Life on the Farm in Late Medieval Jerusalem

Monographs in Islamic Archaeology

Series Editors

Monographs in Islamic Archaeology is a series dedicated to the promotion of innovative and state-of-the-art scholarship on Islamic societies, polities, and communities, from an archaeological perspective. The volumes are problem-oriented, data-rich, theoretically sound, and methodologically innovative studies of archaeological sites and corpuses. The range of studies span the Islamic periods (7th century CE until today), and represent the Islamic world on its global scale.

Published

Archaeology, Politics and Islamicate Cultural Heritage in Europe
Edited by David Govantes-Edwards

Life on the Farm in Late Medieval Jerusalem

The Village of Beit Mazmil, Its Occupants and Their Industry over Five Centuries

Edited by
Bethany J. Walker

equinox

SHEFFIELD UK BRISTOL CT

Published by Equinox Publishing Ltd.

UK: Office 415, The Workstation, 15 Paternoster Row, Sheffield, South Yorkshire S1 2BX

USA: ISD, 70 Enterprise Drive, Bristol, CT 06010

www.equinoxpub.com

First published 2025

British Library Cataloguing-in-Publication Data

A catalogue record for this book is available from the British Library.

ISBN-13 978 1 80050 554 4 (hardback)
978 1 80050 555 1 (ePDF)
978 1 80050 663 3 (ePub)

Library of Congress Cataloging-in-Publication Data

Names: Walker, Bethany J. editor
Title: Life on the farm in late medieval Jerusalem : the village of Beit Mazmil, its occupants and their industry over five centuries / Edited by Bethany J. Walker.
Description: Bristol, CT : Equinox Publishing Ltd, 2025. | Series: Monographs in Islamic archaeology | Includes bibliographical references and index. | Summary: "This book offers a rare glimpse into the daily life of a single rural household and its intimate, but ever-evolving, relationship with Jerusalem from the 14th through the early 20th centuries"-- Provided by publisher.
Identifiers: LCCN 2025011730 (print) | LCCN 2025011731 (ebook) | ISBN 9781800505544 hardback | ISBN 9781800505551 pdf | ISBN 9781800506633 epub
Subjects: LCSH: Farm life--Jerusalem--History | Agriculture--Jerusalem--History | Archaeology and history--Jerusalem | Mamelukes--History | Jerusalem--History
Classification: LCC S522.J4 L54 2025 (print) | LCC S522.J4 (ebook) | DDC 338.1095694/42--dc23/eng/20250711
LC record available at https://lccn.loc.gov/2025011730
LC ebook record available at https://lccn.loc.gov/2025011731

Typeset by Scribe Inc.

Printed and bound by CPI Group (UK) Ltd, Croydon, CR0 4YY

Contents

List of Figures

— 1 —

On Farmsteads and Terraced Fields

The Origins of the Medieval Jerusalem Hinterland Project

Bethany J. Walker

Studies of al-Quds al-Sharīf/Jerusalem in the post-classical periods have traditionally centered, unsurprisingly, on the Old City, isolating it from its hinterland and regional setting. That Jerusalem was part of a larger social world was obvious to contemporaries, who readily recognized the interconnections of satellite settlements, farms and farmsteads, and seasonal settlements that formed a totality which was "the Hill Country of Jerusalem (Jabal al-Quds)." Pilgrimage reinforced this sense of oneness, as did the structure of imperial administration. The agricultural hinterland of Jerusalem—comprising a network of smaller settlements, agricultural terraces, fields, cisterns, water systems, roads, watch towers, and local marketplaces that together fed the city—have not been a focus of archaeological research until very recently.

The following presents the final results of a tightly integrated, multidisciplinary study of a single farmstead (and village) in its agricultural landscape, how both settlement and farmland developed over time, and how these changes impacted the socio-economic development of Jerusalem in the medieval era. Khirbet Beit Mazmīl—perched on one of the highest hills of the Jerusalem area—is the best-preserved medieval farmstead in the immediate outskirts of the Holy City, and one that has been occupied, reused, and remodeled more-or-less continuously since the Late Byzantine period. The region around it has been intensively surveyed and subject to numerous salvage excavations by the Israel Antiquities Authority (IAA). While modern development has cut the site off from the lands that its residents once cultivated, the region around the site of Khirbet Beit

Figure 1 Aerial view of site of Khirbet Beit Mazmīl, view west to Ein Karem (photo: O. Ze'evi).

Mazmīl was traditionally part of nearby Ein Karem, a village that has retained much of its historical landscape of agricultural terraces. The excavations are one component of a larger collaborative research project—the Medieval Jerusalem Hinterland Project—focused on the study of Jerusalem's agricultural landscapes, and which includes the documentation, excavation, and OSL (Optically Stimulated Luminescence) dating of relic agricultural terraces in the lands that we can historically attest were cultivated in the medieval period by the residents of Beit Mazmīl.

What kind of place was Khirbet Beit Mazmīl?

The site under investigation is Khirbet Beit Mazmīl, located on the border between the modern neighborhoods of Bayit Va-Gan and Kiryat HaYovel in southwest Jerusalem, seven kilometers southwest of the New Gate of Jerusalem's Old City. It is one of the fastest-developing neighborhoods in Municipal Jerusalem, and it is real estate of high market value, which explains the numerous salvage excavations (and construction projects) at the site (Figure 1). Perched on the summit of one of the highest hills in the Jerusalem municipality—on an outcrop of Mount Herzl and at 819 meters above sea level—the 2.8 dunam (0.28 hectare) site commands a view to the Old City, Hebron, and the village of Ein Karem. During the medieval period this region constituted the gardens of Jerusalem and took the form of terraced fields largely devoted to intensive cultivation of fruit trees. When we began fieldwork at the site in 2015, however, the terraced fields had been long since covered by concrete, and the part of the site at the southernmost extent (the Southern Sector) had been demolished and a new apartment building, mall, and school had been built. Our focus of excavation—the Northern Sector, on the top of the hill—was obscured from view by the jungle of a wild almond orchard and high grass. Local people knew something was there—some sort of ruined building and urban rubbish deposited by the neighbors—but nothing more. One of the vaulted buildings was used for drying clothes, another for depositing household trash, and the water system occasionally visited by children as a small swimming pool. In the late spring, the almond trees are in full bloom, and the orchard is filled with the singing of birds, which nearly block out the noise of the highway below and the nearby construction that continued for all but our last field season. Our choice to excavate in late spring was deliberate: working in a garden in bloom in Jerusalem was a magical experience.

As for the architectural remains, what was standing above ground looked, from a distance, like a pile of heaped fieldstone (Figure 2). They were, in fact, the remains of a Late Ottoman *'ezbeh*, a drystone farmstead overlooking terraced fields and occupied by peasant families in the harvest seasons. This 19th century building was built directly on, and incorporated the structural ruins of, a Mamluk and Early Ottoman farmstead (Figure 3). The site, in short, represented three layers of building and reuse dating to the Mamluk (13th to early 16th centuries), Early Ottoman (late 16th/early 17th to early 18th centuries), and Late Ottoman/

Figure 2 Khirbet Beit Mazmīl in 2015, before excavations (photo: R. Havilio, Ein Karem).

Tanzimat/British Mandate (late 19th century to the 1920s) periods. These three phases of occupation represented three different ways of living in a village—developing from an amiral estate to a farmstead—and, as we would see from our later fieldwork in the relic terraces nearby, also different phases of development of the agricultural land immediately west of Jerusalem.

Was Khirbet Beit Mazmīl the Arab village of Beit Mazmil?

You will read in the work that follows reference to the site of Khirbet Beit Mazmīl as a "place." That is quite deliberate, as it reflects an ambiguity of placename that complicates the archaeological and historical study of rural settlement. Different genres of textual sources use their own terminology for types of places and for the names of those places; archaeologists have not yet been able to relate the physical remains of sites to these indigenous concepts of place (Walker 2011, 100, 129, 142–143; Walker 2022, 325, note #1). We do not know what this archaeological site was called before the 14th century. On the basis of the architectural remains alone, we cannot say how large a complex this was in its latest phase of occupation, what was near it, and whether it was part of a larger settlement. Was it simply a standalone farmstead? Was it part of a village or hamlet (terms that we, as archaeologists, tend to banter about without clearly understanding how they function in that time and place)? The site that we excavated was cut off from anything historical/archaeological. The north end of the site appears to have continued underneath the current cement-covered parking lot. The south end of the site is likewise covered by modern construction. The high-rises of the neighborhoods of Kiryat HaYovel and Bayit VaGan hedge it in on both sides.

From the mid-19th century, the site was called Khirbet Beit Mazmīl, the "ruins of the House of the Flute."[1] When Guérin, who surveyed the region in the 1860s, called it "Khirbet Ein Karem," his Arab guide corrected him, telling him it was known locally as "Khirbet Beit Mazmīl" (Guérin 1868, 82–83). The site is listed under the name of "Khŭrbet Beit Mizmir" in the Palestine Exploration Fund's Survey of Western Palestine two decades later (Conder and Kitchener 1883, 108, Sheet XVII).[2] The site, described in some detail by Bagatti during his survey of the Ein Karem region in the 1940s, refers to it as "Khirbet Beit Mazmīl" (Bagatti 1948, 16). The Arabic term *khirbah* has many meanings, which change with time and differ from one part of the Arab world to another.[3] While it generally refers to a "ruin," the Palestinian usage, as a placename, is rich in social meaning that goes well beyond the physical state of architectural preservation. In the 19th century, both seasonal farmsteads and villages could be called *khirāb*. The one form of settlement could morph into the other, reflecting either growth or decline. As a placename it was widespread throughout Palestine. Its physical form and function were fluid, which complicates defining the function of an archaeological site such as ours.

The textual sources surveyed and analyzed in Chapter 8 refer consistently to a village called Beit

Figure 3 View north of 19th century architecture built on medieval walls, Northern Sector of site (photo: B.J. Walker).

Mazmil. The first textual reference to it comes from a 14th century manuscript, and documentation regarding the village, its lands, and people living there, continue to the British Mandate period. The modern neighborhood of Qiryat HaYovel in Jerusalem was built "on the land of the Palestinian village of Beit Mazmil" in the early 1950s. The village was depopulated in 1948.[4] Is our archaeological site a remnant of that village, or something else, such as part of an offshoot, or "daughter village" that was created through growth of the village, an independent farmstead built on the outskirts of the village of Beit Mazmil, or, alternatively, a seasonally-occupied *'ezbeh* built in terraced fields to protect the harvest?[5]

Maps produced by the Palestine Exploration Fund in the 1870s (Figure 4) and the British Mandate authorities in the 1940s (Figure 5) specify two interrelated places—an area of orchard cultivation called "Beit Mazmil" and something built on top of a hill at the northernmost end of that land called "Khirbet Beit Mazmīl" (our site) (Abu-Sitta 2010, Map 426). What the maps actually portray is a built area (a settlement) with its cultivated land: they are the same place.

Early drawings of the Ein Karem region by foreign travelers illustrate what is likely the village of Beit Mazmil. It appears as a conglomeration of buildings at the top of the highest hill between Ein Karem and the Old City of Jerusalem. One of

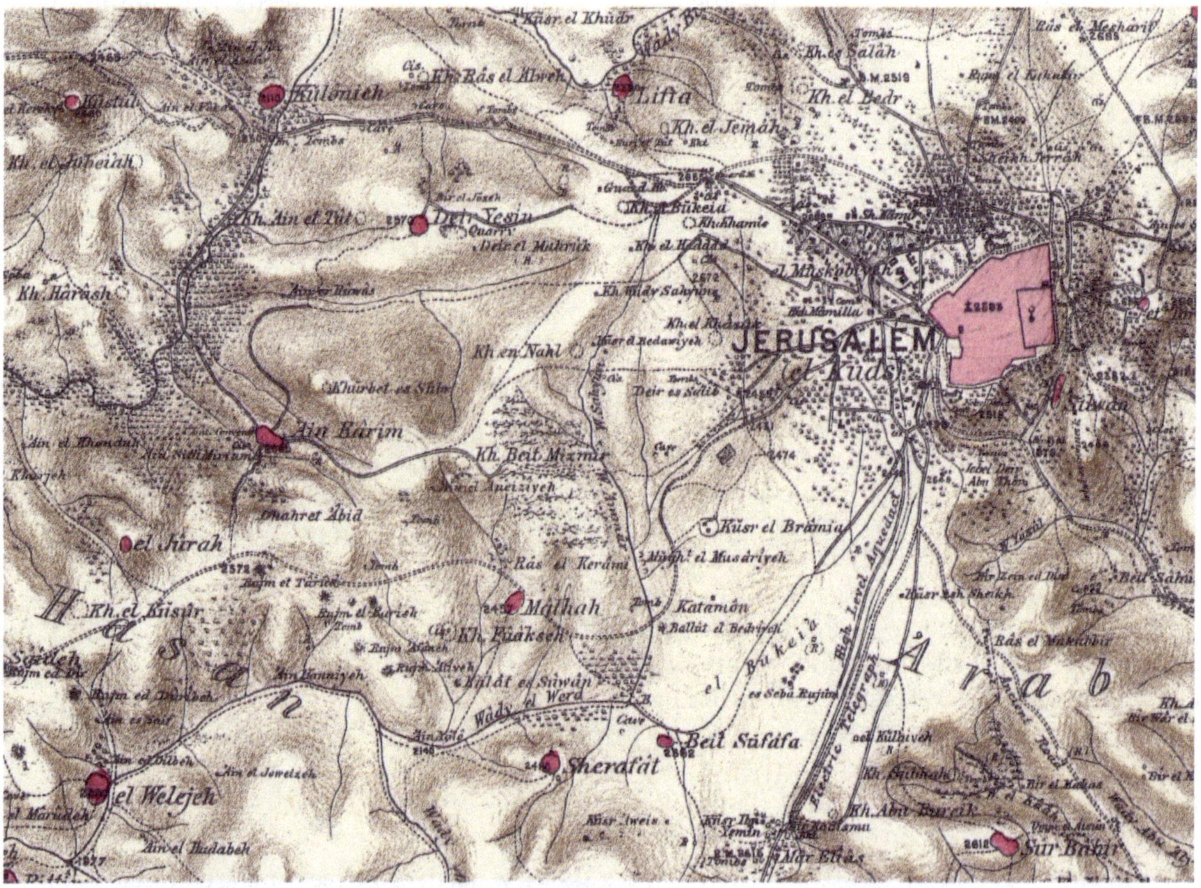

Figure 4 Palestine Exploration Fund Survey of Western Palestine map (9-1-Sheet 17, May 1878, Ein Karem, Jerusalem).

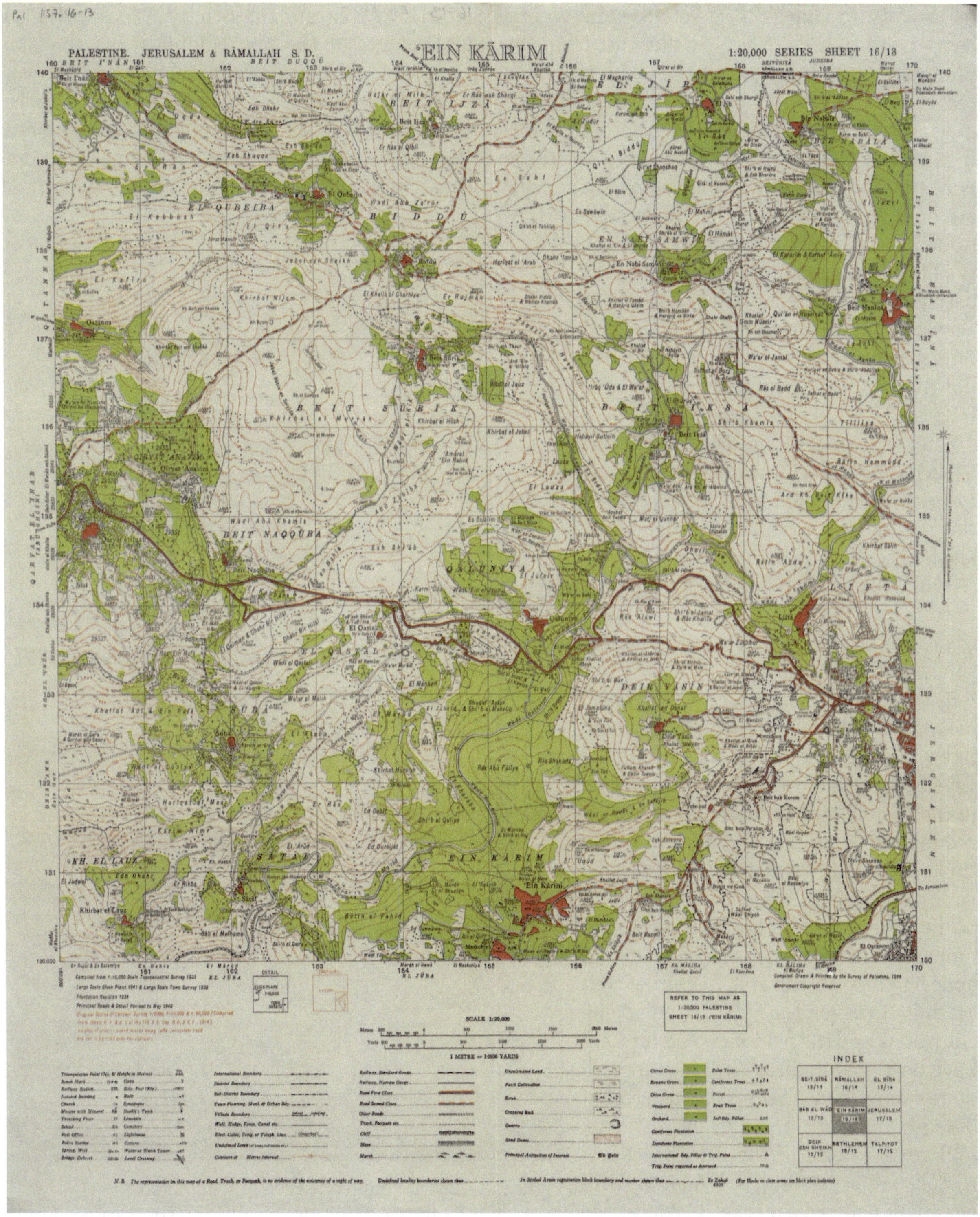

Figure 5 British Mandate survey map of 1944 (16–13, Ein Karim, 1944).

de Bruyn's woodblock image of 1681, drawn in a short-hand fashion, is a view due east from the Church of St. John, showing a hill with two clusters of buildings on its summit, and terraced fields down their slopes (Figure 6).[6] The masonry buildings are flat-roofed (from the exterior), and there is heavy use of vaulting. The rooms (or houses) are built against one another to create a walled space. A photograph preserved in the Matson Collection in Washington, DC, dating sometime between 1898 and 1914, shows a similar scene from a more distant viewpoint, with the Musqubiya and downtown Ein Karem in the foreground (Figure 8).[7]

When enlarged, this image clearly shows a series of agricultural watch towers (*manāṭīr*) in a line (or circle) on two hills: the one on the left is Khirbet al-Ḥamāmah, and that to the right is Khirbet Beit Mazmīl (Fig. 8). What is visible on the right hilltop are three, and possibly six, *manāṭīr*. It is not clear whether these towers were part of a single or multiple compounds.

By the mid-20th century, only one walled compound (the *'ezbeh*) remained. A photo likely taken in the early 1950s shows the developing neighborhood of Kiryet HaYovel in the center, with Ein Karem in the background and a ruined drystone farmstead in the foreground (Figure 9).[8] The farmstead is surrounded by a wall, while *manāṭīr* occupy the slope and are interspersed in numerous field walls and terraces. The site appears to be in ruins: it is Khirbet Beit Mazmīl. However, the fields are still being cultivated, and the field walls and terraces have been maintained. The trees are still quite young, having been planted in recent decades.[9]

In 2022, as part of the Palestine Rural History Project (PRHP), Roy Marom (University of California, Berkeley) conducted interviews with Palestinians who used to live in Ein Karem and the vicinity.[10] These included descendants of the Abu Hudhūds and the Abu Ḥāmeds of Mulabbis.[11] During his fieldwork, he was referred to vital sources, including identifying the residents of Beit Mazmil through local narrations that will be discussed in Chapter 5. One such conversation should be highlighted here. The narrator

Figure 6 Woodblock print drawing of Ein Karem in 1681 (de Bruyn 1715, Vol. I: fig. 126).

Figure 7 Ein Karem from 1898–1914, Matson Collection (Birthplace of John the Baptist, Ain Karim–LOC_matpc.11376).

Figure 8 Detail of Figure 7, showing Khirbet Beit Mazmīl on a hilltop east of Ein Karem.

Figure 9 Kiryat HaYovel neighborhood in the 1950s (photo: A. Kestler, https://www.flickr.com/photos/125243454@N08).

is Fuʿād Abū Hadhūd, who was born in 1935 and is a descendent of the Abū Hadhūd family. The clan Abū Hadhūd lived in Khirbet Beit Mazmīl during its last phase of occupation, namely in the 19th and early 20th centuries. Fuʿād described how his extended family lived together "in one place surrounded by a simple wall with a door." Fuʿād also recalled how it was "the same place where we used to spend our summers after it became known as 'al-Khirba.'" He estimates that the walled area was no bigger than six dunams.[12] The family lands extended from the land endowed to the shrine of Sheikh Jaljīl to the territory of the adjacent village of al-Malḥa. At some point between the 1840s and the 1870s, there was a family dispute that displaced the family to the safety of the larger village of Ein Karem, which absorbed most of the land of Beit Mazmil.[13] The buildings were probably then looted. From that point on, the place where they lived was called "al-Khirba," but they still cultivated its land and orchards, and visited the house complex in the summers.

In previous project publications, we speculated that the archaeological site of Khirbet Beit Mazmīl was located outside of the historical village, and that these were different places (Walker and Dolinka 2020, 193). Marom suggested a different interpretation, tying the earlier phases of the site to Beit Mazmil village of the Jerusalem courthouse records. Therefore, the site is likely what is left of that village which occupied, at one time, the entire hilltop and valley below. The slopes were terraced for cultivation. It is the remnant of a village that has a long history and once was, second only to Ein Karem, the largest and most prominent village immediately west of Jerusalem.

The archaeological site that one sees today is preserved in the Beit Mazmil Archaeological Park in the Bayit VaGan neighborhood of Jerusalem. The park presents the site during its final phase of occupation, from the Late Ottoman era, when it was an *ʿezbeh*. An *ʿezbeh* was the most characteristic rural settlement form of historical Palestine, and, from Ottoman times until the mid-20th century, could be seen on hilltops throughout the Jerusalem highlands. The term refers to more than a form of architecture: it was a kind of settlement, and a way of life. In English, we might consider it a "hamlet," in terms of size and function. As a piece of property, it corresponds to a *dhayʿa* in Syrian Arabic, which refers to a farm devoted to a particular form of cultivation, such as vineyards or palm farms (Maḥjūb 1997, 420; al-Rāzī 1999, 186). In terms of a form of settlement, it was used in Palestine to refer to either a small village or a farm that enclosed the house of the landlord (Omar 2008, vol. 2, 491). The *ʿezbeh* could also be found in Egypt, where it was a low-density settlement that might be larger than a village (Ḥawlīyyāt ādāb ʿAyn Shams 2007, vol. 35, 1248). In both Bilād al-Shām and Egypt, an *ʿezbeh*, if it started out as a farmstead, could grow into a village. Places like this continued to be called *ʿezbeh*s, rather than *qarya*s (villages). In Egypt, the term initially referred to vast contiguous farmlands owned by a single family, but over time it came to be associated with any farm with a mansion or a relatively low-density rural settlement (Barakāt 1987, 12–28). The same was true for Palestine, where the terms *qarya al-kursīyya* ("throne village"), *khirbah* (technically "ruins"), and *ʿezbeh* often merged to describe either different stages in the development of the same place or different aspects of the same settlement. The term *ʿezbeh* was thus an ambiguous one that reflected a fluidity of settlement form and function, as populations grew and declined, the laws and traditions of land tenure changed, and cultivation practices adapted to the market agriculture that developed in town and countryside over the course of Ottoman rule.

At Khirbet Beit Mazmīl, underneath this drystone complex, are two earlier settlements: a family farmstead of Early Ottoman date (late 16th to early 18th centuries) built into a medieval rural estate of the Mamluk era (13th to early 16th centuries). They were all part of the historical village of Beit Mazmil. The Mamluk and Early Ottoman structures are no longer visible to the visitor of the site. The book that follows attempts to reconstruct what all three settlements may have looked like, and how the people residing there lived their lives on a daily basis.

Introducing the Medieval Jerusalem Hinterland Project

History of the project

Though locals may not have been aware of the archaeological and historical significance of this site, the place was known to archaeologists. The excavations that stand at the center of this monograph followed an IAA salvage excavation of the southern part of the site, led in 2012 and 2013 by Benyamin Storchan and Benjamin Dolinka. There were other IAA projects before that, and several surveys, documentation, and mapping efforts made by foreign archaeologists, the Palestine Exploration Society, and the Mandate authorities before that, dating back to the mid-19th century.

The Medieval Jerusalem Hinterland Project was launched in 2015 as a partnership between the University of Bonn and the Israel Antiquities Authority, to document, in as much detail as possible, the contours of the daily lives of the peasants who lived there.[14] I visited the site in 2013, after a field season at Tall Ḥisban in Jordan, and welcomed the opportunity to co-direct the excavation of a site contemporary to Mamluk Ḥisban, in order to better understand demographic trends, peasant practice, and the structure of family and communal life in two different highland landscapes. We conducted five three-week seasons of fieldwork at the site between 2015 and 2019.[15]

In 2017 the project expanded to include the study of the site's ancient fields, specifically the relic agricultural terraces that, according to historical documents, once either belonged to the medieval village of Beit Mazmil or were located in its vicinity. Our research questions grew accordingly, as we aimed to explore the development of Jerusalem's farmland in the later Islamic periods, and to study the site as part of a rural landscape. We wanted to see if the farmstead—that is, the built space of the village—developed, physically and functionally, in tandem with its fields. We were, moreover, interested in peasant decision-making in the past: what were the triggers for changes in land use and resources exploitation by peasant communities, and what defined agricultural practice. We did so by focusing on a single agro-regime: terraced agriculture. This work was led by Tel Aviv University and continued through January 2022.

Methodology

Field recording was done on paper and followed an adaptation of the IAA locus and basket system. After each season this data was digitized and archived in Dropbox folders and eventually the University of Bonn cloud. All soil was sifted. Samples of charcoal from soil deposits and plaster were collected for radiocarbon dating, done at the Institute of Archaeology labs at the University of Cologne in Germany.[16] Our macrobotanist was on site for every field season, and collected soil samples for both macrobotanical and phytolith analysis, conducted in the Netherlands and Greece, respectively, and applying sampling strategies developed in consultation with our phytolith specialist during our previous seasons of fieldwork at Tall Ḥisban. Those soils were floated at "camp"—the Albright Institute in Jerusalem—in the afternoons. Animal bones were collected from select contexts, using a sampling strategy developed by Corbino at Tall Ḥisban; the faunal analysis was done in Italy. Drone photography and photogrammetry of standing architecture was performed every season, the data used to create three-dimensional models and interpret architectural phasing.

All sherds were collected in the field. Due to staff shortages during the first four seasons, individual sherds were pulled for publication. They were washed either at site or at the Albright Institute, and selected sherds taken to the IAA for drawing and photography. Methods of processing pottery were altered in 2019, as our fieldwork was bolstered by the participation of my doctoral students. That season, all pottery was collected, washed, read, and quantified at the Albright Institute, using protocols for form, fabric, ware, periodization, and fragmentation developed for our work at Tall Ḥisban. This kind of quantification has made it possible for us to reconstruct patterns of midden development in the Ottoman period, describe site formation processes over time, and define household assemblages in a way we could not have without this holistic approach to pottery recoding. The drawing and photography of the pottery was initially done in Bonn, with the help of a short-term export loan, and preliminary materials analysis done (organic residue analysis, thin-section petrography, XRF, XRD, and glaze probe analysis). The results of this lab work, however, was not finished by the time of writing, and will appear in two doctoral dissertations and a ceramics monograph.

The field and lab methods used in the terrace part of the project is described in detail in Chapter 6. It is important to mention here, however, that the excavations of the terraces and the farmstead were carefully coordinated. Many members of the team floated between the two excavations, and soil samples for OSL dating were taken from both. Initial survey of relic terraces in the Jerusalem Forest and the region between Ein Karem and our site were conducted with members of both teams during the summer of 2017. Soil samples were taken from the terraces for both pollen and phytolith analysis. No pollen was preserved, unfortunately, and the phytolith analysis was incomplete by the time of the completion of this book. Any information about what was cultivated on these terraced plots comes from the farmstead itself—the botanical record of which becomes a proxy for the fields—and the court registers, which describe exactly what was planted on the lands and orchards being bought and sold.

One of the methods that makes this project unique is the intensive and integrated use, and critical reading, of written sources. Jerusalem is richly documented in Arabic and Ottoman Turkish texts of these periods. What appears in Chapter 8 is only a sampling of the volumes of manuscripts available with information concerning the medieval and Ottoman village of Beit Mazmil. There are 1500 entries alone in the Jerusalem Sharia court registers of the Ottoman period that reference Beit

Mazmil. The Jerusalem series is made up of 400 registers that span 388 years (1529–1917 CE). I began to work in Bonn on these *sijills* several years ago; that work continues today. They provide detailed information on what the land was worth (in the prices of the day), what was cultivated on them, and who bought and sold the land. They are also our only textual source for the names of the villagers in the pre-modern period. The village of Beit Mazmil also appears in six Ottoman tax registers of Jerusalem from the 16th century, spanning 73 years; we have relied on these for general information on cultivation and estimates of numbers of households residing there. In addition to court and tax documents, we have at our disposal endowment records mentioning the site, and fatwa entries documenting conflicts over terrace-building and maintenance, mostly of the Mamluk period. A range of narrative sources (chronicles, travelers' accounts, administrative manuals, and farming manuals) provide background information on Jerusalem's rural hinterland and its administration, as well as peasant knowledge of the day. Beit Mazmil is one of the best-documented places outside Jerusalem in the Ottoman period. The textual study has been an important part of this project from the start and has been invaluable in providing information on cultivation practices, land and produce prices, changes in land tenure, labor organization, village relations, the emergence of market agriculture, and the role of the people of Beit Mazmil in this process. From them we also get names: who lived in this place in times long gone.

For a multi-disciplinary project like this to be successful, you need chronological control, a tightly integrated data set, robust botanical analysis of well-preserved botanical remains, and a well-stratified site. The tight chronology provided by the OSL dates from the terraces and farmstead; long series of radiocarbon dates from key stratigraphic contexts in the farmstead; coins from sealed contexts; excellent stratification by floors, pavements, and collapse (in an otherwise heavily spoliated site); and a textual record that, in the case of the *sijills*, provides us with the exact date of document-filing, has allowed us to correlate very different data sets and identify patterns of change in all of them. This is one reason why we opted for a different kind of final report. We are in the position to tell a story about a place and its people, as we can trace changes with some chronological specificity.

A field report of a different kind

The monograph offered here situates the study of Khirbet Beit Mazmīl and its agricultural land in the larger scholarship of peasant studies, through its focus on peasant decision-making, indigenous knowledge about natural resources and agrarian matters, and communal sovereignty over food production and marketing. In concept, it follows in the steps of the Annales School, by tracing the long history of a place, a landscape, and the people who lived and cultivated there. It is the first study of its kind to approach the landscape history of medieval Jerusalem through an equal measure of archaeological, natural science, and textual analyses. The integration of a multi-proxy botanical analysis and landscape archaeology forms a key component of a project focused on the ways a rural settlement has developed in tandem with its agricultural landscape. The systematic survey and study of medieval and Ottoman-era Arabic texts gives voice to medieval peasants in the Islamic world and enriches the interpretation of the archaeological record.

Life on the Farm in Late Medieval Jerusalem surveys the evolution of a village and its farmland from the Early Islamic era—the Abbasid Caliphate, namely the 9th and 10th centuries—to the mid-20th century, through a combination of archaeological, botanical-chemical, and historical approaches. The focus of the work, however, is on the 14th to 19th centuries, during which time the walled farmstead took on its current form, physically and functionally adapting in step with changes in land use and in response to the vagaries of Jerusalem's food markets. In keeping with long-accepted periodization used in Israeli archaeology, we refer to this period as "Medieval" and "Ottoman." "Late Byzantine" is, for this region, the 5th and 6th centuries. We opted to use the political periodization for the Abbasid strata, as we had pottery and small finds that could be attributed to this dynasty. (It is the equivalent of the Early Islamic 2 period—a cultural periodization used at other sites in Israel/Palestine and Jordan.) The nine chapters of the monograph follow the development of the site and its fields spatially, focusing on the ways that they developed in tandem with one another. Rather than tabling most plans, drawings, charts, and so on, to appendices, all discussion of small finds, architecture, and ceramics are incorporated into each chapter, as part of a synthetic presentation on development of the site over time. In the appendices one will only find the site plans (by occupational phase), a summary of representative entries on Beit Mazmil in the Jerusalem *sijills*, a chart of radiocarbon and OSL dates, a list of critical loci, and instructions on how to use the online 3D modelling program—all for the reader's reference. This organization is a deliberate effort to write a biography of place, trying, as possible, to tell the stories of the people who lived there.

The book is organized into nine chapters. Chapter 2, co-written by Yuval Gadot (director of the terrace work of our larger project) and Gideon Avni (my co-Principal Investigator from the IAA for the GIF grant that funded this project from 2017), situates our project into current landscape studies in the Jerusalem region, which emphasize peasant decision-making. The collaboratively written Chapters 3 through 5 constitute the stratigraphic analysis, written as a historical narrative of the development of the site through the Late Antique and Early Islamic; Mamluk; and Ottoman and British Mandate periods, respectively. The study of the relic agricultural terraces that we believe belonged, at one time or another, to the people of Beit Mazmil, is the focus of Chapter 6, which includes the results of the OSL analysis. The study of foodways and land use through the archaeobotanical and faunal records is found in the subchapters of Chapter 7. A synthesis of the results of all these chapters, read against the very rich, multi-century textual record about this place and its relationship with Jerusalem, is the topic of the historical study in Chapter 8. Chapter 9 summarizes what we believe we have learned about the people who lived at Khirbet Beit Mazmīl, as peasants and businessmen, and why they were able to survive the vicissitudes of politics and economic life.

As for spelling and date formulae, we follow the style guide of the Journal of Islamic Archaeology for the transliteration of Arabic terms. Most Arabic placenames, particularly those of villages and archaeological sites, make use of diacriticals. Therefore, in this volume the placename "Khirbet Beit Mazmīl" which refers throughout to the ruins of the archaeological site (which we excavated) maintains the use of the diacritical. In referring to the village, including the village in the occupational phase following the mid-19th century communal conflict described by Fuʿād Abu Hudhūd, we adopt the local spelling use "Beit Mazmil," without diacritical marks. As for dates, the Gregorian calendar is adopted. Only in Chapter 8, which deals directly with Mamluk and Ottoman-era documents dated according to the Ḥijrī calendar, do document entry dates appear in either Ḥijrī-only or a combined AH/CE form.

The reader accustomed to the traditional final volume format will immediately notice a difference in this work. As a research team—our collaborative efforts representing many different disciplines—we aimed to produce in the end, in written form, what has been our philosophy of fieldwork from the start: interdisciplinarity. To achieve a narrative flow, stay true to our goal of learning about the details of daily life at this place in the past, and to best represent to the reader the carefully orchestrated integration of different data sets collected over the lifetime of the project, our writing is largely a group effort. As an archaeologist, historian, and editor, I have aimed for years at this kind of collaborative writing. The following attempts on a monograph-scale the kind of field reports that I and my colleagues have been writing together in article form for years. We hope that this could be a model for multidisciplinary writing of final reports in the future.

The three stratigraphic chapters (3, 4, 5) have been collaboratively written in every sense. There is a lead author for each, but the co-authors have contributed large parts of the text. This was accomplished in a painstaking way, through monthly Zoom meetings over the course of a year and half with those authors, who live in three different continents, and the sharing of draft texts through the University of Bonn cloud (Sciebo). Each of our texts (our interpretations) were adjusted after reading those of our colleagues and working through stratigraphic problems together online. As this is a group effort, we have deliberately not acknowledged who wrote which part of the text. However, it is appropriate here to give credit to our authors for their special expertise. The formal stratigraphic interpretations were the result of consultations between the two co-directors (Benjamin Dolinka and myself), and were initially based on our unpublished, internal project reports (Dolinka from 2015–2018 and Walker for the 2019 season), as well as the previously published preliminary and interim reports in article form (Storchan and Dolinka 2014; Walker and Dolinka 2020; and ʿAdawī and Dolinka 2021). We divided the tasks for artifact and building analysis, with Dolinka being responsible for ceramics and Walker for vernacular architecture; we both relied on internal, unpublished IAA reports for the analysis of the glass, metals, and stone. Nicolò Pini was the project architect and did all work related to interpretation of architectural phasing, photogrammetry, three-dimensional modelling, and most drone work. Zubayr ʿAdawī, who directed the IAA salvage excavation of the Lower Site in 2017, produced the raw data for work there, which was written-up by Dolinka and made available to all of us for all three chapters. We acknowledge the many contributions of Benyamin Storchan, archaeologist with the IAA, who co-directed with Dolinka the IAA salvage excavations in 2012 and 2013 and wrote the report on the archaeological park in Chapter 5. At a late stage of writing, we were very fortunate to have the historical geographer Roy Marom join us, whose interviews with local peoples who once lived at this place in the early 20th century, and whose communal memories reach back to the 19th century, have allowed us to "put faces on places" and bring the history of this place to modern times. His work is the basis of the section "The People of the Place" in Chapter 5 (and partially based on Abū Hadhūd 2011). All work with Mamluk and Ottoman-era texts and the narration of Mamluk and Ottoman history was the responsibility of Walker, who also took over the ceramics processing in 2019.

A different approach was taken for Chapter 7, where the results of the botanical and faunal analyses were written as three separate subchapters. Our project phytolith specialist (Sofia Laparidou, assisted by Georgia Kasapidou to process the 2019 samples) and macrobotanist (Annette Hansen) shared their draft texts with each other and me, adjusting interpretations on the basis of the emerging results of the other. While these are separate studies, it was necessary to integrate the macro- and microbotanical readings. Our project zooarchaeologist, Chiara Corbino, meanwhile provided an assessment of the site's faunal assemblage.

As this volume was a collective effort, we have many people to thank. In addition to the committed work of our project staff, we have been supported institutionally by the Research Unit of Islamic Archaeology and Annemarie Schimmel Kolleg of Mamluk Studies at the University of Bonn, the Institute of Archaeology at Tel Aviv University, Hebrew University, and the Israel Antiquities Authority, and Matt Adams and the Albright Institute of Archaeology in Jerusalem. Oren Gutfeld (Hebrew University) and his Israel Archaeological Services provided all logistical support, including our team of professional excavators from Ramla and Ramallah. Many people living and working near the site and in Ein Karem were instrumental in facilitating our terrace study; we want to particularly recognize Ron Havilio, Mike Leiter, and the Zippori Center in Jerusalem.

Supporting our excavation of the *khirbah* in 2015 were enthusiastic volunteers from Ein Karem and the neighborhood of Qiryat HaYovel, who helped with excavation, pottery washing, and drawing, and kindly brought tea to the site; they made us all feel welcome. I also want to thank my Erasmus, MA, and PhD students from Bonn, as well as former students from Missouri State University—a group of experienced field archaeologists—who joined us in the field for excavation in 2019, making that final field season particularly special. My doctoral students also contributed to the production of this monograph. Ahmad Ghizawat, an epigrapher who is also well-trained in Arabic manuscript use, assisted me in finalizing the summary of a series of Ottoman court records, while I took over the tax registers; the completion of Chapter 8 would not have been possible without dividing that load. To prepare the manuscript for copy-editing, Nikolaus Hochstein Cox did preliminary editing of each chapter, produced the Index, and designed the basic design of the book cover, which was further refined by Equinox. I am, as always, ever-grateful to Russell Adams, our copy editor at Equinox, and the entire Equinox staff for promoting this book project and making it a reality.

On a final note, we gratefully acknowledge the financial support we have received for these many seasons of excavation. The IAA salvage projects were funded through contractors before the Southern Sector was demolished for modern construction. The scientific project that began in 2015, which guaranteed the preservation of the rest of the site, has been financed through several sources. The 2015 spring survey and architectural documentation was financed through the University of Bonn. The 2015 fall excavation season would not have been possible without the combined funding from the Barakat Trust, the German Society for the Exploration of Palestine, and the Palestine Exploration Society. From 2017 until the completion of fieldwork in 2019, excavation of the *khirbah* and the terrace study were generously funded by the German-Israeli Foundation for Scientific Research and Development (GIF grant no. G.1354–108.4/20176). The architectural analysis by Pini, in the field and back "home" in the computer lab, was made possible through multiple sources: a post-doctoral grant from the Annemarie Schimmel Kolleg in 2017–2018; project grants from Marie Curie Action IF@ULB and the Belgian F.R.S.-FNRS; the logistical and financial support of Université libre de Bruxelles, with a special thanks to Professor Agnès Vokaer; and Henry-Louis Guillaume and his team at Panorama (ULB) for their help in the processing and optimization of the 3D models. We heartily thank Thomas Flynn, Cultural Heritage Lead at Sketchfab, for offering a free Premium subscription to their platform. This author would also like to acknowledge the constant support of Prof. Emeritus Yehoshua Frenkel, who recommended sources, led me to archives in Israel, and made so many useful suggestions during the writing phase. Final production of this publication was made possible by the Deutsche Forschungsgemeinschaft (DFG, German Research Foundation), through project number WA 2525/4-1.

Notes

1. It is known in Palestinian colloquial Arabic today by some other names, such as Beit Muzammil and Beit Madmīl. The residents of the nearby village of al-Malḥa, some of whom came originally from Beit Mamil, often pronounce it as "Mazmīr." A placename "Farfīr" also appears in the Ottoman *sijillāt*; this may or may not be a scribal error, a misspelling of "Mazmīl," but we do not include it in our study.
2. The placename "Beit Mizmir" is also found in 16th-century Ottoman tax registers.
3. For a fuller discussion on the social meaning of the term in different periods of Islamic history, see Walker 2021.
4. The story of what happened to its residents is told through an oral history project, the Palestine Rural History Project (PRHP), led by Roy Marom and summarized at the end of Chapter 5. At the time of writing, we have been unable to get any population figures concerning the village or details of the demographics before 1948, other than what was revealed by the PRHP.

5. For more on the "daughter villages" of Late Ottoman Palestine, see Kark and Oren-Nordheim (2001).
6. In his travelogue, de Bruyn refers to the village of Ein Karem as the "village of St. John," and suggests that the ruins on the top of the hill opposite of the church were the tombs of the Maccabees (de Bruyn 1715, Vol. I, 273).
7. The Musqubiyyah is the Arabic name for all of the churches west of the village and east of the Hadassah Ein Karem Hospital. It includes the Church of the Visitation and the Church of Princess Elizabeth.
8. I am grateful to Benyamin Storchan for sending me this image. The photo is credited to Ayelet Kestler (https://www.flickr.com/photos/125243454@N08/).
9. After geo-rectifying the photograph and overlaying it with the British Mandate map in Figure 4, the field walls of both images align with one another. I am grateful to Benjamin Storchan for this information.
10. Marom curates the Palestinian Rural History Project (PRHP), an oral history archive documenting rural Palestinian life before 1948 (Marom 2023a). For the site gazetteer, see Marom 2023 (*The Palestinian Rural History Project (PRHP): Mission Statement*, last updated October 2022: DOI: 10.13140/RG.2.2.31021.77285).
11. On the history of the village of Mulabbis, located 10.6 kilometers east of Tel Aviv, see Marom 2023b.
12. The size of the space inside the final-phase enclosure wall of the Northern Sector of the site is roughly three dunams. If one includes the estimated size of the Southern Sector, which would be the entire walled *'ezbeh*, the space is roughly six dunams.
13. These dates are based on the interviewee's narrative, which relates the passing of four generations since the incident.
14. The 2015, 2017, and 2018 seasons were co-directed by myself (University of Bonn) and Benjamin Dolinka (IAA). I led the final season in 2019.
15. The March 2015 survey was authorized by IAA license S-555/2015, and the excavations of August/September 2015, March 2017, March 2018, and March 2019 under licenses G-73/2015, G-18/2017, G-24.2018, and G-29/2019, respectively.
16. One should note that while the radiocarbon results gave us a narrow range of dates for most periods, for the Ottoman they were quite general (usually 300–600 years before present). This allowed us to attribute loci to the Ottoman period in a general sense, but not to assign them to any specific Ottoman phase (at this site 16th, 17th, or 19th centuries).

References

Abū Hadhūd, F.

2011. "ʿUṣūl ʿĀʾila-t-Abu Hudūd." *Ein Karem (Occupied Palestinian Village)*, 07.22.23 Website: http://ainkarem.blogspot.com/2011/01/blog-post.html?m=1.

Abu-Sitta, S.H.

2010. *Atlas of Palestine 1917–1966*. London: Palestine Land Society.

ʿAdawī, Z., and B. Dolinka

2021. "Qiryat Ha-Yovel," *Hadashot Arkheologiyot: Excavations and Surveys in Israel* 133 (04.05.2021).

Bagatti B.

1948. *Il Santuario della Visitazione ad ʿAin Karim (Montana Judaeae)*. Jerusalem: Tip. dei PP. Francescani.

Barakāt, ʿĀ.

1987. *Taṭawwur al-mulkīyya al-zirāʿīyya fī miṣr wa atharuhu ʿalā-l-ḥarak als-siyāsīyya 1813–1914*. Cairo: Dār al-thaqāfa al-jadīda.

de Bruyn, C.

1715. *Voyage au Levant, c'est-à-dire, dans les principaux endroits de l'Asie Mineure, dans les isles de Chio, Rhodes, et Chypre etc.* Paris: Cavelier.

Conder C.R. and H.H. Kitchener

1881. *The Survey of Western Palestine III: Judaea*. London: Palestine Exploration Fund.

Guérin M.V.

1868. *Description géographique, historique et archéologique de la Palestine I: Judée*. Paris: Imprimé par autorisation de l'empereur à l'Impr. impériale.

Ḥawlīyyāt ʿādāb ʿĀyn Shams

2007. Vol. 35. Cairo: Kullīyyāt al-ʿādāb, Jāmiʿay-ʿĀyn Shams.

Kark, R. and M. Oren-Nordheim
2001. *Jerusalem and Its Environs: Quarters, Neighborhoods, Villages 1800–1948*. Israel Studies in Historical Geography. Jerusalem: Magness Press.

Maḥjūb, M.
1997. *Dirāsāt fī al-mujtamiʿ al-badawī*. Alexandria: Dār al-maʿrifa al-jāmiʿīyya.

Marom, R.
2023a. *Palestinian Rural History Project (PRHP): Gazetteer of Documented Villages*. Online Edition: https://www.academia.edu/89656575/Palestinian_Rural_History_Project_PRHP_Gazetteer_of_Documented_Villages.
2023b. "The Abu Hameds of Mulabbis: An oral history of a Palestinian village depopulated in the Late Ottoman period." *British Journal of Middle East Studies* 50.1: 87–106.

Omar, A.M.
2008. *Muʿjam al-lugha al-arabīyya al-muʿāṣira*. Cairo: Ālim al-kutub.

al-Rāzī, Abu ʿAbdallah Muḥammad ibn Abu Bakr, Shams al-Dīn
1999. *Mukhtār al-suḥāḥ—muʿjim lughawī ʿarabī/ ʿarabī*. Beirut: Dār al-kutub al-ʿilmīyya.

Storchan, B. and B. Dolinka
2014. "Jerusalem, Bayit Va-Gan." *Hadashot Arkheologiyot* 126 (09.10.2014).

Walker, B.J.
2011. *Jordan in the Late Middle Ages: Transformation of the Mamluk Frontier*. Chicago Studies on the Middle East Monograph Series. Chicago: Middle East Documentation Center, University of Chicago.

Walker, B.J.
2021. "Khirba," *Encyclopaedia of Islam* III, Part 2021.1: 73–75. Internet Edition: https://referenceworks.brillonline.com/entries/encyclopaedia-of-islam-3/khirba-COM_35538?s.num=32&s.start=20.

Walker, B.J.
2022. "The changing face of agricultural 'estates' in 15th and 16th-Century Palestine: The commercialization of Khirbet Beit Mazmīl" in *The Mamluk-Ottoman Transition: Continuity and Change in Egypt and Bilad al-Sham in the Sixteenth Century*, Vol. II, edited by S. Conermann and G. Şen, 325–357. Bonn: V&R Press.

Walker. B.J., and B. Dolinka
2020. "*Ḫirbet Bēt Mazmīl*. Investigations of Medieval Jerusalem's Hinterland: Interim report on the 2015 and 2017 Field Seasons." *Zeitschrift des Deutschen Palästina-Vereins* 136.2: 191–234.

Wilson, C.W.
1882. *Picturesque Palestine, Sinai, and Egypt*. London: J.S. Virtue.

About the author

Prof. Dr. Bethany J. Walker (PhD 1998, University of Toronto) is Professor of Islamic Archaeology at the University of Bonn, the Director of the Research Unit of Islamic Archaeology at the same institution, and former Co-Speaker of the Annemarie Schimmel Kolleg of Mamluk Studies in Bonn. A historically trained archaeologist, Walker directs field projects in Israel and Jordan. Among her monographs, edited, and co-edited works are *Jordan in the Late Middle Ages: Transformation of the Mamluk Frontier* (Chicago 2011), *Reflections of Empire: Archaeological and Ethnographic Studies on the Pottery of the Ottoman Levant* (Boston 2009), *Living with Nature and Things: Contributions to a New Social History of the Middle Islamic Periods* (Bonn 2020), and the *Oxford Handbook of Islamic Archaeology* (Oxford 2020). She is the senior and founding editor of the Journal of Islamic Archaeology, and her research focuses on peasant society, land use and tenure, and landscape transformation.

— 2 —

Peasant Decision-Making in the Jerusalem Highlands

A Landscape Perspective

Yuval Gadot and Gideon Avni

Beit Mazmil is located within the heart of Jerusalem's agricultural lands (Edelstein and Kislev 1981; Gadot 2015; Gibson and Edelstein 1985). As such, the site functioned throughout the ages as a component within a wider system that ensured regular supply of food and other products into the city. At the same time, the site is a reflection of how the urban elite ruled over and exploited the lands surrounding the city. In light of the above, it is clear that an archaeological study of Beit Mazmil should not end at the settlement's outer fence or last building. Rather, it should continue and include the agricultural lands that surround the site. This approach, advocated for over four decades by landscape archaeologists (Bradford 1957; Gibson 1995, 295–298; Gibson 2003), is the one that guided us when planning this project. In what follows we wish to describe the shortcomings of the settlement-oriented approach and summarize the methodological breakthrough that guided us in designing the coordinated study of the relic terraces and the archaeological site of Khirbet Beit Mazmīl.

Traditionally, and with a few notable exceptions, archaeologists working in and around Jerusalem were more interested in the excavation of public and religious buildings that dominated the local landscape, and practically neglected the study of settlements in their agrarian context. Thus, despite the extensive research conducted in Jerusalem and its surroundings, almost no attention was dedicated to the study of agricultural fields, their accurate dating, and their environmental context. Nevertheless, agricultural systems and their related water management installations are highly visible in the agrarian landscapes that surround the Old City of Jerusalem, just as they are visible everywhere else around the Mediterranean. Many rock-cut or built features form an intensely exploited landscape. It seems that the most prominent feature is terrace walls used for dry farming that are built along the slope of almost all mountainsides. Coupled with fences that parcelized the lands into definable plots, the two features mirror past land management and land tenure policies. According to the estimation made by Zvi Ron, as much as 60 percent of the highlands of Jerusalem are covered by built terrace walls (Ron 1966). Using terrace walls for artificial creation of arable plots was a major technological innovation that has led to the complete alteration of the natural terrain (Bevan and Conolly 2011; Davidovich *et al.* 2012). It transformed valleys and slopes into flat arable plots while reducing soil erosion, increasing infiltration, and dramatically expanding the carrying capacity of the land. Furthermore, these walls were built and managed by the population that lived in the supporting settlements and are, therefore, interconnected. Thus, it is unfortunate that until the past decade the intensive terracing of the Judaean Highlands was studied by archaeologists to only a limited extent.

Earlier works conducted in the 1960s, particularly by Zvi Ron (1966), concentrated on the mapping of terraces and agricultural fields in relation to the natural springs prevailing in the region and on determining their relationship to nearby historic villages, and did not aim at accurately dating the systems themselves. The first direct archaeological attempt to study the terraces of the Jerusalem highlands was headed by Shimon Gibson, who chose Mount Eitan as his study area and focused on the land-use patterns of Sataf, a Palestinian village that was built over a multi-period archaeological site (Gibson *et al.* 1991). A second landscape project was initiated by Gershon Edelstein, who documented the cultural landscape around the Refaim Valley at the southwestern outskirts of modern-day Jerusalem (Edelstein and Milevski 1994; Edelstein *et al.* 1998; Gibson and Edelstein 1985). The studies of agricultural hinterlands of Jerusalem were intensified with detailed surveys of the central Palestinian highlands and the Jerusalem region, which were conducted in the framework of the Archaeological Survey of Israel (Kloner 2003).[1] Unlike the two previous projects described, which were limited in their scope, the survey was conducted over a vast area and produced widescale mapping of many features within their context. However, while accurately mapping and dating the remains of villages and farms, these surveys related only in passing to the extensive networks of terraced agricultural fields in the region.

It seems that one of the major factors that deterred archaeologists from investigating these landscapes in depth was the inability to date many of the features in general, and the construction of terrace walls in particular. In many regions of the southern Levant, the tentative dating of these extensive terraced fields correlated them with the peak settlement periods, usually associating them to the "glorious periods" of the past, mainly from the Hellenistic to Late Antique times (Bar 2008; Dar 1986; Frankel *et al.* 2001; Safrai 1994; Wilkinson 2003). This perception of the "ancient lands" was very much enhanced by thousands of drawings and early photographs from the second half of the 19th century and the early 20th century, which depicted the traditional landscapes of Syria-Palestine in the late Ottoman period. The connection between these landscapes and the old "Biblical," Roman, and Late Antique sceneries prevailed in many

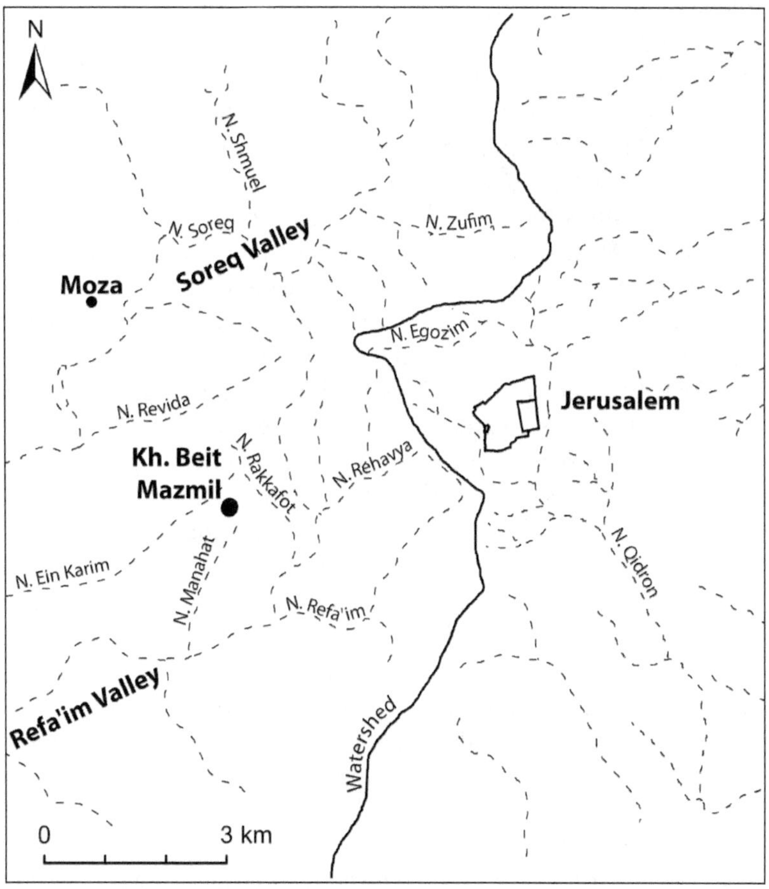

Figure 1 Map of agricultural lands of Jerusalem.

scholarly studies up to the end of the 20th century and impacted traditional archaeological research (de Geus 1975; Stager 1985).

A development of this approach can be seen with several regional projects that emphasized a bottom-up approach and included a systematic study of fields in their rural context, through excavation and documentation of terrace walls. The dating of the walls was usually based on stratigraphy and/or material culture (Edelstein and Milevski 1994, 9–10; Gibson *et al.* 1991, 35–37; Gibson 1995, 175–182). However, most of these studies still connected the rise and decline of agricultural exploitation of the countryside with the intensification and abatement of nearby settlements (Finkelstein 1996; Gibson 2001). The common paradigm was that terrace walls made settlement possible on these inhospitable slopes. The rise in the number of settlements over the course of the Middle Bronze and Iron I periods was therefore seen as proof that terrace walls were constructed. A second assumption, which had been adopted in many regions around the Mediterranean, suggested that the rise of agricultural regimes began in the Hellenistic period with the mass expansion of urban settlements and the growing need for regional "food baskets" to supply their growing demands; this was further intensified in the Roman and Byzantine periods.[2] The establishment of new cities in the southern Levant and the development of vast hinterlands around them expanded the cultivated areas in the countryside. It was believed that these settlements declined in the 7th century when the Sasanian and Arab conquests marked the collapse of the extensive agricultural systems in the Near East.

Consequently, the intensive construction of terraced fields was always related to the expansion of settlements. The earliest peak of agricultural expansion around Jerusalem was dated by some to the Middle Bronze and Iron II periods. Later, much more substantial demographic peaks were dated either to the early Roman period, or to the 5th and 6th centuries (CE). These periods were also considered the zenith of agricultural productivity in Palestine.[3] The regional studies conducted in the central highlands north of Jerusalem, for example, associated fields to nearby settlements from the Roman and Byzantine periods (Dar 1986). A similar methodology was adopted in the Negev Highlands, where the fields were first dated to the Nabataean period and later associated with the extensive settlements of the Byzantine period (Avni 2022; Mayerson 1960; Rubin 1990). However, new archaeological research, which showed increasing evidence for settlement continuity into the Early Islamic period (Avni 2014; Magness 2003; Walmsley 2007) and questioned the nature of settlements in the Medieval period, called for a re-evaluation of the traditional dating of agricultural fields and related irrigation systems, looking at aspects of continuity of previous traditions, together with the penetration of new agricultural technologies (Avni 2020). Recently some attempts have been made at the direct dating of terrace walls, using radiocarbon dating of charcoal found within terrace fills, though not in the dry farming zones of the South Levantine Highlands.[4] These attempts have suffered from the unclear source of the charred samples within the soils, and the processes that led to their deposition—such as the re-working of old charcoal and the penetration of roots—as well as from other possible problems, such as old wood effect.

A major leap forward in the detailed study of agricultural terraced fields and their relation to historical villages and farms west of Jerusalem was conducted in the past decade, with the systematic study of fields in Ramat Rahel, south of Jerusalem (Davidovich *et al.* 2012; Davidovich and Porat 2020). The settlement's main periods of habitation span the 8th century BCE and the 10th century CE. It consisted of a palatial complex in the Iron Age and Persian periods, which was transformed into a village in the Roman period and functioned as one of the rural settlements around Jerusalem. The village was abandoned in the 11th century, but this area continued to serve as the "food basket" of Jerusalem in the Mamluk and Ottoman periods (Lipschits *et al.* 2020,

488–489). The immediate surroundings of Ramat Rahel consist of hillside terraces and agricultural installations. Several stages of terrace construction were identified and dated by OSL samples, showing three main periods of use: the first between the 6th and 9th centuries (CE), the second intermediate phase was dated to the 12th–13th centuries; and the youngest and most extensive terracing was formed in the early Ottoman period.

In contrast to radiocarbon dating of organic materials, OSL dates minerals that were directly affected by the action that is being dated. This method measures the time elapsed since the last exposure of mineral grains to sunlight (Aitken 1998; Avni *et al.* 2013, 333–335; Davidovich *et al.* 2012; Wintle 2008; and Chapter 5 this volume). It uses quartz grains as dosimeters, recording the accumulation of environmental radiation over time. A signal accumulates within the quartz as a result of the ionizing radiation. When exposed to sunlight, this signal is reset to zero, and following the covering of soils deposits the signal builds up again and its intensity is proportional to the time spanned since the covering. This signal can be measured in the laboratory and converted into a "burial age."

Following the pioneering work at Ramat Rahel, the study of the agricultural hinterland of Jerusalem was extended into a larger project titled "The Formation of Terraced Landscapes in the Judean Highlands." It was based on the systematic sampling of a diversity of agricultural landscapes around Jerusalem.[5] Four study areas were defined, three within the valleys of Soreq and Refaim and one at Mount Eitan—an isolated, hilly spur in the center of the region. In each study area, selected terrace systems, ranging from well-preserved standing terraces to barely visible deteriorated remnants, were documented in detail and then excavated. We used the stratigraphical relations to suggest a relative sequence for each wall and plot. We then extracted in dark conditions soil samples that were taken to the lab for dating and for further analysis.

The results of this work, published in several articles, demonstrate that most of the terrace walls seen today around Jerusalem date to the Late Mamluk and early Ottoman periods (Elgart-Sharon *et al.* 2020; Gadot *et al.* 2016a; Gadot *et al.* 2016b; Gadot *et al.* 2018). At Mount Eitan, which is located approximately five kilometers west of Beit Mazmil, more than half of the 42 soil samples date between 1400 and 1750 CE. Two villages are known to have existed on the mountain during the Ottoman period: Khirbet el Loz and Sattaf. We have collected all available data regarding these two settlements in early 20th century maps, tax registers, and aerial photos. Using GIS, we have also mapped and calculated all available flat plots of land that can be potentially cultivated without using terraces. Our calculations show that the 450 registered residents of Khirbet el Loz in 1945 had intensely cultivated 765 dunams of their approximately 4,500 dunams territory and that practically no parts of their cultivated plots are located on sharp terraced slopes. However, within their territory there are over 800 dunams available with a slope of less than 10 degrees, which can be cultivated even if not terraced. Furthermore, if one ignores the village border as marked on the 1945 map and searches for available cultivated land that is located within an hour's walk from the village, more than 3,000 dunams of flat arable land could potentially be exploited (Elinson 2016).

The outcomes of the terrace-dating project are clear and prove that the extensive terracing operations on Mount Eitan (echoed by similar results from other case-studies in the Refaim and upper Soreq valleys) should not be interpreted as a reflection of demographic pressure to conquer previously uncultivated areas. Instead, the reasons behind the construction (and maintenance) of terraces reflect choices made by peasants and landowners, based on a range of considerations (Elgart-Sharon *et al.* 2022).

By conducting excavations at selected points in the agricultural landscape around Beit Mazmil, the current project was designed to continue exactly from the point reached by the "terrace dating project:" to further explore the reasons that motivated peasants living in the hinterland of Jerusalem to adopt terracing as their main economic strategy and to invest the energy and labor needed for their arduous construction and on-going maintenance (Walker *et al.* 2019). This is further emphasized by the fact that the decision-making process and the mechanisms of intensification and abatement of agricultural fields in general, and terraces in particular, have not been systematically addressed to date. Thus, the Medieval Jerusalem Hinterland Project includes three distinct aspects: excavations of Khirbet Beit Mazmīl; a study of archival materials related to land tenure and construction of terrace walls; and a study of the field systems around the site. The latter study included a detailed survey, followed by probes in specific locations. The application of OSL dating within the regional project provided an accurate dating for the construction and use of the terraced fields. The results of these studies enable a refined reconstruction of the duration of agricultural exploitation of specific areas around the site of Khirbet Beit Mazmīl.

Notes

1. See also the data that has been gathered on the Archaeological Survey of Israel website (https://survey.antiquities.org.il/index_Eng.html#/).
2. See Olus *et al.* (2016) for a discussion on the intensification of urban settlement in the Hellenistic period.
3. This paradigm was also applied to other regions such as Northern Syria and North Africa. See, for example, Barker *et al.* (1996) for North Africa; Decker (2009) for Syria-Palestine; Wilkinson (2003) for the Near East in general; Tchalenko (1953–58) and Tate (1992) for Northern Syria; Dar (1986) for the Samarian Hills; Mayerson (1962) and Rubin (1990) for the Negev Highlands; Dagan (2010; 2011) for the Judaean Lowlands; and Ron (1966) and Seligman (2011) for the Judaean Hills.
4. For discussions of the runoff farming systems in the Negev Highlands see Bruins and van der Plicht (2005, 359–362; 2007: 489–492); for dry farming terraces in the Aegean see Krahtopoulou and Frederick (2008).
5. This methodology has been used in several archaeological and geomorphological studies in the last decade. For updated summaries see Srivastava *et al.* (2023).

References

Aitken, M.
1998. *An Introduction to Optical Dating.* Oxford: Oxford University Press.

Avni, G.
2014. *The Byzantine-Islamic Transition in Palestine: An Archaeological Approach* (Oxford Studies in Byzantium). Oxford: Oxford University Press.

Avni, G.
2020. "Terraced fields, irrigation systems and agricultural production in Early Islamic Palestine and Jordan: Continuity and innovation." *Journal of Islamic Archaeology* 7: 111–137.

Avni, G.
2022. "The Nabataeans in the Negev—on archaeological research, historical interpretation and modern Israeli society." *Cathedra* 181: 9–30.

Avni, G., N. Porat, and Y. Avni
2013. "Byzantine-Early Islamic agricultural systems in the Negev Highlands: Stages of development as interpreted through OSL dating." *Journal of Field Archaeology* 38: 329–343.

Bar, D.
2008. *'Fill the Earth'—Settlement in Palestine during the Late Roman and Byzantine Periods 135-640 C.E.* Jerusalem: Yad Ben-Zvi.

Barker, G., D. Gilbertson, and D. Mattingly
1996. *Farming the Desert—The UNESCO Libyan Valleys Archaeological Survey*. Paris: UNESCO.

Bevan, A., and J. Conolly
2011. "Terraced fields and Mediterranean landscape structure: An analytical case study from Antikythera, Greece." *Ecological Modelling* 222: 1303–1314.

Bradford, J.
1957. *Ancient Landscapes. Studies in Field Archaeology.* London: Bell.

Bruins, H.J., and J. van der Plicht
2005. "Desert settlement through the Iron Age: Radiocarbon dates from Sinai and the Negev Highland."

In *The Bible and Radiocarbon Dating: Archaeology, Text and Science*, edited by T.E. Levy and T. Higham, 349–366. London: Equinox.

Bruins, H.J., and J. van der Plicht
2007. "Radiocarbon dating of the 'Wilderness of Zin.'" *Radiocarbon* 49: 481–497.

Dagan, Y.
2010. *The Ramat Bet Shemesh Regional Project: The Gazetteer* (IAA Reports 46). Jerusalem: Israel Antiquities Authority.

Dagan, Y.
2011. *The Ramat Bet Shemesh Regional Project: Landscapes of Settlement: from the Paleolithic to the Ottoman Periods* (IAA Reports 47) Jerusalem: Israel Antiquities Authority.

Dar, S.
1986. *Landscape and Pattern: An Archaeological Survey of Samaria, 800 BCE—636 CE* (British Archaeological Reports International series 308). Oxford: British Archaeological Reports.

Davidovich, U., N. Porat, Y. Gadot, Y. Avni, and O. Lipschits
2012. "Archaeological investigations and OSL dating of terraces at Ramat Rahel, Israel." *Journal of Field Archaeology* 37: 192–208.

Davidovich, U. and N. Porat
2020. "Area T: Survey and excavations on the Western Slope." In *Ramat Rahel IV: The Renewed Excavations by the Tel-Aviv-Heidelberg Expedition (2005-2010). Stratigraphy and Architecture*, edited by O. Lipschits, M. Oeming, and Y. Gadot, 406–466. Tel Aviv: Eisenbraun, Monograph Series of the Institute of Archaeology of Tel Aviv University.

Decker, M.
2009. *Tilling the Hateful Earth: Agricultural Production and Trade in the Late Antique East* (Oxford Studies in Byzantium). Oxford: Oxford University Press.

Edelstein, G., and M. Kislev
1981. "Mevasseret Yerushalayim. The ancient settlement and its agricultural terraces." *Biblical Archaeologist* 44: 53–56.

Edelstein, G. and I. Milevski
1994. "The rural settlement of Jerusalem re-evaluated: Survey and excavations in the Rephaim Valley and Mevasseret Yerushalayim." *Palestine Exploration Quarterly* 126: 2–23.

Edelstein, G., I. Milevski, and S. Aurant
1998. *Villages, Terraces and Stone Mounds: Excavations at Manahat, Jerusalem, 1987-1989*. Jerusalem: IAA Reports 3.

Elgart-Sharon, Y., N. Porat, and Y. Gadot
2020. "Land management and the construction of terraces for dry farming: The case of the Soreq catchment, Israel." *Oxford Journal of Archaeology* 39: 274–289.

Elinson, R.
2016. "The physical potential and the subsistence economy in the Jerusalem Hills." Tel Aviv: MA thesis; Tel Aviv University.

Finkelstein, I.
1996. "Ethnicity and origin of Iron I settlers in the Highlands of Canaan: Can the Real Israel stand?" *Biblical Archaeologist*. 59: 198–212.

Frankel, R., N. Getzov, M. Aviam, and A. Degani
2001. *Settlement Dynamics and Regional Diversity in Ancient Upper Galilee: Archaeological Survey of Upper Galilee* (IAA Reports 14). Jerusalem: Israel Antiquities Authority.

Gadot, Y.
2015. "In the Valley of the King: Jerusalem's rural hinterland in the 8th–4th Centuries BCE." *Tel Aviv* 42: 3–26.

Gadot, Y., U. Davidovich, G. Avni, Y. Avni, M. Piasetzky, G. Faershtein, D. Golan, and N. Porat
2016a. "The formation of a Mediterranean terraced landscape: Mount Eitan, Judean Highlands, Israel." *Journal of Archaeological Science: Reports* 6: 397–417.

Gadot, Y., U. Davidovich, G. Avni, Y. Avni, and N. Porat
2016b. "The Formation of terraced landscapes in the Judean Highlands, Israel, and its implications for Biblical history." *Hebrew Bible and Ancient Israel* 5: 437–455.

Gadot, Y., Y. Elgart-Sharon, N. Ben-Melech, U. Davidovich, G. Avni, Y. Avni, and N. Porat
2018. "OSL dating of pre-terraced and terraced landscape: Land transformation in Jerusalem's rural hinterland." *Journal of Archaeological Science: Reports* 21: 575–583.

de Geus, C.H.J.
1975. "The importance of archaeological research into the Palestinian agricultural terraces, with

an excursus on the Hebrew word gbî." *Palestine Exploration Quarterly* 107: 65–74.

Gibson, S.

1995. "Landscape Archaeology and Ancient Agricultural Field Systems in Palestine." Unpublished PhD dissertation, University College London.

Gibson, S.

2001. "Agricultural terraces and settlement expansion in the highlands of Early Iron Age Palestine: Is there a correlation between the two?" In *Studies in the Archaeology of the Iron Age in Israel and Jordan*, edited by A. Mazar, 113–146. Sheffield: Sheffield Academic Press.

Gibson, S.

2003. "From wildscape to landscape: Landscape archaeology in the Southern Levant—methods and practice." In *The Rural Landscape of Ancient Israel*, edited by A.M. Maeir, S. Dar, and Z. Safrai, 1–25. Oxford: Archaeopress.

Gibson, S., and G. Edelstein

1985. "Investigating Jerusalem's rural landscape." *Levant* 17: 139–155.

Gibson, S., B. Ibbs, and A. Kloner

1991. "The Sataf Project of landscape archaeology in the Judaean Hills: a Preliminary report on four seasons of survey and excavations (1987–1989)." *Levant* 23: 29–54.

Kloner, A.

2003. *Survey of Jerusalem. The Northwestern Sector, Introduction and Indices*. Jerusalem: Israel Antiquities Authority.

Krahtopoulou, A., and C. Frederick

2008. "The stratigraphic implications of long-term terrace agriculture in dynamic landscapes: Polycyclic terracing from Kythera Island, Greece." *Geoarchaeology: An International Journal* 23(4): 550–585.

Lipschits, O, Y. Gadot, and M. Oeming

2020. "Deconstruction and reconstruction: Reevaluating the five expeditions to Ramat Rahel." In *Ramat Rahel IV: The Renewed Excavations by the Tel-Aviv-Heidelberg Expedition (2005-2010). Stratigraphy and Architecture*, edited by O. Lipschits, M. Oeming, and Y. Gadot. 476–491. Monograph Series of the Institute of Archaeology of Tel Aviv University. Tel Aviv: Eisenbraun.

Magness, J.

2003. *The Archaeology of the Early Islamic Settlement in Palestine*. Winona Lake IN: Eisenbraun.

Mayerson, P.

1960. *The Ancient Agricultural Regime of Nessana and the Central Negev*. London: British school of Archaeology in Jerusalem.

Ron, Z.

1966. "Agricultural terraces in the Judean Mountains." *Israel Exploration Journal* 16: 33–49, 111–122.

Rubin, R.

1990. *The Negev as Settled Land—Urbanisation and Settlement in the Desert in the Byzantine Period*. Jerusalem: Yad Ben Zvi.

Safrai, Z.

1994. *The Economy of Roman Palestine*. London: Routledge.

Seligman, J.

2011. "The rural hinterland of Jerusalem in the Byzantine period." Unpublished PhD dissertation, University of Haifa.

Srivastava, A., T. Kinnaird, C. Sevara, J.A. Holcomb, and S. Turner

2023. "Dating agricultural terraces in the Mediterranean using luminescence: Recent progress and challenges." *Land* 12.3: 716. Internet Edition: https://doi.org/10.3390/land12030716.

Stager, L.E.

1985. "The archaeology of the family in Ancient Israel." *Bulletin of the American Schools of Oriental Research*. 260: 1–35.

Tate, G.

1992. *Les campagnes de la Syrie du Nord du IIe au VIIe siècle*. Paris: Geuthner.

Tchalenko, G.

1953–58. *Villages antiques de la Syrie du Nord—Le massif du Bélus a l'époque romaine*. Paris: Paul Geuthner.

Walker, B.J., Y. Gadot, Y. Elgart-Sharon, and O. Ze'evi

2019. "Agricultural terracing and rural revival in Late Medieval Palestine." In *Proceedings of the 11th International Conference on the Archaeology of the Ancient Near East*, edited by L. Korn, 651–661. Wiesbaden: Harrassowitz Verlag.

Walmsley, A.

2007. *Early Islamic Syria—An Archaeological Assessment*. London: Duckworth.

Wilkinson, T.J.
2003. *Archaeological landscapes of the Near East*. Tucson: University of Arizona Press.

Wintle, A.G.
2008. "Fifty years of luminescence dating." *Archaeometry* 50: 276–312.

About the authors

Prof. Yuval Gadot (PhD 2004, Tel Aviv University) heads the Sonia & Marco Nadler Institute of Archaeology and is the Vice Dean at Tel Aviv University. He was formerly the Head of the Department of Archaeology and Ancient Near Eastern Cultures at Tel Aviv University (2018–2023). Currently Gadot serves as lead archaeologist at the "City of David" excavations project, co-directs the Givati Parking Lot Excavations, and leads an environmental study of the rural landscape surrounding the city. He has been affiliated with numerous archaeological projects in Israel focusing on agricultural terraces and employing laboratory science methods, including the excavations at the Ramat Rachel site (2005–2010, as Field Director), and the "Ancient Israel through the lens of natural and exact sciences." Between 2004–2008 he was a research fellow at the Nelson Glueck School of Bible and Archeology at Hebrew Union College in Jerusalem (2004–2008). Gadot has authored seven archaeological books and over 100 scientific articles.

Prof. Gideon Avni is the Chief Archaeologist for the Israel Antiquities Authority and Professor of Archaeology at the Hebrew University in Jerusalem. His academic interests focus on various aspects of Classical, Late Antique, Early Islamic, and Medieval archaeology; the archaeology of Jerusalem and its hinterlands; nomads and sedentary societies in the desert areas of the Near East; Mediterranean connectivity in Late Antique and Early Islamic times; and the diffusion of technologies and movement of peoples in Eurasia. He has authored six monographs and numerous articles, and his recent books are *The Byzantine-Islamic Transition in Palestine: An Archaeological Approach* (2014), and *A New Old City—Jerusalem in the Late Roman Period* (2017).

— 3 —

The Earliest Settlement on Telegraph Hill

Benjamin J. Dolinka, Nicolò Pini, and Zubair ʿAdawī

In 1944, British real-estate magnate Lord Harold Samuel was quoted as saying, "There are three things that matter in property: location, location, location" (Safire 2009). This sentiment can also correlate to the placement of many ancient sites. Location in regard to ancient settlement consists of the following three components: the site has to be positioned on a high place that is easily defensible; it must be situated along a major road system in order to ensure that it can be supplied; and it must be in close proximity to a good source of water.

Khirbet Beit Mazmīl fits well within all of these parameters. First, the site is located on the summit of Mount Herzl (819 meters ASL), the second highest point in the Jerusalem region. The strategic importance of the site was clearly recognized by those who settled there for the past two thousand years, even playing a vital role in more recent history, during the First World War, when the British fought Turkish forces near the site in December 1917 (Grainger 2006, 207–208, Map 5). Second, its presence was not only located along a major hinterland road leading to the Holy City, which lies only seven kilometers to the southeast of the site, but it was also on the Christian pilgrim route that connected Jerusalem, Bethlehem, and Ein Karem. Third, and most importantly, was the access to water, a theme which will pervade throughout this chapter and the rest of this volume. Because the site is positioned prominently along the western watershed that provides runoff water to Jerusalem to the east and Ein Karem to the west, the location of Khirbet Beit Mazmīl was perfectly suited for the collection and storage of water, and for the cultivation of a wide variety of crops and other foodstuffs.

The purpose of this chapter is threefold: to provide a brief overview of the archaeological surveys and excavations conducted at Khirbet Beit Mazmīl during the past century-and-a-half; to present an occupational history of the site that can be divided into 10 strata dating from the Early Roman to Late Ottoman/British Mandate eras and covering a period spanning nearly 2,000 years; and to examine in detail evidence from each of the strata (Stratum IX–VI) that pre-date the feudal floruit of the site under the Mamluks. With regard to the pottery reported upon in this volume, unless otherwise specified, it is not discussed within the traditional typo-chronological framework accompanied by detailed analysis and numerous parallels for each vessel examined; rather it consists of a series of assemblages from individual contexts taken together as a whole unit, in order shed greater light upon the relationships among form, function, and findspot (the "three Fs") for the pottery from the site. To that end, discussion is limited in nature and parallels (if even found) are kept to a minimum.

Previous research at Khirbet Beit Mazmīl

Although the site of Khirbet Beit Mazmīl was referred to in medieval sources, it completely escaped the notice of the classical authors, who fail to mention it either directly or indirectly. The site remained essentially *terra incognita* to westerners until the first surveys of the site in the 19th century, described in Chapter 1. The most direct and extensive description of the ruins at Khirbet Beit Mazmīl were provided by Bagatti (1948), during his discussion on the regional environment surrounding the Sanctuary of the Visitation in Ein Karem. He noted that:

> "At Khirbet Beit Mazmīl . . . the remains of a wall are preserved in three visible courses characteristic of the Crusader [period]. It is a large wall of rough stones [running] from east to west. The land, used as an orchard, has a great amount of mosaic tiles, some very large and many small white ones and occasionally blue or red [in color], and numerous pottery. We noticed handmade pottery decorated in brown on the bottom [most likely casseroles or frying pans], reddish or greenish glazed ceramics, with or without graffiti, [and others in] blue, green and yellow with various incisions . . . there was also unglazed pottery: basin rims, also thin-walled vessels . . . but always with rather weak firing . . . The numerous tesserae make us think of a Roman-Byzantine occupation, but the walls and the pottery are of the Crusader period . . ." (Bagatti 1948, translated by Dolinka 2017, 16)

Gazetteers on the medieval pottery of the region (Pringle 1981, 53) and secular Crusader buildings in the vicinity of Jerusalem (Pringle 1997, 28, no. 36) also list "Khirbat Bait Mazmīl" and make mention of surface pottery collected there dating to the 12th century, which supports the chronological framework suggested by Bagatti with regard to the site. During the comprehensive survey of Jerusalem's northwest sector conducted by the IAA, the area is named "Qiryat Ha-Yovel" (Site 162) and described as a "ruin on a hill [with] remains of walls, cairns and quarry debris" (Kloner 2003, 130). Finally, fieldwork conducted in the vicinity of Qiryat Yovel during February 2011 reported a building complex and its associated pottery, dating from the Mamluk and Ottoman periods, that was briefly surveyed during the course of salvage excavations conducted there (Levi 2012, fig. 3–6).

The more recent archaeological fieldwork at the site was conducted by the IAA in 2012 and 2013, where it was referred to as "Bayit Vagan / Giv'at Antennot." Excavations in two areas revealed the presence of several loci sealed by floors that yielded an abundance of datable ceramic material. Four main occupational phases were identified: Byzantine, dating from the 5th to 6th centuries; Crusader/Ayyubid, dating from the mid-12th to mid-13th centuries, which contained imported pottery from Greece and northern Syria; and two Mamluk phases, one from the 14th century and another from the 14th to 15th centuries, the latter of which produced a remarkable amount of imported pottery from northern Italy.

During the course of the first season of excavations, the site was visited by B.J. Walker (University of Bonn) and E. Stern (IAA), and a preliminary investigation of the adjacent archaeological site was conducted. This initial and informal survey indicated substantial architectural remains that were well-preserved but in imminent danger of being destroyed and formed the genesis for the Medieval Jerusalem Hinterland Project, which ran from 2015 to 2019.

Early Roman period (Stratum IX, 1st century BCE–1st century CE)

The first chapter in the biography of Khirbet Beit Mazmīl demonstrates the beginning of the site's occupational sequence and corresponds to the Early Roman period in Judaea. This period covers the time between the first centuries BCE and CE. During this era, the agricultural hinterland surrounding Jerusalem was characterized by several installations hewn into bedrock. Important amongst these were the Jewish ritual baths (*miqva'ot*) and stone quarries utilized for the building of structures in the Holy City and beyond.

Jewish ritual bath (miqveh)

One of the primary features from Early Roman Khirbet Beit Mazmīl is the Jewish ritual bath (*miqveh*), which represents the earliest known manifestation of a water-collection system at the site. According to the archaeological record, there does not appear to be any associated settlement on the summit of Mount Herzl dating from this period. However, the *miqveh* was most likely associated with an agricultural installation (for example, an olive or wine press), to ensure ritual purity for the production of these items not only for the laity, but perhaps for use as offerings in the temple in Jerusalem as well (Adler 2008). The presence of this ritual bath also clearly demonstrates the Jewish nature of the site during the Early Roman period and is a poignant example of the importance of ritual purity to the people who worked the fields around Khirbet Beit Mazmīl (Snyder 2016). These installations are ubiquitous in the countryside surrounding Jerusalem, and over a decade ago it was noted that more than 850 ritual baths had been documented (Adler 2011, *ii*; Lawrence 2006, 155–183; Magness 2011, 16–17). Since that time *miqva'ot* continue to be discovered in estates, farms, villages, and agricultural lands in the Jerusalem hinterland on a regular basis over the course of excavations.

There has been some scholarly debate regarding the chronology for the presence of *miqva'ot* in Early Roman Judaea. While the initial consensus dated the appearance of these installations from the mid-2nd century (BCE) to the destruction of the Second Temple in 70 CE (Magness 2021, 143) there is a growing body of evidence that suggests continuity of use into the 2nd century (CE), and as late as the Bar Kochba Revolt of 132–135 CE, due to the fact that they were such a fundamental element of Jewish religious and community life (Zissu 2001). A recent study observed that, with regards to the use of *miqva'ot*, "Jews were aware of ritual purity issues during their day-to-day life, while growing crops, preparing food, and celebrating holidays and religious festivals, especially the pilgrimage to the Holy City of Jerusalem, and most of all while entering the temple courts" (Zissu 2014, 246). This phenomenon explains why there was continuity of use for these ritual baths, even after the destruction of the Second Temple and the termination of cultic activities which took place there, because of their importance to daily life (Adler and Amit 2010, 142).

Comprehensive and detailed archaeological studies on *miqva'ot* are relatively new, few in number, and mostly published in the Hebrew language. The primary resource among these is the work of Reich (1990), whose PhD dissertation from the Hebrew University of Jerusalem included an archaeological catalogue of 300 ritual baths, most of which date from the Second Temple Period (Zissu 2001). That work was more recently revised, updated, and published as a Hebrew-language monograph (Reich 2013). Following the work of Reich, Amit (1996) wrote an MA thesis at Hebrew University on the *miqva'ot* from the Hebron hinterland, which is unpublished and also written in Hebrew. The work of Adler has also made a significant impact upon *miqva'ot* studies. Although his doctoral dissertation from Bar-Ilan (2011) was written in Hebrew (with an English abstract), he has since published several articles on the subject in English (see Adler 2008), that deal with both literary sources and archaeological perspectives. Finally, Magness (2021, 168–195) has recently produced an excellent study on the ritual baths from Qumran.

All *miqva'ot* have essentially the same basic components: they are usually hewn directly into the bedrock; have a narrow, stepped passageway that leads down into an inner chamber which has rounded corners (Reich 1984, 220) "to prevent injury to the barefoot bathers" (Magness 2021, 144); and the steps, chamber floors, walls (and usually the roof) are covered with several layers of gray hydraulic plaster (Reich 1990, 53–54; Porat 1989, 74). The number of steps varies from four, as in the *miqveh* at Beit Jala (Hizmi and Sion 2009, 335), to seven, as noted in the one at Khirbat Ka'kul (Seligman 2006, 3).

The *miqveh* at Khirbet Beit Mazmīl is located in Area D1, in the southeast part of the Northern Sector excavated between 2015 and 2019. It was originally documented in 2011, during a survey of the area conducted by the IAA (Levi 2012), which produced the only plan, section, and photos of the installation (Figure 1). The stepped passageway of this *miqveh* is located on the southeast of the installation, is roughly square in shape, measures 2 × 2 meters, and consists of four steps that measure between 30 and 50 centimeters. The inner chamber measures 2 × 1.5 meters and is covered in its entirety by a thick layer of gray plaster. There seems to be another set of four steps on the northern end of the inner chamber. Although these were badly damaged during a later period, they are clearly seen in the plan. This would suggest that this *miqveh* is a variation on the type identified by Reich (1984, 220) as having a double entrance, by which "one path served for an entry and the other for an exit, in order to avoid contact between the impure person entering and the pure [person] emerging from it" (Reich, 222).

In one of the fundamental studies of Jewish ritual baths associated with agricultural installations, it was noted that, "The presence of these ritual baths adjacent to wine-presses and olive-presses suggests that those who made the use of the *miqva'ot* did so immediately prior to commencing work in these installations" (Adler 2008, 72), because ". . . the most immediate potential threat to the purity of wine and oil during the various stages of production was the ritual impurity of the agricultural labourers in these industries . . . [for which there were numerous possible contaminants]" (Adler, 68).

Several *miqva'ot* have been discovered in association with agricultural installations, such as oil and/or wine presses (Adler 2008, 63–65). In the Jerusalem hinterland, *miqva'ot* located adjacent to presses have been found

Figure 1 a) Plan and Section (by: A. Hajian) of the Early Roman *miqveh* first documented in February 2011 (courtesy of D. Levi, IAA); b) Photo of the Early Roman *miqveh* at Khirbet Beit Mazmīl; view to the east (photo courtesy of D. Levi, IAA).

at many places (see Avni and Gudowitz 1987, 6–12; Magen 2004, 29–45; Seligman 1999, 140–141, 148–152, 161–162). Other *miqva'ot* have been found in the Jerusalem countryside located in open fields, in the absence of presses, and with no known settlement nearby. These likely either represent agricultural land holdings of aristocratic families from Jerusalem, or the entities with which they were associated have been lost to the archaeological record. Examples of this phenomenon are known from north of the Old City (see Magen 1984, 64; Magen 1985) and to the south of Jerusalem (see Hizmi and Zion 2009; Reich 1998, 122–123). Taken together, the *miqveh* at Khirbet Beit Mazmīl provides the earliest archaeological evidence for agricultural land-use at the site, which likely engaged in the seasonal harvests of grapes and/or olives and required that the workers be ritually pure in order to perform their duties.

Stone quarries

Stone quarries are another type of feature scattered throughout the Jerusalem hinterland and dating from the Early Roman period. These utilized the natural bedrock to provide both Jerusalem and its hinterland with building stones for domestic, administrative, and religious structures. The presence of quarries at Khirbet Beit Mazmīl was noted in the Survey of Western Palestine conducted by Conder and Kitchener (1883, 108), during the Government of Palestine's investigations in the 1920s (Israel Department of Antiquities 1976, 156), as well as by the survey of Jerusalem's northwest sector conducted by the IAA (Kloner 2003, 130, Site 162). These quarries consisted of the negative impressions of removed rectangular stone building blocks and the presence of detachment or severance channels, which were utilized to remove the stones.

During the 2012 excavation season, evidence for quarrying activities was uncovered in the western portion of Area A in the Southern Sector. In the southernmost area of the quarry, numerous carved rectilinear lines outlining the negative of building blocks removed from the site were found. To the north, the quarrying activity was found to have been more intensive, leaving a deeper carved exposure of the bedrock (Figure 2). The quarry is tentatively dated to this phase due to the fact that it clearly predated all of the later constructional activities in the area (Storchan 2020).

The remains of the Khirbet Beit Mazmīl quarry bear striking similarities to other Early Roman examples found both within and outside of the Old City of Jerusalem. It is comparable in most respects to a terraced stone quarry uncovered in the Yemin Moshe neighborhood to the southwest of the Jaffa Gate. Like Khirbet Beit Mazmīl, the quarry is also situated on a hillside overlooking a valley below. In addition, it had the negatives of removed building stones that had an average size of 0.45 × 1.50 meters (Balila 2020), which corresponds to the evidence from our site.

Figure 2 View of the Early Roman quarry marks cut into the bedrock found in Area A during the 2012 excavations at Khirbet Beit Mazmīl; view to the south (photo: B. Storchan, IAA Photo Archive No. B-297809).

Farther afield from the Old City, excavations conducted in Emek Lavan ("White Valley") revealed the presence of numerous quarrying areas scattered throughout the slopes leading down into the valley. The location of these quarries is only approximately three kilometers south of Khirbet Beit Mazmīl, in the Ir Ganim B neighborhood. These quarries, like most of those that are found, produced no dating evidence to shed light upon when they were in use. However, excavations along Shonei Halachot Street in the Jewish Quarter of the Old City (Raphael 2018, 2) have finally demonstrated that the quarry found there dated from the 1st century (BCE), as the fill layer that covered its floor yielded Hasmonaean ceramics and coins.

Early Roman material culture and concluding remarks

There was a decided paucity of artifactual evidence from the Early Roman period at Khirbet Beit Mazmīl. During the course of the 2012 excavations at the site, topsoil and subsoil fill layers yielded a few sherds of the well-known cooking pot forms (not illustrated here) produced at the major ceramic workshop that was located where the modern Crowne Plaza Hotel (see Levi and Be'eri 2011) and Jerusalem Convention Center currently stand (Magness 2005; Rosenthal-Heginbottom 2005). These cooking pots conform to the narrow-mouthed globular type, which are dated to the third quarter of the 1st century (CE) (Rosenthal-Heginbottom 2005, 243, cat. nos. 14–17). The only other Early Roman artifact to be found at Khirbet Beit Mazmīl, also in a fill level, was a coin known as a *prutah*, which was made of bronze and represented a low denomination of currency (Kool 2014, 1, cat. no. 1). It dated from the years 24–12 BCE, during the reign of Herod the Great, and conforms to Type 15 from the typo-chronology developed by Ariel (2012, 63).

While the evidence from the Early Roman period is indeed limited in nature, it offers a glimpse into the landscape environment of the Jerusalem hinterland. The presence of the *miqveh*, quarry, and cooking pots from this era point to its use as a farming settlement where laborers ritually purified themselves before work began, where building stones were cut from the bedrock for use in now long-gone structures, and where the workforce prepared and ate their meals in the field, just as the archaeologists do today.

Late Roman period (Stratum VIII, 2nd–3rd century)

One feature at the site dating to the Late Roman period was identified: a roughly square cistern (Levi 2012) hewn into the bedrock (Figure 3). Only the bottom half of it was preserved, and this was covered with two distinct plaster layers—one with a gray color that had large inclusions of gravel in the bottom layer, and another that had a lighter gray and whitish color with quartz and limestone in its matrix. Levi had the plaster analyzed (Levi 2012) and after comparison with that of other

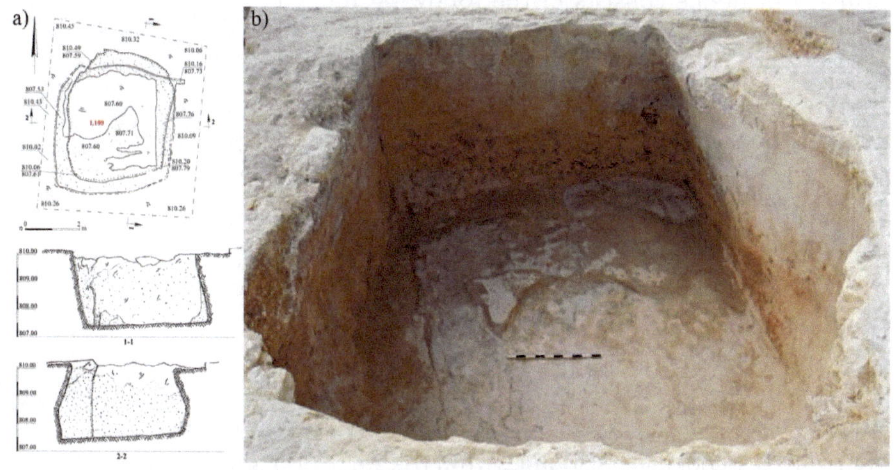

Figure 3 a) LR cistern plan and section (photo courtesy of D. Levi, IAA); b) LR cistern photo (photo courtesy of D. Levi, IAA).

installations from the Jerusalem region, and also due to the lack of crushed pottery sherds in its matrix, suggested that it was pre-Byzantine in date. Only two of the sherds discovered in the cistern were diagnostic, with the latest represented by a Late Roman cooking pot rim and shoulder (Figure 4). A close parallel for this vessel was recovered from a late 2nd-early 3rd century context at Mampsis in the central Negev highlands (Erickson-Gini 1999, fig. 13.3:2, 4), while the exact rim form is known from the Roman Legionary kiln works at the Jerusalem International Convention Center excavations (Magness 2005, fig. 15:7). The cistern at Khirbet Beit Mazmīl "seems to have been part of a complex that was destroyed in development activities of the neighborhood" (Levi 2012). The presence of this installation demonstrates the continuity of water collection on the eastern lower summit of the hill.[1]

Late Byzantine period (Stratum VII, 5th–6th century)

The earliest clear evidence for settlement at the site dates from the 5th to 6th century and may have been associated with a monastic complex. Information for this era corresponds to Stratum VI from the 2012 season, with additional data provided by the IAA salvage excavation in 2017. The primary evidence consists of the following four distinct elements: architectural remains of a large Late Byzantine structure; a sealed plaster floor from within that structure; preserved sections of a mosaic floor and numerous small white tesserae that were scattered throughout the site; and the artifacts that can be attributed to this period.

Structure A1 from the Southern Sector

Remains dated to the Late Byzantine period from the 2012 season were exposed in Area A only (Storchan 2015; Storchan 2021), where the bedrock is much higher than that in Area B. One of the primary features from this era is a large building designated Structure A1 (Figure 5). Only the southeastern corner of the structure was uncovered, and this was formed by wide walls (W111 and W143), which measured approximately 1.7 to 2 meters wide. The walls were constructed from two rows of large stones with a densely packed core composed of small stones and soil fill and that were set directly upon the bedrock within a foundation trench that cut the surrounding reddish earth matrix (Figure 6). The foundation trench (0.30 meters wide) was dug down to the bedrock to ensure the stability of the structure and then later backfilled with a darker-colored earth.

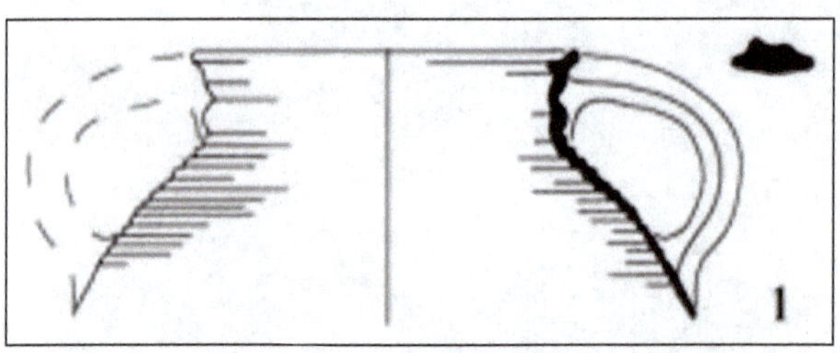

Figure 4 LR cooking pot found within the cistern (drawing: I. Lidsky-Reznikov, courtesy of the IAA).

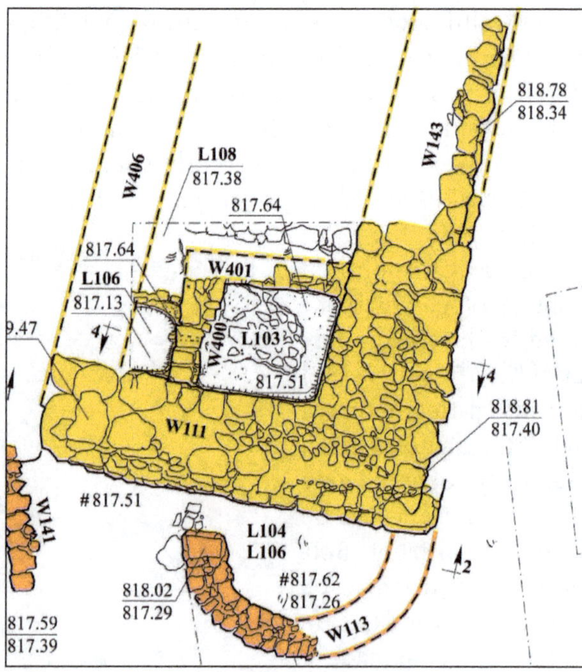

Figure 5 Plan of the walls from the large Structure A1 in Square A8 (courtesy of the IAA).

Figure 6 Photo of the foundation trench for Structure A1; view to the north (photo: B. Storchan, courtesy of the IAA, photo: B-331336).

Excavations in the central portion of Structure A1 revealed the remains of a crushed-lime and beaten earth-floor (L.134), which had flat-lying Late Byzantine sherds embedded in the floor itself. This floor no doubt was part of an associated structure that was later damaged and cut on at least three sides by foundation trenches (L.133) for the construction of the Stratum III Mamluk walls. Sealed under the floor was a soil layer (L.135) that included ceramics dating from the 5th to 6th centuries.

Mosaic floors from the Southern Sector and Lower Site

The presence of numerous small tesserae from Late Byzantine mosaic floors was noted in earlier surveys of the site. According to Bagatti:

> "At Khirbet Beit Mazmīl . . . a great amount of mosaic tiles, some very large and many small white ones and occasionally blue or red [in color were found] . . . The numerous tesserae make us think of a Roman-Byzantine occupation . . ." (Bagatti 1948, translated by Dolinka, 16)

The evidence from recent excavations corroborates this observation. A good example of this was the existence of a dense concentration of mosaic tesserae found within the foundation trench for Structure A1 (Storchan 2015). Additionally, excavation of only half of Square A9, located immediately to the east of the structure, yielded over 3,000 medium-sized plain white tesserae (Storchan 2021). This dense concentration would seem to suggest that a mosaic floor existed in the area that was later dismantled and discarded. These tesserae were of the same size and manufacture as those found on a mosaic floor in Square 1 from the Lower Site in 2017 (see below). It is suggested here that this "dump" of tesserae is the result of the presence of a workshop that produced them, which served the Late Byzantine settlement and probably the surrounding areas (as needed). There is a direct parallel for this scenario: a similar quantity of tesserae was recovered from the Byzantine monastic complex at Khirbat 'Adassa excavated by Zelinger (2015); and it should be noted that this monastic complex is located only about 150 meters away from the Early Islamic and Mamluk settlements at the site. It should also be noted that, among the glass finds from the 2012 excavation (Katsnelson 2014, 1) were colored tesserae, which were most likely meant for use in mosaic floors.

The remnants of a mosaic floor were found during the 2017 excavation at the Lower Site of Khirbet Beit Mazmīl ('Adawī and Dolinka 2021). It was recovered from Square S1, located in the eastern sector of the excavation area, immediately to the south of and below the A1 tower. This area was heavily disturbed by bulldozing activities while leveling the ground for the construction of the community center that now sits there. This small segment of a mosaic floor measured 1.2 × 1 meters and was made of coarse tesserae stones (1–3 × 1–2 centimeters), with a concentration of about 25 tesserae per square decimeter. The foundation of the mosaic floor consisted of two layers: a layer of gray cement (1–2 centimeters thick) beneath the tesserae that was in turn founded on a layer of brown soil (3–5 centimeters thick). This mosaic floor has been dated to the Late Byzantine period because it was built directly on bedrock, and Byzantine pottery was recovered from the excavation area ('Adawī and Dolinka 2021, 11).

Late Byzantine pottery from the Southern Sector and Lower Site (Figure 7)

The primary stratigraphic context for the Stratum IV ceramic assemblage from the Southern Sector was the L.134 and L.135 sequence, which originated from either directly on or directly below, respectively, the plaster floor in Square A4. It should be noted that a small amount of contemporaneous Byzantine pottery represented by the same vessel forms was also uncovered from below the flagstone floor (L.138) in the southern sector of Square A6; it is not illustrated due to the fact that it came from a context that was mixed with the later Crusader/Ayyubid pottery. The Late Byzantine ceramic corpus consists of well-known forms of bowls, basins, and store jars typical of the Jerusalem region during the 5th and 6th centuries. Among the assemblage was a Fine Byzantine Ware Form 1A bowl with a light reddish-brown (5YR 6/4) fabric and an incised wavy band on the vessel exterior below the rim zone (Figure 7:1), which Magness (1993, 193–194) dated from the mid-6th to late 7th centuries. Also present were two examples of the Form 1 Arch-Rim Basins with a reddish-yellow (5YR 6/6) fabric (Figure 7:2–3), which date from "mostly before the mid-6th century" (Magness 1993, 204–206), and a Form 6A store jar with similar fabric (Figure 7:4), which, according to Magness (1993, 227–229), makes its first appearance in the late 6th century. Of particular interest was the presence of a Form 4B store jar (Magness 1993, 224–225) that dates from the 5th to 6th century and was found directly embedded into the matrix of the plaster floor. Although destroyed during the process of removing the floor, and therefore not illustrated, the excavators were able to document a series of in situ photos of the sherd in question. A final Late Byzantine vessel was recovered from L.402 at the Lower Site (Figure 7:5). It is a Holemouth Jar Form 2 (Magness 1993, 234, no. 4) with a reddish-yellow (5YR 6/8) fabric fired that is fired to pink (7.5YR 7/3) on the exterior, which dates to "before the mid-6th century" (*Ibid*, 235).

Abbasid period (Stratum VI, 8th–10th century)

During the course of the 2012 excavations in the Southern Sector, a relatively large quantity of body

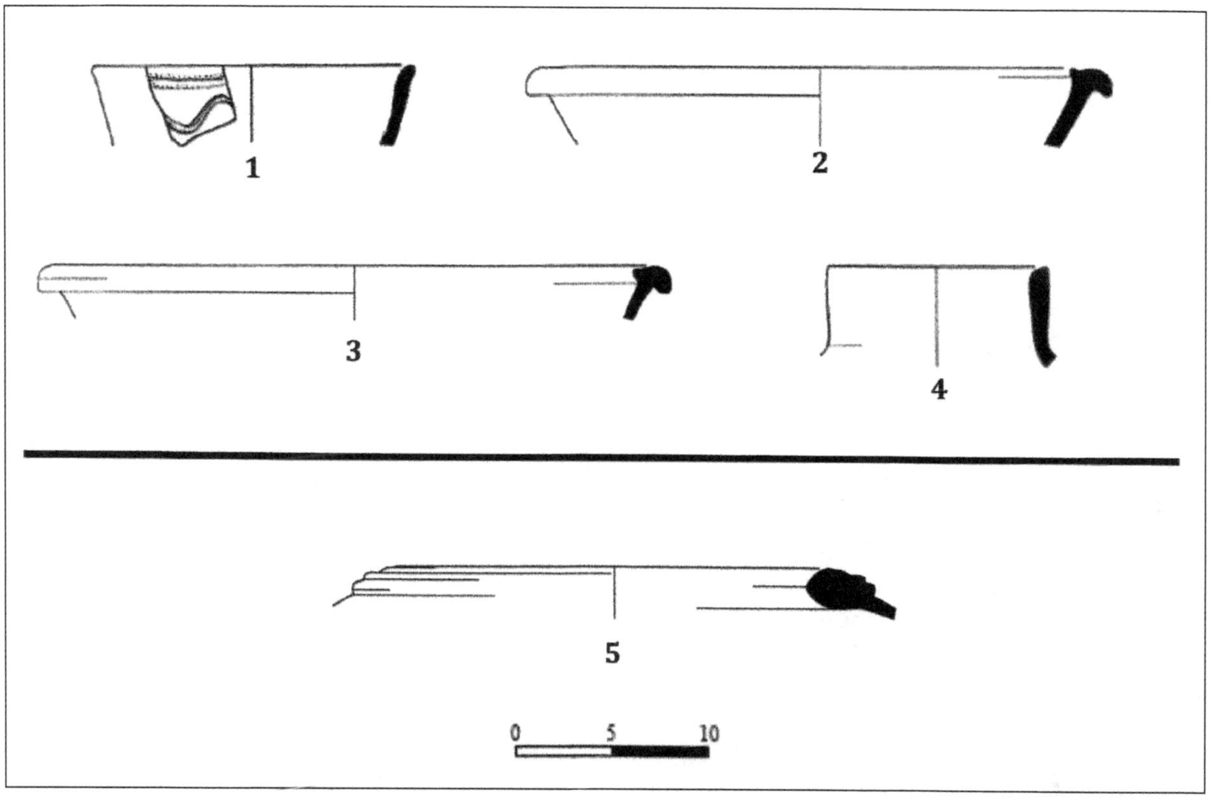

Figure 7 Late Byzantine pottery from the Southern Sector and Lower Site.

sherds from storage jars dating to the Early Islamic period were discovered; these were often mixed with Early Roman cooking pot rim fragments. However, none of these were associated with any architectural features or found in stratigraphic contexts. In 2015, fieldwork inside of the vaulted structure from C4 in the Northern Sector provided the first true evidence for the Early Islamic era in general, and the Abbasid period (9th through 10th centuries) in particular. Modified bedrock in Areas C (Northern Sector) and A (Southern Sector) have been identified as Abbasid constructions on the basis of associated pottery and radiocarbon dates.

The robber's trench in C4 (Northern Sector)

The work conducted in Room C4 during the 2015 season provided some important insights into the occupational history of the site. After removal of the deep subsoil layer (L.301) in the building the entrance and threshold (L.302) were cleared. A stairway (L.305) of three steps was uncovered leading from the threshold in the entrance of C4 down to the floor of the room. The occupational levels inside the room comprised of the following stratigraphic sequence: a beaten earth floor (L.306), which was directly under the L.301 subsoil; below this was a very thin white plaster layer (L.310) which probably represented another floor; and a floor that had been hewn into and modified the bedrock (L.307). In the northeast corner of C4, a pit (L.308) was cut directly into the modified bedrock floor; this was excavated and cleared.

Of interest was the pottery (Figure 8) that was retrieved from the pit contents (L.309): there were neither any Ottoman nor medieval sherds recovered. All of the latest diagnostic ceramics dated

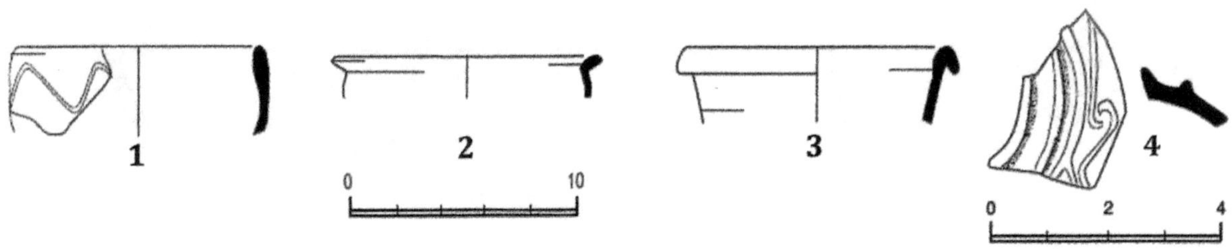

Figure 8 Abbasid pottery from the robber's trench in C4.

from the Abbasid period (9th–10th century). The first three are from a well-known class of vessels referred to as Fine Burnished Ware, which is essentially a continuation of the Fine Byzantine Ware tradition with different forms but a similar surface treatment. One of these is a bowl (Figure 8:1) which conforms to Magness Fine Byzantine Ware Form 1E (Magness 1993, 194). Despite the nomenclature utilized in the classification provided by Magness, this bowl is Early Islamic and a close parallel for it is attested at Ramla from an Early Abbasid context dated to the 8th–9th century (Tal and Taxel 2008, fig. 6.78:1). Another Fine Burnished Ware vessel, this one an ansate mug (Figure 8:2), was also found in L.309. Two parallels for this mug are attested from the excavations in the *Givati* Parking Lot, just outside of the Dung Gate of the Old City, the first of which was found in Stratum IV and is firmly dated to the 9th century (Cytryn-Silverman 2013, figs. 7.2:3, 7.7:6). A final Fine Burnished Ware specimen uncovered from L.109 is a jar with an everted rim (Figure 8:3). While this vessel is similar to the earlier Fine Byzantine Ware Form 1C, which Magness dates from the mid-6th to early 8th century (1993, 237, 239 nos. 1–2), the type is known from Abbasid contexts in the *Givati* Parking Lot (Reuven 2020, fig. 13.60:1). The final sherd from L.309 is that of an Abbasid lamp (Figure 8:4). These lamps are known from the typo-chronologies provided by both Hadad (2002, 95–106, Type 37) and Arnon (2008, 220–222, Medieval Type C), all of which were found in post-749 CE earthquake contexts. Parallels for the sherd come from Caesarea (Arnon 2008, fig. 22:108), Bet She'an (Hadad 2002, 98 no. 434), and Ramla (Cytryn-Silverman 2010, pl. 9.7:3).

Cisterns from Square A11 (Southern Sector)

A water collection system of two interlinked rectangular cisterns, designated Cisterns I and II (Figure 9), was uncovered during the 2013 excavation at the site (Storchan and Dolinka 2014). The inner faces of the two walls associated with these cisterns (W111 and W143) were covered with a layer of pinkish colored hydraulic plaster containing many ground ceramic inclusions. The plaster in the central area of the Cistern I floor was not preserved; however, the foundation layer was. It consisted of a layer of packed earth and flat stones located just above the bedrock. At the base of Cistern I a foundation course for wall W400 with a built-in channel was uncovered (Figure 9). The channel was plastered and led to a plaster spout that drained into Cistern II. The walls of Cistern II were also coated with the same hydraulic plaster as Cistern I, but differed in that they were hewn into the bedrock. The cisterns were both robbed out and intentionally dismantled during the Mamluk period, as indicated by the ceramic assemblage from the fill of the cistern and the robber's trench. At the time that the report for the 2013 season at Khirbet Beit Mazmīl was published, the construction for the cisterns was attributed to the Byzantine period, although no

Figure 9 Clockwise from top: Cistern I after excavation (photo: B. Storchan, courtesy of the IAA); Cistern II after excavation (photo: B. Storchan, courtesy of the IAA); Plan and section of Cisterns I and II from the 2013 season at Khirbet Beit Mazmīl (courtesy of the IAA).

clear, dateable material had been found within either the content or makeup of the cisterns. Given the stratigraphic nature of Cistern I, with its packed-earth and flagstone layer directly on top of modified bedrock, it fits the overall construction technique attested in the C4 vaulted structure of the Northern Sector.

Arch springer platforms for the monumental building in C6 and B5 (Northern Sector)

Dated primarily on the basis of construction technique and stratigraphic associations to the pre-Mamluk period, are a set of three roughly square piers that were utilized as platforms for arch springers from a monumental, vaulted building. Originally uncovered during the 2017 (Figure 10) season, these were thought to date from the Late Mamluk period, due to their stratigraphic positioning below the Early Ottoman building (Walker and Dolinka 2020, 203). Further excavation of the two piers (W.124/1–2) from Square B5 in 2019, revealed that they were constructed on a thin, compact soil layer (L.164) that covered the modified bedrock (L.163), much like the cisterns in A11.

Figure 10 The W.124/1–2 platforms for the monumental building, while excavation was in-progress during the 2017 season (view to the east).

The C6 press room

Of a similar construction technique to the above, a "press room" was identified in 2019 in C6 (Figure 11). It had a flagstone and packed earth floor (L.250) immediately above the modified bedrock (L.252). A charcoal sample taken from the preparation layer on the bedrock yielded a radiocarbon reading of between 891 and 1030 CE, which dates the original construction of that room to the Abbasid period.

Pottery from levels above bedrock in C6 and E1, Northern Sector (Figure 12)

A small sample of pottery dating to the Abbasid period was uncovered in fill levels above the bedrock in C6 and E1, where they were mixed with later Crusader/Ayyubid sherds, which suggests major clearing activities in each respective period at the site. An Abbasid fine-burnished ware bowl (Figure 12:1) was recovered from L.250, the flagstone floor on modified bedrock in C6. A similar example (albeit in a buff fabric) was found at the White Mosque excavations in Ramla (Cytryn-Silverman 2010, pl. 9.11.16). The rim and neck fragment of an Abbasid store jar with a pointed rim (Figure 12:2) was recovered from L.447, the lowest stratified context that was excavated in E1 and which was sealed under flagstone floor L.445. This jar has very thin vessel walls and a well-levigated buff fabric, and a comparable example was found at Tiberias, where it was attributed to Stratum III, which was dated to between 880 and 980 CE, or the Late Abbasid period. The final specimen from the Northern Sector originated from L.428, the soil fill inside of the circular wall W.140. It is a lamp (Figure 12:3) that conforms to Hadad Type 37 (Hadad 2002,

Figure 11 The C6 press room, which had a flagstone and packed earth floor that was constructed immediately above the modified bedrock (view to the north).

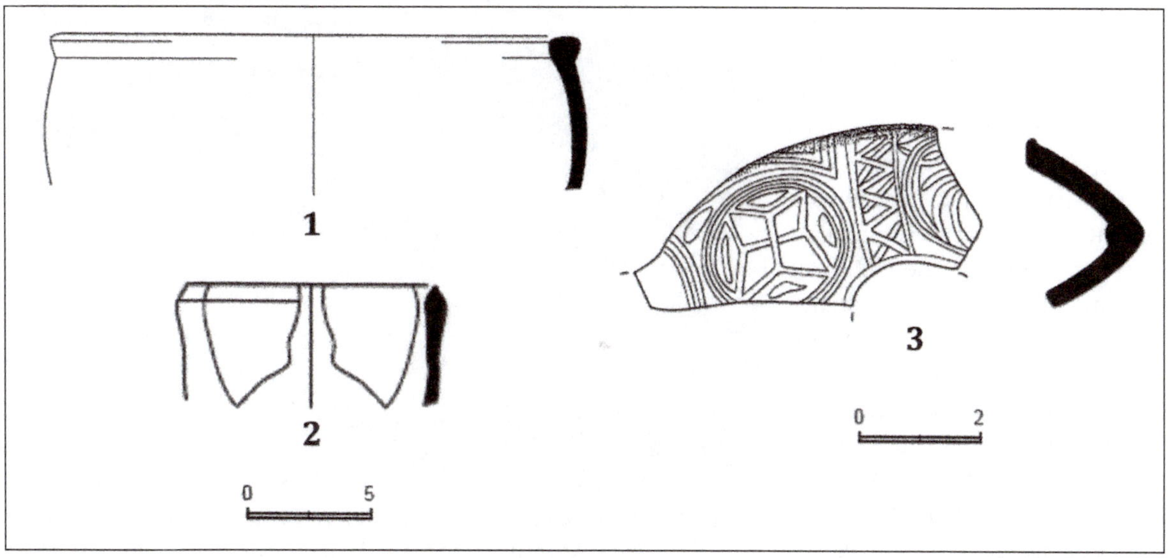

Figure 12 Abbasid pottery from levels above bedrock in C6 and E1.

105), which represents the most common Abbasid lamp form and had a long period of use from the late 8th/early 9th century through the 11th century. Although no parallels could be found for this interesting piece, its decorative schema includes the geometric designs and spirals, surrounded by three-lined medallions, that are typical of one of the sub-types for this form.

Taken together, although there was limited evidence for the Abbasid period, it does appear to have been a transformative one at the site, during which time the hilltop was cleared for the construction of a monumental barrel-vaulted building that continued to be reused in subsequent phases of occupation.

The Crusader/Ayyubid period (Stratum V, mid-12th–mid-13th century)

The Crusaders wrested control of Jerusalem from the Fatimids after a weeklong siege in July 1099. Within a short period of time, the Franks established a series of farmsteads scattered throughout the hinterland of the Holy City, one of which was likely Khirbet Beit Mazmīl. When Pringle visited the site during the 1980s, while investigating Crusader secular buildings and associated ceramics in the Jerusalem region (Pringle 1981, 53; Pringle 1997, 28, no. 36), he not only noted the presence of surface pottery collected there dating to the 12th century, but also described a particular wall in the C4 vaulted structure at Khirbet Beit Mazmīl as being of Crusader construction. The finds from all field seasons, and particularly those in 2012 and 2019, have confirmed what appears to be a small lens of Crusader/Ayyubid occupation. A modest but significant amount of evidence dating from this period, or the mid-12th through mid-13th centuries, was recovered from the lowest occupational levels and just above bedrock in both the Northern and Southern Sectors of the site.

The stratigraphic contexts of the Northern Sector

There were four primary contexts, three of which were uncovered in E1, that provide information about the Crusader/Ayyubid period in the Northern Sector. The first was L.447, a soil layer that was sealed under the L.445 Mamluk flagstone floor in the northern part of E1, where it continued into C6 as flagstone floor L.267. A charcoal sample taken from an associated locus (L.448)—which was the soil layer covering bedrock and sealed by the last flagstone floor at the bottom of the doorway—produced a radiocarbon date between 1040 and 1220 CE. The second important findspot, also from the lowest levels of E1, consisted of two successive stratified layers: the upper was the L.433 fill found on the L.424 burnt plaster floor (Figure 13), sealed by the L.429 Mamluk flagstone floor in the southern part of E1; and the lower was L.430, sealed by a burned plaster floor (L.424) and

Figure 13 The L.429 burnt-plaster floor, after removal of the L.424 final Mamluk flagstone floor in the southern part of E1 (view to the south).

just above the bedrock. A third context from E1 was L.434, the ashy soil layer sealed below the pebbled bottom of stone-lined pit L.140. The final findspot that yielded evidence from the mid-12th through mid-13th century was the soil deposit removed along the south face of W118 in B4, which was built directly on bedrock.

Crusader/Ayyubid pottery from the Northern Sector (Figure 14)

The Northern Sector yielded a small but significant corpus of well-stratified ceramics dating from the Crusader/Ayyubid period, or the mid-12th through mid-13th centuries. All the sherds were recovered from the lowest occupational levels at the site and just above bedrock. The assemblage presented here consists of 10 vessels represented by only bowls, cooking pots, and store jars. The bowls include those with gritty glaze that were either locally produced or imported from the Aegean and Lebanon; the cooking pots were of Beirut manufacture; and the store jars were all well-known forms that were produced locally in the Jerusalem region.

Three vessels were found in L.447. The first specimen is a glazed bowl with a slightly out-turned rounded rim that is covered with a gritty yellow glaze on the interior and dripping over onto the rim exterior (Figure 14:1). It conforms to Avissar/Stern Type I.1.2.1 Bowls with Gritty Yellow Glaze (Avissar and Stern 2005, fig. 2:4), which "... are characteristic of the Crusader period and do not continue into the Mamluk period" (Avissar and Stern 2005, 8) and are only found in 12th century contexts.

Another ceramic ware type found in the Jerusalem region that may truly be characterized as Crusader is the Aegean Coarse Incised Ware (Avissar and Stern 2005, 46, fig. 17:3–4, Type I.5.3). A body sherd from one of these bowls was also recovered from L.447 (Figure 14:2). It has a thick white slip that is covered with a transparent yellow lead glaze and is decorated with broadly incised linear sgraffito on the vessel interior. This ware corresponds to Stern Type GR.GL.6 and dates from the late 12th to the early 13th centuries (Stern 2012, 65–69, pl. 4.49:5–6). This pottery was brought to Jerusalem by the Franks and is found in small quantities from the Old City of Jerusalem, usually in areas associated with Christian and/or Frankish domestic dwellings and religious structures. It is also known from Crusader settlements in the hinterland surrounding the Holy City, where sherds of this type were uncovered during the 2012 excavations (see Southern Sector pottery below).

The final vessel from L.447 is a Beirut Cooking Pot (Figure 14:3). These are also known in small numbers from excavations in the Old City, although in the hinterland surrounding Jerusalem they are less common, as the cooking pots from these rural sites overwhelmingly consist of handmade local types. While the Beirut cooking pots occur in Ayyubid ceramic assemblages from Jerusalem—such as the Armenian Garden (Tushingham 1985, fig. 36:2)—in Akko they appear in both the Early (12th century) and Late (13th century) Crusader assemblages, with the exemplar from Khirbet Beit Mazmīl dating from the latter (Stern 2012, 41–44, pl. 4.16:3–11, Type BE.CW.2). These cooking pots are very distinct, act as a good chronological indicator that can be used for external dating of a ceramic corpus from a locus, and are a standard type found at Crusader sites throughout the region. Their production in Beirut is supported by the discovery of pottery kilns there that were uncovered at excavations in the modern city center.

The next four vessels were found in successive stratified layers from the lowest levels of E1: the upper sequence was recovered from the L.433 fill found on the burnt plaster floor (L.429), which was sealed below the final Mamluk flagstone floor (L.424); the lower sequence from L.430 was sealed below the burnt plaster floor (L.429) just above bedrock. Two glazed bowls were found in the lower sequence. The first (Figure 14:4) is a gritty glazed bowl with a rounded, externally thickened rim. The streaky glaze covers the interior of the vessel and drips over the rim and its exterior has tiny calcite inclusions. The rim form is very atypical for

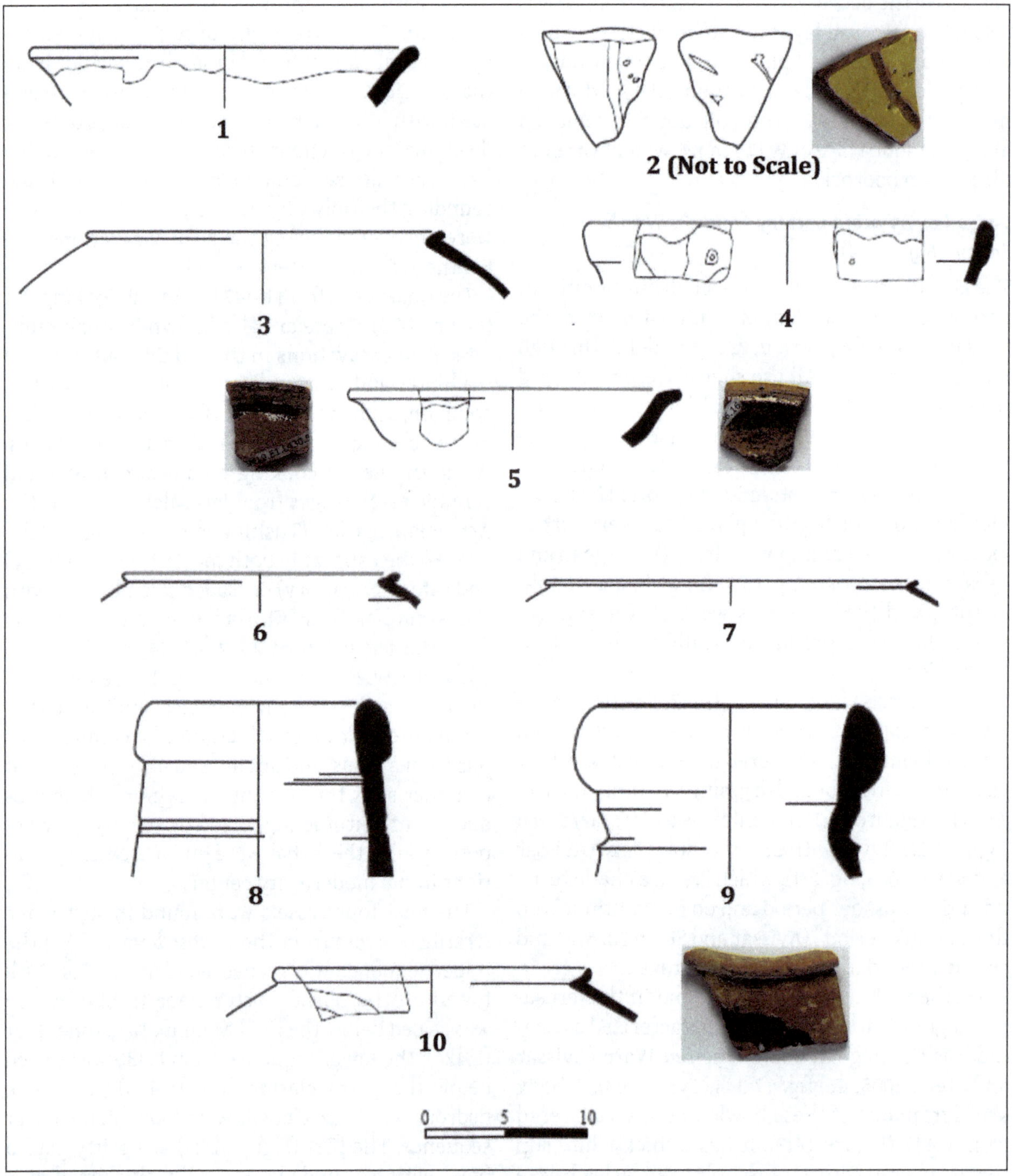

Figure 14 Crusader/Ayyubid pottery from the Northern Sector.

this ware type and no parallel could be found. The other bowl (Figure 14:5) has a small ledge rim that is covered with a splashed and mottled gritty yellow glaze that covers the vessel's interior, rim top and rim exterior. A dark brown band of paint was added horizontally along the lower end of the rim exterior. The fabric color, gritty glaze, and added brown paint suggest that this bowl is an import from Beirut, and the form is known from the Late Crusader assemblage (13th century) at Akko (Stern 2012, 44–47, pl. 4.19:14, Type BE.GL.2). The upper sequence from L.433 yielded two more examples of the Beirut Cooking Pots (Figure 14:6–7), both of which are attested in the Late Crusader assemblage at Akko, where they find good parallels from the BE.CW.2 corpus (Stern 2012, pl. 4.16:6 and 9, respectively).

In a second key stratigraphic context, L.434, the ceramic finds included two variants of the Jerusalem Ridged Neck Jars, a term coined by Boas (2006, 83, fig. 8) in his study on the materials from Khirbat Ka'kul, located in the hinterland to the north of the Old City of Jerusalem. The first (Figure 14:8) has a collared rim with two horizontal incised lines on the upper neck exterior, and the second (Figure 14:9) is more typical of the archetype for the form and has the commonplace buff exterior. Parallels for both of these jars are attested at Emmaus el-Qubeibeh.

The final vessel from the Crusader/Ayyubid assemblage in the Northern Sector is yet another example of an early 13th century version of the Beirut Cooking Pot (Figure 14:10), which is also paralleled in the Akko corpus (Stern 2012, pl. 4.1:7, Type BE.CW.2). What differentiates this one from the rest is that it has splashes of transparent glaze on the exterior of the upper shoulder, which is fairly commonplace for this form.

Stratigraphic contexts of the Southern Sector

The 2012 excavations in the Southern Sector produced further support for Crusader/Ayyubid occupation. Architecturally, this phase is represented in Area A by wall W405, which runs east-west and spans the northern limit of Squares A2/4/6, and the north-south oriented wall W143 in Square A2, which presumably abutted W405. In Area B, W414, located in the southwest sector of Square B8, was assigned a Crusader/Ayyubid date, although only partially excavated. Stratigraphically, evidence from this phase came from the following contexts: L.136 in Square A6, which was the makeup of plaster floor L.114 associated with W405; L.138, in the southern sector of Square A6, which represents the makeup of flagstone floor L.128 and was directly above the bedrock (Figure 15); the stratified fill (L.132) above the bedrock in Square A2; and floors L.543 and L.544, which were associated with W414 in the southwest sector in Square B8,

Figure 15 The L.138 fill, situated between the L.128 Mamluk flagstone floor and the bedrock in A6, produced Crusader/Ayyubid sherds; view to the north (courtesy of the IAA, photo: B-297802).

and were just above the bedrock, which is considerably lower than in Area A, due to the southward slope of the hill.

Crusader/Ayyubid pottery from the Southern Sector

A larger corpus of Crusader/Ayyubid ceramic material than was found in the Northern Sector was recovered from the Southern Sector at the site. The evidence comes from the stratified contexts attested in both Areas A and B, with additional pottery that was recovered from secondary deposits found on top of the L.114 plaster floor in Square A6, as well as from the L.510 subsoil in Square B1. Of importance are the well-known imported Aegean and North Syrian wares from these loci, which provide good external dating for the contexts from whence they came. The small assemblage consists mostly of bowls, with a store jar, jugs, and lamps completing the corpus.

Glazed bowls (Figure 16:1–8)

A variety of glazed bowls were recovered from Stratum III contexts. Noteworthy among them are the two fragments (the rim and ring-base, respectively) from an Aegean Coarse Incised Ware bowl (Figure 16:1–2). Both sherds have a reddish-brown (5YR 4/3) fabric with a thin white slip and are covered with a transparent yellowish lead glaze. The bowl is decorated with a broadly incised sgraffito consisting of linear motifs that are dark yellowish brown (10YR 4/6) in color and emanate from the center interior of the bowl. The ware is representative of the well-known variety (see Avissar and Stern 2005, 46, fig.17:3, Type I.5.3) which conforms to Stern Type GR.GL.6 and dates from the 13th century (Stern 2012, 65–69, pl. 4.49:5–6). Of interest is the bowl with a polychrome (green/yellow/brown) decoration and double slip with a gritty glaze on its interior and exterior (Figure 16:3). It has a light reddish-brown (5YR 6/4) fabric and a slightly upturned and flattened rim. It conforms to Avissar/Stern Type I.1.1 (Avissar and Stern 2005, 6, fig. 1:3), which dates from the mid-11th to the mid-12th century. The polychrome decorative schema on these bowls is reminiscent of its Fatimid forbears and appears to be a local Jerusalem production during the Crusader/Ayyubid period. Two slip-painted ware bowl fragments typical of the mid-12th through mid-13th centuries were recovered from Area A. The first is a very-low ring-base (Figure 16:4) with a reddish-brown (2.5YR 6/4) fabric, white slip-painted decoration, and a transparent green lead glaze which corresponds to Avissar/Stern Type I.1.6.2 (Avissar and Stern 2005, 19, fig. 7:5–8). Parallels for this bowl base are attested in the Jerusalem hinterland at Emmaus El-Qubeibeh and Belmont Castle, which also has a green glaze. The second sherd is from a bowl with an everted rim (Figure 16:5) and an upright wall that is broken just above where the carination point would have been. It has a red (2.5YR 5/6) fabric and the white slip-painted decoration is covered with the more common transparent yellow lead glaze (Avissar and Stern 2005, 19, fig. 7:1–4, Type I.1.6.1). A good parallel for this vessel was recovered from the Frankish farmhouse at Har Hozevim. The next three vessels are monochrome glazed bowls that are homogenous in both their fabric and surface treatment and exhibit similar rim forms as well. They have a reddish-yellow (5YR 6/6) fabric with tiny calcite inclusions, a thin white interior slip, and a well-adhering dark green interior glaze dripping over rim. The first two examples (Figure 16:6–7) have a flat rim with a beveled exterior and are similar to a vessel from Area X-5 from the Cardo excavations (Avissar 2012, fig. 10.9:9), while the last bowl in this series (Figure 16:8) has a slightly incurved rim that is beveled on the interior and has a parallel from the Damascus Gate (Wightman 1989, 75, pl. 64:3). Taken together, both the dark-green well-adhering glaze on these vessels and their rim forms are representative of the earlier medieval assemblage from the site (M. Avissar *personal communication*), and were all found in association with the lowest floor level (Loci 543 and 544) uncovered in Area B.

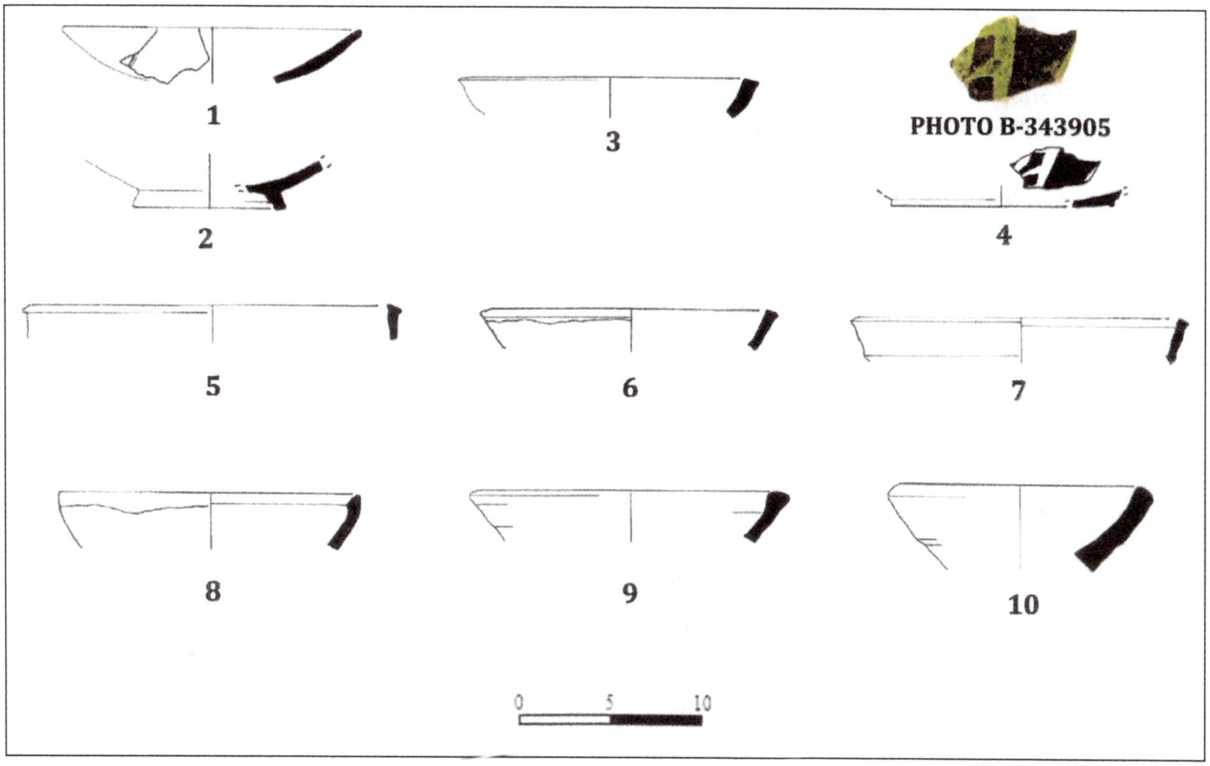

Figure 16 Crusader/Ayyubid glazed and plainware bowls from the Southern Sector (Figure 16:4 courtesy of the IAA, photo: B-343905).

Plainware bowls (Figure 16:9–10)

Plainware bowls with neither glazed nor decorated surfaces were under-represented in the ceramic assemblage from Stratum III, and only two vessels are presented here. The first example (Figure 16:9) is handmade and has a very coarse light reddish-brown (5YR 6/4) fabric, numerous grits, and calcite inclusions of varying sizes. The rim is rounded and thickened on both the interior and exterior. These bowls usually had a flat base, as in the parallel from the Armenian Garden (Tushingham 1985, fig. 41:14). The second piece (Figure 16:10) is also handmade, has an inverted conical body with thick vessel walls and a flat rim. The fabric is a light reddish-brown (5YR 6/4), with numerous grits, large calcite inclusions, and burned-out organic temper. A parallel or this vessel was uncovered at Emmaus-El Qubeibeh where it was dated from the mid-12th to mid-13th century.

Handmade decorated bowls (Figure 17)

The next group of bowls is relatively homogenous in size, shape, and surface treatment. They are all handmade and all but one (Figure 17:7) are burnished either on the interior, top of the rim, or both. In well-stratified assemblages from the Old City of Jerusalem, handmade bowls with burnishing first appear in contexts dating from the mid-12th to mid-13th centuries (M. Avissar *personal communication*). Many of the bowls are painted with either simple linear or geometric designs on the interior (Figure 17:1–5) and on the top of the rim (Figure 17:3–5), and several from this phase are characterized by appliqué finger-impressed "rope" decoration on the exterior just below the rim zone (Figure 17:4–7), a trend which continues well into Phase IIA, or Mamluk II, which dates from the 14th through 15th centuries. The first three bowls (Figure 17:1–3) have relatively thin vessel walls when compared to their later counterparts and have rims that are beveled slightly on the exterior, interior, or a simple ledge that is slightly rounded, respectively. The examples with the rounded rim and impressed "rope" decoration on the exterior (Figure 17:4–6) are typical forms of the 13th century from Jerusalem (M. Avissar *personal communication*). The final bowl from this group (Figure 17:7) is unique: it has much thicker walls, a rim that is thickened on the interior, and a vestigial horizontal strap handle. The coarse fabric of these vessels ranges from pink (5YR 7/4) (Figure 17:1, 3), through reddish-yellow (5YR 6/6 or 7/6) (Figure 17:2, 4–5), to light reddish-brown (5YR 6/4) in color (Figure 17:6–7), and contains numerous calcite inclusions and occasionally has burned-out organic temper (Figure 17:5, 7) in its matrix. The painted examples are dark reddish-brown (5YR 3/2 or 3/3) or reddish-brown (5YR 5/4), and rarely have red (10R 5/8) paint. Of interest is the burnished and painted bowl found below the lowest floor in Square B8 (Figure 17:2), which has a rim form and pressed-on handle reminiscent of the frying pans and/or cooking bowls with glazed interiors that are typically found in Crusader/Ayyubid assemblages. Although there were no exact parallels for any of the examples presented here, they are similar in form and decoration to those published from the Armenian Garden and the Damascus Gate in the Old City (Tushingham 1985; Wightman 1989) as well as those from Khirbat Ka'kul in northern Jerusalem (Boas 2006). It should be noted that all of

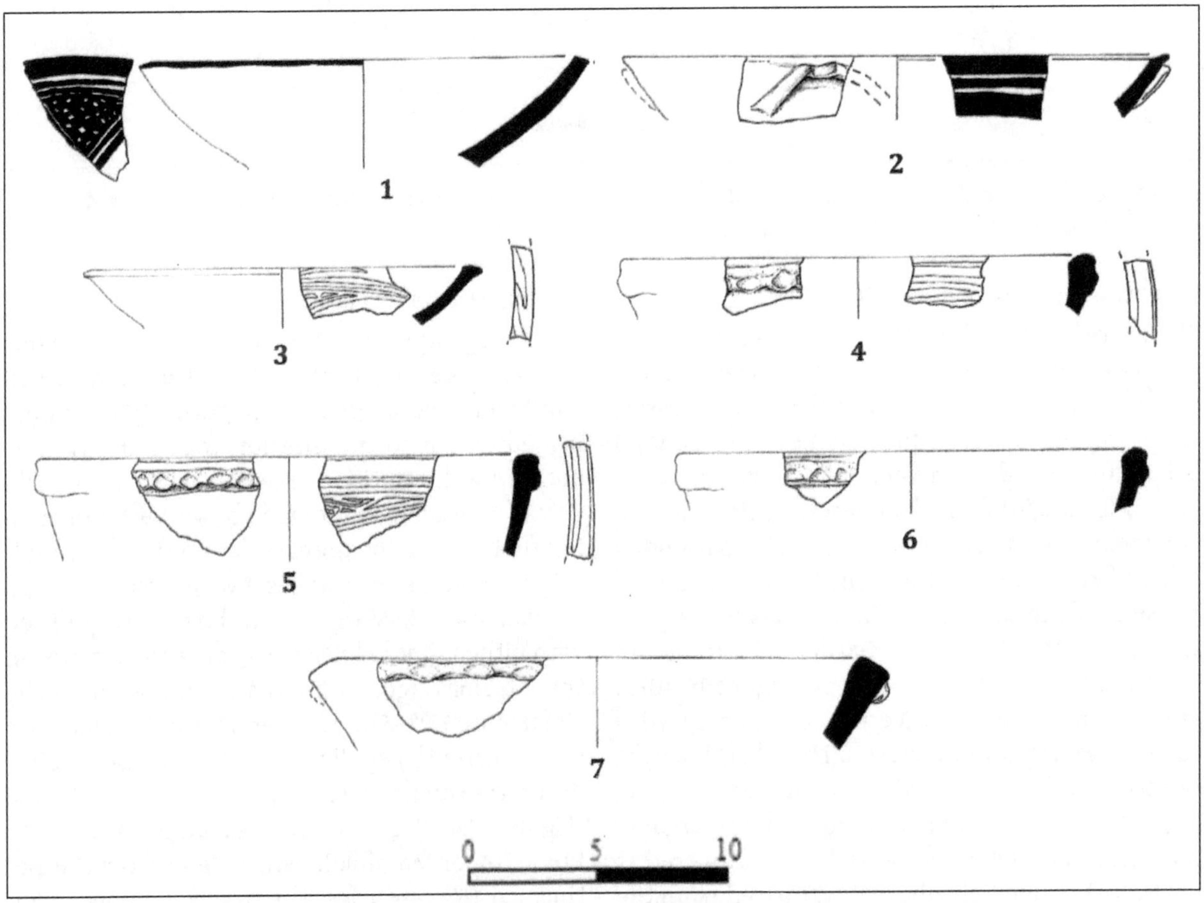

Figure 17 Crusader/Ayyubid handmade decorated bowls from the Southern Sector.

the bowls from this small corpus originated from good, sealed contexts, such as the makeup of floors and directly above the bedrock.

Store jars, jugs, and lamps (Figure 18)

Vessels with closed forms, such as store jars and jugs, as well as lamps, were also represented amongst the Phase III ceramic corpus from Khirbet Beit Mazmīl. An example of the ubiquitous Jerusalem Ridge-Neck Jar (Figure 18:1) was recovered from the makeup of the L.114 floor in Square A6. It has a very pale brown (10YR 8/3) fabric and a collared incurved rim that is slightly thickened on the exterior. The vessel conforms to Avissar and Stern's "Store Jars with Thickened Rim and Ridged Neck" (Avissar and Stern 2005, 100, fig. 4:24, Type II.3.1.3) and was "probably manufactured in and around Jerusalem" (Boas 2006, 83). This family of jars first appears in a buff ware form during the mid-12th to mid-13th century at Emmaus/El-Qubeibeh and continued production into the 14th century, although the later examples are in a pink ware. Good parallels for the Khirbet Beit Mazmīl jar were uncovered during recent excavations in the Old City of Jerusalem, from Phase IIIc at the Austrian Hospice (Dolinka 2017, fig. 3) and from the lowest floor at Beit Musa Effendi, where they were found with wares dating from the Crusader/Ayyubid period.

Of great importance for dating was the presence of two Port St. Symeon Ware sherds (Figure 18:2–3), both of which come from the long and narrow necks of the globular jugs in this ware type (Stern 2012, 56). They have a pink (7.5YR 7/4) fabric with calcite inclusions, a thick white exterior slip covered by a light yellow transparent and shiny lead glaze, and incised decoration painted with dabs of yellowish-brown, dark brown, and green glaze. Port St. Symeon Ware has been referred to as "the Crusader pottery *par excellence*" by Pringle (1986, 458). The ware conforms to Avissar/Stern Type I.7.4 (Avissar and Stern 2005, 54–56, fig. 22, pl. XVII:2–4) and to North Syria Glazed Ware NSY.GL.4 from Akko, where it dates from the late Crusader assemblage, or the 13th century (Stern 2012, 55–58). While the first sherd (Figure 18:2) from Khirbet Beit Mazmīl was recovered from a secondary deposition, the second example (Figure 18:3) was found in the makeup of the lowest floor (L.544) from the southwestern sector of Square B8, which provides a good relative dating for the construction of that floor. Closed forms in the Port St. Symeon Ware are a rarity, as most of the finds are limited to bowls, but a similar jug neck sherd was uncovered from the Hospitaller Compound at Akko (Stern 2012, 86, pl. 4:39:3).

A small, decorated body sherd from a strainer jar (Figure 18:4) was recovered from the makeup (L.136) of the Phase III floor (L.114) in Square A6. It has a very pale brown (10YR 8/4) fabric with pinpricked, incised and gouged decoration on its exterior. This jar sherd conforms to Hama Category D.XVIII (Poulsen 1957, 258–260), as well as Avissar/Stern Type II.4.3.2 (Avissar and Stern 2005, 111, fig. 46:3–6), which was assigned a broad date range from the 13th to 15th centuries. Similar decorated strainer jars are known from Ayyubid and Mamluk contexts in Jerusalem, from the Armenian Garden (Tushingham 1985, 149, fig. 43:1–8) and Damascus Gate (Wightman 1989, pl. 61), as well as from Emmaus/El-Qubeibeh in the Jerusalem hinterland, where they are dated from the mid-12th to mid-13th century. Good parallels for the Khirbet Beit Mazmīl jars have been found at both the City of David and the Visitor's Center in Jerusalem (Dolinka 2021b, 431–432, fig. 21.2:4–5). Other jug sherds were uncovered in Area A. The first is a spout (Figure 18:5) that was uncovered in L.138 from the makeup of the flagstone floor in the southern part of Square A6. It has a pink (5YR 7/3) fabric with a few tiny calcite inclusions that is fired to a very pale brown (10YR 8/3) in color, and is typical of examples found on the buff ware jugs with stamped decoration on the neck (Avissar and Stern 2005, 110–111, fig. 45:6, Type II.4.2.2) that are dated exclusively to the 13th century. The second sherd is from the neck of a jug with a collared everted rim that is triangular in section (Figure 18:6). This vessel was over-fired and most likely meant to be buff in color on the exterior, but instead has a light reddish-brown (5YR 6/4) fabric fired to reddish-gray (2.5YR 6/1). Although no parallels could be found for this jug, it is clearly of a 13th century type (M. Avissar *personal communication*).

Completing the Phase III ceramic assemblage are two examples of the Ayyubid mold-made slipper lamps, which were almond shaped, featured a high tongue handle and were decorated with linear patterns, arabesques, and calligraphic designs rendered in a relatively fine, high, and thin relief when compared with their later Mamluk counterparts, which had a much thicker and lower relief and were limited to geometric and linear designs. The first lamp (Figure 18:7) has a very pale brown (10YR 8/3) fabric throughout and the second one (Figure 18:8) is of a pink (7.5YR 7/4) fabric fired to very pale brown (10YR 8/3) color. They conform to Avissar/Stern Type II.2.1.1 (Avissar and Stern 2005, 126–128) and are limited to the first half of the 13th century in date. Although there are no exact parallels for these lamps, they are similar to those from Abu Ghosh (de Vaux and Stève 1950, pl. XIX:1), Emmaus/El-Qubeibeh (Bagatti 1993,

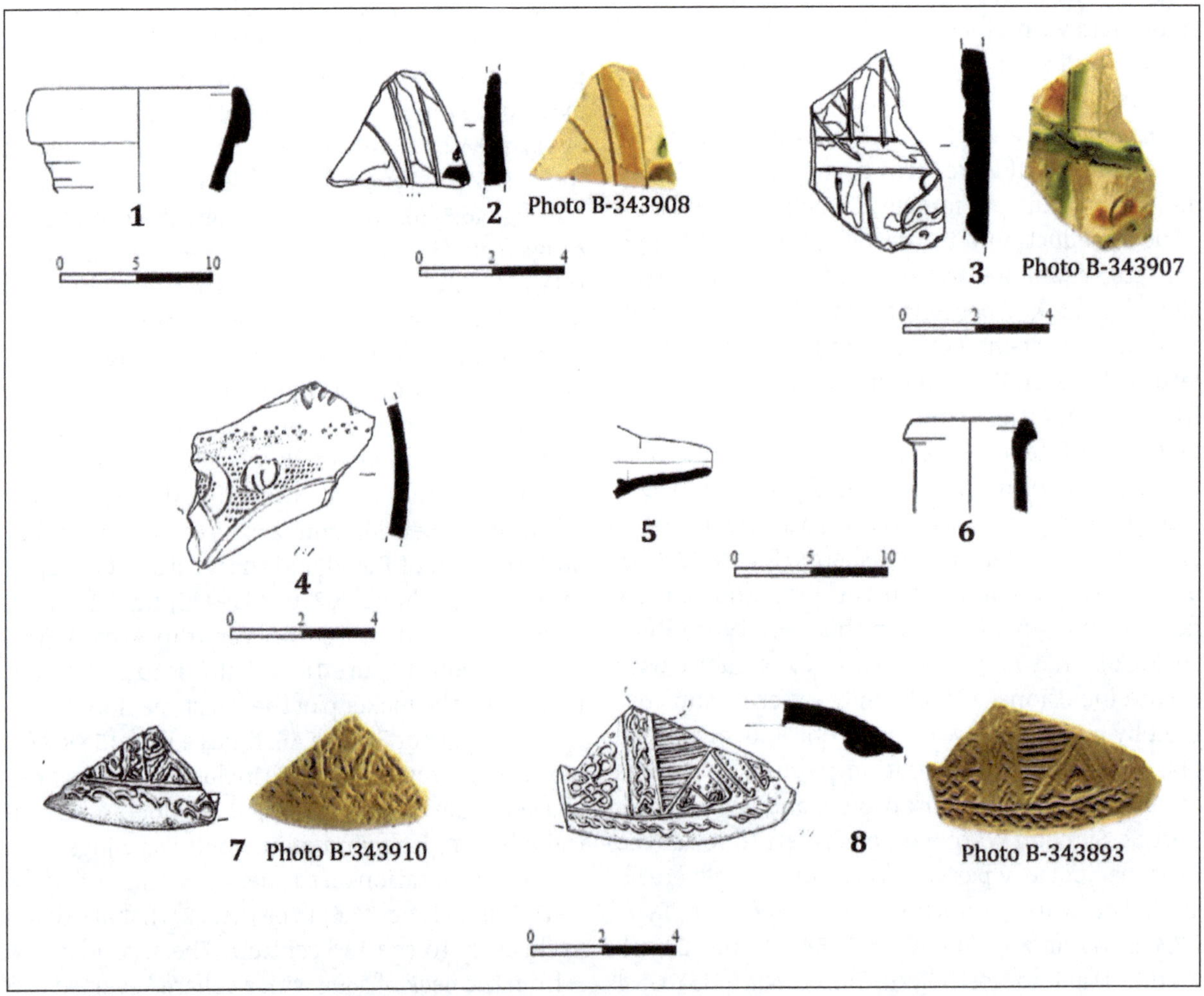

Figure 18 Crusader/Ayyubid store jar, jugs, and lamps from the Southern Sector (Figures 18: 2–3, 7–8 courtesy of the IAA, photos: B-343908, B-343907, B-343910, B-343893, respectively).

fig. 34:3), the Damascus Gate (Wightman 1989, 73, pl. 62:7–8) and the Armenian Garden (Tushingham 1985, 151, figs. 34:41, 37:6, 38:14, 17), while the decorative border on the top of them is reminiscent of contemporaneous lamps uncovered at Ramla (Cytryn-Silverman 2010, pl. 9.33:3, photos 9.44–45).

Concluding remarks on the Crusader/Ayyubid pottery

In her groundbreaking work on the Crusader pottery from Akko, located on the northern littoral of Israel, Stern demonstrated that there was indeed a locally produced ceramic repertoire that was produced during the Crusader period (Stern 2012, 34–38, pls. 4:1–11). This was proven by both petrographic and chemical analysis. The Jerusalem area, however, presents an entirely different situation. Regarding the Crusader ceramic repertoire from Jerusalem, the following points made by Boas (2005, 141–143) provide the best synopsis: the vast majority of vessels were manufactured by local, non-Frankish potters. These included store jars and water jugs with a buff-colored fabric which continued traditions established long before the arrival of the Franks. Considerable quantities of unglazed/glazed ceramics reached Jerusalem from Fatimid Egypt, and these included a variety of lamps as well as buff-ware water jugs. Ayyubid Syria supplied high-quality pottery which consisted of alkaline glazed bowls, jugs, and lamps decorated with underglaze-painted designs in blue and black. The third source for ceramics were countries under Christian rule, such as Lebanon and Greece, from the late 12th through early 13th centuries.

The pottery assemblages from Khirbet Beit Mazmīl attributed to this period seem to have confirmed the model provided by Boas. For Jerusalem there was no pottery made by the Crusaders *per se*. The vast majority of it was locally produced by the region's Muslim inhabitants, accompanied by imported glazed wares from Islamic lands, and supplemented by a relatively small number of glazed bowls and cooking pots from Beirut and a limited number of glazed bowls from the Aegean that the Franks brought with them. This phenomenon presents a vastly different picture when compared to the northern and coastal sites of Israel that were under Crusader dominion for nearly a century longer than Jerusalem, which fell to the Ayyubids in 1187.

The Crusader-period "Seal of Mar Sabas" (Figure 19)

Without question, the most important find from Khirbet Beit Mazmīl dating from the Crusader period was the discovery of a lead seal uncovered from the Southern Sector in A6 during the 2012 season at the site. This seal was studied in great detail by Kool (2014), who revealed just how significant the find was, and the unearthing of this rare artifact even made international news (Gannon 2014). The obverse depicts a portrait bust of St. Sabas (Figure 19a), who founded an eponymous monastery on a cliff overlooking the Kidron Valley in the Judaean Desert sometime during the late 5th or early 6th century. The monastery (*laura*) of Mar Saba is only one of two places—the other being St. Catherine's monastery on Mount Sinai—to have a continuous monastic community since its founding (Kool 2014, 8). The image of St. Sabas on the obverse portrays him as a bearded figure, who holds a cross in his right hand and possibly a New Testament in his left and bears his holy name (Kool 2014, 8). The reverse of the seal (Figure 19b) bears a Greek inscription that translates as "the seal of the *laura* of Holy Sabas" (Kool 2014, 8). It functioned as a *bulla*, which consisted of a cord placed between

Figure 19 a) Obverse of the Mar Saba *bulla*; b) Reverse of the Mar Saba *bulla*.

the obverse and reverse discs that was then hammered together and was used to seal an important document (Kool 2014, 7), much like sealing wax. It appears then, that the *bulla* sealed either an important document or deed that was dispatched from the monastery of Mar Sabas to some type of Frankish official residing at the site, which may have been either the property of the monastery or perhaps even of the Church of the Holy Sepulcher (Bresc-Bautier 1984, 259–260), "which owned numerous rural estates around Jerusalem" (Kool 2014, 13).

Reconstructing building practice in the pre-Mamluk period

The architectural remains present numerous challenges that exacerbate the problems of studying the pre-Mamluk site. Modern urban expansion has limited our study area to only a small fraction of what the site once was. The irreversible destruction of evidence due to modern building activities and the fact that four archaeological projects (three of them salvage excavations) were conducted separately and without coordination has created gaps in the architectural picture of the site. With the physical remains scattered among different non-contiguous areas, it is often impossible to clearly establish the functional and spatial connections between them.

In addition to these modern factors, one needs to consider one further historical process: a continuous reuse of building materials and reoccupation—with or without major modifications—of earlier structures. This practice, which is of course not limited to Khirbet Beit Mazmīl (for a theoretical overview on the subject, see Pini 2021), prevents us from having a complete idea of the kind of buildings that once stood here, as well as of their chronology. Structures were rearranged and/or demolished to accommodate new functional or spatial needs and building materials from demolished or ruined earlier buildings were often reutilized several times before entering their final phase of use. Each reuse might have made use of ever smaller building blocks, requiring modification and adjustments by the builders, and making the task of recognizing their original date even harder. This is particularly true for the earliest phases of the site, those pre-dating the Mamluk occupation.

Before looking more into the architectonical evidence, some words about the criteria for the identification of reoccupied structures and reused materials are necessary. One should first consider that all buildings appear to be built using locally quarried limestone, and there are no visible traces of materials other than stone or of imported building materials. The stonework rarely carries diagnostic toolmarks to allow a precise dating of their original construction. Some information about the architectural development comes from the degree of homogeneity of building materials: within certain areas of excavation, in fact, materials appear more homogenous than in others, indicating that the structures were either only slightly modified after their original construction or that the source for these reused materials was the same and, possibly, not far away. This homogeneity is generally one of the major criteria for dating structures prior to the Mamluk phases, when building materials reused in single structures tend to come from a multiplicity of sources and therefore appeared extremely diversified. However, one needs to consider that reuse processes which took place already during earlier, pre-Mamluk phases (the later Byzantine or Early Islamic periods) might, in certain cases, present rather variegated building materials.

In general, delineating generalized and universally valid criteria is impossible and one should be cautious in the interpretation of materials remains on a case-to-case base. In certain cases, the use of a distinctive building technique helps to identify chronologically significant reuse episodes, and this is sometimes helpful to recognize certain patterns of reuse and reoccupation. Still, most chronological information comes from stratigraphic relationships. Even if it is almost impossible to offer a precise date for these pre-Mamluk structures, it is nonetheless possible to assess their anteriority in terms of a relative chronology. Given these premises, if pre-Mamluk materials reused in later structures are documented almost everywhere on site, only an extremely limited portion of structures appears to have been left in place throughout its entire history of occupation.

In fact, standing remains of the pre-Mamluk period are documented exclusively in six areas, scattered across the different areas of the site (see Appendix 3). The Northern Sector is where most evidence is still visible: a wall between C4 and C5, almost entirely embedded into the most recent drystone structures (W120); two low platforms/pillars in C6 and B5 (W124/1 and W124/2), also heavily impacted by later construction phases; a cloister vault in C6/C7 (composed by a platform W123 and the superstructure L111), also almost entirely covered by 19th century drystone walls; and finally, the *miqveh* complex in Area D (Levi 2012).

The interpretation of the evidence in Areas C4 and C5 is particularly problematic since all the structures have been covered and/or modified by the later drystone buildings (W105). However, an extremely limited portion of an earlier structure is clearly visible in the northeastern corner of C5 as well as through some holes inside the vaulted chamber C4 (W120) (Figure 20). The parts of wall still accessible are characterized by blocks markedly bigger than in any other stonework at the site, finding as direct comparison only the cornerstones of W111 and W143 of Area A and some reused blocks in Area B of the most recent excavation.

The wall in C5 seems to form a corner (Figure 21) and continue northwards, running more-or-less parallel to the later vaulted rooms—making it visible through a series of preexisting holes/niches in the eastern side of C4 (Figure 22). Dating this structure is, at present, impossible, since there are no technical features or associated materials offering us any precise indication. The large, well-cut and squared appearance of the blocks do not necessarily imply an earlier dating, or a direct connection with a Roman quarry found in Area A of the Southern Sector. It may indicate a specific static requirement of the wall or be related to the function of the building itself: possibly a structure with a defensive character which would require a more massive construction. It is difficult to determine whether this structure was already built reusing earlier materials. This would affect our understanding of specific occupation dynamics during the early

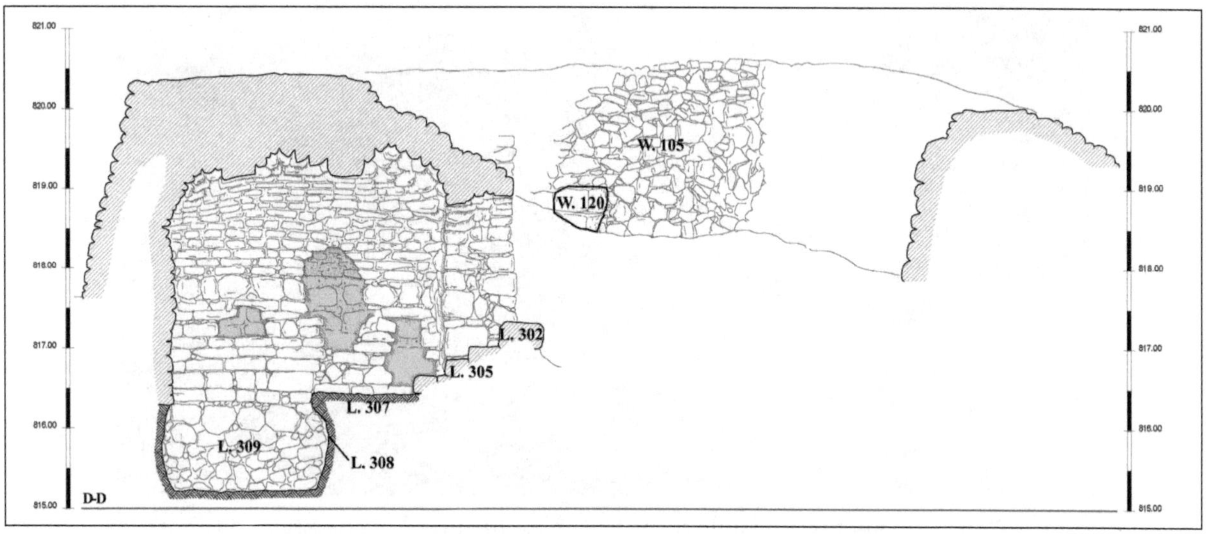

Figure 20 Section of Chamber C4 and the north-eastern corner of C5 (elaboration by N. Pini on S. Pirsky).

Figure 21 Visible portion of the ancient wall in C5, under W105 (photo: N. Pini).

Figure 22 The ancient wall visible through the eastern wall of chamber C4 (photo: N. Pini).

phases of the site. Notably, the first five courses of the eastern wall of C4 are quite different from the rest of the masonry of the chamber, being regular and made by oblong and well-cut blocks, closely recalling the ones used for the pillars (W124/1 and W124/2). Yet there is no certainty they are in their original place and their stratigraphic relation with the previously described wall, which appears to run right behind them, remains unclear.

A chronological clue comes from a robber's trench excavated in the northeastern corner of C4, where another part of a pre-Mamluk structure was documented (Figure 23). The trench exposed some further lower courses under the eastern wall of C4, which look like a foundation level—with roughly-cut blocks of large size but regularly layered—and were associated with a securely dated Abbasid locus. As mentioned, with the evidence at our disposal, it is impossible to determine whether these foundations were contemporary either with the lower courses directly above them or with the wall hidden behind the eastern wall of C4. What is certain is that this foundation level refers to the first documented architectonical phases in this part of the site, and pre-dates all the remaining structures. Therefore, the evidence indicates that this area was occupied already during the Early Islamic period by some structures, which were partially maintained and later reoccupied, certainly in the Mamluk and Ottoman periods. Absolutely no element gives us any indication of the possible function or scale of this building or any other of the pre-Mamluk structures at the site.

Another example of extant, pre-Mamluk construction are two rectangular platforms found in areas C6 and B5 (W124/1 and W124/2) (Figure 24).

Figure 23 Robbers' trench in C4, exposing the foundation level under the eastern wall of C4 (photo: N. Pini).

Determining the kind of complex they were originally part of is impossible, as the later interventions too-heavily transformed the built environment. In any case, the two constructions appear to be contemporary, built with the same set of building materials and with the same building techniques, both quite distinctive and different from almost all other structures on site. Both platforms present one face of regular rows of well-cut, prevalently oblong blocks (average height of 17 centimeters; average length between 30–50 centimeters, with numerous blocks reaching 60–70 centimeters), laid on 1 centimeter-thick layers of grey mortar. The fill, partially hidden by later building phases, appears to be of earth and rubble, similar to the "sack masonry" found elsewhere at the site (as in Area A in the Southern Sector).

It is impossible to determine the function of these structures, nor to clearly date them. The archaeological record in C6/E1 documents abundant Abbasid materials in layers certainly earlier than the creation of the room east of pillar W124/1 and the installation filling the gap between it and W123 (a possible third pillar to the north). The installation was built directly on top of a pebble floor dated to the Abbasid period by radiocarbon analysis (L.245). The installation necessarily required the existence of the pillar, so it is certain that it was built before the Mamluk period, when the entire area was transformed, and the productive facility likely built.

Contemporary with "pillar" W124/1 is a channel (W127), abutting its southern face: the building materials and the way the courses are laid are

Figure 24 Aerial view of pillars W124/1 and W124/2 from C7, looking south (photo: N. Pini).

identical to the faces of the pillar, suggesting that they were built at the same time. The channel was probably cleaned in the Early Mamluk period, when the area was reoccupied, and maintained at least until the Late Mamluk phase, when it was once again filled with debris.

The southern pillar (W124/2) was highly impacted by later building activities, being partially dismantled in its southern half, and embedded in a later building in Area B. This area appears to be completely transformed during the Early Mamluk period, and rubble was dumped in B5, between the two pillars. In this deposit, several materials dating to the Abbasid period were found, suggesting that Area B was occupied during the Early Islamic period as well. The fact that the structures dating to the Mamluk period clearly abut and transform the pillar is clear proof that the latter was already existing at the time of the 13th to 14th century reoccupation of the site. It is the only evidence of a pre-Mamluk structure that was, otherwise, completely destroyed.

In general, both pillars clearly pre-date the Early Mamluk reoccupation of the Northern Sector. The fact that the Abbasid evidence is so widespread across the areas of this sector of the site and is found in the immediate proximity of the few surviving pre-Mamluk structures, might indicate that the pillars were part of a large Abbasid structure. However, it is hard to determine if their construction first dated to the Early Islamic period or were part of an earlier building later re-occupied under the Abbasids.

A third pillar (W123), located north of W124/2, supports one of the most indicative pre-Mamluk features, a cloister vault (L111), unfortunately visible only from the outside through a gap in the collapsed later masonry (Figures 25 and 26). Both pillar and vault were, in fact, partially obliterated by the later Mamluk and Ottoman constructions (W121 and W126); the collapse of the late Ottoman drystone structure partially exposed the vault. It presents the same features as the previously described platforms, but differs in the upper portions of its masonry, namely the presence of a springer of a barrel vault made of thin slabs of well-cut blocks. The vaulting technique and the materials employed are markedly different from anywhere else on site, and it might well be one of the few surviving examples of original, pre-Mamluk roofing left in place (Figure 27). The vault itself is built with rather small, well-cut blocks, carefully laid and set in place with regular layers of whitish mortar. From the limited visible portion of the vault, the materials might be of reuse, but their reemployment certainly followed a careful selection and adaption process. The most direct comparison of the materials is with the original vaulting of the *miqveh* in D5, although the typology of the vault is entirely different. No other examples of cloister vaults are to be found on site.

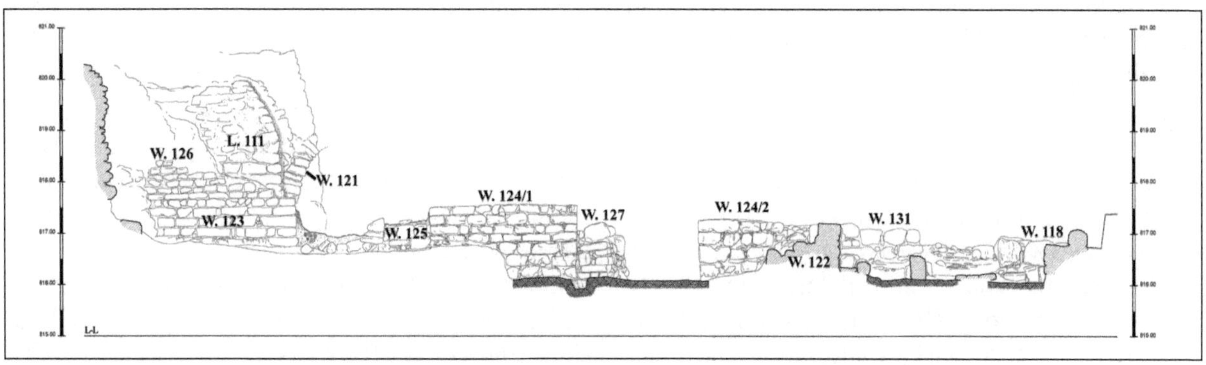

Figure 25 North–South section of Area C6 and B5, with all three pillars, the cloister vault and the Mamluk barrel vault visible (drawing: S. Pirsky).

Figure 26 Cloister Vault L111 and the supporting pillar W123, with later Mamluk constructions (photo: N. Pini).

As we were unable to investigate the inside of the room extending eastward, we have only a partial picture of the structure, as well as limited chronological references. Notwithstanding, it appears that at least a second, similar vaulted chamber stood in the neighbouring room C3, where we possibly have a second surviving arch springer in the northwestern corner, whose features closely resembles those of the cloister vault in C6. Looking at the dimensions of the spaces between C7 and C5, it might be likely to have one single room covered by two vaults or two adjacent cloister-vaulted rooms. The space was later modified to accommodate two smaller vaulted rooms (C3 and C4) in the eastern part of the building, where almost nothing of the previous construction was left in place.

One further element of interest is the structural relation between the vault and the platform underneath it. Indeed, the association of a potential third pillar with a cloister vault may be an intriguing possibility to explain the presence also of the other two pillars, which appears to share with it the same set of materials and the same building technique. There can be no certainty at present that all the pillars were associated and supported cloister vaults: however, if this general reconstruction of the site would be confirmed, one should start wondering whether the two other platforms were also connected to the same vaulting system. In principle, cloister vaults do not need massive pillars, since they convey the stress primarily of the walls, rather than on the corners. In our case, the walls in phase with the pillars and the cloister vaulting are either not visible or destroyed. The platforms still preserved in place were clearly compatible with different and multiple vaults or arches. An example of this can be seen in C6, where one arch springing from the southern side of the third pillar (W121)—the same still supporting the cloister vault—is related to a new vaulting system set in the Mamluk period, clearly different both in terms of building materials and technique.

In any case, even imagining that the pillars originally supported all cloister vaults, it does not seem likely that they belonged to the same vault system, running north to south from C6, since the span appears to be too limited. If they were indeed part of the same building, it is more reasonable to imagine a different type of roofing; otherwise, we might consider the presence of two neighboring and contemporaneous structures covered by the same type of vault.

Clearly, later building phases deeply impacted the original structures, leaving us with few chronological indications, with the exception of an almost certain pre-Mamluk construction. The only way to more precisely date these structures is through regional comparisons, which are unfortunately not so numerous. This is partially due to the fact that cloister vaults are often mistaken with cross vaults, but mainly because of their less-frequent construction. In general, however, establishing comparisons for the entire complex, regardless of the specific matter of the cloister vault, is a particularly hard task: the evidence is too fragmentary to offer clear indications. In the case of some potentially interesting parallels, the issue of a

Figure 27 The Roman quarry in Area A2 of the southern sector (courtesy of the IAA, photo: B-297809-15071515519).

precise chronology for these other buildings is a major concern, especially in the absence of archaeological excavations conducted together with an architectural survey. We are lucky to have access to *Riwaq's Registry of Historical Buildings in Palestine* (2006): the architects compiling the registry did survey some buildings that might offer hints to the pre-Mamluk phase at Khirbet Beit Mazmīl. One recurring type of structure appears to present some striking similarities with the "fragments" of architecture documented at the site, including a direct comparison for the pillars.

Some of the surveyed buildings took the form of rather squared structures with one or two arches per side. The most interesting example is from Ein Yabrud, near Ramallah (RIWAQ 2006, 326), where the typology of the building seems to differ slightly from other structures in this village, both in terms of size and architectural technique. Here, the corners of the structure have the appearance of large pillars supporting two-barrel—or sometimes pointed—arches forming a 90-degree corner. The pillar located in the middle of each wall appears to be more modest in size, just offering a support for the two subsequent arches. The massive corner pillars also support the roofing of the structure. From the outside, when the stonework is well-preserved, the walls seem to continue straight up to the roof. On the inside, on the other hand, it is possible to have cross-vault, but nothing prevents it from having cloister vaults as well, especially in association with central domes.[2]

Unfortunately, it was impossible to gather more information on the structure through published materials, especially in relation to the kind of vaulting and, more importantly, to the period of construction. Similar massive structures are often described as "Crusader" in fortresses and hilltop sites—an example being Khirbet Jiddin, northeast of Akko (Pringle 1997, 80–82)—even though extensive and systematic archaeological and architectonical investigations have rarely been conducted in rural sites. To this concern, even if specific technical elements—most notably typologies of vaults—might be found in better-studied urban centers, such examples hardly represent useful comparisons for our site, because of their different context, and because certain constructive elements continued to be employed over long periods.[3]

The last area in the Northern Sector where pre-Mamluk structures can be identified is Area D. Here, a *miqveh*—with later modifications—was found. It is likely the best-preserved early architectural remain of the site, dating to the Early Roman period. It is significant because, due to the assumption that it is the first trace of occupation at the site, it shows how water collection and management played a key and continuous role at the site. Water management will continue to be a recurring theme throughout the entire life of Khirbet Beit Mazmīl.

In the same phase of occupation, the *miqveh* was converted into cisterns and on the surface a series of structures might relate to this newly established water system. Wells and channels were created using extremely well-cut, polished stones that are still partially visible in the eastern wall of Area D (W110). It is unclear if this water system was in communication with the channel found in Area C (W127) or with what the excavations exposed in the Southern Sector of the site.

The earliest phase in Area A appears to be a quarry found underneath the Mamluk structures (Storchan and Dolinka 2021) (Figure 27). This quarry, dated tentatively to the Early Roman period, still shows the outlines of masonry blocks, and possibly supplied the first construction activities at the site. Interestingly, the dimensions of these blocks do find parallels with materials reemployed in other structures at later phases.

The only standing structures of the period found in the area, however, are a series of walls (W111 and W143) connected to a two-basin water system identified in A11 (Storchan and Dolinka 2014), which was heavily affected by later Mamluk interventions. The corner created by W111 and W143 might have been one of the perimetral walls of a first Byzantine construction, as suggested by a series of beaten earth floors found immediately on the outside, associated with small field walls. The thickness of the walls (1.7–2 meters) reinforces this impression. The structure sat directly on bedrock, in a trench filled with dark soil with a high concentration of mosaic tesserae. The masonry is built in "sack technique," composed by two rows of large blocks and filled with rubble soil. However, the two walls W111 and W143 do not appear to belong to the same phase.

W111 is certainly built in "sack technique." Its faces are made of quite mixed building materials, alternating between more-or-less regular rows of larger, yet clearly not properly well-dressed, blocks, and smaller stones (Figure 28). These mixed materials appear to have been reused and come from different sources, with a limited use of more-carefully squared, larger blocks used as corner stones or door jambs, possibly originally from the nearby Early Roman quarry. In contrast, the internal parament presents a more homogeneous set of larger, regularly cut and squared blocks disposed in regular lines, probably in their original location. The fill between the two paraments is an extremely compact layer of earth mixed with small stones. Considering the differences in building materials and techniques, it can be suggested that the exterior face of W111 is the result of a later repair. It would not come as surprise that the original appearance would have been as W406, against which the exterior face of W111 abuts, and which uses the same materials as the interior face (Figure 29).

A marked difference in techniques and materials is appreciable in the northernmost portion of W143, where we have a simple "two-face" wall made by larger blocks, which is compatible with the modification of the installations found on the inside of these walls. Interestingly, farther north and to the east of Area D, another wall (W107), which seems to be the continuation of W143,

Figure 28 W111 (courtesy of the IAA, photo: B-297800-15071515340).

Figure 29 W111, abutting W406 (courtesy of the IAA, photo: B-297804-15071515414).

presents a rather similar technique to the exterior face of W111, even if it is more regular in its use of alternating rows of larger stones and flat stones.

The two connected basins are another interesting feature found in this part of the excavation. It is not entirely surprising to find water-related installations here, given the high elevation of this part of the site, but it is not entirely clear what function this installation played on a more general level: if it was connected into a broader hydraulic or productive system, or was instead a smaller, isolated catchment facility. A pinkish plaster covering the inside of the cisterns was recognized by the excavators as typical for waterworks from the Byzantine period in the Jerusalem area (Storchan and Dolinka 2021).

In this area in 2019, construction work also brought to light a lime kiln (L.300). Unfortunately, the structure was already damaged, so it was impossible to conduct a systematic analysis of the features. Its construction was initially dated to the Crusader period. It fell out of use with the expansion of the Mamluk site (Storchan and Dolinka 2021). This might make sense, since the use of mortar at the site appears to end with the Early Ottoman period, but this area at this point seems to have already been entirely converted into terraced space and field plots. Nonetheless, formal similarities have been noted with a Mamluk kiln excavated north of Jerusalem (B. Dolinka *personal communication*). The stratigraphic relationship between the kiln and the Mamluk walls (especially W140) is unclear, and the installation might either abut or cut the walls.

The last example of pre-Mamluk structures found in a primary context comes from the Lower Site, where a mosaic floor was discovered in Area S. The well-laid floor is built in coarse tesserae and was discovered under a Late Mamluk or Early Ottoman construction, later covered by a 19th century *manṭarah*. No surviving structure appears to be associated with this floor, but its preparatory levels rested directly on bedrock. In light of the few archaeological materials found in the preparatory layers of the mosaic, a tentative date to the Byzantine and/or Early Islamic period was suggested by the excavators (ʿAdawī and Dolinka 2021). One could suppose that it relates to an initial occupation of this part of the site, and certainly indicates the presence of some structure prior to the later construction of the tower building, but no further element for discussion is available.

In addition to these five areas, where proper pre-Mamluk structures can be clearly identified, it is also possible to recognize "clusters" of reused materials in later construction. These concentrations of materials reflect possibly the destruction and scavenging of the same earlier building or from different buildings constructed at the same time. These clusters might offer important indications on the development of the site at the later phases. Before considering the evidence from the site, though, a first question to be answered is whether these clusters reflect a selection of building materials dictated by functional purposes or by practical reasons. Among practical explanations, the most common is generally the proximity of the destructed or already-ruined building to the specific construction area of the site. Gathering materials from the same source might result, in certain cases, in a higher degree of homogeneity in the reused materials. It could also be the case that the ruined or destroyed building was already built with reused materials: in this case the materials would obviously be more mixed in their secondary or later reuse.

The most-recognizable cluster of materials was found in B4. This is one of the few examples onsite where well-cut and polished boulders are reused in later constructions. The blocks are considerably larger than the rest of the building materials used anywhere else at the site (spanning from approximately 80–110 centimeters in length, 35–50 centimeters in width, and 25–30 centimeters in height), closely resembling the materials of the early standing wall in C5. It is impossible to state whether these materials indeed originate from the same complex whose remaining walls are under the more recent structures in Area C. However, the fact that there is no comparable stonework anywhere else leads to the assumption that these materials come from a single source that was partially dismantled or already in ruins at the moment of the Mamluk development of this part of the site. Other blocks of the same fashion could be found in tertiary (if not more) reuse in the drystone Ottoman building in B3.

Other clusters of reused materials are documented in different structures across the site, although not presenting such distinctive features as the previous example. These are smaller blocks, still well-cut and polished but are clearly different than the larger boulders found in B4. These materials closely resemble the stonework around the *miqveh* in Area D and can be found in the two pillars in B5 and in the earlier standing structures in C. The platforms were likely built with reused materials, since no attempt has been made in utilizing stones of the same dimensions. However, the walls of the pillars, as previously described, were carefully and regularly laid, despite the rather miscellaneous building materials. Similarly, smaller blocks used for vaulting are quite clustered. This is not surprising: where rooms are preserved, notably in C3 and partially in C4 and C5, earlier masonry has often been reused in later vaults, and even reemployed in the drystone Ottoman building.

To summarize, the architectural remains offer little information on the pre-Mamluk occupation and organization of the site, due to subsequent, widespread, and highly destructive, construction activity. Combining the archaeological record with the dispersed architectural evidence, we can suggest the following scenarios. First, the site was certainly occupied, quite intensively, before the Mamluk period. The scattered evidence documents a widespread—and yet not extremely visible—occupation during the Roman and Byzantine periods, and a more evident Early Islamic phase. The Abbasid period is particularly well-represented in the Northern Sector, when there might have been a large complex or multiple smaller buildings, which likely reused materials from preexisting structures. Second, the reuse of materials was a constant practice throughout the entire occupational history of the site. Third, from a very early stage of occupation, water collection and management was a focus, along with industrial activities. The two aspects likely connected to one-another. These water systems and installations were regularly maintained and, in certain cases, partially repurposed.

Summary and concluding remarks

Taken together, there was a significant, albeit relatively brief, Crusader/Ayyubid occupational period at Khirbet Beit Mazmīl, which is demonstrated by the presence of architectural elements and domestic living spaces and is documented in the ceramic record as well. Both the Northern and Southern Sectors at the site yielded evidence for this era, almost all of which was recovered from contexts that were just above the bedrock or directly on it. The presence of the Mar Sabas seal would seem to indicate that the Frankish site was more than a simple farm or small rural site during this period: rather it was likely an important estate that had ties with the religious leadership and infrastructure in Jerusalem.

Notes

1. It represents just one of many cisterns that dotted the landscape in this area, as shown from combined maps of the 1930s and 1940s produced by Palestine Open Maps: https://palopenmaps.org/en/maps/beit-Mazmīl?basemap=9&overlay=pal1940&color=status&toggles=places|year#14.00,35.1764,31.7649.
2. All these features—the dome supported by different typologies of vaults, including the cloister vault, the single multi-functional room, and upper floor accessible either from the courtyard or from the single room—are listed among characteristic features of the traditional Palestinian house (Fuchs 1998).
3. There are several Mamluk and Ottoman sites reoccupying earlier settlements that might offer some parallels both in terms of specific constructive techniques and site history: Yoqne'am I (Ben-Tor *et al.* 1996); the Ottoman throne village at Majdal Yābā (Taxel 2017; Tsuk *et al.* 2016); Nabi Musa (Taragan 2007), especially for the vaulting system; Nahal Haggit (Seligman 2010); and Qālūnyā (Wachtel *et al.* 2019). Unfortunately, not one of these sites appears to exactly match the typology of Khirbet Beit Mazmīl, nor do they provide real architectural parallels.

References

ʿAdawī, Z., and B. Dolinka.
2021. "Jerusalem, Qiryat Ha-Yovel: Preliminary report." *Hadashot Arkheologiyot: Excavations and Surveys in Israel* 133 (04.05.2021). Internet Edition: https://www.hadashot-esi.org.il/report_detail_eng.aspx?id=25962&mag_id=133.

Adler, Y.
2008. "Second Temple period ritual baths adjacent to agricultural installations: The Archaeological evidence in light of Halakhic sources." *Journal of Jewish Studies* 59: 62–72.

Adler, Y.
2011. "The archaeology of purity: Archaeological evidence for the observance of ritual purity in Erez-Israel from the Hasmonean period until the end of the Talmudic era (164 BCE–400 CE)." Ramat Gan: PhD dissertation, Bar-Ilan University.

Adler, Y., and D. Amit
2010. "The observance of ritual purity after 70 CE: A reevaluation of the evidence in light of recent archaeological discoveries." In *Follow the Wise: Studies in Jewish History and Culture in Honor of Lee I. Levine*, edited by Z. Weiss, O. Irshai, J. Magness, and S. Schwartz, 121–143. Winona Lake: Penn State University Press, Eisenbrauns.

Amit, D.
1996. "Ritual baths (miqva'ot) from the Second Temple period in the Hebron Mountains." Jerusalem: Unpublished MA thesis, Institute of Archaeology, Hebrew University of Jerusalem.

Ariel, D.T.

2012. *The Coins of Herod: A Modern Analysis and Die Classification*. Leiden: Brill.

Arnon Y.

2008. "The ceramic oil lamps of the transitional and Medieval periods (640–1300): A chronological and typological study." In *Caesarea Reports and Studies: Excavations 1995–2007 within the Old City and the Ancient Harbor. BAR International Series No. 1784*, edited by K.G. Holum, J.A. Stabler, and E.G. Reinhardt, 213–269. Oxford: Archaeopress.

Auld, A.G.

2011. *I and II Samuel: A Commentary*. Louisville: Westminster John Knox Press.

Avissar, M.

2012. "Pottery from the Early Islamic to the Ottoman Period from the Cardo and Nea Church." In *Jewish Quarter Excavations in the Old City of Jerusalem, Volume V: The Cardo (Area X) and the Nea Church (Areas D and T)—Final Report*, edited by H. Geva, 301–345. Jerusalem: Israel Exploration Society.

Avissar, M., and E.J. Stern

2005. *Pottery of the Crusader, Ayyubid, and Mamluk Periods in Israel. IAA Reports* 26. Jerusalem: Israel Antiquities Authority.

Avni, G., and S. Gudowitz

1987. "The oil press and the subterranean storage complex at Achuzat Chazzan." In *Olive Oil in Antiquity: Israel and Neighbouring Countries from Neolith to Early Arab Period*, edited by M. Heltzer and D. Eitam, 6–12. Haifa: University of Haifa.

Bagatti, B.

1948. *Il Santuario della Visitazione ad ʿAin Karim (Montana Judaeae). Esplorazione archeologica e ripristino*. Jerusalem: Gerusalemme Tipografia dei Francescani (Studium Biblicum Franciscanum 5).

Balila, M.

2020. "Jerusalem, Émile Botta Street: Final report." *Hadashot Arkheologiyot: Excavations and Surveys in Israel* 132 (09.11.2020).

Ben-Tor, A., M. Avissar, and Y. Portugali

1996. *Yoqne'am I: The Late Periods. Qedem Reports* 3. Jerusalem: Israel Exploration Society.

Boas, A.

2006. "The Medieval Ceramics from Khirbat Ka'kul." *ʿAtiqot* 54: 75–104.

Bresc-Bautier, G.

1984. *Le cartulaire du Chapitre du Saint-Sépulcre de Jérusalem*. Paris: P. Geuthner.

Conder, C.R., and H.H. Kitchener

1881. *The Survey of Western Palestine III: Judaea*. London: Palestine Exploration Fund.

Cytryn-Silverman, K.

2010. "The ceramic evidence." In *Ramla: Final Report on the Excavations North of the White Mosque. Qedem 51*, edited by O. Gutfeld, 97–212. Jerusalem: Israel Exploration Society.

Cytryn-Silverman, K.

2013. "The Islamic period pottery." In *Jerusalem. Excavations in the Tyropoeon Valley (Giv'ati Parking Lot), Volume I. IAA Reports* 52, edited by D. Ben Ami. Jerusalem: Israel Antiquities Authority.

Dolinka, B.

2017. *The Middle and Late Islamic Pottery from the Excavations at the Austrian Hospice, 2011-2012*. Internet Edition: https://www.academia.edu/10347780/THE_MIDDLE_AND_LATE_ISLAMIC_POTTERY_FROM_THE_EXCAVATIONS_AT_THE_AUSTRIAN_HOSPICE_2011_2012.

Erickson-Gini, T.

1999. *Mampsis: A Nabataean-Roman Settlement in the Central Negev Highlands in light of the Ceramic and Architectural Evidence found in Archaeological Excavations during 1993–1994*. Tel Aviv: Unpublished MA thesis, Tel Aviv University.

Fuchs, R.

1998. "The Palestinian Arab House and the Islamic 'Primitive Hut.'" *Muqarnas 15*: 157–177.

Gannon, M.

2014. "'Extraordinarily rare' Crusade-Era seal discovered in Jerusalem." *Livescience.com*, 05.27.2014. Website: https://www.livescience.com/45893-rare-lead-seal-discovered-ancient-jerusalem.html.

Grainger, J.D.

2006. *The Battle for Palestine 1917*. Rochester: Boydell & Brewer Ltd.

Guérin, M.V.

1868. *Description géographique, historique et archéologique de la Palestine I: Judée*. Paris: E. Leroux.

Hadad S.
2002. *Excavations at Bet She'an I: The Oil Lamps from the Hebrew University Excavations at Bet She'an. Qedem Reports 4.* Jerusalem: Israel Exploration Society.

Hizmi, H., and S. Zion
2009. "An oil press and ritual bath at Kh. El-Najar in Beit Jala." In *Oil and Wine Presses in Israel from the Hellenistic, Roman and Byzantine Periods. BAR International Series No. 1972*, edited by E. Ayalon, R. Frankel, and A. Kloner. Oxford: Archaeopress.

Israel Department of Antiquities.
1976. *Geographical List of the Records Files, 1918-1948.* Jerusalem: Israel Department of Antiquities.

Katsnelson, N.
2014. "The glass finds from Bet She'an (Youth Hostel)." *Atiqot* 77: 23–57.

Kool, R.
2014. *The Coins and Lead Seals from Khirbet Beit Mazmīl.* Jerusalem: Unpublished report on file at the Israel Antiquities Authority.

Kloner, A.
2003. "Map of 'En Kerem (101; Sheet 16–13)." In *Survey of Jerusalem: The Northwestern Sector*. Jerusalem: Israel Antiquities Authority.

Lawrence, J.
2006. *Washing in Water: Trajectories of Ritual Bathing in the Hebrew Bible and Second Temple Literature. Academia Biblica.* Atlanta: Society of Biblical Literature.

Levi, D.
2012. "Jerusalem, Qiryat Ha-Yovel." *Hadashot Arkheologiyot: Excavations and Surveys in Israel* 124 (01.08.2012). Internet Edition: http://www.hadashot-esi.org.il/Report_Detail_Eng.aspx?id=1999&mag_id=119.

Levi, D., and R. Be'eri
2011. "Jerusalem, Binyene Ha-Umma: Preliminary report." *Hadashot Arkheologiyot: Excavations and Surveys in Israel* 123 (29.09.2011).

Magen, Y.
1984. "Kalandia: A vinyard farm and winery of Second Temple times." *Qadmoniot* 17: 61–71.

Magen, Y.
1985. "The Miqva'ot in Kedumim and the purification standards of the Samaritans." *Cathedra* 34: 15–26.

Magen, Y.
2004. "Qalandiya: A Second Temple period viticulture and wine-manufacturing agricultural settlement." In *The Land of Benjamin*, edited by I. Magen, and N. Haimovich-Carmin, 29–144. Jerusalem: Israel Antiquities Authority.

Magness J.
1993. *Jerusalem Ceramic Chronology, circa 200–800 CE.* Sheffield: Sheffield Academic Press.

Magness, J.
2005. "The Roman Legionary pottery." In *Excavations on the Site of the Jerusalem International Convention Center (Binyanei ha Uma). JRA Supplement 60*, edited by B. Arubas and H. Goldfus, 69–191. Portsmouth: Journal of Roman Archaeology.

Magness, J.
2011. *Stone and Dung, Oil and Spit: Jewish Daily Life in the Time of Jesus.* Grand Rapids: Eerdmans.

Magness, J.
2021. *The Archaeology of Qumran and the Dead Sea Scrolls. 2nd edition.* Grand Rapids: Eerdmans.

Palestine, Government of.
1922. *Official Gazette of the Government of Palestine, No. 62* (1.3.1922). IAA Archives (Mandate File 32).

Palestine, Government of.
1929. "Provisional schedule of historical sites and monuments." In *Official Gazette of the Government of Palestine, Gazette Extraordinary, Supplement 9*, edited by E.T. Richmond (15.6.1929). IAA Archives (Mandate File 32).

Pini, N.
2021. "The different fates of architectures: preliminary thoughts on reuse and reoccupation of vernacular architecture from Late Antiquity to the Middle Islamic period in the Near East." *Journal of Islamic Archaeology* 8.1: 23–51.

Porat, Y.
1989. "Hydraulic plaster of aqueducts as a chronological indicator." In *The Aqueducts of Ancient Palestine*, edited by D. Amit, Y. Hirschfeld, and Y. Patrich, 69–76. Jerusalem: Journal for the Study of Jerusalem.

Poulsen, V.
1957. "Les poteries." In *Hama. Fouilles et récherchés, 1931–1938, IV.2: Les verreries et poteries médievales,*

edited by P.J. Riis and V. Poulsen, 117–283. Copenhagen: Nationalmuseet.

Pringle, D.
1981. "The Medieval pottery of Palestine and Transjordan (A.D. 636–1500): An introduction, gazetteer and bibliography." *Medieval Ceramics* 5: 45–60.

Pringle, D.
1986. "Pottery as evidence of trade in the Crusader States." In *I Comuni Italiani nel Regno Crociato di Gerusalemme*, edited by G. Airaldi and B.Z. Kedar, 451–475. Genoa: Collana storica di fonti e studi.

Pringle, D.
1997. *Secular Buildings in the Crusader Kingdom of Jerusalem: An Archaeological Gazetteer*. Cambridge: Cambridge University Press.

Raphael, K.
2018. "Jerusalem, the Old City: Second-Temple Period Quarry and a Medieval Building." *Hadashot Arkheologiyot: Excavations and Surveys in Israel* 130.

Reich, R.
1984. "A Mikweh at ʿIsawiya near Jerusalem." *IEJ* 34/4: 220–223.

Reich, R.
1990. *Miqwa'ot in the Second Temple Period and Period of the Mishna and Talmud*. Jerusalem: Unpublished PhD dissertation, Institute of Archaeology, Hebrew University of Jerusalem.

Reich, R.
1998. "The *MIQWEH* (ritual bath)." In *Villages, Terraces and Stone Mounds: Excavations at Manahat, Jerusalem, 1987–1989. IAA Reports* 3, edited by G. Edelstein, I. Milevski, and S. Aurant, 122–124. Jerusalem: Israel Antiquities Authority.

Reich, R.
2013. *Ritual Baths during the Second Temple Period and in the Mishnah and Talmud Periods*. Jerusalem: Yad Itzhak Ben-Zvi and Israel Exploration Society.

Reuven, P.
2020. "The Islamic period pottery." In. *Jerusalem. Excavations in the Tyropoeon Valley (Giv'ati Parking Lot), Volume II, Part 2. IAA Reports* 66/2, edited by D. Ben Ami and Y. Tchekhanovetz, 375–521. Jerusalem: Israel Antiquities Authority.

RIWAQ
2006. *Riwaq's Registry of Historical Buildings in Palestine (Riwaq's Monograph Series on Architecture History of Palestine 9.1)*. Ramallah: RIWAQ—Centre for Architectural Conservation.

Rosenthal-Heginbottom, R.
2005. "The 1968 excavations." In *Excavations on the Site of the Jerusalem International Convention Center (Binyanei Ha'Uma): The Pottery and Other Small Finds. Journal of Roman Archaeology Supplementary Series No. 60*, edited by B. Arubas and H. Goldfus, 229–282. Portsmouth: Journal of Roman Archaeology.

Safire, W.
2009. "Location, location, location." *New York Times Magazine*, 06.26.2009.

Seligman, J.
1999. "Agricultural complexes at Ras Abu Maʻaruf (Pisgat Ze'ev East A), north of Jerusalem." *ʿAtiqot* 38: 137–170.

Seligman, J.
2006. "Jerusalem, Khirbat Ka'kul (Pisgat Ze'ev H): Early Roman farmsteads and a medieval village." *ʿAtiqot* 54: 1–73 (cf. pp. 3–5).

Seligman, J.
2010. *Nahal Haggit—A Roman and Mamluk Farmstead in the Southern Carmel*. Jerusalem: Israel Antiquities Authority.

Snyder, B.
2016. "s.v. Mikvah." In *The Lexham Bible Dictionary*, edited by J.D. Barry, D. Bomar, D.R. Brown, R. Klippenstein, D. Mangum, C.S. Wolcott, and L. Wentz. Bellingham: Lexham Press.

Stern, E.
2012. "Akko I. The 1991–1998 Excavations. The Crusader period pottery, Part 1: Text, Part 2: Plates (Israel Antiquities Authority Reports 51/1)." Jerusalem: PhD dissertation on file at the Israel Antiquities Authority.

Storchan, B.
2015. "Jerusalem, ʿEmek Lavan." *Hadashot Arkheologiyot: Excavations and Surveys in Israel* 127 (04.10.2015).

Storchan, B.
2020. *A Crusader- and Mamluk-Period Farmstead in the Southwestern Hinterland of Jerusalem: Excavations at Khirbet Beit Mazmīl*. Jerusalem: Unpublished report on file at the Israel Antiquities Authority.

Storchan, B.
2021. "Jerusalem, Bayit Va-Gan: Preliminary report." Submitted for publication in *Hadashot Arkheologiyot: Excavations and Surveys in Israel* (25.02.2021).
Storchan, B., and B. Dolinka
2014. "Jerusalem, Bayit Va-Gan." *Hadashot Arkheologiyot: Excavations and Surveys in Israel* 126.
Storchan, B., and B. Dolinka
2021. "Excavations at Kh. Beit Mazmīl: A Crusader and Mamluk period farmstead in the southwestern hinterland of Jerusalem." Unpublished project report.
Tal, O., and I. Taxel
2008. "Pottery: Typological discussion." In *Ramla (South). An Early Islamic Industrial Site and Remains of Previous Periods. Salvage Excavation Reports 5*, edited by O. Tal and I. Taxel, 125–165. Tel Aviv: Israel Antiquities Authority.
Taragan, H.
2007. "Holy place in the making: Maqam al-Nabi Musa in the Early Mamluk period." *ARAM* 18–19: 621–639.
Taxel, I.
2017. "Majdal Yābā: More insights on the site in Medieval and Late Ottoman to Mandatory Times." *Journal of Islamic Archaeology* 4.1: 49–86.
Tsuk, T., I. Bordowicz, and I. Taxel
2016. "Majdal Yābā: The history and material culture of a fortified village in Late Ottoman- and British Mandate Palestine." *Journal of Islamic Archaeology* 3.1: 37–88.
Tushingham, A.D.
1985. *Excavations by K.M. Kenyon in Jerusalem 1961–1967, Volume I: Excavations in the Armenian Garden of the Western Hill.* Toronto: Oxbow Books.
Walker, B.J., and B. Dolinka.
2020. "*Ḫirbet Bēt Mazmīl.* Investigations of Medieval Jerusalem's Hinterland: Interim report on the 2015 and 2017 Field Seasons." *Zeitschrift des Deutschen Palästina-Vereins* 136.2: 191–234.
Wachtel, I., E. Ayalon, Y. Gadot, and A. Nadan
2019. "An interdisciplinary study of Qālūnyā." In *New Studies in the Archaeology of Jerusalem and its Region, Special Issue: Proceedings of the International Conference on Motza*, edited by Peleg-Barkat, O., Y. Zelinger, J. Uziel, and Y. Gadot, 381–405. Jerusalem: Israel Antiquities Authority.
Wightman, G.J.
1989. *The Damascus Gate, Jerusalem: Excavations by C.M. Bennett and J.B. Hennessy at the Damascus Gate, Jerusalem, 1964–66. BAR International Series No. 519.* Oxford: British Archaeological Reports Oxford Ltd.
Zelinger, Y.
2015. "Horvat 'Adassa: Changes in settlement patterns in the agricultural hinterland of Jerusalem." *New Studies in the Archaeology of Jerusalem and Its Region: Collected Papers*, Volume 9: 107–117.
Zissu, B.
2001. *Rural Settlement in the Judean Hills and Foothills from the Late Second Temple Period to the Bar-Kokhba Revolt.* Jerusalem: Hebrew University.
Zissu, B.
2014. "A classification of the Second Temple Period Judean Miqwa'ot (ritual immersion baths)." In *Speleology and Spelestology to the Centenary of A.V. Ryumin's Birth*, 246–261. Naberezhnye Chelny: UIS Commission on Karst Hydrogeology and Speleogenesis UIS Artificial Cavities Commission IGU Commission Karst (C12–23).

About the authors

Dr. Benjamin J. Dolinka is an archaeologist and ceramics specialist who has conducted research and fieldwork in Greece, Jordan, and Israel during the past 31 years. He received his PhD in Archaeology from the University of Liverpool in 2007 with a dissertation on Nabataean pottery and worked for the Israel Antiquities Authority from 2010 to 2018. Mentored on the Islamic pottery by the late Dr. Miriam Avissar (IAA), he has come to be one of the leading specialists for the Jerusalem area. His forthcoming *The Pottery of Medieval Jerusalem, ca. 1050–1550 CE* will be published by Equinox as part of its *Monographs in Islamic Archaeology* series.

Dr. Nicolò Pini (PhD in Classical Archaeology at the University of Cologne, Germany) is an archaeologist, specialized in ancient urbanism, architecture, and social structures in the Roman and Byzantine periods. Over the last years he has ventured into the Early, Middle, and even Late Islamic periods. He currently conducts a postdoctoral research project at the CReA-Patrimoine, Université libre de Bruxelles (Belgium), funded by the Belgian Fonds de la Recherche Scientifique—FNRS (CR—Chargé de recherches). He was previously awarded a COFUND Marie Skłodowska-Curie Individual Fellowship at the same institution. He collaborates on several projects in the Near East, in particular with the Islamic Archaeology Research Unit at the University of Bonn and is also external research associate of the Panorama Platform (based at the Universitè libre de Bruxelles), specialized in acquisition and digitization of objects and architectural surveys.

Zubair ʿAdawī is an archaeologist with the Israel Antiquities Authority, operating in the Jerusalem area, and has conducted a number of development surveys and participated in and led dozens of small- and medium-scale excavations, both independently and together with his colleagues from the Jerusalem Regional Department—and from other divisions of the Antiquities Authority—in coordination with development work. He has been involved with many notable excavations including: Roman and Byzantine burial caves on the Mount of Olives; a Roman cemetery on Salah a-Din Street near the Old City; Khirbet Umm Tuba, at the foot of the Tel of Ramat Rachel, between Jerusalem and Bethlehem—a rural settlement that originated during the Iron Age and was continually occupied through almost all subsequent periods; and the tomb of a monk, discovered bound in iron rings, below the church at Ramat Shelomo in northern Jerusalem.

— 4 —

Remnants of a "Feudal" Past

The Mamluk Estate

Benjamin J. Dolinka, Nicolò Pini, Bethany J. Walker, and Benyamin Storchan

Historical background

The Mamluks were a Muslim dynasty of emancipated military slaves that controlled much of the eastern Mediterranean from their capital in Cairo in the 13th to early 16th centuries. For the first century of their rule, their priority in Egypt and Syria was security of the pilgrimage route and defense from internal and external threats. In Bilād al-Shām this meant the investment in travel infrastructure; development of the pigeon post; pacifying rebels (particularly Bedouin); and expanding its network of castles and garrisons. After signing a peace treaty with the Il Khānid Mongols in 1323, which secured the eastern border, the Mamluk sultans shifted much of their attention to domestic concerns: lavish building projects and the arts; the provision of public services through endowments (*awqāf*); and the construction of roads, water systems, and caravanserais.

Jerusalem presented a special case to the Sultanate. It was a pilgrimage town—mostly for Muslims and Christians in this period—and hosted numerous religious hostels for both faiths. As a result, it was the recipient of religious donations, and under the Mamluks the *waqf* system expanded exponentially, resulting in many building projects on the Ḥaram al-Sharīf and the marketplace, and devouring the revenues of the villages that ringed the town. Politically, Jerusalem was a quiet backwater, the place where officers retired and disgraced members of the ruling elite were exiled. At the same time, the state built *quṣūr* (walled estates) for officers in active service in the green zones beyond the Old City, on hilltops and in orchards. Jerusalem was also an international town. Foreign merchants and diplomats were welcome and lived in their own *wikālas* (urban caravanserais) and consular houses. Muslim and Christian poor and widows had housing there as well, funded through generous endowments.

The countryside was impacted by the constant movement of peoples and the diversity of the marketplace. The villages closest to Jerusalem benefitted from a thriving network of markets, which gave them ready access to goods from other villages and from town. Ceramic imports, among other goods, infiltrated the countryside. In this sense, Beit Mazmil, as a village, was representative of a small community in Jerusalem's orbit. In another sense, it sheds new light on the way that the Mamluk state penetrated rural society through these amiral estates, embedded in local communities but serving state interests.

The Mamluk period in the Northern Sector at Khirbet Beit Mazmīl

Introduction

The 2017–2019 excavation seasons conducted in the Northern Sector at Khirbet Beit Mazmīl yielded a modest but significant amount of information about the site during the Mamluk period. Two primary zones of occupation provided the necessary data. The first encompassed both C6 and E1, with the former consisting of a plastered press installation with a channel on its floor that led to a settling basin, and the latter being a storeroom that was entered from a corridor with steps that led into it. The second context originated from a domestic space in the northwest area of B4, where cooking installations and spaces for food preparation were uncovered. The associated ceramics and finds from these two spheres provide valuable information about lives of those who inhabited and worked at the site between the 13th and 15th centuries.

Square C6

During the 2017 season, the top levels of a large plastered Early Ottoman cooking installation were uncovered from L.225 in Square C6. In 2018 one of the goals was to remove the remnants of the L.225 installation in order to reach the level of the L.229 flagstone floor upon which it presumably rested (Figure 1). This was achieved by dismantling the cooking installation (L.228) and removing the fill on the L.229 floor (L.233), during which time the first few sherds dating from the Mamluk II period in the area were uncovered. The focus of the work was then shifted to the southwest sector of the square in order to reveal more of the L.229 floor. In this part of the square an oval cooking installation (L.238) was revealed (Figure 2). It consisted of burnt beaten earth layers alternating with thin plaster layers and was cut into the L.229 flagstone floor. Unlike the other cooking installations from the site, this one was not very artifact rich, and yielded no coins. The finds were mostly earlier in date and typical of the Mamluk II period, or the 15th century, at the site. These included an almost complete small glass bottle, a glass sprinkler spout, and a green steatite whetstone.

One of the primary goals of the 2018 season was to conduct a probe under the L.229 flagstone floor in C6. Most of the flagstones from the floor were removed and the fill below it (L.244) was excavated to provide datable material for its construction. While only a few artifacts were recovered from

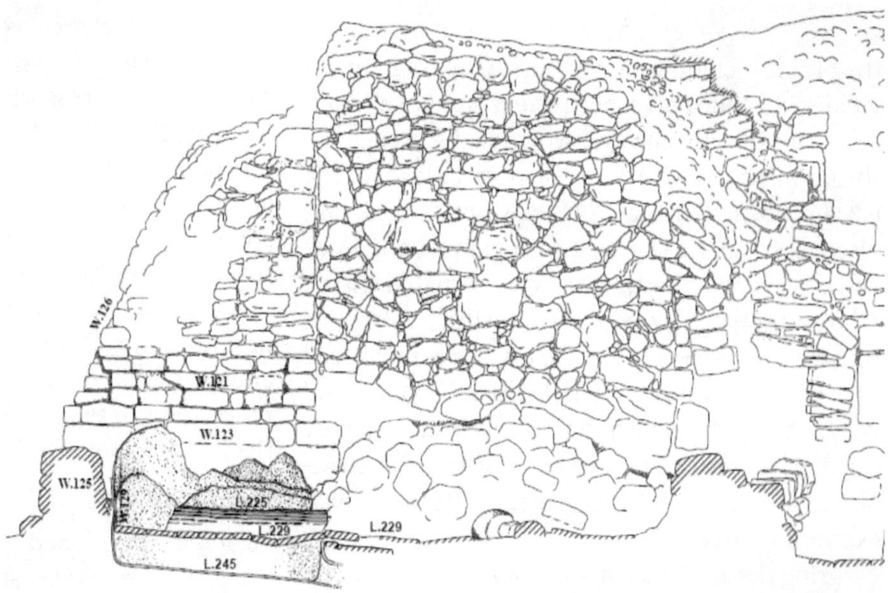

Figure 1 Section drawing showing the L.229 flagstone floor and the L.245 rectangular plaster-lined basin from the press installation in C6 (drawing: S. Pirsky).

Figure 2 View of the oval cooking installation (L.238) in the southwest sector of C6; view to the west (upper left of image).

the probe, including an iron horseshoe fragment, it yielded a good representative example of a Mamluk II ceramic assemblage, which had some earlier sherds mixed in with it. At the bottom level of the L.244 probe under the L.229 flagstone floor, it was discovered that the area encompassing the floor and both cooking installations was situated within a large rectangular plastered basin (L.245) with rounded corners (Figure 3) This basin measured 2.6 × 1.6 meters and was bounded by walls W.128 and W.129. Along the eastern side of this installation, at the floor level of L.245, a small semi-circular channel (L.246) was cut into the plaster at its floor level, extending eastward.

During the 2019 season, work was resumed in the northeast corner of C6 with a goal of completely clearing the remnants of the L.229 flagstone floor and L.248 fill above L.245 in order to fully delineate the plastered basin. To the east of this installation in C6 was a corridor which led to the south (L.256) and connected the plastered basin with the E1 storeroom, via a set of stairs (L.247) which led upwards into the room (Figure 4). The L.256 corridor also led eastward, and excavation in that part of C6 revealed that the L.246 circular channel from the press room drained into the L.258 semi-circular basin (Figure 5). A doorway (L.263) to the east that led into another room that was not excavated due to time constraints. By the end of the final field season for the project in 2019, one thing was certainly clear: the plastered installation in the western sector of C6 was a press of some kind, given the presence of the rectangular plastered basin, above which were small, square holes in a preexisting wall where a wooden press beam may have been secured, with a channel at floor level in the basin that led to a settling or storage vat. A good early-modern example of an olive press with such accoutrements, dating from the region during the 1920s and published on the Matson Collection website, bears striking similarities to the press at Khirbet Beit Mazmīl. It should be noted that a radiocarbon sample taken from L.259—the plaster-lined settling basin in C6—yielded a date range of 1233–1391 CE. This data provides a firm Mamluk I date for the construction and use of the entire C6 press system, including the press and the two settling basins, and most likely the E1 storeroom for this phase as well.

Square E1 (2019)

C6 and E1 functioned as one unit during the Mamluk period. While C6 acted as the main industrial area in the Northern Sector, E1 served as a storeroom associated with the press, and continued to do so well into the Early Ottoman period, after C6 was converted into a kitchen. E1 was first revealed during the 2018 season, in what was originally

Figure 3 View of the L.244 probe under the L.229 flagstone floor, which revealed most of the L.245 large rectangular plastered basin in the C6 press room; view to the southwest.

Figure 4 Corridor (L.256) and set of stairs (L.247) that connected the plastered basin (L.245) from the press installation in C6 with the E1 storeroom; view to the east.

thought to be the courtyard of the enclosed area of the Northern Sector. Instead of the relatively sterile fills that were thought to be contained within that new square, it yielded a room delimited by W.136 on the west, which was built abutting the W.124/1 platform; W.133 on the south; and W.134 on the east; the room was entered from the north, via the stairway L.247 in C6. A complete stratified sequence dating from the Early Ottoman period was revealed, which consisted of a series of flagstone and plaster floors. By the end of the 2018 season, excavation in the square had stopped at the top levels of L.420, a plaster floor that was sealed by the L.419 flagstone floor above it. One of the first priorities of the 2019 season was a return to L.420 in order to check if the Early Ottoman sequence continued. Found on that floor, however, was a clear Mamluk II context, as demonstrated by the ceramics retrieved from it (see below), with no evidence for the Early Ottoman period contained therein. Interestingly, the soil layer below this (L.423) yielded a clean Mamluk I ceramic assemblage that was mixed with earlier sherds dating from the Crusader/Ayyubid period. This stratigraphic sequence demonstrated that the area associated with the E1 storeroom was continuously in-use from at least the mid-13th until the early 18th century.

Another excavation strategy for E1 addressed trying to date some of the walls and installations contained therein. An important feature from this fieldwork was the semi-circular stone-lined pit W.140 (Figure 6) in

Figure 5 From bottom to top (C6): eastern part of the plastered basin (L.245), semi-circular channel (L.246), semi-circular plastered settling basin (L.258) and doorway (L.263); view to the east.

Figure 6 Semi-circular stone-lined pit W.140 in the northeastern corner of E1.

the southwest corner of E1, the fill of which (L.428) yielded the fragment of a green-glazed molded relief bowl with part of an Arabic inscription, which dates from the Mamluk I period, or the 14th century. A soil sample from the ashy lower levels of L.428 and tested for radiocarbon dating yielded a date range between 1312 and 1442 CE, which fits well with the ceramic evidence. A similar situation was attested in L.441—a delicate probe into the matrix and makeup of W.137a—which produced another Mamluk I monochrome green-glazed bowl of contemporaneous date to that from L.428.

Square B4

While C6 and E1 provide evidence for industrial and storage functions for the Northern Sector, respectively, the environs associated with B4 were clearly domestic in nature. The B4 domicile was bounded by W.118 on its southern end; W.131 and the southern half of platform W.124/2 on the east; W.122 on the north; and W.117 on the west. In the northwest area of the structure, the primary feature of importance is Hearth 180 (Figure 7). Two other contexts yielded small amounts of Mamluk I pottery: L.161, the soil layer associated with stairs on the final flagstone floor (L.162); and L.165, the soil under W.127, which dates the L.186 water channel (which was not fully excavated before the end of the project). Although the evidence for the Mamluk period in the Northern Sector was limited in nature, it nevertheless provides a glimpse into the domestic and industrial activities that took place at the site during the 13th through 15th centuries.

Figure 7 Hearth Δ180 in the northwest corner of the B4 domestic space.

Mamluk pottery from the Northern Sector at Khirbet Beit Mazmīl (Figures 8–10)

The corpus of Mamluk pottery retrieved from the Northern Sector was recovered from three excavation areas: the C6 Press Room, which was sealed below the Early Ottoman Kitchen; the E1 Storeroom, which continued in use as such into the Early Ottoman period; and the domestic space in the northwest part of B4, which yielded areas for cooking and food preparation. The ceramic assemblage presented here consists of vessels found in sealed contexts such as either directly on or under floors, from installations, and from the soil matrix of walls for the purposes of dating them. The pottery corpus from the Northern Sector includes two distinct sub-phases: Mamluk I (= Stratum IV), dated from the mid-13th to the end of the 14th century; and Mamluk II (= Stratum III), which covers the 15th to early 16th centuries. The ceramics consist of a limited number of forms which include locally produced bowls, basins, cooking bowls, cooking pots, store jars and jugs. It should be noted here that many of the handmade vessels presented here, particularly the undecorated bowls, lack any published parallels, due to the fact that these types are under-represented at sites in the Old City of Jerusalem and its surrounding hinterland as well, which suggests that perhaps many of these forms were produced within the vicinity of Khirbet Beit Mazmīl. The third bowl (Figure 8:2) is paralleled at the Damascus Gate in the Old City of Jerusalem (Wightman 1989, fig. 50:7), and the form is ubiquitous at the site. While the thickness of the vessel walls and diameter size may vary, the form is fairly standardized, consisting of a flat (or occasionally slightly rounded), inverted wedge-shaped rim that is almost triangular in section.

Another small corpus of ceramics that was uncovered in 2018 originated from L.244, the probe below the L.229 flagstone floor down to the bottom of the plaster-lined basin (L.245) in the press room. The first specimen from this assemblage (Figure 9:1) is a flat-bottomed bowl with a slightly out-turned rounded rim, that has correlates with an exemplar from the Coptic Patriarchate at the Church of the Holy Sepulcher (Avissar 2009, fig. 40:1). The next three bowls (Figure 9:2–4) demonstrate the variations for the type that were found at the Damascus Gate, with one of them (Figure 9:3) representing the archetype for the form. The handmade cooking pot (Figure 9:5) is of interest, and has correlates from the typo-chronology of Avissar and Stern (2005, fig. 40:4), for which they offered a broad date range from the 13th to 15th centuries (Avissar and Stern 2005, 94). The final vessel from the L.244 probe is an Handmade Geometric Painted Ware (hereafter HMGP) body sherd from the neck of a store jar (Figure 9:6) that represents a good example of the decorative scheme for the Mamluk II period at the site and is distinguished by its thick buff-colored exterior slip and light reddish-brown painted decoration; the lack of any paint under where the handle was placed is very typical of the later Mamluk assemblages (S. Gabrieli *personal communication*).

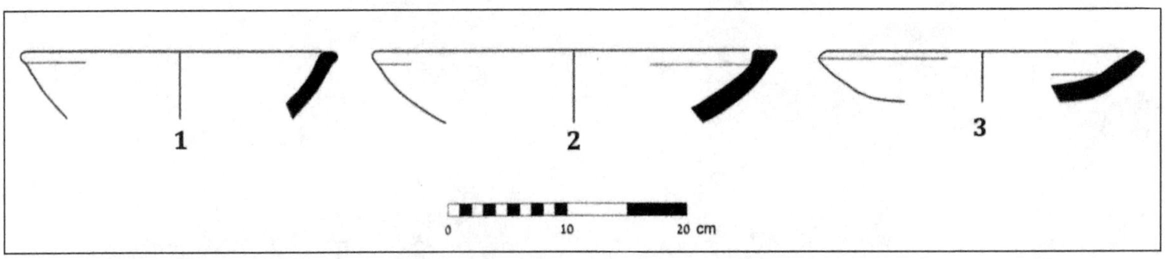

Figure 8 Mamluk II Pottery from above (nos. 1–2) and below (no. 3) the oval cooking installation abutting the north face of the W.124/1 platform.

In 2019, two more contexts yielded a handful of sherds attributable to the Mamluk period. The first corpus consists of three vessels that were sealed under the northeast corner of the L.229 flagstone floor, after it was removed down to the bottom level of the rectangular plaster basin in the press room (L.245). This small corpus of pottery dates to the Mamluk I period, or around the 13th century. The handmade bowl (Figure 10:1) is another, thinner-walled variety of the

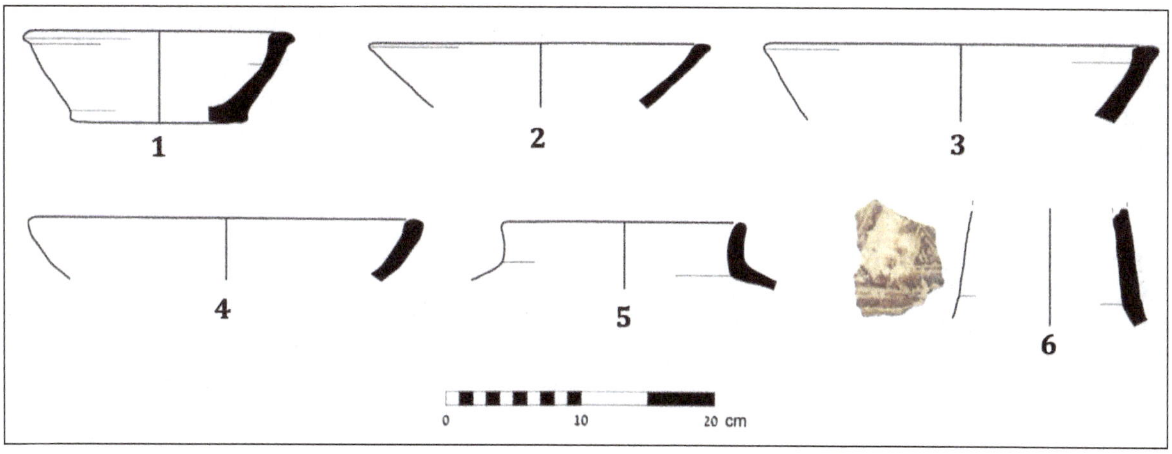

Figure 9 Mamluk II Pottery found below the L.229 flagstone floor in C6.

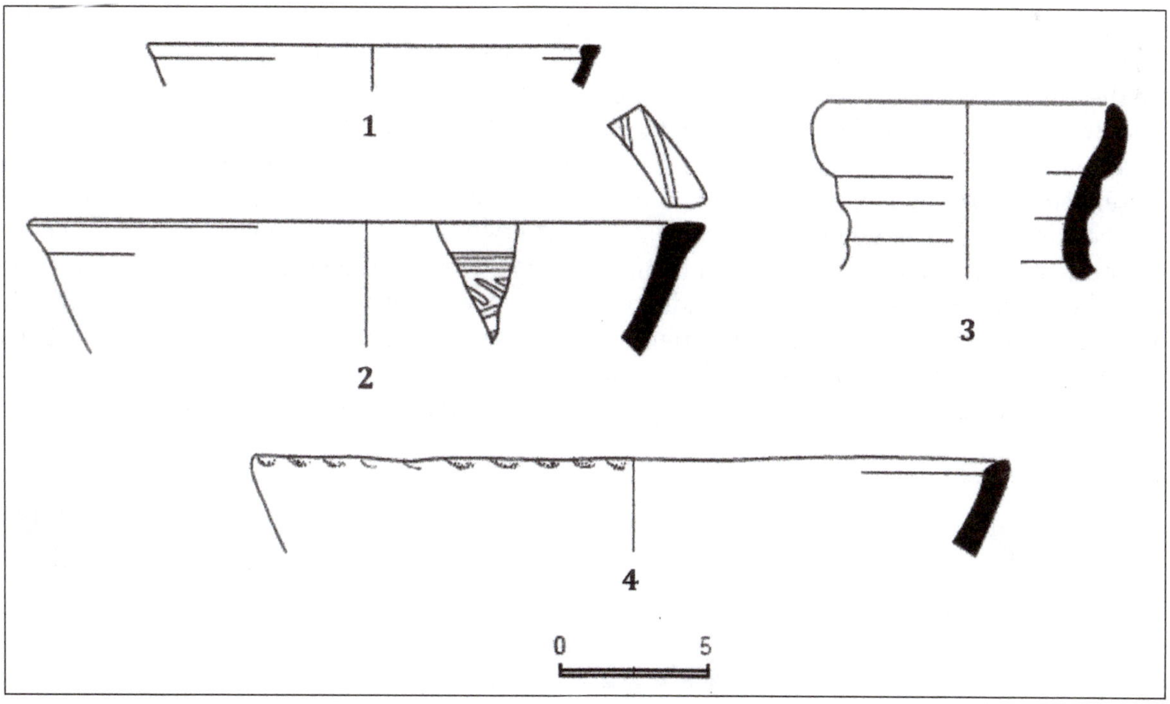

Figure 10 Mamluk I Pottery sealed under flagstone floors L.229 (nos. 1–3) and L.255 (no. 4) from the C6 plastered installation.

aforementioned common type that is found at Khirbet Beit Mazmīl. A contemporaneous HMGP bowl (Figure 10:2) was also uncovered in L.248. It is burnished on the interior, exterior, and rim top, and has linear decorations on the rim top that are painted in a reddish-brown color. The fabric is pale red on the interior and exterior with a dark gray core, and there are several medium-sized (and some larger) calcite inclusions. A very close parallel to this vessel was found in a sealed context under the Mamluk I floor from Square A6 during the 2012 excavations (see Figure 22:10 below). The rim, neck, and partial shoulder of a Jerusalem Ridged-Neck Jar (Figure 10:3) was also found in the L.248 corpus. The name for this jar type was coined by Boas (2006, 83, 85, fig. 8), and it is one of the most common jar forms from the medieval Jerusalem corpus. This form first appears in Ayyubid contexts during the early 13th century, where they occur in a buff fabric and continue to be produced well into the Mamluk I period. A final handmade bowl (Figure 10:4) was recovered from the L.256 fill, which was sealed below the L.255 lower flagstone floor, upon which the stairs into the E1 storeroom were constructed. This was the only diagnostic sherd from this context and dates from the Mamluk II period. It is of a much coarser reddish-yellow fabric than its Mamluk I predecessors and has large inclusions of calcite and chert and a gray core. Its decoration is similar to the earlier type with the finger-indented appliqué on the exterior rim zone but differs in that the decoration is directly on top of the rim instead of below it.

Pottery from the E1 storeroom (Figures 11–13)

There were three ceramic sequences that were recovered from good contexts in the E1 storeroom during the 2019 season: one dating from the Mamluk II period and the other two with vessel forms characteristic of the Mamluk I era. The first assemblage originates from the L.420 plaster floor, which was sealed by the L.419 flagstone floor above it. The following three examples are represented from this context: a small handmade bowl with a rounded, slightly flared rim (Figure 11:1), which is known from the Church of the Holy Sepulcher (Avissar 2009, fig. 4:1); the standard handmade bowl form found at Khirbet Beit Mazmīl (Wightman 1989, fig. 50:7) mentioned above (Figure 11:2), for which two other examples were found from this locus; and a handmade basin with a rounded rim and a horizontal band of finger-indented appliqué just below the rim (Figure 11:3) that is comparable to an example from the Armenian Quarter in the Old City of Jerusalem (Seligman 2001, fig. 8:6). This corpus of bowls illustrates three different sizes, from small to medium to large, which could easily be stacked together in order to conserve space in the E1 storeroom.

The next assemblage consists of a diverse group of vessels that originated from L.423, which was sealed under the L.419 lower flagstone floor. The latest pottery from this context dates from the Mamluk I period. The first example from L.423 is a ring-base and sidewalls of a monochrome glazed bowl (Figure 12:1). This is of interest, despite the fact that there is no rim to indicate what exact type it is. It has a reddish-yellow fabric that is fired to a pinkish-buff,

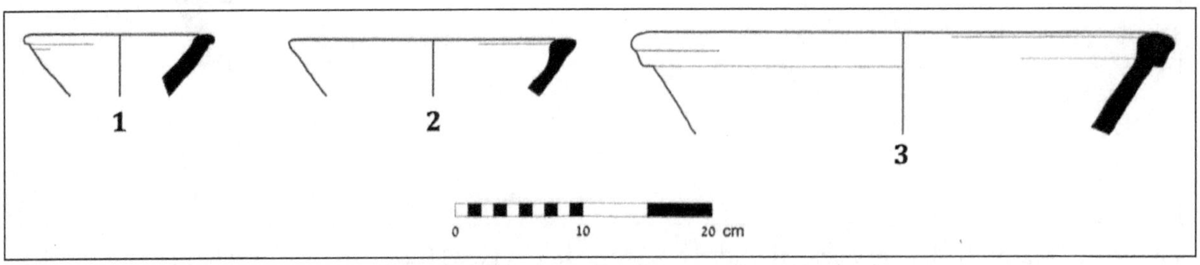

Figure 11 Mamluk II pottery sealed under the L.419 flagstone floor in E1.

it is reasonably well-levigated and contains mostly small calcite inclusions. The interior is covered with a thin white slip, upon which the dark yellow glaze was added. The glaze itself is well-adhering and somewhat streaky in appearance. On the interior of the vessel two out of the three circular impressions from a tripod stand that was on it when it was fired in the kiln are clearly visible. Remnants of green glaze are present on both the interior and exterior of the ring-base, which suggests that different colored glazed bowls of the same type were fired at the same time.

HMGP and handmade bowls were well-represented in the L.423 corpus, and an example of each is presented here. The HMGP bowl (Figure 12:2) has a fairly coarse reddish-yellow fabric with small- and medium-sized calcite inclusions and a flat rim. The interior and top of the rim are burnished and decorated with a dark reddish-brown paint. The fact that the decorative scheme exhibits the *horror vacui* technique, whereby the entire interior of the vessel is covered with painted decoration, indicates a likely date in the 14th century for this bowl. Most of the profile from a handmade bowl of Mamluk I date was also present in the assemblage (Figure 12:3). It has a coarse pinkish-gray exterior with a light gray core, and a burnished interior that is reddish-yellow in color. In addition to the calcite inclusions of various sizes, there was a lot of burned-out organic

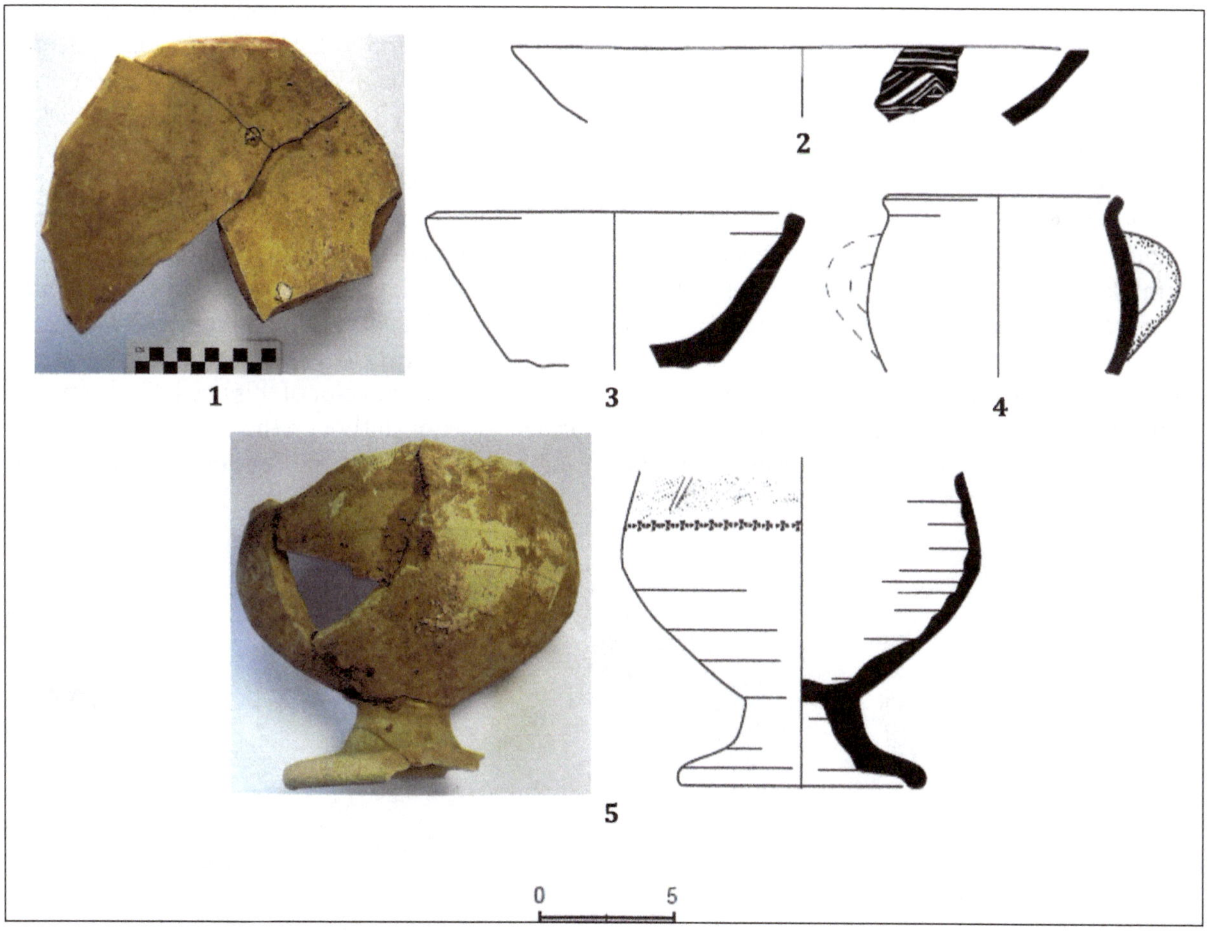

Figure 12 Mamluk I pottery sealed under the L.420 plaster floor in E1.

temper, and a good parallel for this vessel is known from the Damascus Gate excavations (Wightman 1989, fig. 50:1).

Cooking pots were not very well represented in this corpus. The example from Locus 423 (Figure 12:4) is small, handmade and has a cyma-recta body contour with an out-turned rounded rim. It has a reddish-yellow fabric that is fired to very pale brown. This cooking pot type is found in small numbers from both the hinterland of Jerusalem as well as the Old City (see Avissar 2012, fig. 10.6:10, from Cardo Area X-4).

Of importance is the high pedestal base and lower half of the body from a strainer-neck jug with incised and pinpricked decoration (Figure 12:5) with a pink fabric that is fired to very pale brown on the interior and exterior. It conforms to the typo-chronology of Avissar and Stern (2005, 111, fig. 46: 3–4, Type II.4.3.2), which has been assigned a broad chronological range spanning the 13th to the 15th centuries. The bases of this type most often consist of a depressed ring-base that is concave in section; however, the example from Khirbet Beit Mazmīl has a high pedestal base typical of the Mamluk I period that is known from the Old City of Jerusalem (Avissar and Stern 2005, fig. 46:5–6). It has been previously noted that it is "... not clear where these vessels were made. There were probably local workshops ..." (Avissar and Stern 2005, 111). Given the fact that all of the examples noted in Avissar and Stern's corpus come from the Old City of Jerusalem, have a homogenous buff fabric, and are only decorated on the upper shoulder, it is suggested here that this strainer jar is from a local production.

The final two ceramic examples from E1 consist of well-known and well-dated forms of green-glazed bowls dating from the Mamluk I period. The first one (Figure 13:1) was recovered from the soil fill (L.428) inside the semi-circular stone-lined pit W.140. It is a glazed molded relief bowl with part of an Arabic inscription on its exterior. This well-known class of vessels conforms to Avissar/Stern Type I.1.7 and are good chronological indicators (Avissar and Stern 2005, 22, fig. 8: 8–11, Pl. VII). These vessels all have a very thin white slip to which the thick, shiny, and well-adhering glaze is added on both the interior, exterior, and base of the vessel. The final specimen from E1 was found in L.441 (Figure 13:2), the soil within the matrix of W.137a. This vessel conforms to Avissar/Stern Type 1.1.4 Monochrome Glazed Bowls II (Avissar and Stern 2005, 12–15), which were ubiquitous in the Jerusalem region during the 14th century. It has a yellowish-red (5YR 5/6) fabric that contains a few small and medium calcite inclusions. The interior and rim are covered by a thick white slip and a shiny, well-adhering green glaze, both of which drip over the rim onto the vessel exterior. The fabric color of this bowl is indicative of local production, although the best parallel for this example was found at the Bet She'an Youth Hostel (Avissar 2014, fig. 37:5).

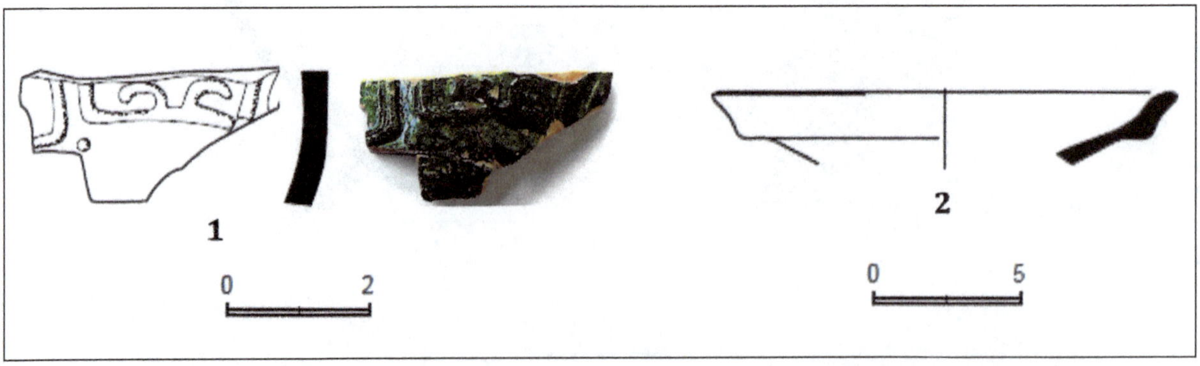

Figure 13 Mamluk I pottery found in the matrix of W.140 (no. 1) and W.137a.

Pottery from B4 (Figure 14)

A very small assemblage of Mamluk I ceramics was retrieved from the domestic space in B4, which consisted of separate living, cooking and food preparation areas; only three vessels are presented here. The first (Figure 14:1) is a handmade cooking bowl that was recovered from L.161, which was the soil layer on top of the L.162 flagstone floor. It has a medium-coarse pinkish-gray fabric with a few small calcite and chert inclusions, and its exterior is heavily carbonized. No parallel could be found for this specimen. A handmade bowl that, due to its find context, most likely dates from the Mamluk I period, was also recovered from L.161 (Figure 14:2). It has a reddish-yellow coarse fabric with a few inclusions of calcite and small voids that represent burned-out organic temper (such as straw) and is burnished in the interior of the vessel and the top of the rim. A good parallel for this vessel was found on Shonei Halakhot Street in the Jewish Quarter of the Old City of Jerusalem (Raphael 2011, fig. 12:3). The final piece of ceramic evidence from B4 was uncovered in L.165, the soil layer beneath W.127 associated with the L.186 water channel that was not fully excavated by the end of the 2019 season. It is the complete upper profile from another example of the buff ware Mamluk strainer jars (Figure 14:3) with incised and pinpricked decoration that has already been discussed above, for which no parallel was discerned.

Mamluk clay rattle and iron knife from the E2 midden in the Northern Sector (Figure 15)

During the 2019 season, Square E2 yielded two exceptional finds: a clay rattle and an iron knife. The stratigraphic context for both objects was L.435, a thick deposit of subsoil associated with the midden located between W.135 and W.138. While the midden was used in the Early Ottoman period, the deposit is very mixed, and included Mamluk and some earlier materials. The formation process for this midden began when the Early

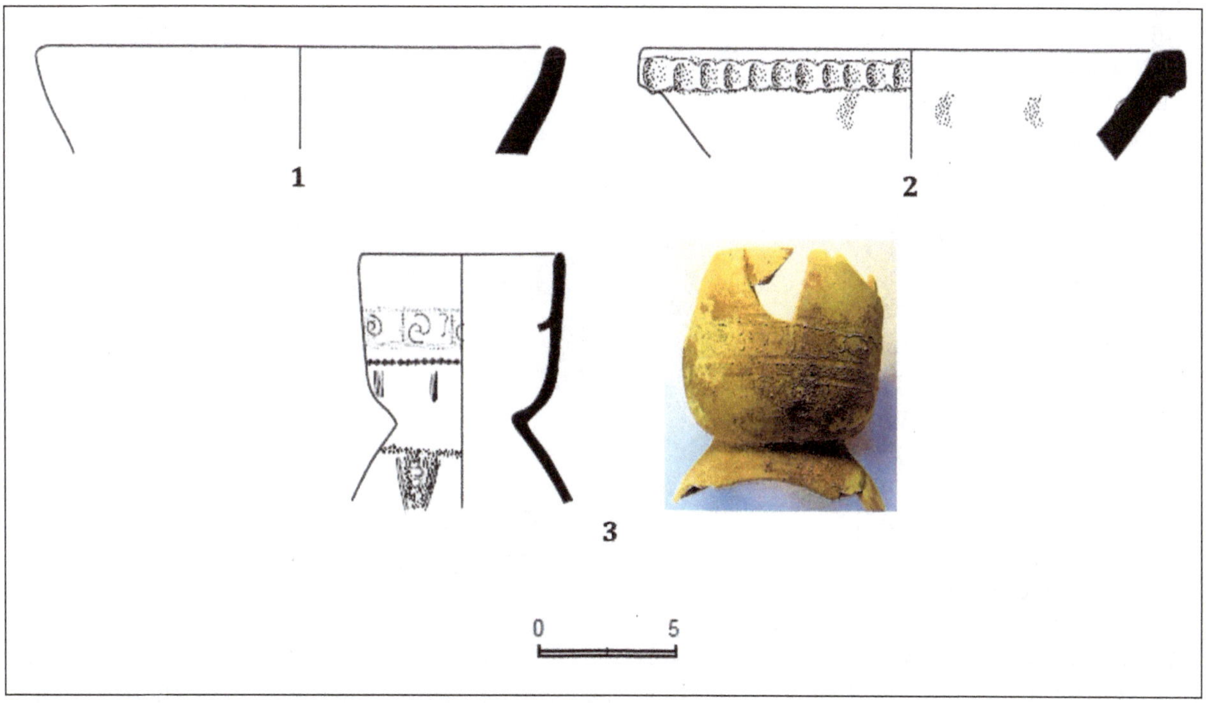

Figure 14 Mamluk I pottery from the B4 domestic space.

Ottoman period newcomers cleared out the Mamluk II remains from all the rooms in the Northern Sector, in order to repurpose them for their own use, and dumped the debris from those rooms in the midden—which they continued to use as such throughout the Early Ottoman period.

The clay rattle (Figure 15:1) is of interest, because it represents an artifact type that is becoming an increasingly more important part of Mamluk material-cultural studies in the region. The rattle from Khirbet Beit Mazmīl is relatively round in shape and a very coarse fabric that is very pale brown in color, has inclusions of mica and small black grits, and voids in the vessel surface from burned-out organic temper (such as straw).

A recent and important study of these objects (Taxel 2022) examined an assemblage consisting of almost 60 fragments of rattles (Taxel 2022, 155, fig. 11), which were found at the Netafim 2 campsite near Eilat, which was located along the Darb al-Hajj al-Maṣri: the pilgrim route from Cairo to Mecca via Aqaba. Petrographic analysis of these objects revealed an Egyptian origin for the rattles (Taxel 2022, 157), which makes sense due to their findspot. The rattle from Khirbet Beit Mazmīl is clearly not a local production from the Jerusalem region, although examples are known from the Damascus Gate excavations in the Old City of Jerusalem (Tushingham 1985, 151, figs. 41:44–45), where they were dated to the 14th century.

With regards to the function and/or use for these rattles, it was suggested that the Netafim finds "served apotropaic purposes, expelling the evil eye" (Taxel 2022, 163; Taxel *et al.* 2013, 224) and that they were used by sorcerers in magic rituals (Taxel 2022: 145). That being said, the presence of this rattle at a village settlement—that is, not a campsite—suggests that they were most likely used to amuse or pacify the babies who were born at the site.

The other object recovered from the E2 midden from L.435 was an almost complete and intact blade and tang from an iron knife (Figure 15:2), and a similar example was found amongst the metals assemblage from the Southern Sector (see Figure 31:1 below). The presence of this weaponry in the E2 midden, combined with ceramic finds such as the glazed molded relief bowls with Arabic inscriptions, would seem to suggest a Mamluk military presence at Beit Mazmīl.

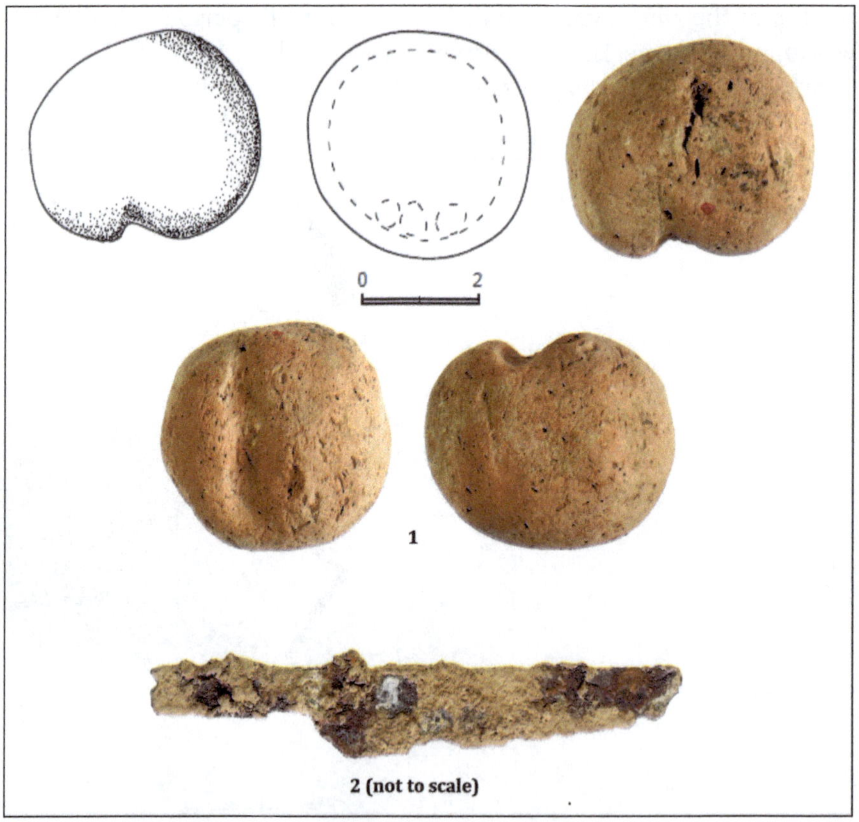

Figure 15 Mamluk Clay Rattle and Iron Knife from the E2 Midden in the Northern Sector (photo: D. Gazit, IAA).

Concluding remarks: Northern Sector

Taken together, the stratigraphy, pottery and other finds from the Northern Sector that were excavated in 2018 and 2019 have revealed valuable insights into the Mamluk-period settlement, highlighting its industrial activities, as substantiated by the press complex and associated storeroom from C6 and E1, respectively, as well as its domestic occupation, as demonstrated in the B4 domicile, during the period covering the mid-13th to the end of the 14th centuries. When combined with the more extensive evidence from the Southern Sector, a clearer picture begins to emerge, shedding light on the lifeways of the people who lived and worked at Beit Mazmil during its existence as a farming estate in the Mamluk period.

The Mamluk period in the Southern Sector at Khirbet Beit Mazmīl

The settlement at Beit Mazmil flourished during the Mamluk period. A great deal of evidence for this era was provided by the 2012 excavation conducted in the Southern Sector by the IAA, under the direction of Storchan and Dolinka (License No. A-6507). The material remains for the Mamluk period from this part of the site were recovered from Area A, which was situated in the northern part of the excavation zone, and Area B, which was located to the south of the Cellcom service road that separated the two precincts by 12 meters. The Mamluk-period remains from the site are divided into two distinct phases based on careful examination of the ceramic assemblages and their associated find contexts. This analysis allowed for the clear separation of an early Mamluk occupational horizon at the site dating from the mid-13th to the end of the 14th century (Mamluk I = Stratum IV) and a later phase that covered the period from the 15th to early 16th century (Mamluk II = Stratum III). During the transition to the Mamluk period a number of changes occurred at the site, most notably an extension of the settlement further to the south beyond the limits of the Byzantine, Early Islamic, and Crusader/Ayyubid-period remains, and the rise of a Mamluk farmstead with agricultural and industrial installations.

Stratum IV (Mamluk I, mid-13th to mid-14th century)

While the remains and ceramic materials from Strata VII–V, or the Late Byzantine through the Crusader/Ayyubid periods, were limited in nature, those from Stratum IV were more considerable. This phase dates to the Mamluk I period (MM-I) at the site, or around the mid-13th to the end of the 14th century. Although this initial era saw the first major constructions that extended the settlement to and beyond the south of walls W.111 and W.450, the Mamluks made direct use of nearly all the pre-existing architectural elements.

Area A

In Area A, the Mamluk I period is marked by the enlargement and expansion of architectural remains from Stata VII–V and the construction of new walls to the south. In order to prepare the eastern part of the area, the Mamluk builders created a leveling course of flat stones, placed directly above the Early Roman quarry in A2 (Figure 16). Architecturally, Stratum IV is represented by a series of walls encompassing Squares A4 and A6 that make up a room bounded by W.112 on the east; W.140 on the south; and W.144 on the west. Abutting W.144 from this room on its east is W.139, which is the primary architectural feature in Square A2. In the eastern sector of Square A6 there was a series of poorly constructed spur walls (Figure 17) which included W.141, W.142, and W.140 from north to south, respectively, while in the central portion of Square A8 there was a semi-circular coping wall W.113, which was most likely a storage installation or small animal pen. Stratigraphic evidence from Stratum IV originated from the following contexts in Area A: L.104 and L.106, from on and below the plaster floor in Square A8; L.114, on the plaster floor in the northwest sector of Square A6; L.117, on the plaster floor in the southwest sector of Square A6;

Figure 16 Mamluk I leveling course of flat stones, placed directly above the Early Roman quarry on the bedrock of A2; view to the north (IAA Photo Archive No. B-322995).

Figure 17 Series of poorly constructed Mamluk I spur walls in the eastern sector of Square A6; view to the southwest (IAA Photo Archive No. B-297803).

L.126, the makeup of plaster floor L.116 in A6; L.121, on the plastered floor in Square A4; and L.133, the Mamluk robber's trench cut into the Byzantine plaster floor in Square A4.

Area B

During the earliest stages of the Mamluk period, the eastern sector of Area B seems to have been an open industrial area at the site, due to the presence of three circular installations (Figure 18) whose functions at present are unknown, but may have been kilns, large ovens (*tannurs*) or perhaps granaries. Similar installations were uncovered in great numbers at the Netafim 2 campsite (Taxel 2022, fig. 6), which is located along the Darb al-Hajj al-Maṣri, where they are clearly tent bases. However, their find context at Khirbet Beit Mazmīl is clearly industrial, especially given the large amount of ash associated with them. They were simply referred to as "Circular Installations" in the original field reports. These were built as a wall composed of a single row of stones which range in diameter from 1.5–2.5 meters. During the course of their excavation, a layer of collapsed stones was uncovered both inside and outside of these circular features, which would seem to indicate that they were capped by stone dome-shaped superstructures. Evidence for the intensive firing that occurred within these installations was found in the eastern sector of the excavation area, as attested by the presence of a dense layer of charcoal and ash. This black layer can be attributed to the routine cleaning of the installations, at which time their waste was dumped in the area to the south. It should be noted here, that within the soil matrix of this ashy layer were several burnt olive pits, which indicates the fuel type that was used for these installations. Immediately to the west of these, remains of a clay *ṭābūn* were also uncovered. Another exact example of these circular installations was also uncovered in the nearby Lower Site (ʿAdawī

Figure 18 Three Mamluk I circular installations in the eastern sector of Area B; view to the south (IAA Photo Archive No. B-297820).

and Dolinka 2021, fig. 10), where it is referred to as a "*ṭābūn*."

In Area B, Stratum IV, the primary features which yielded both stratigraphical and material evidence included the following: L.512, the fill above Installation 1 in Squares B1/2; L.534, the *ṭābūn* contents from Installation 2; L.539, the Installation 3 contents in Squares B2/3, yielded by its sectioning and probe; and L.515, the fill above Installation 2 in Squares B2/3.

Stratum III (Mamluk II, 15th to early 16th century)

As the settlement was continuously inhabited into the 14th and 15th centuries, once again structural changes to previously existing architectural elements, as well as new additions, both indicate another separate archaeological stratum within the Mamluk period at the site. During this period a large structure was built in Area B (Figure 19). The building's northeastern section was composed of two rooms (Rooms 507 and 535), with a large courtyard (Room 532) situated to the southwest.

Room 507

The main entrance to Room 507 was located in the northern area and central portion of wall W.560, and its entrance was indicated by the presence of a built step. The room forms a generally rectangular plan measuring 5.5 × 3 meters and was found completely filled with a dense layer of collapsed stones measuring approximately one meter in depth. After the removal of the layer of stone collapse, a natural step in the bedrock bisecting the room into eastern and western sectors was uncovered. In the upper eastern sector of the room, a plaster floor (L.507) abutting W.554 was unearthed. Upon closer examination, this plaster floor revealed a succession of multiple plaster surfaces. These were homogenous, composed of a simple thin white plaster layer and placed directly upon the previous layer with no substantial fill between surfaces, indicating that the room was routinely cleaned and replastered. Within the western sector of the room, below the bedrock step, a partial flagstone floor (L.528) was uncovered. The eastern closing wall of the structure was not preserved, however a likely reconstruction can be postulated based upon its roughly symmetrical plan.

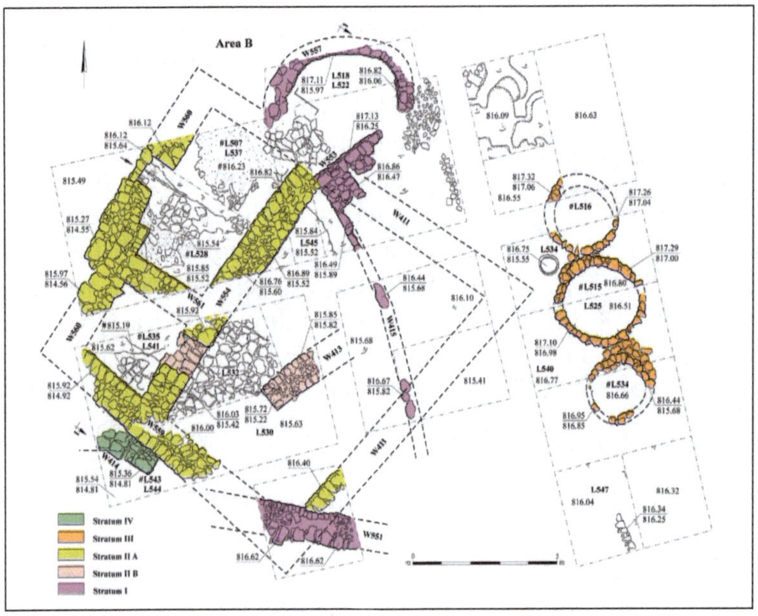

Figure 19 Plan of the Mamluk II structure in Area B, with Room 507 (northeast), Room 535 (southeast) and the courtyard.

Room 535

To the southwest of Room 507 was another, smaller square-shaped room measuring 2.5 × 2.6 meters that was separated from the other room by wall W.561. At the beginning of excavation in this room, a dense layer of fallen stone collapse was uncovered, which sealed the stratigraphy below it. The entrance to this room was in the south and was found to have been blocked by two rows of stones. Below the collapse, as was the case in Room 507, a partial flagstone floor and flat bedrock floor were found and abutting the western wall of the room (W.559) was the base of an arch springer.

Courtyard (Room 532)

The rectangular-shaped courtyard of the structure was located to the south of Rooms 507 and 535, measured 9.5 × 6.5 meters and was separated from the rooms by wall W.554. It was framed by walls W.559 to the west, W.411 to the south, and W.412 to the east. The courtyard was only partially excavated down to the floor levels in the area abutting Room 535. In the easternmost area of the courtyard, the bedrock served as the floor, and in its western sector was the well-preserved, roughly triangular-shaped remnant of the L.537 flagstone floor (Figure 20). Within the open space of the courtyard, an inner division wall with an offset orientation (W.413) was built directly upon the stone pavement.

Stratigraphically speaking, the primary features in Area B were comprised of the following: L.528, under the flagstone floor and above the bedrock in the south-central sector of Square B7; L.530 = 532 = 541, the flagstone floor which covered most of the building in Square B8; L.536, the makeup of the flagstone floor; L.507 and L.537, above and below the plaster floor in Square B7; and the fill layer above the bedrock along the southern face of W.554.

Mamluk pottery from the Southern Sector

The Mamluk ceramic assemblages from the site were recovered from two excavation areas (A-B) and consisted of vessels found in sealed contexts such as either directly on or under floors, from installations, and above the bedrock. The pottery corpus presented here encompasses Stratum IV (Mamluk I) and Stratum III (Mamluk II) from the Southern Sector, which together date from the mid-13th to the end of the 15th century. The assemblages are arranged according to typo-chronological considerations, with the majority of the examples for each stratum originating from sealed contexts and supplemented by other relevant exemplars from secondary and tertiary depositional contexts. This method was chosen in order to provide a basic framework for interpreting similar ceramic assemblages from the Jerusalem region in order to highlight forms and decorative motifs that are representative and typical for each of the strata whence they originate. The pottery presented here consists of locally

Figure 20 Roughly triangular-shaped remnant of the L.537 flagstone floor in the courtyard (Room 532); view to the southeast (IAA Photo Archive No. B-323039).

produced and imported bowls, basins, cooking bowls, cooking pots, store jars, jugs, and lamps. Parallels and suggested dates are contained within the text below, and not listed in the tables for the catalogue; instead, the stratigraphic context for each vessel appears in the far-right column of the tables.

The pottery from Stratum IV (Mamluk I, mid-13th to the end of the 14th century)

Introduction

While the architectural remains and ceramic materials from the Late Byzantine and Crusader/Ayyubid periods were limited in nature, those from Stratum IV were more considerable. This phase dates to the Mamluk I period at the site, or around the mid-13th to the end of the 14th century. The Stratum IV ceramic assemblage presented here is divided into four assemblages based upon typological considerations. A total of 31 vessels are illustrated herein, and these consist of glazed bowls, bowls with burnished interior, bowls with impressed rope decoration, plainware bowls, cooking pots, store jars, and lamps.

Mamluk I glazed bowls from the Southern Sector (Figure 21)

There was a limited quantity of glazed ceramics recovered from Stratum IV deposits, and these few sherds were represented by monochrome slip-glazed wares as well as glazed molded relief bowls. The glazed pottery presented here was covered with a thin white slip before the addition of the green, yellow, or polychrome glaze, which was fairly thickly applied, well-adhering, and shiny. The first bowl (Figure 21:1) has an out-turned rim, a lip that is beveled on the interior and a carinated body. The fabric is a light reddish-brown (5YR 6/3) with tiny calcite inclusions and a green glaze on the interior which drips over the rim onto its exterior. A good parallel for this vessel was found at the Youth Hostel excavations in Beit She'an (Avissar 2014, fig. 37:5), also with a green glaze, where it was dated to the 13th century. Another green monochrome slip-glazed bowl (Figure 21:2) was found above the plaster floor in Square A6 (L.117). It has a direct rim with a slightly offset collar that is beveled on the exterior and a reddish-yellow (5YR 6/6) fabric with a green glaze on its interior and rim top. There are no published parallels for this vessel. An interesting incurved glazed bowl (Figure 21:3) was found in L.106 (the makeup of the plaster floor from the easternmost sector of Square A8). The form, with

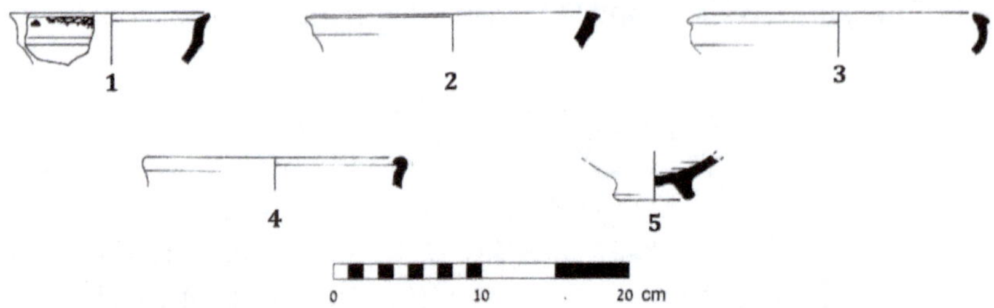

6

Photo B-343888

(after Avissar and Stern 2005: Fig. 8 – not to scale)

7

No.	Locus / Basket	Ware Description	Context
1	528 / 5184_2	Light reddish brown (5YR 6/3) fabric, tiny calcite inclusions, thin white interior slip, green glaze on interior dripping over rim onto exterior	Under floor in B7
2	117 / 1068_5	Reddish yellow (5YR 6/6) fabric, thin white slip, green glaze on interior and rim top	Above floor in A6
3	106 / 1018_3	Pink (5YR 7/3) fabric, thin white slip, pale yellow and green mottled glaze with dark-brown gritty spots	Below floor in A8
4	539 / 5164_1	Reddish yellow (5YR 7/8) fabric, think white interior/exterior slip, dark yellow glaze covering entire surface	Kiln 3 contents in B2/3
5	534 / 5110_8	Light reddish brown (5YR 6/4) fabric, tiny calcite inclusions, thin white interior slip, well-adhering green interior glaze	*Ṭābūn* contents in B2
6	106 / 1043	Reddish yellow (5YR 6/6) fabric, yellowish-brown glaze, vegetal motifs in molded relief on exterior	Below floor in A8
	117 / 1073	Fabric as no. 6, green glaze, vegetal and geometric molded relief on exterior	Above floor in A6
	512 / 5050a	Red (2.5YR 6/6) fabric, dark brown glaze, UD molded decoration on exterior	Above Kiln 1 (B1/2)
	512 / 5050b	Fabric as no. 6, green glaze, UD molded decoration on exterior	Above Kiln 1 (B1/2)
	534 / 5110_5	Fabric as no. 6, yellow glaze, molded Arabic inscription in relief on exterior	*Ṭābūn* contents in B2
	128 / 1104	Light reddish brown (5YR 6/4) fabric, yellowish-brown glaze, vegetal design in molded relief on exterior	Secondary deposition
	532 / 5187_13	Fabric as no. 6, brownish-yellow glaze, molded geometric decoration in relief on exterior	Secondary deposition
	513 / 5052_2	Pink (5YR 7/3) fabric, yellow interior glaze, mottled yellow/green exterior glaze, Arabic inscription in raised molded relief on exterior	Tertiary deposition
	513 / 5052_6	As no. 13 above	Tertiary deposition
	513 / 5052_3	As no. 13 above	Tertiary deposition

Figure 21 Mamluk I glazed bowls from the Southern Sector.

its collared and everted rim that is almost triangular in section, is well known from Jerusalem. It conforms to Avissar/Stern Type I.1.4.3 (Avissar and Stern 2005, 14, fig. 5:10), which was recovered from the Christian Quarter of the Old City. Similar bowls to the example from Khirbet Beit Mazmīl have been found from Area X-6 of the Cardo (Avissar 2012, fig. 10.13:1) and the Armenian Garden (Tushingham 1985, fig. 41:10). What differentiates the present bowl is its surface treatment. Its pink (5YR 7/3) fabric is covered with a pale yellow and green mottled glaze with dark-brown gritty spots and a rather low firing which is similar to the gritty glazed bowls of the preceding Crusader/Ayyubid period. The last two monochrome slip-glazed bowls are important due to their find contexts. The first has a dark-yellow glazed with an incurved rim (Figure 21:4) that is similar in most respects to those found on the glazed molded relief bowls (see discussion below) and was the only datable pottery revealed from the sectioning and probe of Installation 3 in Square B2/3. This type dates to the 14th century (M. Avissar *personal communication*) and is known from the Armenian Garden (Tushingham 1985, fig. 41:25). The second noteworthy bowl fragment is the monochrome green-glazed ring-base (Figure 21:5), which was the only diagnostic sherd recovered from the *ṭābūn* in Square B2. The base form of this vessel has parallels from Area X-4 of the Cardo (Avissar 2012, fig. 10.5:6) and is of a type that is common in the 14th century.

The final group of glazed bowls from Stratum IV at Khirbet Beit Mazmīl is represented by another example of the aforementioned type with molded relief designs (Figure 21:6) consisting of either floral or geometric decorations, or Arabic inscriptions on the vessel exterior (see Figure 14:1 above). While only a handful of sherds from these bowls were recovered from the excavation, the complete form is fairly standardized (Figure 21:7), consisting of almost hemispherical vessel walls that terminate in an incurved rim that is thickened on the interior and exterior, with a high, flaring trumpet-shaped base.

Mamluk I bowls with burnished interiors and impressed rope decoration from the Southern Sector (Figure 22)

The bowls with a burnished interior first appear in Stratum V (Crusader/Ayyubid), where they are represented in small numbers. They continued to be produced throughout Stratum IV at Khirbet Beit Mazmīl (Figure 22:1–9), or the mid-13th to the end of the 14th century. In addition, bowls with impressed rope decoration on the exterior of the vessel just below the rim zone also continue production into Stratum IV, either burnished (Figure 22:10) or un-burnished (Figure 22:11–12). These bowls form a rather homogenous group: all are handmade and have a coarse fabric with calcite inclusions and numerous grits; many have tiny pockets of burned-out organic temper (such as straw); the fabric color is predominantly light reddish-brown (5YR 6/3–4) or reddish-yellow (5YR 7–6/6); most have either simple linear or geometric painted decoration rendered in a reddish-brown color on the interior and (less frequently) on the top of the rim; and the most typical rim form is rounded and slightly out-turned with thickening on either the exterior or interior. Although there are no parallels for some of them (Figure 22:2, 8–9, 11–12) similar examples for the others are known from the Jerusalem region, including the Armenian Garden (Tushingham 1985, fig. 45:11) and Khirbat Ka'kul (Boas 2006, figs. 3:20, 4:24). Of interest are the two bowls with only a single (finger, or perhaps thumb) indented impression on the exterior below the rim zone (Figure 22:5–6). This type is occasionally found in the Jerusalem hinterland (see Boas 2006, fig. 4:23) but so far not reported from excavations in the Old City. As these bowls do not appear in the Mamluk II assemblage from Stratum III at the site and are limited to the 14th century, as are their counterparts from Khirbat Ka'kul, it would seem that these unique vessels could be used as a chronological indicator for the local Jerusalem ceramic repertoire.

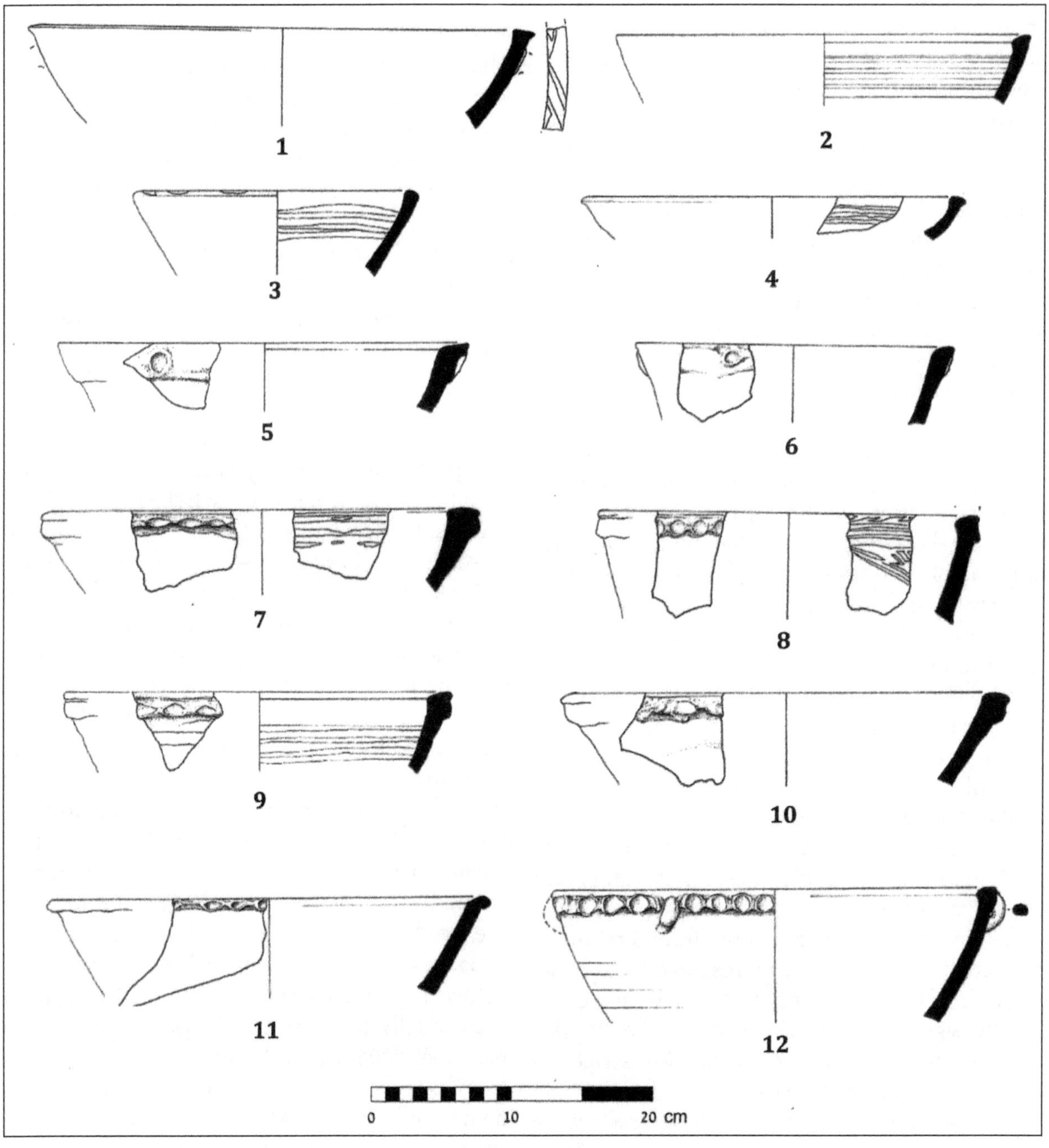
1
2
3
4
5
6
7
8
9
10
11
12
0
10
20 cm

No.	Locus / Basket	Ware Description	Context
1	126 /1087_2	Pink (5YR 7/4) fabric, pale red (10R 6/4) burnished interior, large calcite inclusions, traces of reddish brown (2.5YR 5/4) painted geometric decoration on rim top	Below floor in A6
2	512 / 5207_2	Reddish yellow (5YR 6/6) fabric, numerous grits, calcite inclusions, burnished interior, reddish brown (5YR 4/3) painted horizontal linear design on vessel interior	Above Kiln 1 in B1/2
3	104 / 1011_1	Pale red (10R 6/4) fabric, very dark gray (2.5YR N/3) core, calcite inclusions, burned-out organic temper, burnished interior painted with weak red (10R 4/4) geometric decoration	Above floor in A8
4	121 / 1086_1	Light reddish brown (5YR 6/4) fabric, numerous grits, calcite inclusions, burned-out organic temper, burnished interior, reddish brown (2.5YR 5/4) painted geometric decoration on interior	Above MM floor in A4
5	512 / 5192_2	Very coarse reddish yellow (5YR 7/6) fabric, numerous grits, medium to large calcite inclusions, burnished interior	Above Kiln 1 in B1/2
6	126 / 1087_1	Pinkish gray (5YR 7/2) fabric with pinkish gray (5YR 6/2) burnished interior, calcite inclusions	Below floor in A6
7	117 / 1068_2	Light red (10R 6/6) fabric with numerous grits, calcite inclusions and burned-out organic temper, reddish gray (2.5YR 6/1) core, burnished interior, traces of reddish brown (5YR 5/4) painted geometric design on interior	Above floor in A6
8	126 / 1087_3	Pink (5YR 7/4) fabric, burnished interior, reddish brown (5YR 5/3) painted geometric decoration on interior and rim top	Below floor in A6
9	512 / 5219_2	Light reddish brown (5YR 6/3) fired to pink (5YR 7/4), numerous grits, calcite inclusions, burnished interior, reddish brown (5YR 5/4) painted decoration as above	Above Kiln 1 in B1/2
10	121 / 1086_3	Light reddish brown (5YR 6/4) fabric, numerous grits, calcite inclusions, burned-out organic temper, burnished interior	Above MM floor in A4
11	539 / 5164_2	Coarse reddish yellow (5YR 6/6) fabric, dark reddish gray (10R 4/1) core, large calcite inclusions, burned-out organic temper	Kiln 3 Contents (probe)
12	117 / 1073_1	Light reddish brown (5YR 6/3) fabric with reddish gray (5YR 5/2) core, numerous grits	Above floor in A6

Figure 22 Mamluk I bowls with burnished interior (1–6) and impressed rope decoration (7–12) from the Southern Sector.

Mamluk I plainware bowls, cooking pots and store jars from the Southern Sector (Figure 23)

A variety of other vessels were recovered from Stratum IV contexts at Khirbet Beit Mazmīl. These included plainware bowls, cooking pots, and store jars, all of which were only represented in small numbers. The number of wheelmade plainware bowls from the Mamluk I period was few, and one is presented here. It is a carinated bowl with an everted rim that is grooved on top and slightly offset (Figure 23:1). It has a light reddish-brown (5YR 6/3) fabric fired to weak red (2.5YR 6/4) and tiny calcite inclusions. It was found in the makeup of the flagstone floor in Square B7. The rest of the plainware bowls from this stratum were handmade and consisted of a coarse, mostly reddish-brown fabric with numerous grits, calcite inclusions, and burned-out organic temper. Most of them have direct, simple rims that are either flat or slightly rounded. Two of these bowls originated from the contents of installations in Square B2: the first was the complete vessel profile from the contents of Installation 3 (Figure 23:2), and the second (Figure 23:4), was recovered from contents of the *ṭābūn*; parallels could not be found for either of these. A large handmade plainware bowl (Figure 23:5) was found with the wheelmade one in the makeup of the flagstone floor from Square B7, described earlier. Two handmade plainware bowls were also recovered from Mamluk I plaster floors in Area A. The first (Figure 23:3) was found in Square A6 and the second (Figure 23:6) is the thick HMGP base found on the plaster floor in Square A4. It has reddish-brown (5YR 5/4) fabric fired to reddish-yellow (5YR 6/6) and dark reddish-gray (2.5YR 4/1) painted geometric decoration on central interior of the base finds.

Cooking wares were not well-represented in the ceramic assemblage from Stratum IIB, and only two are illustrated here. The first is represented by the rim and shoulder of a cooking pot with a flared and rounded rim (Figure 23:7). It has a weak red (10R 5/4) fabric fired to light reddish-brown (5YR 6/4), with numerous grits, as well as large calcite and crushed quartzite inclusions. Found within the content of Installation 3 from Square B3, the form is typical for the 14th century, and it conforms to Avissar/Stern Type II.2.2.2 (Avissar and Stern 2005, 94–95, fig. 40). Similar examples have been uncovered in the Old City, from Cardo Area X-4 (Avissar 2012, 306, fig. 10.6:13). The other handmade cooking pot fragment presented here is a large and complete pulled-up strap handle (Figure 23:8) with a series of incised dots and finely incised double lines on either side of the dots. It has a pale red (10R 6/4) fabric fired to pink (5YR 7/3), with inclusions of calcite and crushed quartzite. This type of incised decoration is typically found on Mamluk cooking pots of the 14th century and is “usually confined to the handles only” (Avissar and Stern 2005, 94). Although no exact parallels for this handle have been found, the only examples attested locally were uncovered from Cardo Area X-4 (Avissar 2012, pl. 10.6:12) and at Khirbat Ka’kul (Boas 2006, fig. 2:9, 11). As with the cooking wares, there was a paucity of jugs and jars recovered from the Stratum IIB ceramic assemblage. Both of the examples illustrated here were found in the makeup of the flagstone floor from Square B7, and both exhibit the same light reddish-brown (5YR 6/3) fabric. The first is the rim and neck from a large HMGP store jar with a wide neck, and a flared and slightly pointed rim (Figure 23:9). It has a light pink (7.5YR 7/3) exterior slip and reddish-brown (5YR 5/3) painted geometric decoration on its exterior. The vessel conforms to Avissar/Stern Type II.4.4.1 (Avissar and Stern 2005, 113). HMGP store jars with a similar rim and neck form have been uncovered in and around Jerusalem (see Tushingham 1985, fig. 42:16; Boas 2006, fig. 3:15) while the closest parallel to its painted decoration is attested at Banias (Avissar and Stern 2005, fig. 47:4). Of interest is the store jar handle with an almost excessively gouged imprint (possibly from a fluted instrument) on its outer portion (Figure 23:10). It has a light reddish-brown (5YR 6/4) fabric that is fired to pink (5YR 7/3), with numerous grits, as well as small and medium calcite inclusions. No parallels for this unique handle could be found.

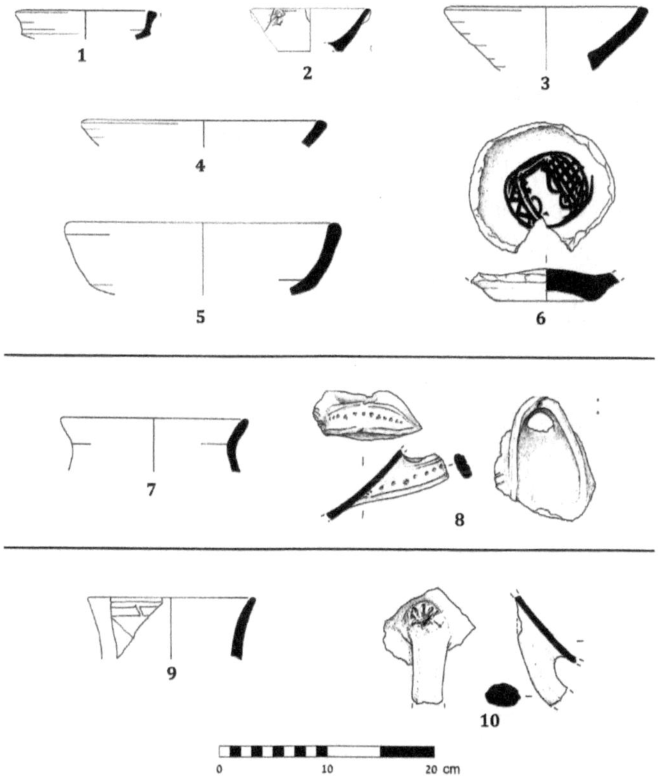

No.	Locus / Basket	Ware Description	Context
1	528 / 5184_6	Light reddish brown (5YR 6/3) fabric fired to weak red (2.5YR 6/4), tiny calcite inclusions	Under flagstone floor in B7
2	539 / 5164_3	Coarse light reddish brown (5YR 6/3) fabric, gray (5YR 5/1), large calcite inclusions, burned-out organic temper	Kiln 3 contents in B2/3
3	117 / 1068_4	Pale red (2.5YR 7/4) fabric, dark gray (5YR 5/2) core, calcite inclusions, numerous grits, burned-out organic temper	Above floor in A6
4	534 / 5110_1	Light reddish brown (5YR 6/4) fabric, dark gray (5YR 4/1) core, numerous grits, large calcite inclusions, burned-out organic temper	*Ṭābūn* contents in B2
5	528 / 5174_2	Coarse reddish yellow (5YR 6/6) fired to pink (5YR 7/3), tiny calcite inclusions, burned-out organic temper	Under flagstone floor in B7
6	121 / 1117_1	Reddish brown (5YR 5/4) fabric fired to reddish yellow (5YR 6/6), numerous grits, large calcite inclusions, dark reddish gray (2.5YR 4/1) painted geometric decoration on central interior of the base	Above MM floor in A4
7	539 / 5164_4	Weak red (10R 5/4) fabric fired to light reddish brown (5YR 6/4), numerous grits, large calcite and crushed quartzite inclusions	Kiln 3 contents in B2/3
8	106 / 1018_2	Pale red (10R 6/4) fabric fired to pink (5YR 7/3), inclusions of calcite and crushed quartzite	Below floor in A8
9	528 / 5184_4	Light reddish brown (5YR 6/3) fabric, pink (7.5YR 7/3) exterior slip, reddish brown (5YR 5/3) painted geometric decoration on exterior	Under flagstone floor in B7
10	528 / 5184_1	Light reddish brown (5YR 6/4) fabric fired to pink (5YR 7/3), numerous grits, small and medium calcite inclusions	Under flagstone floor in B7

Figure 23 Mamluk I plainware bowls (1–6), cooking pots (7–8), and store jars (9–10) from the Southern Sector.

Mamluk I lamps from the Southern Sector (Figure 24)

A small number of moldmade Mamluk lamps were found amongst the Stratum IV pottery, and they form a rather homogenous group that typically dates from the mid-13th to the end of the 14th century. Their form represents a development of the Ayyubid almond-shaped slipper lamp with the high tongue handle but differs in that the handle is bent over so that it is either touching the rim of the lamp, or completely pressed down onto the rim. Instead of having the decorative scheme consisting of arabesques, floral patterns, and pseudo-calligraphy rendered in a thin, fine relief which typically characterize their chronological predecessors, they are most often limited to geometric designs in a thicker relief. Typologically, the group conforms to Avissar/Stern Type III2.1.2 (Avissar and Stern 2005, 128, fig. 53:2–4), Hadad Type 45 (Hadad 2002, 112–114), and Type 2 from the Damascus Gate (Tushingham 1985, 151). A number of these lamps, along with the molds in which they were made, were found at Nebi Samwil, not far from Jerusalem (Magen and Dadon 1999, 76). The first lamp from the Mamluk I period at Khirbet Beit Mazmīl (Figure 24:1) has a very pale brown (10YR 8/3) fabric and was uncovered in the Mamluk robbers trench cut into the western sector of the Byzantine floor in Square A4. Although there are no exact parallels for the decorative scheme on this lamp, with its spade-shaped motifs on the rear of the lamp near the handle, similar examples were uncovered from the Armenian Garden in Jerusalem (Tushingham 1985, fig. 45:2–3). The other two lamps were found in secondary depositional contexts: the first (Figure 24:2) from the Stratum IV flagstone floor in Square B8 (L.530); and the second example (Figure 24:3) from the contemporaneous Mamluk II flagstone floor in the southern sector of Square A6 (L.128). Their stratigraphical context fits well with the fact that they are both so worn that there is a decided lack of any discernable remaining decoration on either of them.

The pottery from Stratum III (Mamluk II, mid-14th to end of 15th century)

Introduction

The next occupational level at Khirbet Beit Mazmīl is represented by Stratum III, which dates from the Mamluk II period, or the mid-14th to the end of the 15th century. There was only limited evidence for new architectural features and related stratigraphy in Area A, all of which originated in Square A6. To the south, however, Area B witnessed considerable construction during Stratum III with the building of a roughly square structure exhibiting a north-northeast/south-southwest orientation. The Stratum III ceramic assemblage presented here is also divided into four typological assemblages. A total of 34 vessels are illustrated herein, and these consist of glazed bowls, imported north Italian glazed bowls, handmade decorated and undecorated bowls, cooking pots, jars, jugs, and a lamp.

Mamluk II glazed bowls from the Southern Sector (Figure 25)

An interesting corpus of glazed bowls was recovered from the Mamluk II ceramic assemblage at Khirbet Beit Mazmīl, and it is highly representative of the typical vessel forms and decorations characteristic of the Jerusalem region during the late-14th and 15th centuries. These bowls are either rounded or slightly carinated, have a thin white slip and a shiny and well-adhering green interior glaze that often drips over onto the exterior rim zone, and the fabric color ranges from reddish-yellow (5YR 6/6) to light reddish-brown (5YR 6/4–5). Typologically, these vessels conform to Avissar/Stern Monochrome Glazed Bowls II (Avissar and Stern 2005, 12–15, figs. 4–5), which have been assigned a date range from the late 13th to the 15th century. The majority of these bowls (Figure 25:1–6) are of the Avissar/Stern Type I.1.4.3 Bowls with Out-Folded or Everted Rim (Avissar and Stern 2005, 14, fig. 5:7–10), which "seems to be unique to Jerusalem where it appeared in

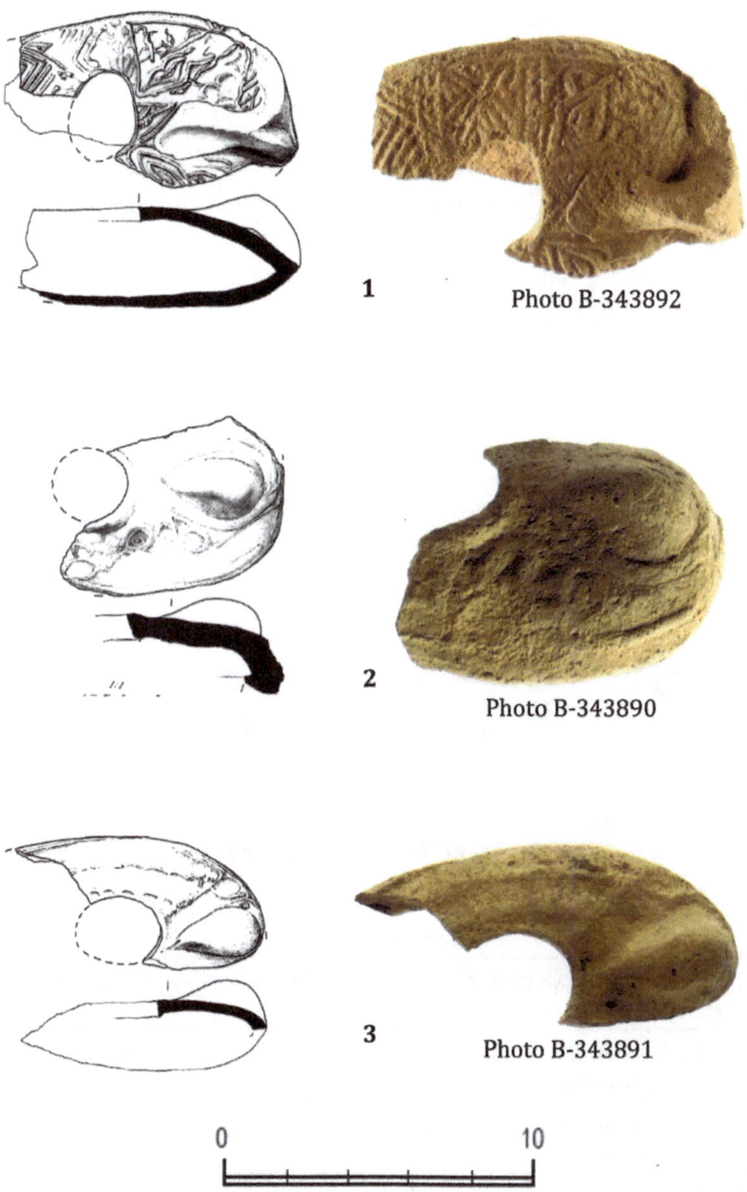

No.	Locus / Basket	Ware Description	Context
1	133 / 1125	Very pale brown (10YR 8/3) fabric	MM Robber's Trench in A4
2	530 / 5176	Very worn. Pinkish gray (5YR 7/2) fabric	Secondary Deposition MM II flagstone floor
3	128 / 1106	Light reddish brown (5YR 6/3) fabric fired to pinkish gray (7.5YR 7/2)	Secondary Deposition MM II flagstone floor

Figure 24 Mamluk I lamps from the Southern Sector.

several, so far unpublished, excavations" (Avissar 2012, 310). The rim form for these bowls is very distinct: it has an upturned stance with a profile that is either a small internal or external slightly thickened ledge or T-shaped, with either a flat (Figure 25:1–2, 4–6) or rounded (Figure 25:3) lip. Several examples of these bowls were recovered from the Jerusalem region (see Avissar 2012, fig. 10.5:6–7; Boas 2006, fig. 12:82–83; Tushingham 1985, fig. 41:1–21). Taken together, this glazed bowl type seems to be a strictly local production, as attested by one example recently uncovered from the Muslim Quarter of the Old City (S. Kisilevitz *personal communication*), which had the tripod-stand from where it was stacked in a kiln still attached to and melted into the interior base of the vessel. Stratigraphically, all the aforementioned bowls were found on either the plaster or flagstone floors from Squares A6 and B7–8, which were laid during the Mamluk II period.

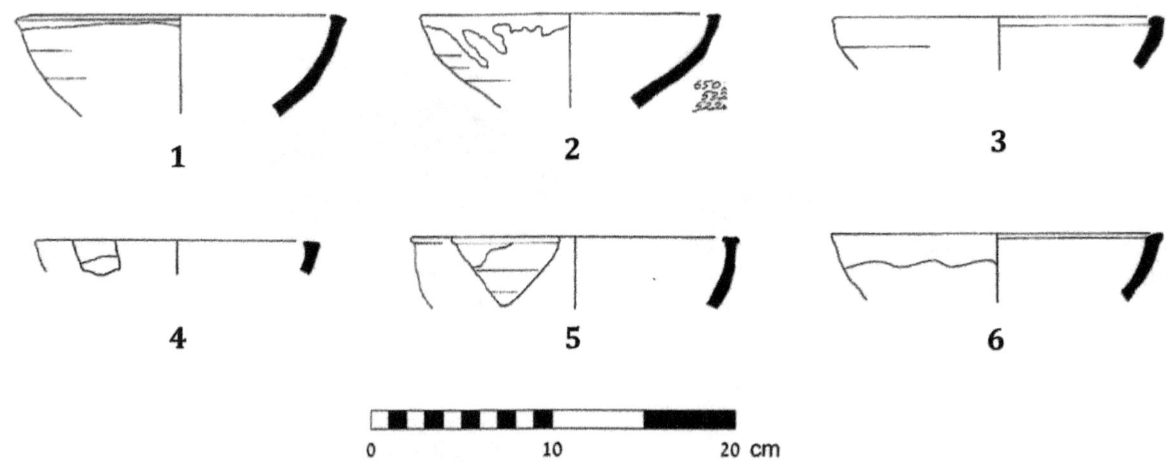

No.	Locus / Basket	Ware Description	Context
1	128 / 1104_15	Light reddish brown (5YR 6/5) fabric, thin white slip, green glaze on interior dripping over rim	Above floor in A6
2	532 / 5220_6	Reddish yellow (5YR 6/6) fabric, thin white interior slip, light green interior glaze dripping onto exterior	Floor in B8 NE
3	532 / 5187_7	Reddish yellow (5YR 6/6) fabric, thin white slip, green interior/ exterior glaze	Floor in B8 NE
4	507 / 5122	Reddish yellow (5YR 6/6) fabric, tiny calcite inclusions, thin white slip, green glaze on interior, rim, and exterior rim zone	Above floor in B7
5	545 / 5212_2	Reddish yellow (5YR 6/6) fabric, tiny calcite inclusions, thin white interior slip, dark green high-quality well-adhering glaze	B4/7 Fill along W.554 down to bedrock
6	128 / 1104_3	Light reddish brown (5YR 6/4) fabric, thin white slip, thick shiny well-adhering dark green glaze on interior dripping over onto exterior	Above floor in A6

Figure 25 Mamluk II glazed bowls from the Southern Sector.

Imported north Italian glazed bowls from the Southern Sector (Figure 26)

Perhaps the most significant group of ceramic finds from the excavations at Khirbet Beit Mazmīl is the corpus of imported north Italian glazed bowls recovered from Area B. Small quantities of bowls and basins from northern Italy are known from sites in the north of the country (Stern 1999, 129–131, fig. 2), from the Old City of Jerusalem (Johns 1950, 189, pl. LXIII:3, 5; Pringle 1984; Wightman 1989, pls. 67:7–8, 68:4). The north Italian finds from Khirbet Beit Mazmīl fall into the following three categories, all of which were of Venetian and/or Paduan production and date from the late 14th to the end of the 15th century: monochrome sgraffito wares; glazed carinated bowls; and *Graffita Arcaica*, which is also referred to as *Rinascimentale* (or "Rennaisance") ware. The monochrome sgraffito bowls were the most common type of north Italian wares found at the site, and four are illustrated here (Figure 26:1–4). All of them have a light red fabric with tiny mica flecks, a thin white interior slip and a well-adhering shiny dark green glaze that covers the interior of the vessel, occasionally dripping over onto the rim. The interior is incised with sgraffito using simple floral designs, and the very distinct exterior is decorated with a series of two or three protruding ridges located just below the rim zone which are rouletted vertically in an even fashion across the circumference of the bowl. They conform to Avissar/Stern Type I.9.5 (Avissar and Stern 2005, 73, fig. 31:4–6) and have been found at other places in Jerusalem (see Johns 1950, 189, pl. LXIII:5; Wightman 1989, pl. 68:4), and at their production center (Saccardo 1993, 365, pl. III:42) in Venice. While two of these bowls originated from secondary depositional contexts, one (Figure 26:3) was found on the L.541 flagstone floor in Square B8, and another (Figure 26:4) was discovered in the makeup of that flagstone floor beneath L.530 in the same square.

The next group of imported north Italian pottery from the site consists of the glazed carinated bowls that correspond to Avissar/Stern Type I.9.6 (Avissar and Stern 2005, 73–74, fig. 31:7–8). Two examples were recovered from the excavation, and one is presented here (Figure 26:5). It has the same light red fabric as the aforementioned monochrome sgraffito bowls and is covered with a shiny, brown well-adhering glaze on the interior and exterior. The primary morphological characteristic of this bowl type is its pronounced horizontal ridge at the carination point of the vessel.

The final imported bowl sherd from northern Italy is represented by the ring-base from a *Graffita Arcaica/Rinascimentale* bowl (Figure 26:6) produced in Venice during the 15th century (Saccardo 1993; Saccardo 1996; Siviero 1986; Avissar and Stern 2005, 72–73, Type I.9.4, Fig. 31:1–3). This bowl sherd has a light red (10R 6/6) fabric with tiny mica flecks and is covered with a thick white interior slip and a transparent white lead glaze. The incised interior decoration is highlighted with splashes of brown, green, and yellowish glaze. The central interior of the bowl has a distinctive raised and protruding ring around its circumference, a characteristic trait found on many of the Venetian *Graffita Arcaica* bowls (see Alvera-Bartolotto 1981, tab. XVI; Conton 1981, 164; Wightman 1989, pl. 67:8). Taken together, the fact that this small but significant amount of imported pottery from northern Italy was found in a rural context well outside of Jerusalem is noteworthy.

Mamluk II handmade bowls from the Southern Sector (Figure 27)

The overwhelming majority of the ceramic assemblage from Stratum III consisted of handmade decorated and undecorated bowls of varying sizes. While most of them bear similarities to the pottery recovered from sites in Jerusalem and the surrounding hinterland, they are markedly different in a few key aspects of form and decoration, especially with regard to their incurved rather than upright and direct vessel walls. Exact parallels for these vessels were therefore difficult to discern.

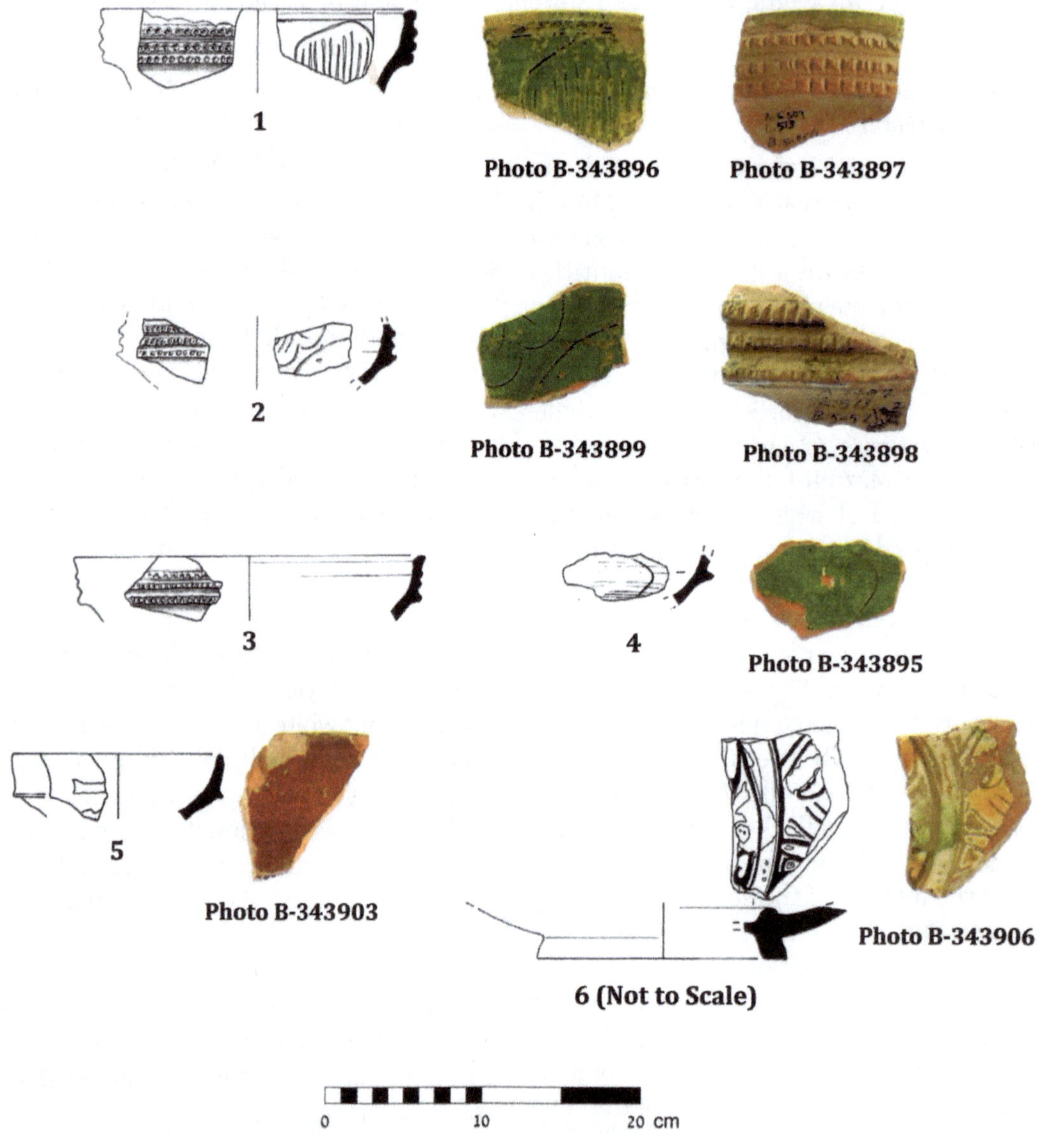

No.	Locus / Basket	Ware Description	Context
1	513 / 5066_1	Light red (10R 6/6) fabric, thin white slip, shiny dark green well-adhering glaze on interior dripping onto rim and exterior, incised decoration on interior	Subsoil in B5
2	513 / 5052_2	As no. 1 above	Subsoil in B5
3	541 / 5189_3	As no. 1 above	Above floor in B8 NW
4	536 / 5119_1	As no. 1 above	Makeup of L530 floor in B8
5	513 / 5061_2	Light red (10R 6/6) fabric, shiny brown well-adhering glaze on interior and exterior	Subsoil in B5
6	503 / 5015_13	Light red (10R 6/6) fabric, tiny mica flecks, thick white interior slip, transparent white lead glaze, incised interior decoration highlighted with splashes of brown, green and yellowish glaze	Topsoil in B5

Figure 26 Imported north Italian glazed bowls from the Southern Sector.

It is suggested here that perhaps some of these handmade vessels were produced locally—that is, in the vicinity of the site. The following corpus presented here consists of plainware bowls, those with either simple linear or complex geometric painted designs on the interior and (less frequently) on the top of the rims, and several examples with impressed rope decoration on the exterior just below the rim zone. The first two bowls (Figure 27:1–2), both found on Mamluk II floors, are small and have simple, direct rims that are somewhat rounded. Similar examples have been uncovered at the Damascus Gate (Wightman 1989, pl. 50:1) in the Old City of Jerusalem. The next three vessels (Figure 27:3–5) are important for two reasons: first, they all originate from the makeup (L.537) of the plaster floor in the western sector of Square B7, and therefore come from stratigraphically sealed deposits. All of them are characterized by a slightly incurved body. The first bowl from this group (Figure 27:3) has a pale red (2.5YR 7/4) fabric, dark gray (5YR 4/1) core, numerous grits, medium calcite inclusions, and burned-out organic temper in its matrix. Unlike most of the vessels from this assemblage, it has a vestigial pressed-on handle just below the rim zone. The next bowl (Figure 27:4) is a shallower version of the preceding one but has an interior slip and is decorated with a worn and barely visible reddish-brown (5YR 4/3) paint on its interior and rim top. The final example from this group (Figure 27:5) uncovered from the makeup of the plaster floor in Square B7 is similar in most respects to its predecessor, although has a pink (7.5YR 7/4) fabric and a gray (5YR 5/1) core, and its painted decoration (which is much better preserved) is horizontal and linear in nature. The next two bowls bear striking similarities to those uncovered from the French excavations in Abu Ghosh (de Vaux and Stève 1950, pls. F8 and 16, respectively), and are the first representatives from this group which bear the impressed rope pattern on the vessel exterior, a decorative motif which forms the core for this ceramic repertoire. The first bowl (Figure 27:6) is slightly smaller and has thicker vessel walls, while the second example (Figure 27:7) is larger and deeper. Both vessels have a light reddish-brown (5YR 6/3–4) fabric with numerous grits and calcite inclusions, and the larger of the two also contains burned-out organic temper in its matrix. A plainware bowl with rope decoration (Figure 27:8) was uncovered from above the flagstone floor in Square B8. It has a reddish-yellow (5YR 6/6) fabric that was fired to a pale red (2.5YR 7/3) and a reddish-gray (10R 5/1) core. An HMGP bowl with rope decoration (Figure 27:9) was found above the flagstone floor in the southern sector of Square A6. It is made from a coarse, very pale brown (10YR 7/3) fabric with calcite inclusions and burned-out organic temper, and is decorated with a pinkish gray (5YR 6/2) painted geometric design on the rim top. A similar example was recovered from Cardo Area X-4 and dated to the 14th and 15th centuries (Avissar 2012, 306, pl. 10.6:7). The next three vessels (Figure 27:10–12) are representative of what have been termed either "workshop" or "atelier" bowls, that seem to have been produced in the vicinity of Khirbet Beit Mazmīl during the 14th and 15th centuries. This corpus of relatively thin-walled HMGP bowls all seem to have been made by the same hand(s). They essentially have the same buff fabric, although there are a few examples with light reddish-brown fabric, and usually have a white (or pinkish-white) interior slip and very dark—almost black—painted decoration. Most of them have a horizontal applied band of finger indentations or rope pattern along the exterior rim zone which is carefully executed and evenly spaced. The painted decorative motif is relatively uniform and consists of a thick band along the interior just below the rim, a zone below that consisting of two or three thin horizontal lines, and a major decorative zone covering the rest of the lower vessel interior to where the vessel walls join the base. This zone usually consists of a series of either pendent superimposed triangles or semi-circles. Most of the vessels have a series of diagonal lines covering the entire rim top, a decorative scheme that has been referred to as the

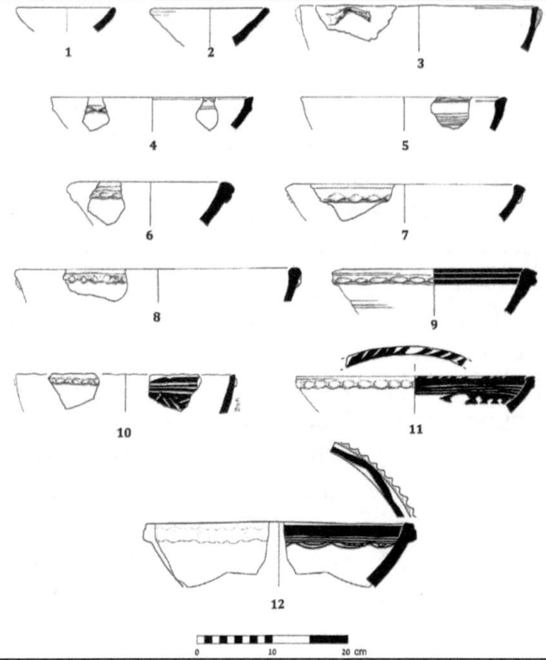

No.	Locus / Basket	Ware Description	Context
1	507 / 5151_1	Pale red (2.5YR 7/3) fabric fired to pink (5YR 7/4), numerous grits, calcite inclusions	Above floor in B7
2	128 / 1104_7	Reddish yellow (5YR 6/6) fabric, calcite inclusions, numerous grits	Above floor in A6
3	537 / 5240_2	Pale red (2.5YR 7/4) fabric, dark gray (5YR 4/1) core, numerous grits, medium calcite inclusions, burned-out organic temper	Below floor in B7
4	537 / 5236_2	Light reddish brown (5YR 6/3) fabric, numerous grits, medium calcite inclusions, reddish yellow (5YR 6/6) interior slip, reddish brown (5YR 4/3) painted decoration on interior, rim top	Below floor in B7
5	537 / 5236_9	Pink (7.5YR 7/4) fabric, gray (5YR 5/1) core, numerous grits, medium calcite inclusions, thin pink (5YR 7/3) interior slip, reddish brown (5YR 5/4) painted horizontal linear decoration on interior and rim top	Below floor in B7
6	507 / 5151_2	Light reddish brown (5YR 6/4) fabric fired to pink (7.5YR 8/4), dark reddish gray (2.5YR 5/1) core, numerous grits, calcite inclusions	Above floor in B7
7	530 / 5176_4	Light reddish brown (5YR 6/3) fabric, numerous grits, calcite inclusions, burned-out organic temper	Floor in South of B8 East
8	541 / 5189_2	Reddish yellow (5YR 6/6) fabric fired to pale red (2.5YR 7/3), reddish gray (10R 5/1) core, large calcite inclusions	Above floor in B8 NW
9	128 / 1103_4	Very pale brown (10YR 7/3) fabric, calcite inclusions, burned-out organic temper, pinkish gray (5YR 6/2) painted geometric design	Above floor in A6
10	545 / 5218_1	Light reddish brown (5YR 6/3) fabric, a few medium calcite inclusions, very pale brown (10YR 8/3) slip, very worn grayish brown (10YR 5/2) painted geometric decoration	B4/7 Fill along W.554 down to bedrock
11	127 / 1111_8+	Light reddish brown (5YR 6/3) fabric, very pale brown (10YR 8/3) interior/exterior slip, very dark gray (7.5YR 3/1) painted geometric decoration on interior and rim top	Door Blockage in A6
12	103 / 1002_2	Reddish yellow (5YR 6/6) coarse fabric, very pale brown (10YR 8/3) interior slip, reddish black (2.5YR 2.5/1) painted geometric decoration on interior and rim top, impressed decoration on rim exterior	Cistern 1 in Square A11

Figure 27 Mamluk II handmade bowls from the Southern Sector.

"zebra crossing" motif (S. Gabrieli *personal communication*), but on rare occasions the rim top is covered with a centrally placed continuous wavy line. These bowls were recovered primarily from Area A (see Figure 27:11–12), but one example was attested in Area B (Figure 27:10) as well. The only other known example of this unique bowl type was uncovered in the Old City of Jerusalem and a sample from this vessel has undergone petrographic analysis (S. Gabrieli *personal communication*) to determine its provenience.

Mamluk II cooking pots, jars, jugs, and lamp from the Southern Sector (Figure 28)

Completing the ceramic assemblage from Stratum III is a small corpus of cooking pots, jars, jugs, and a lamp—all of which are typical of the Mamluk II period, or the 14th through 15th centuries, at Khirbet Beit Mazmīl. Handmade cooking pots with out-turned rims were uncovered from the flagstone floor in Square B8 (Figure 28:1) and the makeup of the plaster floor in Square B7 (Figure 28:2). Their coarse fabric is either red or light reddish-brown and has numerous grits and inclusions of calcite, crushed quartzite and occasionally feldspar in its matrix. Similar examples are known from the Damascus Gate (Wightman 1989, pl. 54:2), and the type with the out-turned rims is a later, 14th to 15th century, manifestation of the earlier form with an upright rim (M. Avissar *personal communication*). An interesting, relatively thin-walled cooking pot handle with a pinkish gray (5YR 6/2) exterior and incised decoration was found in Square B4/7 (Figure 28:3). This handle is typical of the Mamluk ceramic repertoire and is similar in most respects to an example from Khirbat Ka'kul (Boas 2006, fig. 2:9).

Part of the neck and handles from a large HMGP strainer jar (Figure 28:4) was found on the flagstone floor in Square B8. It has a reddish-yellow (5YR 6/6) fabric with numerous grits and calcite inclusions. Its pale red (2.5YR 7/4) exterior slip is covered with a reddish-brown (5YR 4/3) painted decoration. Similar vessels are known from the Old City (Tushingham 1985, fig. 42:16) and the Jerusalem hinterland (de Vaux and Stève 1950, pl. F3). Two handmade store jars were represented in the Stratum III assemblage. The first (Figure 28:5) consists of a rim, neck, and partial handle made from a coarse reddish-yellow (5YR 6/6) fabric with a dark gray (5Y 4/1) core, numerous grits, calcite inclusions and burned-out organic temper. A store jar with the same rim form, but with painted decoration, is attested at the Armenian Garden (Tushingham 1985, fig. 42:16). A thick-walled HMGP store jar with an upright rounded rim was recovered from the fill above Circular Installation 2 in Square B2 (Figure 28:6). It has a light reddish-brown (5YR 6/3) fabric, numerous grits, and calcite inclusions. Both the interior and exterior are covered with a very pale brown (10YR 8/2) slip, and the exterior is decorated with a reddish-brown (5YR 5/4) painted geometric decoration. The presence of the buff slip on this vessel, along with the color of the paint, indicates a production date in the 15th century (S. Gabrieli *personal communication*), and it is similar to an example from the Armenian Garden (Tushingham 1985, fig. 42:17). A few jugs were also recovered from Mamluk II contexts at Khirbet Beit Mazmīl. The first example (Figure 28:7) was found in the makeup of the plaster floor from Square B7. It is a plainware jug with folded-over rim with a light reddish-brown (5YR 6/4) fabric and a few medium-sized calcite inclusions. A good parallel for this vessel is attested from Cardo Area X-3 (Avissar 2012, pl. 10.3:15). Another jug attributed to this stratum is of the well-known variety often referred to as a "trumpet jug" (Figure 28:8). This moniker is due to the shape of its neck, which is long and narrow at the neck/shoulder join and flares significantly outward towards the rim, much like the shape of a trumpet (see Avissar and Stern 2005, pl. XXX:4). The example illustrated here has a reddish-brown (5YR 5/4) fabric and an everted rim that is almost triangular in section, which is characteristic for this vessel form. This jug type is commonplace in the Jerusalem ceramic repertoire (Wightman 1989, pl. 60:12), where it has a long date

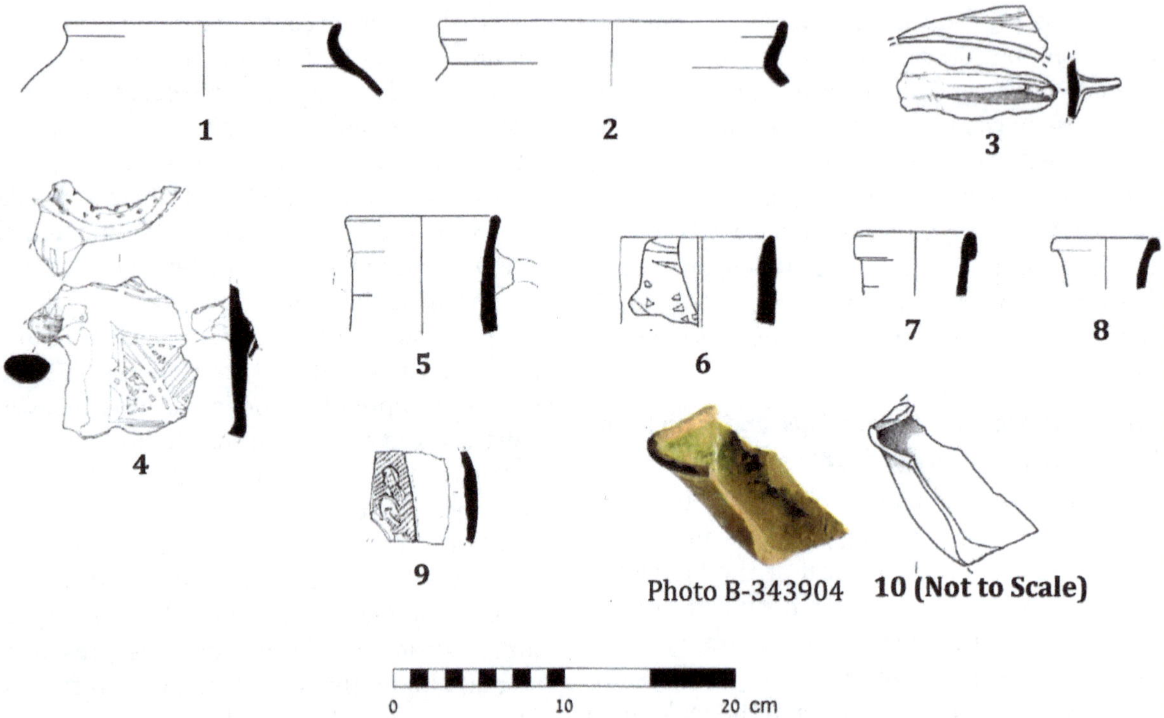

No.	Locus / Basket	Form	Ware Description	Context
1	532 / 5187_01	CP	Red (2.5YR 6/6) fabric, gray (5YR 5/1) exterior, numerous grits, inclusions of calcite, crushed quartzite and feldspar	Floor in B8 NE
2	537 / 5240_3	CP	Light reddish brown (5YR 6/3) fabric fired to pale red (10R 6/2), numerous grits, calcite and crushed quartzite inclusions	Below floor in B7
3	545 / 5212_3	CP	Light reddish brown (5YR 6/3) fabric fired to reddish yellow (5YR 6/6), pinkish gray (5YR 6/2) exterior, numerous grits, calcite and crushed quartzite inclusions	B4/7 Fill along W.554 down to bedrock
4	532 / 5220_9	StrJ	Reddish yellow (5YR 6/6) fabric, numerous grits, calcite inclusions, pale red (2.5YR 7/4) exterior slip, reddish brown (5YR 4/3) painted decoration	Floor in B8 NE
5	128 /1103_7	SJ	Coarse reddish yellow (5YR 6/6) fabric, dark gray (5Y 4/1) core, numerous grits, calcite inclusions, burned-out organic temper	Above floor in A6
6	515 / 5098_1	SJ	Light reddish brown (5YR 6/3) fabric, numerous grits, calcite inclusions, very pale brown (10YR 8/2) interior/exterior slip, reddish brown (5YR 5/4) painted geometric decoration on exterior	Above Kiln 2 in B2
7	537 / 5236_4	JG	Light reddish brown (5YR 6/4) fabric, a few medium calcite inclusions	Below floor in B7
8	530 / 5176_14	JG	Reddish brown (5YR 5/4) fabric, dark gray (10YR 4/1) core, calcite inclusions	Floor in South of B8 E
9	128 / 1104_10	JG	Pink (5YR 7/3) fabric fired to pinkish gray (7.5YR 7/2)	Above floor in A6
10	501 / 5148	LP	Light red (2.5YR 7/6) fabric, dark green interior/exterior glaze	Secondary/Tertiary Topsoil in B4

Figure 28 Mamluk II cooking pots (1–3), jars (4–6), jugs (7–9), and lamp (10) from the Southern Sector.

range covering the 13th through early 16th centuries (M. Avissar *personal communication*). The fabric color indicates a date in the 15th century and this vessel has an exact parallel from Phase II at the Austrian Hospice (Dolinka 2017, fig. 1:13). A unique ceramic find from the Mamluk II assemblage is the body sherd from a jug decorated with incised and pinpricked decoration (Figure 28:9). It has a pink (5YR 7/3) fabric fired that was fired to pinkish-gray (7.5YR 7/2). This sherd was most likely from an imported Syrian strainer-neck jar decorated with thinly incised lines and small floral motifs. It is of the same type discussed above (see Figure 13:5 above), with decorative schemes attested at Hama (Poulsen 1957, 258–260, figs. 943–947, Category D.XVIIIe). Based upon its find context in L.128, above the flagstone floor in Square A6, a date in the late 15th century is suggested here. The final Mamluk II vessel presented here is a small wheel-made Mamluk glazed lamp with a pinched nozzle (Figure 28:10). It conforms to Avissar/Stern Type III.3.1 (Avissar and Stern 2005, 128, fig. 53:5). This lamp has a light red (2.5YR 7/6) fabric with a thin white interior/exterior slip and is covered with a patchy dark green interior/exterior glaze. The production for most lamps of this type has been attributed to a workshop at Emmaus (Gichon and Linden 1984), where numerous examples were uncovered. Unglazed lamps with a similar form have been uncovered in the Old City of Jerusalem (Avissar 2003, pl. 19.2:22; Avissar 2012, pl. 10.19:9), where they are dated from the 14th to 15th centuries: but those examples have a much longer and narrower spout than the glazed example from Khirbet Beit Mazmīl.

In Small Things Forgotten: Material culture down on the farm

In the popular study *In Small Things Forgotten: An Archaeology of Early American Life* by anthropologist James Deetz (1996), emphasis was placed upon how small objects and finds can shed light on the activities of daily life, thereby painting a picture of how these minutiae help bind us to our human past. The same is also reflected in the metal, stone, and glass finds, which provide important information about how the people lived and what type of things they did on a daily basis. Presented here are assemblages of metal (21), stone (10), and glass (18) objects from the Southern Sector of the site during the Mamluk period.

Mamluk metal finds from the Southern Sector

A total of 46 iron, bronze, lead, and copper objects were recovered from the two seasons of excavation at Khirbet Beit Mazmīl. These finds were attested in nearly all areas of the site. Out of the total metal assemblage, only half were selected for publication, as they were either found in good stratigraphical contexts or were important due to their form and function. Taken together, the metal finds can be divided into four main categories: equine-related accessories; nails and farm implements; weaponry; and personal and decorative items. It should be noted that a concentration of the metal objects presented here were recovered from L.537, the makeup of the plaster floor in Room 507 from Structure B1. These included a bridle bit, a cinch buckle from a saddle, a horseshoe, an S-hook, and an arrowhead. Other important find contexts included the following: L.115 and L.130 from Square A1, which produced a bridle/rein buckle, two functional rings, two iron nails and a pruning hook; L.511 from Square B8, which yielded an iron nail and a decorated wedding ring; and L.505 in Square B4, which contained a bridle/rein buckle, a noseband buckle, a dagger, and a cosmetic spoon.

Equine-related accessories (Figure 29)

A small, yet significant, number of metal finds from the site comprised of equine-related accessories, including a bridle bit, bridle/rein buckles, a cinch buckle from a saddle, and a horseshoe. A unique and un-paralleled object is the iron three-link bridle bit (Figure 29:1). Since there is a decorative boss on only one side of the central bar, it was likely used in a paired draught (Jope 1956, fig. 514B). The next four metal objects are buckles that

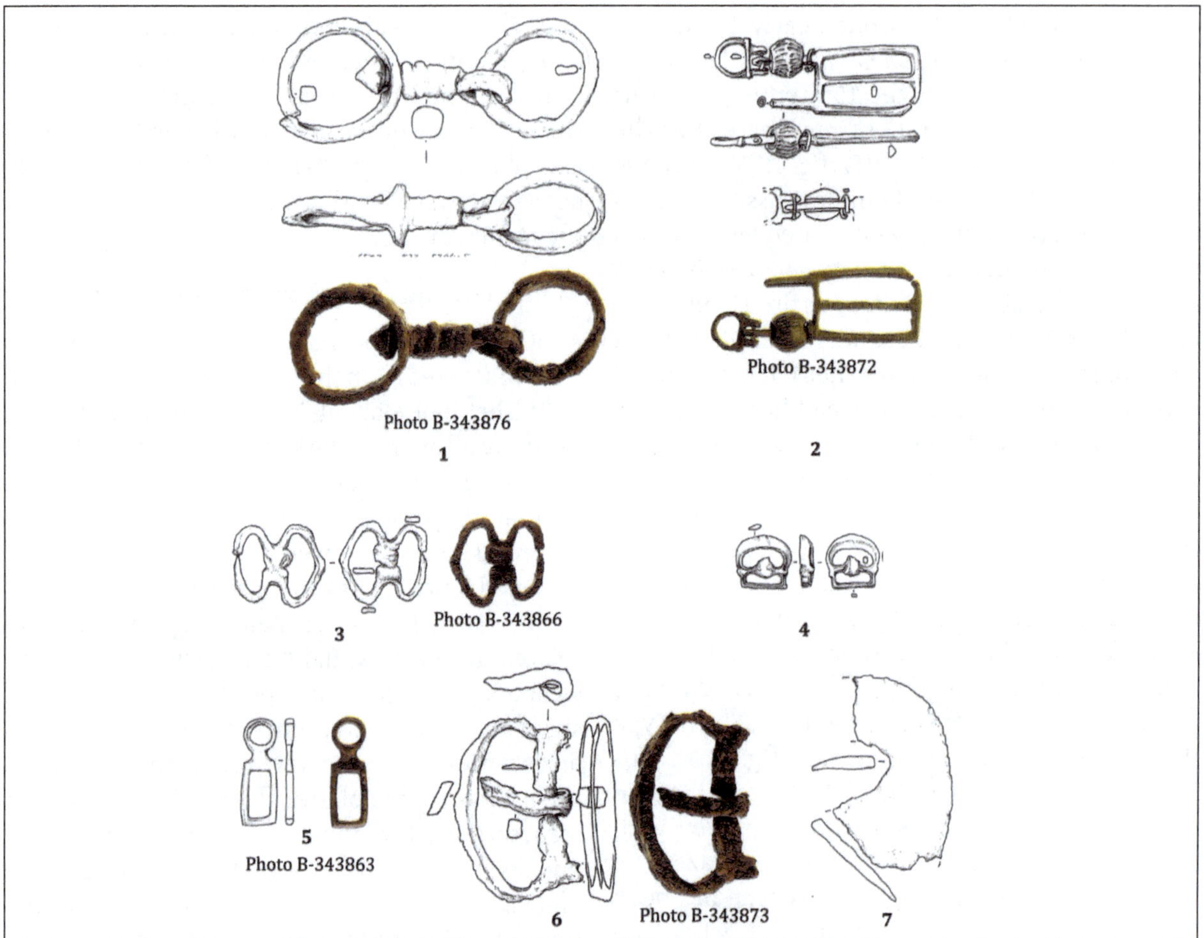

No.	Square	Reg. No.	Description
1	B7	537 / 5244_5	Three-link bridle bit. Iron. Complete. Length: 12.4 cm. Weight: 75.6 gm. Left ring diameter: 4.2 cm. Right ring diameter: 4 cm. Ring thickness (both): 5 mm. Maximum thickness of central bar: 11 mm.
2	B4	505 / 5017	Decorative bridle/rein buckle. Bronze. Almost complete. Weight: 32.6 gm. Length: 7.8 cm. Width of buckle: 2.6 cm. Thickness: 2mm. Bead Diameter: 1.3 cm.
3	A1	115 / 1065_1	Bridle/rein buckle. Bronze. Complete. Weight: 9.9 gm. Length: 3.3 cm. Width: 2.8 cm. Thickness: 4 mm.
4	B4	514 / 5138	Bridle/rein buckle. Bronze. Complete. Weight: 4.1 gm. Length: 2 cm. Width: 1.8 cm. Thickness: 1 mm.
5	B4	505 / 5134	Noseband (*cavesson*) buckle. Bronze. Complete. Weight: 5.4 gm. Length: 4 cm. Width: 1.2 cm. Thickness: 1 mm.
6	B7	537 / 5127	Cinch buckle from a saddle. Bronze. Corroded. Complete. Weight: 47.3 gm. Length: 7.8 cm. Width: 4.4 cm. Thickness: 4 mm.
7	B7	537 / 5244_4	Horse shoe. Iron. Both ends broken. Weight: 52.6 gm. Length: 7.2 cm. Width: 4.4 cm. Thickness: 5 mm.

Figure 29 Equine-related accessories from the Southern Sector.

were used in either bridles or reins. As has been aptly noted, ". . . due to the fact that few clothes had little need for buckles, most must have been used for . . . part of the tack for horses and other burden animals" (Raphael 2008, 447). Another interesting piece is the almost complete bronze decorated bridle/reign buckle (Figure 29:2), which had at least one incised bead on it. Its heavy weight (32.6 grams), when compared with the following two examples (9.9 and 4.1 grams, respectively), suggests that it may have been used as a connector between the crownpiece and one of the cheekpieces. The next two bronze buckles (Figure 29:3–4) are much smaller, but bear striking similarities to those still used on modern bridles, while the final specimen (Figure 29:5) would appear to come from the part of the *cavesson* where the noseband and cheekpieces attached. In addition to those, another much larger buckle (Figure 29:6) was uncovered from the makeup of the plaster floor in Square B7 (L.537). Although highly corroded, it is complete and is clearly the cinch buckle from a saddle. Almost all of the parallels for this object come from contexts dating from the 13th century (that is, Mamluk I), including the Red Tower (Pringle 1986, 164–165, fig. 55:11, from Phase C), 'Atlit (Johns 1936, fig. 16:1, dated from 1217–1291 CE), Montfort Castle (Dean 1927, 38, fig. 53f-j), and Hama (Ploug and Oldenberg 1969, 85, fig. 32:6).

Of perhaps the greatest importance amongst the equine accessories found at Khirbet Beit Mazmīl is the iron horseshoe (Figure 29:7). This medieval type of horseshoe was a departure from the original Roman version, which consisted of a flat piece of iron that was strapped to the hoof (Jope 1956, 515, fig. 469), as well as the local shape which developed from it, characterized by a thin plate which covered the hoof (Johns 1936, 43). Instead, the medieval type followed the contour and profile of the hoof, and was nailed to its bottom, rather than being strapped to it (Jope 1956, 561). While this medieval version was clearly brought to the region from Europe by the Crusaders (Khamis 1996, 220), based upon the smaller size of these horseshoes, ". . . they must have been intended for horses of native rather than European origin" (Johns 1936, 43). With regard to parallels for this horseshoe, the Khirbet Beit Mazmīl specimen is similar in most respects to an example published from a 14th century context in the Museum of London (Ward-Perkins 1940, figs. 36:12, 37:5, both 14th century).

Nails and farm implements (Figure 30)

The second category of metal artifacts consists of a series of nails, either a pruning hook or perhaps a small sickle blade, an S-hook and two functional rings. The majority of these items were recovered from Square A1. Nails are a common artifact from archaeological excavations and are related to constructions made of wood (Raphael 2008, 437). A total of seven iron nails were found during the excavations, three of which (Figure 30:1–3) are presented here. All of the nails have a tapered shaft that is square in section. The heads are either square or round in shape, and flat or domed in profile. The lengths of the nails vary according to their associated function and purpose, with the longer ones being used for fastening of thicker and wider beams or planks. The technology and production methods employed for fashioning these implements is simple and long-lived, which makes dating them problematic at best and, as has been noted, "no comprehensive research on the subject has been undertaken" (Seligman 2010, 194). Nails similar in form to those from Khirbet Beit Mazmīl have been found in medieval and Mamluk contexts at the following sites: Horvat 'Uza (Tatcher 2009, 185, fig. 3.42:5), Abu Ghosh (de Vaux and Stève 1950, fig. 36:6, 9–10) and Yoqne'am (Khamis 1996, 223, no. 31, fig. XVIII.4:14). An iron pruning hook (Figure 30:4) was uncovered at the site and identified as such by its curved iron blade with evidence of sharpening on its inner surface. These tools reached the pinnacle of their development during the Roman period and thus the general shape and function have not changed since (Khamis 2008, 180; Seligman 2010, 193). They were used during routine pruning activities and maintenance of fruit trees as well as for the harvesting of grains. Hooks and functional rings are common features amongst the metal assemblages from rural farmsteads such as Khirbet Beit Mazmīl. These implements vary in size and shape, with the hooks being utilized primarily for the hanging of tools and the functional rings used in a series for connecting chains (Khamis 2008, 178). An iron S-hook (Figure 30:5) was recovered from the makeup of the plaster floor (L.537) in Square B7, a similar example of which was found in a 13th century Crusader stable at 'Atlit (Johns 1936, fig. 15:7). Two functional rings were uncovered from L.115 in Square A1. The larger of the two (Figure 30:6) is made of bronze and is flat in section, while the smaller, thicker ring (Figure 30:7) is iron and has a circular section. No parallels for either of these two objects could be found.

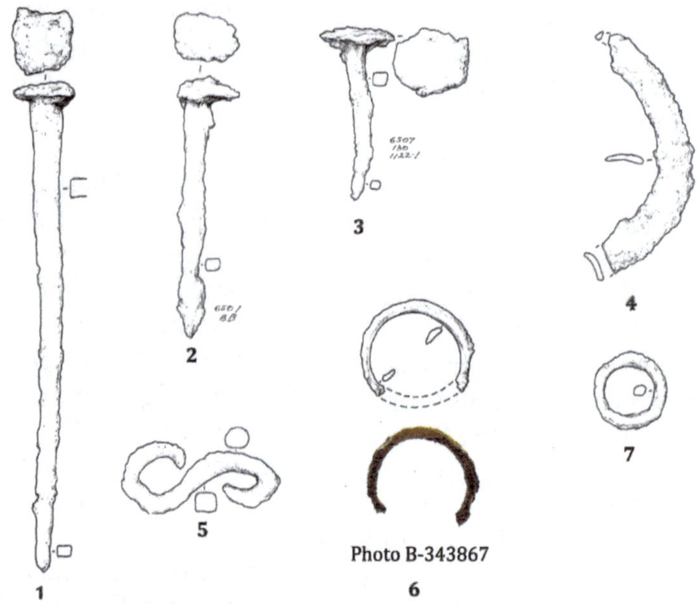

No.	Square	Reg. No.	Description
1	A1	130 / 1122_3	Nail. Iron. Corroded. Complete. Square flat head and square section. Weight: 69.8 gm. Length: 17.4 cm. Head width: 2 cm. Maximum thickness: 6 mm.
2	B8	511 / 5159	Nail. Iron. Complete. Corroded. Square domed head and square section. Weight: 23.8 gm. Length: 9.2 cm. Head width: 2.1 cm. Maximum thickness: 7 mm.
3	A1	130 / 1122_1	Nail. Iron. Complete. Round flat head and square section. Weight: 18.2 gm. Length: 6.3 cm. Head width: 2.6 cm. Maximum thickness: 6 mm.
4	A1	130 / 1122_3	Pruning hook or small sickle blade. Iron. Both ends broken. Weight: 20.1 gm. Length: 8.3 cm. Width: 1.5 cm. Maximum thickness: 9 mm.
5	B7	537 / 5244_3	S-hook. Iron. Complete. Minor corrosion. From square to round in section. Weight: 25.6 gm. Length: 5.8 cm. Width: 2.5 cm. Maximum thickness: 9 mm.
6	A1	115 / 1061_2	Functional ring. Bronze. Flat in section. Weight: 8.7 gm. Diameter: 4 cm. Thickness: 3mm.
7	A1	115 / 1065_2	Functional ring. Iron. Circular in section. Weight: 5.6 gm. Diameter: 2.5 cm. Thickness: 4 mm.

Figure 30 Nails and farm implements from the Southern Sector.

Weaponry (Figure 31)

A small corpus of weaponry was uncovered at Khirbet Beit Mazmīl, including a knife, the base of a scabbard, and a large arrowhead. With regard to knives in general, it has been pointedly noted that "they are difficult to date based on style alone" (Khamis 2008, 179). The knife presented here is made of iron (Figure 31:1), measures 13.6 centimeters overall and has its blade completely preserved, while the tang is broken. Though a similar example to this knife was uncovered at Vadum Iacob in a late 12th century context (Boas 1999, fig. 6.2:3), one must recall the preceding quote in the previous sentence when considering the dating of knives. Also included among the metal assemblage was a well-preserved bronze scabbard base (Figure 31:2) which, due to its size and shape, was most likely used to hold a dagger. It was made from sheet metal that was folded over on its ends into a triangular cone shape. Parallels for this artifact have been found at Mamluk settlements at Banias (Khamis 2008, 184, no. 85) and Hama (Ploug and Oldenburg 1969, fig. 18:8–9), while similar examples are known from the Museum of London (Ward-Perkins 1967, fig. 88:2–3), where they are dated from the 14th to 15th century. The fragment of an iron arrowhead (Figure 31:3) was recovered from the makeup of the plaster floor (L.537) from Square B7. Its tip was broken but the tang was completely preserved. The broken and raised double ridge along the blade represents the remnants of a third wing, meaning that the arrowhead conforms

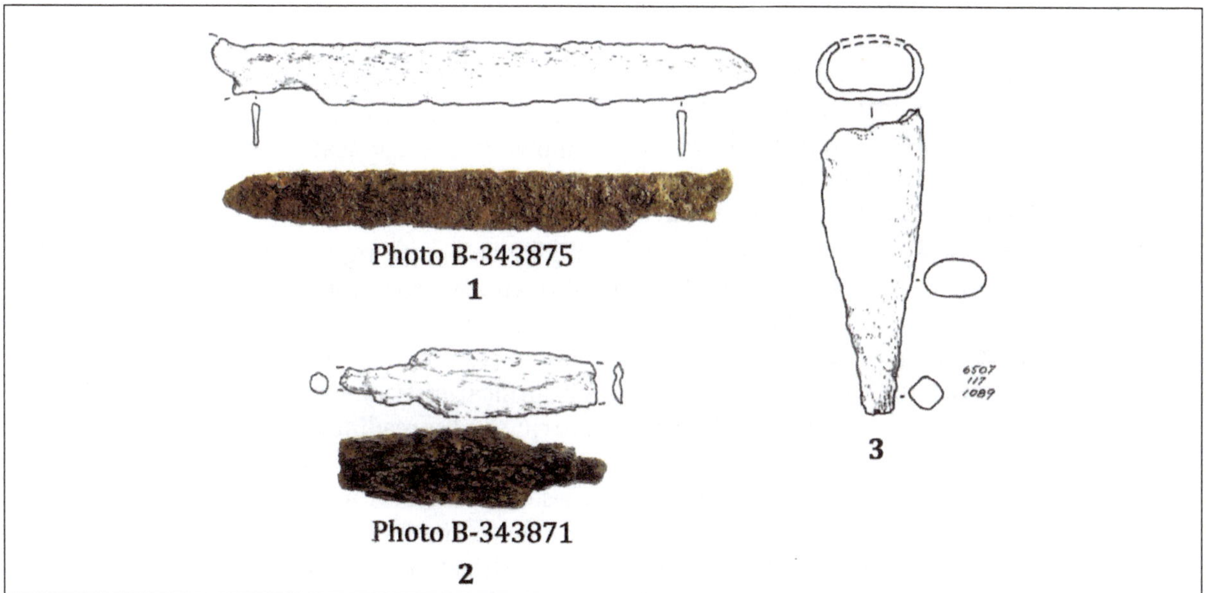

No.	Square	Reg. No.	Description
1	B4	505 / 5024_1	Knife. Iron. Complete blade, tang broken. Weight: 18.9 gm. Length: 13.6 cm. Width: 1.4 cm. Blade thickness: 3 mm.
2	A6	117 / 1089	Dagger scabbard base. Bronze. Lower half, upper part missing. Hollow interior. Elliptical section at top, square section at bottom. Weight: 31.3 gm. Length: 7.1 cm. Width at top: 2.3 cm. Width at bottom: 9 mm. Maximum thickness: 3 mm.
3	B7	537 / 5244_2	Arrowhead. Iron. Tip broken, tang complete Raised double ridge on blade. Weight: 12.8 gm. Length: 6.1 cm. Width: 1.6 cm. Maximum thickness: 5 mm.

Figure 31 Weaponry from the Southern Sector.

to the Type IV "Three-Sided Arrowheads with Tang" from Yoqne'am, which dates from the "Middle Ages" (Khamis 1996, 218). Similar three-winged ("trilobate") iron arrowheads were uncovered at Caesarea, where they were interpreted as being used for either hunting large game or in warfare, but not for armor-piercing (Raphael 2008, 442). A similar example for the Khirbet Beit Mazmīl arrowhead was found amongst the Caesarea assemblage (Raphael 2008, 456, no. 113), but the closest parallel is housed in the Museum of London (Ward-Perkins 1940, fig. 17:13), where it has been assigned the brief date-range of 1242–1263 CE.

Personal and decorative items (Figure 32)

The last grouping of metal objects presented here consist of items for personal adornment, including two finger rings and a complete cosmetic spoon, as well as a decorated hinge, all made of bronze. The first finger ring (Figure 32:1) has all the characteristics typical of this artifact type as noted by Khamis (2008, 178): a fairly uniform shape, full circular loop, a flat and rectangular section, and a diameter of 2.0 to 2.5 centimeters suited to fit a finger. The example from Khirbet Beit Mazmīl measures only 1.9 centimeters in diameter, so perhaps it belonged to a little girl, as the small glass ring from L.517 in Square B4 also may have. Parallels for this ring have been found at Yoqne'am (Khamis 1996, 230, phot. XVIII.34:2) and En-Gedi (Chernov 2007, pl. 4:3). The second bronze finger ring is a decorated wedding band for a male (Figure 32:2). It conforms to Type 2 from En-Gedi, which have "open hoops [that are the] same width throughout with its two ends meeting . . . [is] flat in section . . . [and has] linear patterns along the body" (Chernov 2007, 514). From that report there are parallels for both the form and shape (Chernov 2007, pl. 4:8), as well as the decoration (Chernov 2007, fig. 27). A ring with similar shape but different decoration was uncovered at Yoqne'am (Khamis 1996, 229, photo XVIII.33, no. 61, Type II Band Ring), where it is dated from the Crusader to Mamluk periods. A final example of this general type is also known from the excavations at Hama (Ploug and Oldenburg 1969, fig. 31:2). Of interest are the decorative motifs on this ring, which bear a remarkable similarity to the decorated hinge or box fitting from L.126 in Square A6 (see Figure 32:4 below for comparison), which would seem to indicate that they were made by the same local craftsman.

Another artifact type uncovered at Khirbet Beit Mazmīl is the spatula (Figure 32:3), which consists of a long, thin bronze rod that is circular in section and terminates in a wider flattened and elliptical end that is partially broken. These and similar items are commonplace finds from the Roman (Seligman 2010, 198) through Medieval (Tatcher 2009, 187) periods, where they have been generally classified as "Cosmetic and Surgical Instruments" (Khamis 1996, 225). While many have engraved decorations on them (see Khamis 2008, 173, no. 27), the Khirbet Beit Mazmīl example is plain and undecorated. This class of artifact was used for a variety of purposes, which include the following: the application of cosmetic preparations to the eyes and face (Pringle 1986, 162); as a medical instrument for examining wounds or for mixing and applying medications (Chernov 2007, 509); an ear probe for removing foreign objects; taking liquids, ointments and powders out of their containers (Raphael 2008, 225); a pharmaceutical or surgical instrument; or an item used for the stirring and spreading of ointments (Khamis 1996, 225). As with their manifold functions and uses, these metal objects are also assigned a wide variety of names. The most comprehensive typology for these artifacts, recently provided by Raphael (2008, 445–446), serves as a poignant example. She suggested the following six categories: ligulae, spatulae, tweezers, probes, kohl sticks, and scalpels. That being said, the term "spatula" is employed here to describe the implement from Khirbet Beit Mazmīl, and similar items using the same nomenclature were uncovered at Hama (Ploug and Oldenburg 1969, fig. 25:7, dated from 1260–1401 CE) and Yoqne'am (Khamis 1996, fig. XVIII.6:1). Parallels utilizing different terms for these objects are also attested from the following sites: Horvat 'Uza (Tatcher 2009, fig. 3.42:21, "makeup applicator"); Banias (Khamis 2008, 173, no. 27, "kohl spoon"); Caesarea (Raphael 2008, 466, no. 185, "ligula"); En-Gedi (Chernov 2007, pl. 2:15, "rod"); and the Red Tower (Pringle 1986, fig. 54:5, "cosmetic rod").

The final metal object presented here is a well-preserved decorated copper hinge or box fitting (Figure 32:4), which has a hole drilled in the center of its top portion to attach it to an object. This artifact is important because of its sealed stratigraphical context. It was found in L.126, the makeup of plaster floor L.116 in the northeast sector of Square A6, just south of Wall 111. This locus is phased to the Mamluk I period (Stratum IV), which dates from the mid-13th to the end of the 14th century. Similar decorative bronze hinges with holes pierced into them for use as an attachment have been found in medieval Fustat, where it was noted that they are similar to the hinges found on book covers (Bacharach and Rodenbeck 2002, 199), not unlike those housed in the L.A. Mayer Museum of Islamic Art in Jerusalem (R. Hasson *personal communication*). Good parallels for both the form and the floral/geometric decoration were among the metal assemblage from Caesarea (Raphael 2008, 441, no. 95, 456 nos. 95.15, 20, 21, 463–464, nos. 95.12, 14–21), where they are thought to have been used

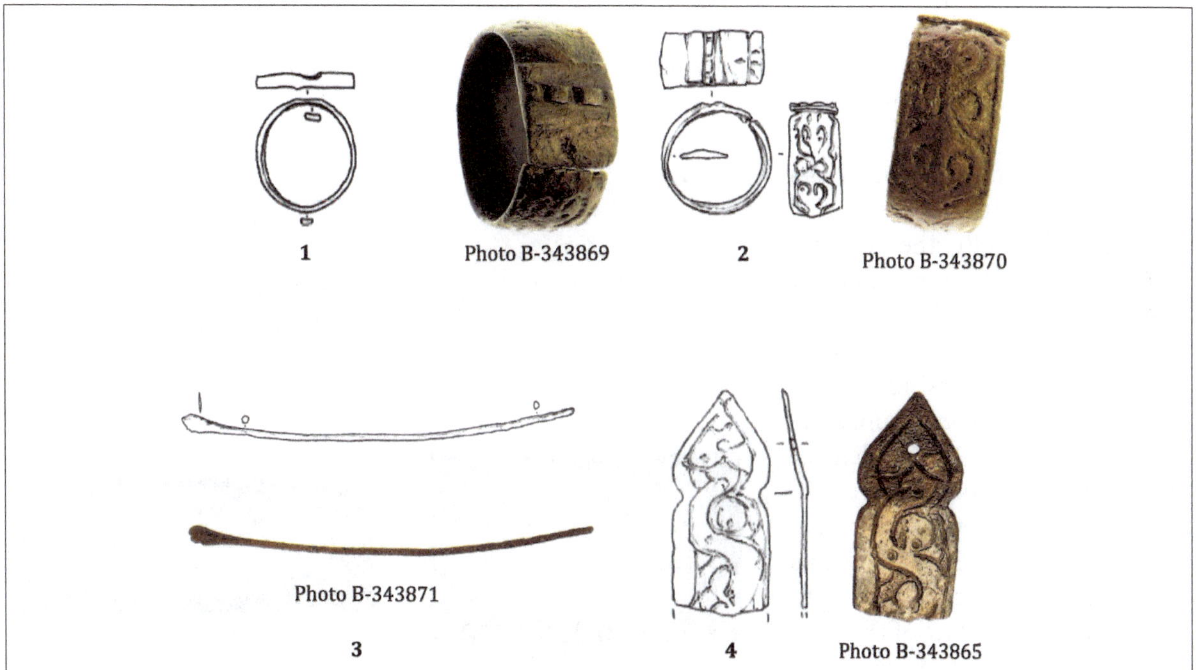

No.	Square	Reg. No.	Description
1	B7	507 / 5129	Finger ring. Bronze. Flat in section. Weight: 1.3 gm. Diameter: 1.9 cm. Thickness: 1 mm.
2	B8	511 / 5095	Decorated wedding band for a male. Bronze. Weight: 12.6 gm. Diameter: 2.2 cm. Thickness: 1 mm.
3	B4	505 / 5024_2	Spatula. Bronze. Circular in section. Terminates in a flattened and elliptical end, partially broken. Weight: 4.6 gm. Length: 13.5 cm. Diameter of handle: 1 mm. Width of spatula: 5 mm.
4	A6	126 / 1088	Decorated hinge or box fitting. Copper. Well preserved. Hole drilled into the center of the top portion. Weight: 3.5 gm. Preserved length: 3.5 cm. Width: 1.5 cm. Thickness: 1 mm.

Figure 32 Personal and decorative items from the Southern Sector.

on either a wooden box or chest. A final example of similar shape was uncovered at Hama (Ploug and Oldenburg 1969, figs. 1:2, 2:10), which are of contemporaneous date (1260–1401 CE).

Concluding remarks on metal objects

The assemblage of metal objects greatly enhances our understanding of the site at Khirbet Beit Mazmīl and its occupational history from the 13th through 15th centuries. Noteworthy among these finds were the equine-related accessories, especially the three-link bridle bit, cinch buckle from a saddle, and the iron horseshoe that were uncovered from the L.537 makeup of the plaster floor in Square B7. The nails and farm implements offer insights into the activities that took place at Khirbet Beit Mazmīl, which included fruit and grain harvesting as evidenced by the presence of the pruning hook. The existence of weaponry at the site would seem to indicate that one of its inhabitants may have had some connection to the military. Finally, the personal and decorative items, particularly the wedding band and the hinge or box fitting, hint at a localized style of metal craftsmanship. Taken together, the metal finds from Khirbet Beit Mazmīl largely supplement and complement the stratigraphic, architectural, and other artifactual finds provided by the excavations at the site, thereby contributing to a better understanding of the lives of those who resided there during the Middle Islamic period.

Stone objects (Figures 33–34)

Mamluk contexts yielded a small assemblage of 13 stone objects made of basalt, limestone, and steatite. The corpus includes grinding slabs, a pestle, and whetstones, which represent a range of daily activities associated with farm life and food preparation at the site, including grinding, bashing, pounding, and sharpening.

Four fragments of coarse-grained basalt grinding slabs were uncovered, and these are classified as lower-idle simple flat slabs with an active upper surface (Figure 33:1–4). An additional find included the bottom disc-shaped fragment of a pestle (Figure 33:5) that was fashioned from a finer-grained basalt and exhibits traces of grinding and bashing activity on its lower active surface. Two well-worked spherical limestone pounding stones (Figure 33:6–7) were uncovered just above a paved floor in Structure B1, and a good parallel for these was found at the Mamluk site of Nahal Hagit (Seligman 2010, fig 7.11: 86). Perhaps the most notable objects among the stone assemblage were four rectangular, green-colored steatite whetstones

Figure 33 Basalt grinding slabs and limestone pounding stones from the Southern Sector.

(Figure 34:1–4). All of these were formed from a generally rectangular block, and each has a perforation on one end. These whetstones show traces of many narrow, grooved striation marks on either one or both faces as a result of sharpening activities. Two similar whetstone fragments fashioned from a greenish steatite block were uncovered in an unclear stratigraphic sequence—and erroneously attributed to the Early Islamic period—at Ramla (Kletter 2009, fig. 12); another example made from a dark grey steatite block is also attested from Crusader-Mamluk contexts at Horvat 'Uza (Tatcher 2009, fig.43: 3–4). In the Northern Sector of the site, L.435 from the E2 midden yielded another two whetstones (not illustrated) as well. It seems that these items were made to be portable, and their perforations indicate that they may have been attached to a cord or thin rope and possibly worn. The whetstones were most likely used for routine sharpening of farm equipment and in some cases may have been part of a soldier's toolkit for maintaining his personal weapons.

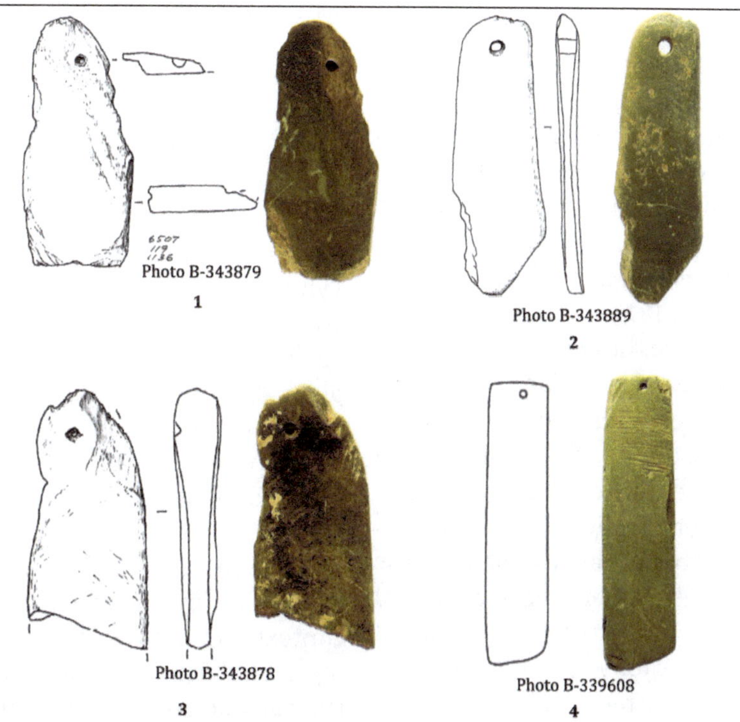

No.	Locus	Reg. No.	Description
1	119 (A6)	1136	Length 7.7 cm, Width 3.4 cm, depth 1.0 cm, Blue-grey, Semi-perforated on one end (Hole 0.4 cm)
2	119 (A6)	1069/2	Length 11.2 cm, Width 3.3 cm, depth 0.6–0.8 cm, Blue-grey, completely perforated (hole diameter 6 cm) broken on opposite end
3	548 (B9)	5232	Length 6.0 cm, Width 3.0 cm, depth 0.8 cm, dark grey, Semi-perforated on one end (Hole 0.4 cm)
4	200 (A12)	2000/4	Length 12.0 cm, Width 3.0 cm, depth 1.25 cm. Light green, completely perforated (hole diameter 0.3 cm) and broken on opposite end

Figure 34 Whetstones from the Southern Sector.

Glass objects (Figures 35–36)

The 2012 excavations in the Southern Sector recovered a small-but-mighty assemblage of glass objects. While more than half of those recovered were bracelets, the rest included blown vessels and a complete finger-ring. These were studied in great detail by Katsnelson (2014b) and the vast majority of them, from the large building in Area B, dated from the Mamluk II period, while only two were found in Area A. The latter two objects were attributed stratigraphically to the Mamluk I period. Seven blown vessels are presented here, and these represent the common forms found in domestic household assemblages (Katsnelson 2014b, 1). Also illustrated is a cache of 10 bracelets from Area B.

The first specimen of a blown-glass vessel is a cylindrical beaker (Figure 35:1) that has two horizontal bands of turquoise trail along the rim zone (Katsnelson 2014b, 3, fig. 1:1). This was a commonplace practice amongst the Mamluk glass makers in Jerusalem (Brosh 2005, 187) and the vessel can be dated by its find context to the Mamluk I period. Of interest is the mold-blown ribbed jar (Figure 35:2) with a tall cylindrical neck and an out-turned rim (Katsnelson 2014b, 4–5, fig. 3:7), the form of which is known from an Ayyubid-Mamluk context at Bet She'an (Katsnelson 2014c, 50, fig. 13:4). A final find that can be attributed to the Mamluk I phase is the complete and intact green finger-ring (Figure 35:3) which was flattened in one part to create a bezel (*Ibid*, 5–6, not illustrated), for which a parallel was found in the Al-Watta Quarter at Safed in a Mamluk I context (Gorin-Rosen, n.d.).

A small corpus consisting of four blown-glass vessels dating from the Mamluk II phase at Khirbet Beit Mazmīl, and it included the following: a small hemispherical blue bowl (Figure 35:4, see Katsnelson 2014a, 3–4, fig. 2:4), which has correlates from a late 14th century context at Hama (Poulsen 1957, fig. 70); a small cylindrical bowl with a folded over rim has a horizontal turquoise trail along its rim (Figure 35:5, see Katsnelson 2014a, 4, fig. 2:5) that is another example of a Mamluk Jerusalem production; a well-known type of large wide-necked jar with an everted rim (Figure 35:6, see Katsnelson 2014a, fig. 2:6) that is commonplace during the Mamluk period (Gorin-Rosen 2004, 60, 132, fig. 1:10); and the nozzle of a Mamluk sprinkler (Figure 35:7, see Katsnelson 2014a, 5, fig. 3:8), which was most likely manufactured in Jerusalem—as the finds from the Jewish Quarter would seem to indicate (Brosh 2005, 188)—and is comparable to examples found in Mamluk contexts at Bet She'an (Hadad 2005, 61, pl. 53:1090, 1091).

The final group of glass finds from the site consists of a group of eleven bracelets (Figure 36) that were part of a larger assemblage of 45 found at the site (Katsnelson 2014a, 6–7, fig. 4: 2–3, 4–11), which is a surprisingly large number for a rural site and much more typical of those found in urban contexts. These conform to the type of twisted bracelets with a circular cross-section noted in the typo-chronologies of both Spaer (1992, 49–50, Type C) and Shindo (2001, 81–82, Type B), which are a hallmark of the Mamluk period. While some of the examples from Khirbet Beit Mazmīl were plain, those illustrated here are decorated with closely spaced spiral bands in a variety of colors, and most—if not all—were likely produced by the Mamluk glass workshops of Jerusalem (Brosh 2005).

Taken together, the small finds from the 2012 excavation in the Southern Sector fit both the context and lifeways present at a Mamluk farming estate. The metal and stone artifacts reflect the daily activities that occurred there, with the metal implements used for pruning and harvesting and stone objects utilized for the grinding of grain. The presence of both the equine accessories and weaponry would seem to indicate some type of military presence at the site (possibly cavalry), while the glass finds, particularly the bracelets, speak to the dress codes and fashion of the times.

The built environment: A site transformed

The state of architectural preservation pales in comparison to the richness of the site's ceramics

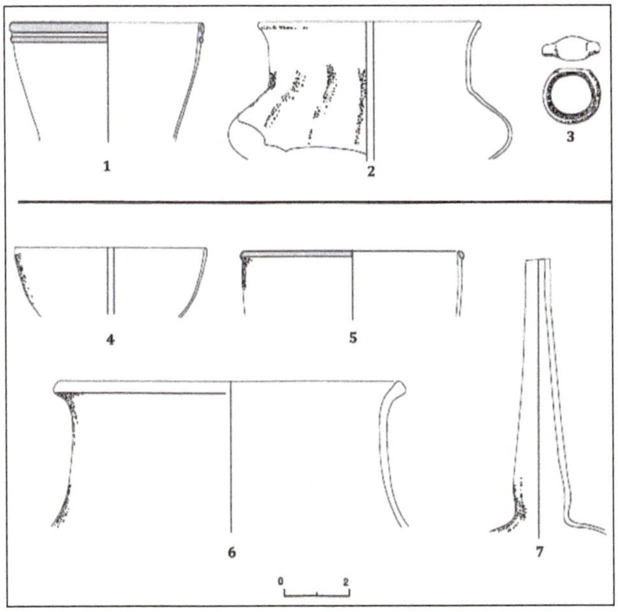

Figure 35 Mamluk I (nos. 1–3) and Mamluk II (nos. 4–7) blown glass objects from the Southern Sector.

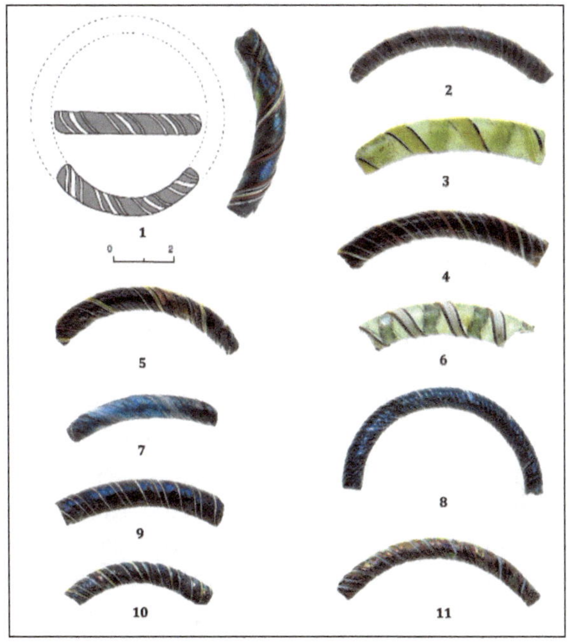

Figure 36 Mamluk glass bracelets from the Southern Sector.

and small finds. Only bits and pieces remain to reconstruct what must have been one of the most ambitious periods of building activity. The site at Khirbet Beit Mazmīl appears to have undergone a radical transformation in the Mamluk period in terms of spatial organization and construction, which were responsible for the almost complete destruction and obliteration of the earliest phases of occupation. Despite the fragmentary nature of the architectural remains, we can identify two interconnected phenomena: the complete destruction of earlier buildings and the reoccupation of the few still-standing structures. Both phenomena imply a large degree of reuse of building materials. As seen in the previous chapter, this is a repeated practice at the site. What appears to be particular to the Mamluk period though, is the extent and frequency of complete rebuilding after demolition and the radical transformation of earlier structures while reoccupying them.

To this concern, some words of caution are necessary before examining more in detail the evidence at our disposal. First, this might not be the first time the site experienced a similar extensive and highly impacting reoccupation, and earlier phases might have been as destructive for the more ancient remains as it was in the Mamluk period. Second, Ottoman-era and modern reoccupations and rebuildings took a toll on the Mamluk structures too, leaving us with highly fragmented evidence and with the impossibility of reconstructing the entire process of resettlement of the site. However, the degree of destruction appears to be lower than the one documented—or supposed—on structures from the pre-Mamluk period. There is, finally, the problem of recognizing and dating reused materials and building techniques. As for the pre-Mamluk period, building materials rarely bear recognizable tool marks or distinguishable features. The same applies to building techniques that are part of long-term, vernacular traditions, and do not change over time. This is due, in part, to the kind of materials at disposal in place, which in a way determine their way of employment.

Ultimately, in the absence of clear interfaces between different stratigraphic units, it is the degree of homogeneity of the materials and the accuracy in the construction that might give some indication of the different phasing of the construction. To this concern, the recurrence of flagstone or pebble floors, which have proven to be sometimes hard to date by stratigraphically associated artifacts, is quite surprising.

The Mamluk-era contexts, however, are much better stratified than in earlier periods, and the association with architecture and architectural phases more secure. This results in a better understanding of the chronology of certain areas as well as of their function. It must be noted, though, that it is not possible to identify in the architecture the two occupational phases that have been distinguished stratigraphically—namely the 13th–15th centuries and the 15th–16th centuries. Only if there is a clear functional or spatial transformation of the built environment can we distinguish these two phases architecturally.

Remains from the Mamluk period are found in all areas of excavation, demonstrating that the entire site was occupied from the end of the 13th century (see Appendix 4). This resettlement was done through both the demolition of earlier structures—or of their ruins—followed by a complete reconstruction; and a reoccupation of still-standing buildings of earlier date, but with important and radical transformations. Only in few areas—notably the former *mikveh* in Area D and some of rooms in Area C in the Northern Sector—were buildings adapted and restored with minimal effort. In all other areas of the site, an intense building activity is documented, with the creation of entirely new buildings that completely transformed the previous built environment.

No clear pattern is recognizable, and the choice of the different resettlement strategies was likely made on a case-by-case basis. The easier explanation could be the conservation status in which earlier structures were at the moment of the resettlement: where the buildings were beyond restoration, or interventions required certain expertise not available at that specific time, demolishing to rebuild anew would likely have been seen as the most practical solution. Given the fact that it is entirely unknown what kind of structure—or structures—stood here prior to reoccupation in the Mamluk period, it is also impossible to say if the complete transformation of the built environment was dictated by a change in site function. Water management was still considered a priority in the "new" site: some installations, such as the *mikveh* and a basin in Area A of the Southern Sector, were kept in place and restored, connected with channels. As it will be described more in detail further on in the chapter, some new installations, also connected with the water system, appear to be built during this macro phase of occupation.

The major challenge is to exactly grasp the relationship—spatial and functional—between the part of site sitting on the top of the hill and the one below. How the topography of the site influenced the built environment is a key question for our understanding how the site physically developed over time. Enveloping the site in its entirety in a terraced system, with towers or accessible flat roofs allowing the circulation between different levels, was the solution adopted in the Late Ottoman period of occupation. As suggested by the survey and the OSL analysis on the valleys surrounding the site, the western Jerusalem hinterland was the subject of intense development already in the Mamluk period, when several terraces appear to have been created, often reusing pre-existing terraced systems (see Chapter 6). Similar slopes and elevation gaps may have been compensated in this manner before the Ottoman period as well, especially since there is direct evidence for the intensive re-occupation of both Upper and Lower Sites. Considering the elevation of dated Mamluk contexts from different areas, the average difference between Area S in the Lower Site and Areas B, C, and D in the Northern Sector is between 2.5 and 3 meters. It is therefore plausible that similar terracing and architectural expedient were also set in place at this stage, to be later either restored or dismantled and totally rebuilt.

A detailed analysis of the wall dividing the Upper and Lower Sites may be the key to understanding the topographical setting of the site, not only during the Mamluk occupation but also for earlier periods. Unfortunately, its dating remains highly problematic, although the existence of different phases for W1, W101, and W102 are easily recognizable (Figure 37). On the northernmost end, the later *manṭarah* (L.110) abuts W102, which appears to be the original portion of the wall. W101 could have been a continuation of W102, but it underwent an almost complete reconstruction in the corner directly under the dwelling B1–B2. The construction of this Late Ottoman complex certainly implied heavy interventions, but it does not seem that this portion of the wall was built specifically to support this new structure, since the lower half of the W101 appears similar to W102, as well as its north–south-running continuation towards Area S. Ultimately, the latter portion of W101 is the same as the northern part of W1. Further south, though, the wall appears to have been radically changed: the original structure is abutted and possibly partially overlayed by another massive,

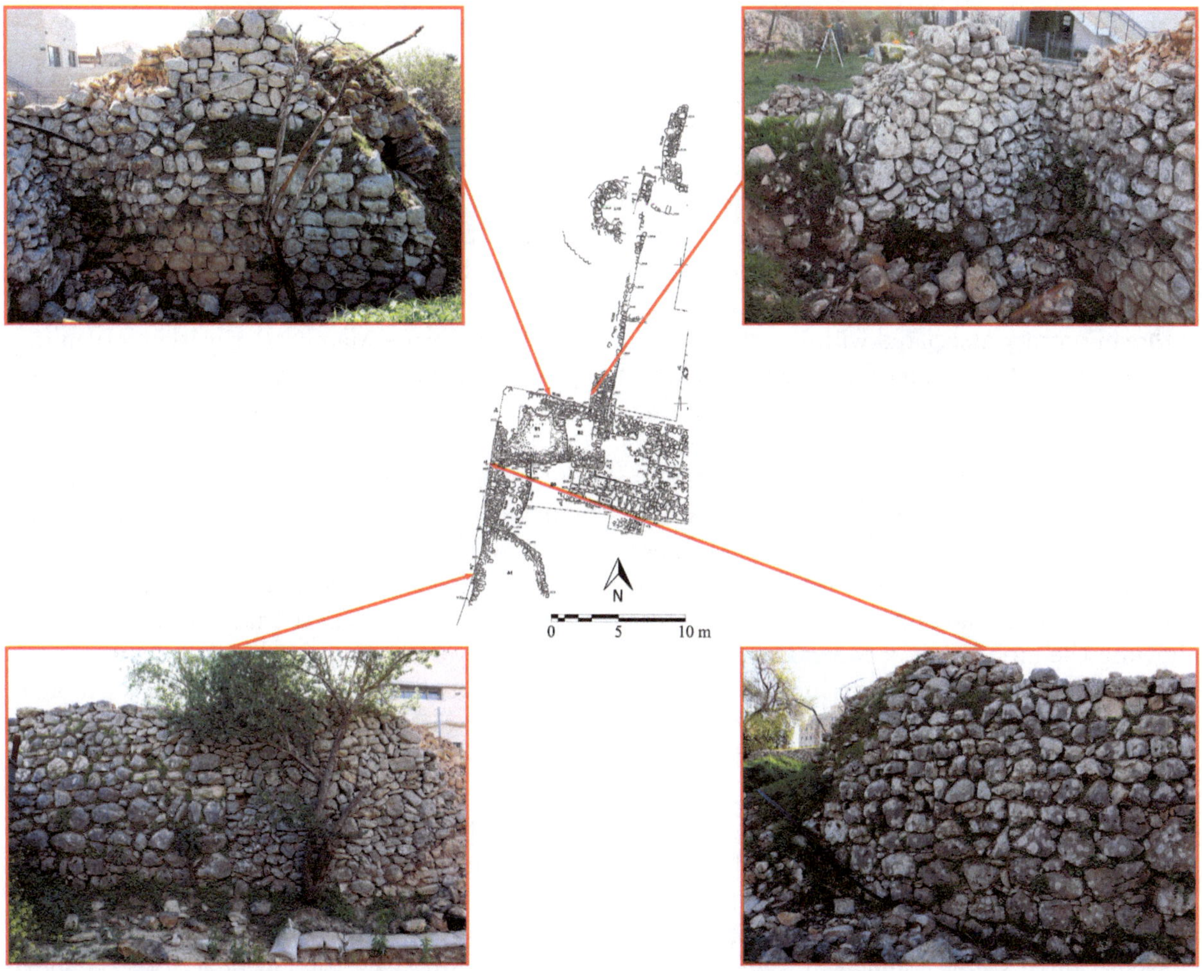

Figure 37 The retaining wall: plan and pictures of its different phases (photo: N. Pini; plan: S. Pirsky).

drystone structure—described as a "stone heap" in the original field report (ʿAdawī and Dolinka 2021), tentatively dated to the Late Ottoman period. The difference between this construction and the rest of the wall is evident not only for its stratigraphic relation with other structures—since it abuts W101—but also in terms of building technique and materials: the stonework appears to be far less regular and carefully built than the original wall, recalling quite closely the more recent drystone architecture of the later Ottoman period on the Upper Site. The stone heap—what remains of a 19th century *manṭarah* (in A1)—covers the remains from the Byzantine-Early Islamic periods in S1 and some later structures of a possible tower building dated to the Early Ottoman period, of which a series of steps are preserved.

The area occupied by the latter Early Ottoman tower is certainly a crucial area that will unfortunately remain largely unknown, due both to the later stone heap and for the modern construction activities. Despite the lack of solid and extensive archaeological data, the presence of pre-Mamluk remains, and Ayyubid materials mixed with Early Ottoman pottery associated with structures overlaying the pre-Mamluk mosaic, suggest that this area was possibly continuously occupied since the early phases of the site. As the architectural remains in the Northern Sector of the Upper Site demonstrated, Early Ottoman structures tended to maintain earlier Mamluk constructions. It is therefore possible that also in this case the Early Ottoman tower in S1 maintained an earlier structure, allowing the circulation between the Lower and Upper Sites. Interestingly, the highest elevation taken in S1, namely on top of one of the tower's wall, is more or less at the same level with the elevations in the Northern Sector (an average of 816.40 meters ASL.).

As described more in detail below, there was important building activity in the area immediately to the west of S1 (S2 and S3), related to the construction of a large complex. It is possible that a part of it was either connected to structures later embedded in or transformed by the Early Ottoman tower. The Late Ottoman drystone wall contributed to covering—if not erasing—any possible trace of a previous building north of S1.

Similarly uncertain remains the exact topography of the Upper Site. Here the elevation gap, if any, was probably far less relevant than in the previously described area, and there are no indication of massive retaining walls predating the drystone architecture. The ground appears to gently slope from north to south, since there is a more or less one meter gap between the Mamluk floors in the Northern Sector (including also Area A of the Southern Sector) and the one in Area B of the Southern Sector.

Regardless of the exact topography of the ground, it is certain that the site was intensively developed during the Mamluk period, and we can grasp some features of the spatial arrangement—as well as of the function of the site. Most information comes from the Northern Sector of the Upper Site. The northernmost part of the Northern Sector (Area C) is where the majority of the pre-Mamluk structures can be still recognized, making it plausible that also the spatial organization was not radically altered. Of the pre-Mamluk (Early Islamic) vaulting system, only one cloister vault was partially maintained (W104). Similarly, the pillars to the south (W124/1 and W124/2) were also kept in place or underwent limited restorations. Channel W127, immediately south of pillar 124/1, was cleaned and maintained at least until the Late Mamluk phase: materials from this period were found blocking the channel and suggest that it went out of use in the Early Ottoman period.

In association with these earlier remains, though, a new and different vaulting system was put in place: it consists of a series of barrel vaults that are still visible in C3 (Figure 38), and likely continued until the last vaulted room (C4), even if later reconstructions prevent us from being entirely certain. It seems plausible that the same roofing system was built towards the south, over C6 and E1, as suggested by the springers (W121) above the northernmost pillar (W123; Figure 39). Less

Figure 38 Vault in C3 from north in 2017 (photo: N. Pini).

Figure 39 Springers (W121) set above the northernmost pillar (W123) (photo: N. Pini).

detectable inside the same room through a hole in the eastern drystone wall. The orientation of this wall is clearly the same as the vaulted rooms, as well as the two pillars W124/1 and W124/2, making us suppose that whatever it was, the pre-Mamluk complex was sufficiently stable and well-preserved to be kept in place, restored, and adapted to the new structure. The wall W120, under the 19th century wall W105 in C5, was possibly the end of the series of vaulted rooms forming the northern wing of the site.

If the earlier architecture influenced the special arrangement of the Mamluk reoccupation of the summit of the site, certain areas were clearly rebuilt. One of the best examples is in C6/E1. Here, a room was created, immediately the south of the cloister vault (L.111), formed by one L-shaped wall closing the space to the south and to the east (W133 and W134, possibly restored in the Ottoman period), abutting against another wall built against pillar W124/1 (W136) and running parallel to it northwards until W104. Between this pillar and the cloister vault (W123/121), an installation in C6—possibly a press (L.245) with a connected round basin (L.258) on the other side of the W133—and an adjacent storeroom with several other installations in E1 were set (Figures 40 and 41).

clear is how the Mamluk reoccupation interacted with the pre-Mamluk structures to the east (especially in C5), where the latest Ottoman occupation totally obliterated earlier structures. As mentioned in the previous chapter, remains of a pre-Mamluk wall are still visible right outside the entrance to the vaulted system in C4, and its continuation is

The stratigraphy of the architectural remains in this area is extremely complex and presents challenges. The presence of pillar W124/1 and the cloister vault demonstrate that this area was intensively occupied in the pre-Mamluk period and that

Figure 40 New room in C6/E1.

Installation L.245 is set on top of a pebble floor, radiocarbon dated to the Abbasid period. However, its original bottom was not preserved, and the plaster covering the sides of it clearly bent towards the centre, suggesting that the plaster was covering the entire surface. The stratigraphical posteriority of this structure as it relates to the Abbasid pebble floor is therefore certain. However, the absolute date is still questionable, and it cannot be entirely excluded that the posterity was not chronological but simply technical (different construction phases but taking place during the same occupational period). The ceramic record suggests that it was likely used in the Mamluk period, and that the flagstone floor marked the end of its use as a productive installation, which is tentatively dated to the Early Ottoman period. Alternatively, the installation could have been constructed before the Mamluk-era reoccupation, and simply maintained (or transformed) thereafter.

the Mamluk reoccupation was to a certain extent influenced by these pre-existing structures. Consistent traces of Abbasid occupation—also supported by radiocarbon dating—were found below a flagstone floor dated to the 14th–15th century; a similar radiocarbon date comes from a pebble floor found below the later installation in C6 (L.245). The press is one critical context that has been heavily transformed over different occupation phases.

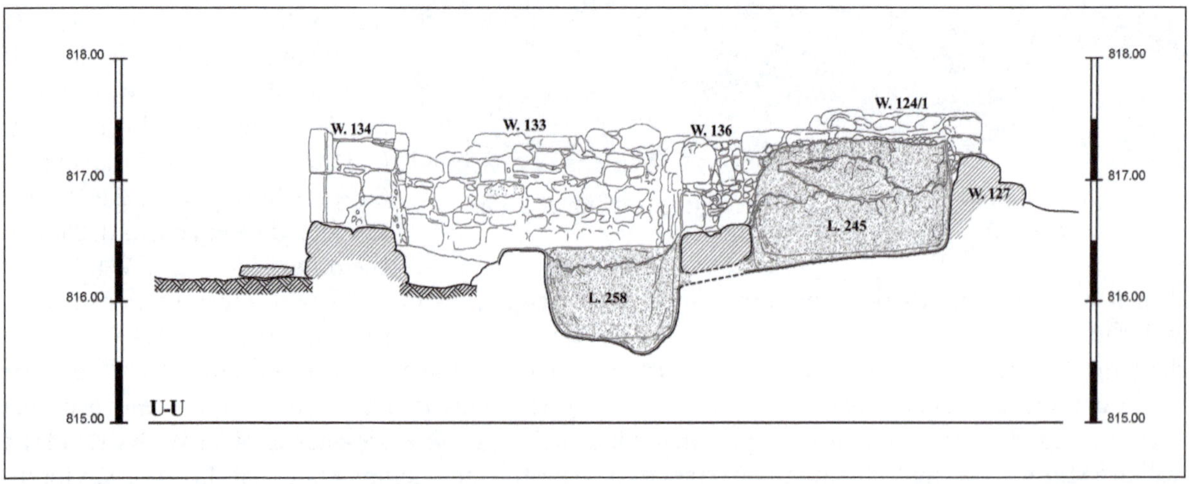

Figure 41 Section of Area E1 and C6, looking south (drawing: S. Pirsky).

The phasing of the walls creating a new room in Area C is equally problematic. There are clear differences in construction between the L-shaped wall formed by W133 and W134—that compose one single wall probably restored in parts in the Ottoman phase—and W136/128, which closed the installation on its eastern side. W133 and W134 are composed of two faces with roughly squared, small and medium-size blocks, laid in regular courses, with a relatively limited core filled with earth and small pebbles. The other wall has one face made of smaller stones and only two large well-cut and dressed blocks in correspondence with the beginning of the installation, and the space separating it from the pillar was filled with rubble stones and small pebbles. The result is ultimately a sack masonry, obtained by constructing a new interior face and reusing the preexisting structure as second one.

W136 breaks in correspondence with the installation, but it is certain that it continued for the entire length of the installation until W123/121 and W104, since its first course, that divides the main, rectangular basin of the installation (L.245) and the smaller, circular one (L.258)—where a drainage hole was carved—is still in place (the second section of the wall after the break was named W128; Figures 42 and 43). Interestingly, this wall disappears under the Ottoman drystone structure at the interface between the latter architectural phase and the original cloister vault (L.111). Also of importance is the fact that the materials employed at least in this first course are exactly the same as the ones used immediately before the break in the elevation of the wall: large squared and polished blocks, carefully laid in extremely regular courses. Since materials worked in the same way are found also elsewhere on the site, in secondary use but in contexts dating to the Mamluk period, it is possible that it was a set of reused materials coming from the pre-Mamluk structures occupying the hill.

Figure 42 Installation L.245, from the east.

Regardless of the difference in building technique and materials, it is clear that the L-shaped wall was built later than the other wall along the pillar, since the first abuts against the second, and the contact interface is constructed with smaller stones to better tie the two structures. Whether this represents a later phase of occupation or was simply a subsequent construction phase is unclear. Nonetheless, the western wall of the room was built immediately before or together with the installation since the latter could not have existed without it. The same applies to the wall on the opposite side of the rectangular basin (W125). Whether the rest of the room was built at the same time is not certain, but the fact that during the Mamluk period a room was established is beyond question. The layer dated to the Abbasid period (L.433) was covered by a series of plaster and flagstone floors; above the first one of them (L.419), a hard-packed surface with plaster (L.418) was dated to the Mamluk II

Figure 43 Installation L.245, from the west.

the former basin L.245, and the construction of which put circular basin (L.258) out of use.

At the same time, the vaulting system of Area C was reconstructed with a low barrel vault or a half vault covering at least the area immediately above the press (Figure 44): one springer of the barrel vault is still in place on top of the northern pillar. The building materials and the technique matches entirely the vault covering room C3: small, well-cut but roughly finished slabs of stone are carefully set on thin layers of mortar. Similar building materials and technique are also documented in the vault covering the upper part of the stairs descending to the *mikveh*, pointing to a restoration done in the Mamluk period to maintain the facility.

It is possible that the pre-existing pillars were maintained from the earlier phase because they were easily adaptable to different vaulting techniques. We could imagine, then, a series of rather low barrel vaults running west-east on the northern side of the complex, as well as another row of covered areas running north–south and forming a 90-degree angle with the northern side. It is also impossible to say if the low barrel vaults also supported a mezzanine or a second floor: this question will remain unanswered also for other phases of occupation, even though there is a clear indication of the existence of an upper floor, at least in the Ottoman and modern periods of occupation, for which parallels are numerous and consistent for the region.

Further information on building practice comes from the phytolith analysis conducted in E1. One sample (MAZ8, see Chapter 7.1) was taken from the fill underneath the floor laid during the first phase of the Ottoman occupation of the site; the

phase and contained several horseshoes and other equine accessories, suggesting the use of this space as storage. Interestingly, this context is mirrored by L.507/537 in the Southern Sector. The installation of a second floor made of flagstones (L.414), directly above L.418, was also dated to the Mamluk period, and it appears to be adapted to all the walls of the rooms. On top of it, a fill and a later beaten floor (L.413) dating to the Ottoman period mark the partial transformation of the area.

All things considered, there is sufficient evidence to suggest that regardless of the kind of structure that was standing here and even without knowing the exact nature of the development of the space, the area was repurposed between the end of the 13th century and the beginning of the 14th century. This period saw the establishment of a new specialized, productive space with associated storage and a paved room that underwent several restorations and re-makings but remained in continual use until the conversion of the area into a dwelling as early as the Early Ottoman period. Possibly some initial major functional transformations already took place in the Late Mamluk period, during the late 15th to 16th century. This was when the first flagstone floor was laid in E1, possibly inside

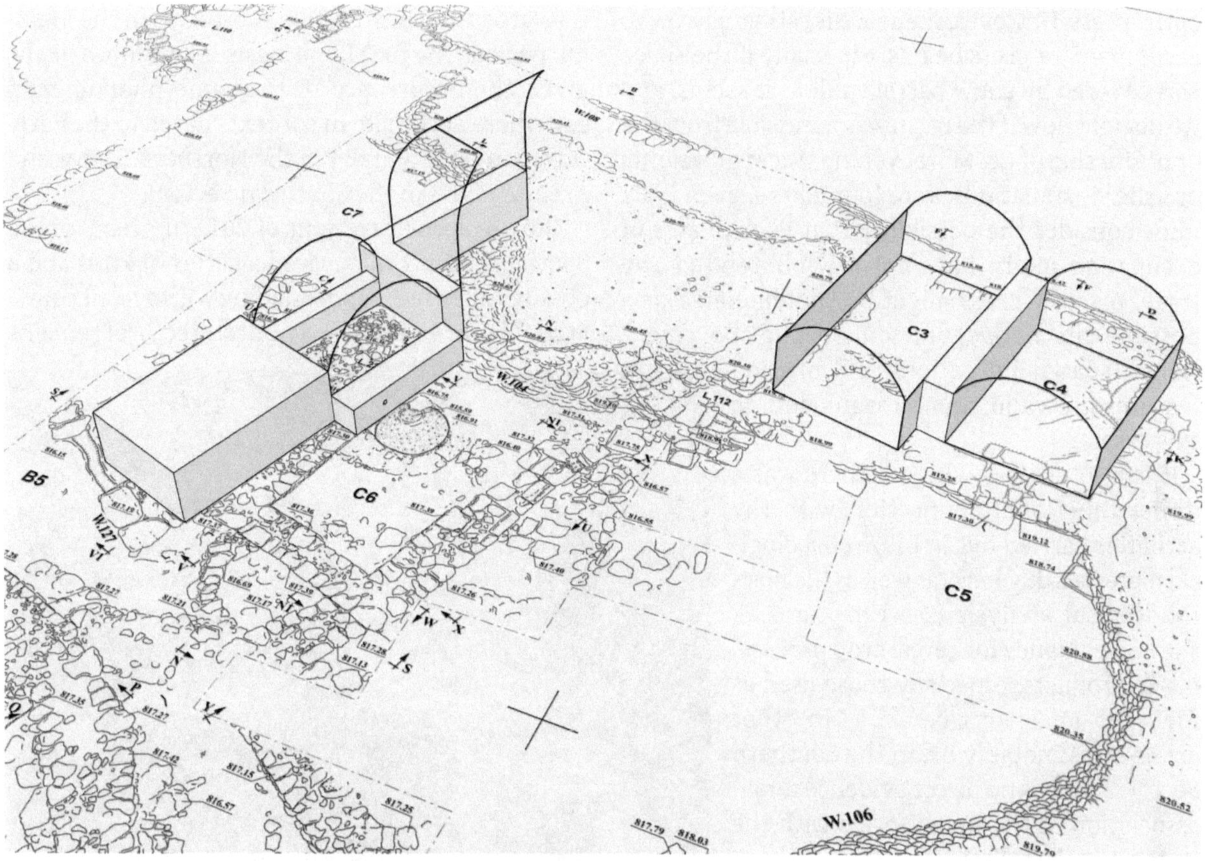

Figure 44 Axonometric reconstruction of Area C.

sample yielded traces of reed grasses, normally used in construction. This is consistent with different ceiling techniques combining stone vaults or arches and weeds and loose materials (frequently soil and small rubbles), attested at several sites of the region and for different time periods (Hirschfeld 1995, 237–244). Among the different possible interpretations,[1] one that is particularly interesting for Khirbet Beit Mazmīl relates to the construction of cross-vaults. In the "traditional Palestinian house," a frame made of reeds, wood, and earth is used to support vaults during their construction, to be afterwards covered with mortar (Hirschfeld 1995, 128–129). As said before, it is possible that the room in C6/E1 was covered by a vaulting system: it is then possible to assume that a similar system might have been put in place.

As for the installation placed between the pillar supporting the cloister vault and pillar immediately to the south (W124/1), two walls were built (or restored): one closed the western side (W125), while the other (W136) blocked the eastern one and continued farther south until the end of the pillar, ultimately abutting against it along its eastern side. A small drain hole was created halfway along the length of the basin and the inside of it was plastered. A second, circular basin was dug just in front of the hole (see Figures 40 and 41), making it clear that the two basins were part of a single installation, probably a press.

If this was the case, it is not entirely clear what kind of press it was. The presence of two small niches in W104 might relate to this construction and be the accommodation for some of the beams of the press. However, the space is rather narrow to accommodate press beams, especially if the space was covered by a low barrel vault as it seems, and we do not know if the basin was accessible from the opposite side of C6. Moreover, no traces of weights or other installations were found here, even if we must consider the possibility that later phases of occupation might have entirely obliterated any trace. A screw press might be contemplated as a possible alternative, but some of the abovementioned spatial limitations and missing features are valid counterarguments in this case.

It is also unclear whether there was a direct functional connection with the activities carried out in E1. Archaeological materials dug here, as well as the macrobotanical analysis (see Chapter 7.2), showed evidence for cereal crop processing byproducts, especially those used as fuel, and *ṭābūn* fragments, as in other areas—most notably B7, in the Southern Sector. There is no direct evidence for an association of the Area C press with the productive installations in the Southern Sector. The Northern Sector might have been a less specialized productive area, where multiple activities were conducted in parallel and not on a properly industrial scale.

It is impossible to determine whether Area C was an isolated productive space in the Northern Sector, or if it expanded also beyond the room in C6. In general, the arrangement of the built environment remains highly unclear: for instance, we do not know if the western side of the Northern Sector was closed to the outside, or if it allowed a passage farther west, possibly offering a direct connection with the Lower Site. Also, it is not entirely clear if the blockage of the eastern door of the room in C6 took place in the Early Ottoman period or if it is related to a constructive sub-phase in the Mamluk period. The first hypothesis seems more likely, given the occurrence of the same phenomenon elsewhere at the site in contexts dated to the Early Ottoman period (in B4 in the Northern Sector and Areas A and B in the Southern Sector).

The salvage excavations of 2012 in Area B of the Southern Sector uncovered a series of kilns and a possibly later two-room building with a front courtyard (Figure 45 and 46). The architectural remains

Figure 45 Area B of the southern sector (courtesy of the IAA, photo: B-323090-150113392615).

Figure 46 The kilns in the southern sector (courtesy of the IAA, photo: B-297820-150715160920).

are extremely poor, but there are some similarities with Area C in the architectural development of the period, and the marked productive specialization of the spaces (even more so for Area B of the Southern Sector than for Area C). There are few traces of occupation in Area B of the Southern Sector prior to the Mamluk period, with the exception of a single wall—built on bedrock—and a plaster floor, dated to the Crusader/Ayyubid period. Development initiated in the Mamluk period in this part of the site benefitted from a rather large, empty space, free from standing structures of days gone by.

The three kilns (L.516, L. 515, and L.534) are the first constructions attested here, dated to the Early Mamluk period, between 13th–14th century. The middle kiln is the largest (2.5 meters in diameter) while the smallest is the third to the south (1.5 meters in diameter), but they all consist of simple, single circles of stone, preserved to only a single course high. The first layer of collapse suggests that they were completely built of stone and possibly domed. Around the kilns hard evidence for intensive firing is attested (charcoal and ash layers), also indicative for periodic cleaning of the installations and dumping of the waste in the immediate surroundings. A *ṭābūn* was also found west of the first and middle kilns.

The association of this area with a larger scale of production, with potentially polluting and loud activities, can also explain its relative spatial isolation from the other structures of the site. In fact, the other building found in Area B of the Southern Sector—the two-room complex with an associated courtyard—appears to have been built later, in the Late Mamluk period, when the kilns had already gone out of use. The construction technique of this building closely recalls the one documented in some walls in Area A of the same sector of the site, and the walls in Area C of the Northern Sector. Another common element is the effort to level the bedrock by setting a flagstone floor: traces of it were found in both rooms as well as in the courtyard.

The circulation within this building can only be partially reconstructed: the larger, northern room (measuring 5.5 × 3 meters) has a door (with built step) to the northwest, but no other apparent openings to the other spaces of the complex.

On the contrary, the other, southern room has only one opening to the courtyard (measuring 9.5 × 6.5 meters), developing along the entire southeastern side of the building, where supposedly there would have been another entrance. In the second room, a base of an arch springer was also identified, though it did not provide decisive information of the ceiling technique. Both rooms show evidence for routine and periodic cleaning and replastering of the floors.

As mentioned earlier, it is unclear if the building had any functional association with the kilns, and if their physical proximity is to any extent motivated by some productive reason. However, botanical analyses from layers in the two-room building found proof for cereal and olive processing byproducts, as well as dung pellets and charcoal, used as fuel, as in Area C and E1, suggesting that some sort of cooking or productive activity was still conducted here during the Late Mamluk period. Similarly, equine accessories were also unearthed, suggesting that certain spaces of the building were used for storage or even as stables.

Outside of these installations or specialized production areas, there is clear evidence for intense activity dating to the Mamluk period across the Northern Sector, as well in Area A of the Southern Sector and in the Lower Site. The architectonical development of Area B of the Northern Sector is less clear than the one in Area C (Figures 47 and 48), and there are no such architectonical features reflecting a particularly specialized function of the space.

Figure 47 Aerial view of field B, with W118 on the left side (photo: O. Ze'evi).

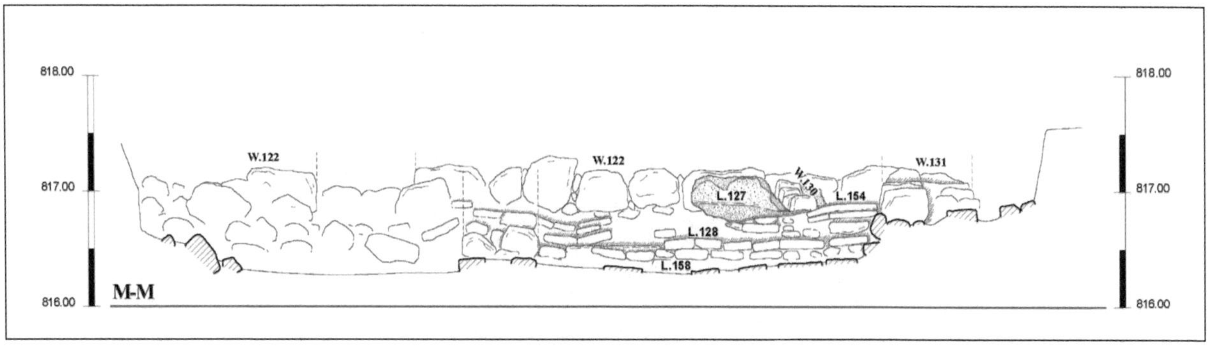

Figure 48 Section of Field B of the northern sector, looking north (drawing: S. Pirsky).

Unfortunately, our understanding of the architectural remains in Area B is highly affected by the construction of the Late Ottoman dwelling B1 and B2, which largely spoliated earlier materials and partially covered previous structures. Therefore, the remains attributable to the Mamluk occupation of the site are quite limited and only a few certain features can be ascertained.

This part of the site was an intensively built area before Mamluk reoccupation and development. It is not only the abundance of reused materials included in the new structures—that due to their homogeneity can be imagined coming from a single earlier building, possibly Byzantine—but also for the maintenance of a few structural elements, most notably the pillar W124/2. The pillar was abutted—and partially destroyed—by a series of walls (W122 from the west, W131 from the south); to the south and to the west two further walls (W118 and W117) enclosed the space, also creating, in this case, a rectangular room. The better-preserved wall of the area is the southern one, a massive wall (W118; one meter thick) built by using a large quantity of reused materials and, most interestingly of all, large, well-squared and polished boulders, normally used as an external face. On the inside, the face was built using smaller stones.

A smaller wall (approximately 80 centimeters thick) was built against the interior face of W118, in a second phase of construction possible intended to reinforce the entire structure. Unfortunately, it is impossible to determine when this buttress was added. Either the outer wall built with larger boulders was already in place in the pre-Mamluk period, and the inner addition is then a Mamluk intervention; both phases are to be dated to the Mamluk period, possibly reflecting two different sub-phases—Early and Late Mamluk—as suggested by archaeological materials at the site; or the outer wall was built in the Mamluk period and the later addition was contiguous to the Early Ottoman transformation of the area.

In any case, it is highly plausible that W118 closed the Northern Sector to the south, although an opening—later blocked in the Ottoman period—was identified, potentially being a direct passage toward the Lower Site. Unfortunately, modern constructions prevented us from further investigating this area, leaving a considerable gap in our knowledge about this crucial part of the site. More certain is that during the Mamluk period Area B was dominated by a single, rectangular room without any clear productive connotation in the architecture.

Even less can be told for Area D (Figure 49), where excavations conducted in 2011 (Levi 2012) also found Mamluk materials. Furthermore, an OSL sample taken underneath the flagstone floor offers a *terminus post quem* for its construction of the 14th century. The *mikveh* continued to be used as a cistern during this period, and part of the preexisting structures were kept in place, especially those in

Figure 49 Area D of the northern sector (photo: N. Pini).

relation to the water system. Among these, most notable is a portion of the eastern wall (W110) in D3: the building materials (well-cut and dressed regular blocks) and the quality of the construction (extremely carefully laid walls, without chip stones) are manifestly different from the rest (Figure 50). The remaining walls enclosing the southern part of Area D appear to belong to the same phase and built abutting these earlier ones: the irregular technique and the extremely diversified materials closely recalls the drystone architecture in the northern part of the area. The only exception is the southern portion of the arch in W110 that shows a similar technique and materials as the barrel vaults in C3 and C6, suggesting that this (re)building activity might be dated to the Mamluk period. Interestingly, the northern half of the same arch presents a similar technique, but the materials are rather irregular and the construction extremely poor. It was probably a restoration—trying to replicate the original aesthetic of the arch—taking place in the Ottoman period, when most of the walls enclosing Area D appear to have been entirely rebuilt. Apart from the presence of cisterns and wells, no further functional indication for this area is available. It is possible that this area in the Mamluk period was closely connected, both physically and possibly functionally, with Area A of the Southern Sector, but such a link is unfortunately unknown and lost.

At the Lower Site, an extensive Mamluk construction phase was documented (ʿAdawī and Dolinka 2021, Stratum II). Remains of a large complex with massive walls was found, built with a technique quite common on site (two outer faces and a harder core made of mortar and rubbles). The considerable thickness of the walls led the excavators to imagine

Figure 50 Detail of W110 in Area D (photo: N. Pini).

the existence of barrel vaults as a ceiling, as well as a second floor, even though the material evidence is poor. Certainly, these appear to be the thickest structures found on site, making it plausible to imagine a structure with multiple storeys. No information on its function or even exact plan of the area is available. The relative stratigraphy of the massive walls suggests the existence of multiple phasing, most notably a later expansion northward (Wall 3) of an original nucleus (Wall 2, 6, and 4), but the exact chronology is unclear. However, it still seems to predate the Ottoman reoccupation of the site when this structure underwent some modifications.

A similar development appears to be attested in Area A of the Southern Sector of the site (Figure 51). In this case, the Mamluk construction reoccupied the area where Roman quarries were found: the bedrock was levelled by setting a flagstone floor, and new walls were built abutting against the perimeter wall of the Byzantine period (W111 and 143). A new room of rather irregular shape was thus created. Connected to it, a series of single course, dry walls were found, possibly facilitating some storage space or pens for animals. Major changes were made to W111 and 143: here, the two interconnected basins (Cisterns 103 and 106), assigned to the Byzantine occupation of the area, were dismantled. This suggests a larger transformation than it was possible to assess archaeologically in the area between the Northern and the Southern Sector—ultimately immediately east of Area D—both in terms of function and spatial organization.

It is impossible to understand the exact function or extent of this new space, and it is likewise impossible to understand its exact physical relationship with the Northern Sector of the site,

Figure 51 Area A of the southern sector (courtesy of the IAA, photo: B-323093–150113393318).

apart from the walls abutting the Byzantine wall. It is certain that the structures in Area A continued farther west, possibly reaching the intermediate ground between the Lower Site and Area B of the Northern Sector: here the corner of a large, plaster-lined rectangular installation (L. 201) was found in A12 (Storchan and Dolinka 2014), associated with two poorly preserved walls (W402, W403). The installation presented three plaster layer floors: underneath the lowest one a layer containing Byzantine sherds was found directly above bedrock. This is a constant throughout the area, where the Mamluk structures appear to have been built on top of this preparatory, cleared level. Therefore, though it is not entirely certain when L.201 was installed, there is a high likelihood that it belongs to the Mamluk reoccupation activity in Area A.

To summarize, during the Mamluk period the site expanded massively beyond its original core, which coincided tentatively with the Northern Sector, starting with a reoccupation at the end of the 13th century. By the end of the Mamluk period all three areas of the site—the Northern and Southern Sectors uphill and the Lower Site—were intensively occupied, which required the construction of new buildings. These buildings, in most cases, used or replaced previous structures: this appears to be the case documented in Area C, where Mamluk structures reoccupied earlier structures and new ones were built above a levelling layer containing abundant Abbasid material. Similarly, in Area A of the Southern Sector, a similar layer, although in this case with materials dating to the Byzantine period, was found directly on bedrock and underneath the new Mamluk constructions. This suggests a general effort at clearing all ruined architecture from the site, maintaining only limited portions of better-preserved ones, and otherwise levelling the ground to build anew, reusing the spoiled building materials. There is also evidence—especially in the Southern Sector—of the occupation of previously untouched areas, or areas that at least do not appear to have been continuously and intensively occupied in earlier times.

The new built environment appears to have been organized organically, with interconnected spaces, some of which bore witness to specialized productive activities, requiring specific facilities and installations. Even if there is no direct evidence for the Mamluk period, the fact that the site extended downhill makes it plausible that the two parts of the site were somehow functionally and physically connected. The existence of at least a portion of the wall and of pre-Mamluk structures where the Early Ottoman building and the Late Ottoman *manṭarah* was built opens the possibility that also during the Mamluk period a similar way to compensate for the difference in elevation between the different areas was adopted.

Finally, it is possible that at a certain point during the late Mamluk period, towards the end of the 15th century, a series of important changes took place at the site, leading to both functional changes of certain areas or installations and to a new spatial arrangement of the built environment. These transformations potentially foreshadow the radical changes of the site taking place in the Ottoman period, when the rural estate was turned into a rural complex of small, independent dwellings.

A synopsis of the Mamluk period at Khirbet Beit Mazmīl

The Mamluk farming estate at Beit Mazmīl flourished throughout that period. It was created through an active process of demolition, repurposing, and demolition again, radically transforming the early medieval site beyond recognition. The Southern Sector of the site provides a wealth of material evidence that sheds valuable light on the lives and activities of those who lived and died there so long ago. As with the Northern Sector, there was clear evidence for both industrial and domestic spheres at the site, as evidenced by the three "Circular Installations" in the eastern part of Area B, with its ashy dump layers that are indicative of some type of industrial production, while the structures in both Areas A and B were clearly domestic in nature. Fortunately, a comparative study between the ceramics and their associated find contexts revealed a clear chronological separation between the Mamluk I and Mamluk II occupational phases at Khirbet Beit Mazmīl, something that until this project was not really attempted at other hinterland sites in the Jerusalem region, where a broader range of dates was provided for the period in question. This period of construction began when the leveling courses of flat stones were laid over the Early Roman quarry in Area A (Southern Sector) and the first flagstone floors put in place in Area C (Northern Sector)—providing the literal foundations for the Mamluk occupation of the site—and culminated with the major constructions in Area B (Southern Sector) and repurposings of vaulted buildings in Area C (Northern Sector). Such activity is informative about local building practices of the day.

Note

1. For instance, at the Negev site of Shivta, within the collapse walls of the Byzantine multistorey dwellings, traces of wooden beams and branches in the ceiling between two floors are still visible. Similarly, wooden branches supported by stone arches are also documented in several photos of traditional Palestinian houses in the Hebron area (see Hirschfeld 1995, 179–184).

References

ʿAdawī, Z., and B. Dolinka

2021. "Jerusalem, Qiryat Ha-Yovel." *Hadashot Arkheologiyot: Excavations and Surveys in Israel* 133. Internet Edition: https://www.hadashot-esi.org.il/report_detail_eng.aspx?id=25962&mag_id=133.

Alvera-Bartolotto, A.

1981. *Storia della Ceramica a Venezia*. Florence: Sansoni.

Avissar, M.

2003. "Early Islamic through Mamluk pottery." In *Jewish Quarter Excavations in the Old City of Jerusalem, Volume II: The Finds from Areas A, W and X-2—Final Report*, edited by H. Geva, 433–446. Jerusalem: Israel Exploration Society.

Avissar, M.

2009. *Pottery from the New Excavations at the Church of the Holy Sepulchre Compound*. Jerusalem: Unpublished manuscript on file at the Israel Antiquities Authority.

Avissar, M.

2012. "Pottery from the Early Islamic to the Ottoman Period from the Cardo and Nea Church." In *Jewish Quarter Excavations in the Old City of Jerusalem, Volume V: The Cardo (Area X) and the Nea Church (Areas D and T)—Final Report*, edited by H. Geva, 301–345. Jerusalem: Israel Exploration Society.

Avissar, M.

2014. "The ceramic finds from the Late Roman to the Ottoman periods at Bet She'an (Youth Hostel)." *'Atiqot* 77: 65–146.

Avissar, M., and E.J. Stern

2005. *Pottery of the Crusader, Ayyubid, and Mamluk Periods in Israel. IAA Reports* 26. Jerusalem: Israel Antiquities Authority.

Bacharach, J.L., and E. Rodenbeck

2002. "Metal objects." In *Fustat Finds: Beads, Coins, Medical Instruments, Textiles, and Other Artifacts from the Awad Collection*, edited by J.L. Bacharach, 198–205. Cairo: The American University in Cairo Press.

Boas, A.

1999. *Crusader Archaeology: The Material Culture of the Latin East*. Cambridge: Routledge.

Boas, A.

2006. "The Medieval Ceramics from Khirbat Ka'kul." *'Atiqot* 54: 75–104.

Brosh, N.

2005. "Islamic Glass finds of the thirteenth to fifteenth century from Jerusalem: Preliminary report." In *Annales du 16 congrès de l'association international pour l'histoire du verre (London 2003)*, 186–190. Nottingham: International Association for the History of Glass.

Chernov, E.

2007. "Metal objects and small finds from En-Gedi." In *En-Gedi Excavations II: Final Report (1996–2002)*, edited by Y. Hirschfeld, 507–543. Jerusalem: Israel Exploration Society.

Conton, L.

1981. *La Antiche Ceramiche Veneziane Scoperte in Laguna*. Venice: Fantoni libri arte.

Dean, B.

1927. "The exploration of the Crusaders fortress in Palestine: A report of explorations made by the Museum, 1926." *Bulletin of the Metropolitan Museum of Art* 22: 5–44.

Deetz, J.

1996. *In Small Things Forgotten: An Archaeology of Early American Life*. New York: Anchor Books.

Dolinka, B.

2017. *The Middle and Late Islamic Pottery from the Excavations at the Austrian Hospice, 2011–2012*. Internet Edition: https://www.academia.edu/10347780/THE_MIDDLE_AND_LATE_ISLAMIC_POTTERY_FROM_THE_EXCAVATIONS_AT_THE_AUSTRIAN_HOSPICE_2011_2012.

Gichon, M., and R. Linden

1984. "Muslim oil lamps from Emmaus." *IEJ* 34: 156–169.

Gorin-Rosen, Y.

2004. "Glass vessels from a salvage excavation at Sarafand el-Kharab, Nes Ziyyona." *'Atiqot* 46: 59–64.

Gorin-Rosen, Y.

n.d. "A fourteenth century ce glass assemblage from the Safed Al-Watta Quarter." *'Atiqot* (forthcoming).

Hadad, S.

2002. *The Oil Lamps from the Hebrew University Excavations at Bet Shean (Excavations at Bet Shean, Volume 1) Qedem Reports 4*. Jerusalem: Institute of Archaeology of the Hebrew University in Jerusalem.

Hadad, S.

2005. *Islamic Glass Vessels from the Hebrew University Excavations at Bet She'an. Qedem Reports 8*. Jerusalem: Institute of Archaeology of the Hebrew University in Jerusalem.

Hirschfeld, Y.

1995. *The Palestinian dwelling in the Roman-Byzantine period*. Jerusalem: Franciscan Printing Press (Studium Biblicum Franciscanum).

Johns, C.N.
1936. "Excavations at Pilgrims' Castle, 'Atlit (1932–3): Stables at the south-west of the suburb." *QDAP* 5: 31–60.

Johns, C.N.
1950. "The Citadel, Jerusalem. A summary of work since 1934." *QDAP* 14: 121–190.

Jope, E.M.
1956. "Vehicles and harness." In *History of Technology II*, edited by C. Singer, 537–562. Oxford: Oxford University Press.

Katsnelson, N.
2014a. *The Glass Finds from Khirbet Beit Mazmīl*. Jerusalem: Unpublished manuscript on file at the Israel Antiquities Authority.

Katsnelson, N.
2014b. "Glass finds from the al-Wata Quarter, Safed (Zefat)." *'Atiqot* 78: 153–157.

Katsnelson, N.
2014c. "The glass finds from the Bet She'an Youth Hostel." *'Atiqot* 77: 23–57.

Khamis, E.
1996. "The metal objects." In *Yoqne'am I: The Late Periods. Qedem Reports* 3, edited by A. Ben-Tor, M. Avissar, and Y. Portugali, 218–234. Jerusalem: Israel Exploration Society.

Khamis, E.
2008. "The metal artifacts." In *Paneas Volume II: Small Finds and Other Studies. IAA Reports* 38, edited by V. Tzaferis and S. Israeli, 177–188. Jerusalem: Israel Antiquities Authority.

Kletter, R.
2009. "Ramla, Ha-Palmah Street." *Hadashot Arkheologiyot: Excavations and Surveys in Israel* 121 (29.7.2009).

Levi, D.
2012. "Jerusalem, Qiryat Ha-Yovel." *Hadashot Arkheologiyot: Excavations and Surveys in Israel* 124. Internet Edition: http://www.hadashot-esi.org.il/Report_Detail_Eng.aspx?id=1999&mag_id=119.

Magen, Y., and M. Dadon
1999. "Nebi Samwil (Shmuel Hanavi—Har Hasimha)." *Qadmoniot* 32: 62–77.

Ploug, G., and E. Oldenberg
1969. "Les objets en métal sauf les monnais." In *Hama IV,3: Les petits objets médiévaux sauf les verreries et poteries*, edited by E. Hammershaimb, E. Oldenberg, G. Ploug, and R. Thomsen, 13–88. Copenhagen: Aarhus University Press.

Poulsen, V.
1957. "Les poteries." In *Hama. Fouilles et récherchés, 1931-1938, IV.2: Les verreries et poteries médievales*, edited by P.J. Riis and V. Poulsen, 117–283. Copenhagen: Nationalmuseet.

Pringle, D.
1984. "Italian pottery from Late Mamluk Jerusalem: Some notes on Late and Post-Medieval Italian tradewares in the Levant." In *Atti del XVII Convegno Internazionale della Ceramica*, 37–44. Albisola: Centro Ligure per la Storia della Ceramica.

Pringle, D.
1986. "Pottery as evidence of trade in the Crusader States." In *I Comuni Italiani nel Regno Crociato di Gerusalemme*, edited by G. Airaldi and B.Z. Kedar, 451–475. Genoa: Collana storica di fonti e studi.

Raphael, K.
2008. "The metal objects." In *Archaeological Excavations at Caesarea Maritima, Areas CC, KK and NN—Final Report. Volume I: The Objects*, edited by J. Patrich, 435–469. Jerusalem: Israel Exploration Society.

Raphael, K.
2011. "Jerusalem, the Old City: Second-Temple Period Quarry and a Medieval Building." *Hadashot Arkheologiyot: Excavations and Surveys in Israel* 130 (29.7.2009).

Saccardo, F.
1993. "Contesti medievali nella Laguna e prime produzione graffite veneziane." In *La ceramica nel mondo bizantino tra XI e XV secolo e I suoi rapporti con l'Italia: Atti del Convegno, Siena 1991*, edited by S. Gelichi, 201–239. Florence: Quaderni del Dipartimento di archeologia e storia delle arti, Sezione archeologica, Università di Siena.

Saccardo, F.
1996. "Nuovi dati sulla ceramica tardomedievale veneziana: il Ritrovamento di Malamocco." In *Atti del XXVI Convegno Internazionale della Ceramica*, 353–372. Albisola: Centro Ligure per la Storia della Ceramica.

Seligman, J.
2001. "Yet another Medieval tower and section of Jerusalem's ancient walls (Armenian Patriarchate Road)." *'Atiqot* 42: 261–276.

Seligman, J.
2010. *Nahal Haggit: A Roman and Mamluk Farmstead in the Southern Carmel. IAA Reports* 43. Jerusalem: Israel Antiquities Authority.
Shindo, Y.
2001. "The classification and chronology of the Islamic glass bracelets from al-Tür, Sinai." In *Cultural Change in the Arab World. Senri Ethnological Studies* 55, edited by T. Nishio, 73–99. Osaka: Osaka National Museum of Ethnology.
Siviero, G.B.
1986. "La ceramica graffita veneta." In *Atti del XIX Convegno Internazionale della Ceramica*, 32. Albisola: Centro Ligure per la Storia della Ceramica.
Spaer, M.
1992. "Ancient glass in the Israel Museum: Beads and other small objects." *Journal of Glass Studies* 34: 44–62.
Stern, E.
1999. "The pottery of the thirteenth–fifteenth centuries from Giv'at Yasaf (Tell er-Ras)." *'Atiqot* 37: 125–136.
Storchan, B., and B. Dolinka
2014. "Jerusalem, Bayit Va-Gan." *Hadashot Arkheologiyot: Excavations and Surveys in Israel* 126. Internet Edition: https://www.hadashot-esi.org.il/report_detail_eng.aspx?id=11622&mag_id=121.
Tatcher, A.
2009. "Miscellaneous finds from Strata 5–1." In *Horbat 'Uza II, The 1991 Excavations Volume II: The Late Periods. IAA Reports* 42, edited by N. Getzov, D. Avshalom-Gorni, Y. Gorin-Rosen, E.J. Stern, D. Syon, and A. Tatcher, 182–188. Jerusalem: Israel Antiquities Authority.
Taxel, I.
2022. "A unique assemblage of Late Islamic magical artifacts from Netafim 2: A Campsite on the Darb al-Hajj, Southern Israel." *Journal of Material Cultures in the Muslim World* 3: 145–173.
Taxel, I., M. Iserlis, and E. Yannai
2013. "The inside out of childhood archaeology: Petrographic and X-ray analysis of Early Islamic ceramic rattles from Mishmar David (Israel) and their cultural implications." *Journal of Eastern Mediterranean Archaeology and Heritage Studies* 1/3: 219–227.
Tushingham, A.D.
1985. *Excavations by K.M. Kenyon in Jerusalem 1961–1967, Volume I: Excavations in the Armenian Garden of the Western Hill.* Toronto: Oxbow Books.
de Vaux, R., and A.M. Stève
1950. *Fouilles à Qaryet el-'Enab, Abu Ghosh, Palestine.* Paris: Etudes archéologiques / Ecole Biblique et Archéologique Française.
Ward-Perkins, J.B.
1940. *Medieval Catalogue. London Museum Catalogues No. 7*. London: The London Museum.
Ward-Perkins, J.B.
1967. *Medieval Catalogue*. London: The London Museum.
Wightman, G.J.
1989. *The Damascus Gate, Jerusalem: Excavations by C.M. Bennett and J.B. Hennessy at the Damascus Gate, Jerusalem, 1964–66. BAR International Series No. 519.* Oxford: British Archaeological Reports Oxford Ltd.

About the authors

Dr. Benjamin J. Dolinka is an archaeologist and ceramics specialist who has conducted research and fieldwork in Greece, Jordan, and Israel during the past 31 years. He received his PhD in Archaeology from the University of Liverpool in 2007 with a dissertation on Nabataean pottery and worked for the Israel Antiquities Authority from 2010 to 2018. Mentored on the Islamic pottery by the late Dr. Miriam Avissar (IAA), he has come to be one of the leading specialists for the Jerusalem area. His forthcoming *The Pottery of Medieval Jerusalem, ca. 1050–1550 CE* will be published by Equinox as part of its *Monographs in Islamic Archaeology* series.

Dr. Nicolò Pini (PhD in Classical Archaeology at the University of Cologne, Germany) is an archaeologist, specialized in ancient urbanism, architecture, and social structures in the Roman and Byzantine periods. Over the last years he has ventured into the Early, Middle, and even Late Islamic periods. He currently conducts a postdoctoral research project at the CReA-Patrimoine, Université libre de Bruxelles (Belgium), funded by the Belgian Fonds de la Recherche Scientifique—FNRS (CR—Chargé de recherches). He was previously awarded a COFUND Marie Skłodowska-Curie Individual Fellowship at the same institution. He collaborates on several projects in the Near East, in particular with the Islamic Archaeology Research Unit at the University of Bonn and is also external research associate of the Panorama Platform (based at the Universitè libre de Bruxelles), specialized in acquisition and digitization of objects and architectural surveys.

Prof. Dr. Bethany J. Walker (PhD 1998, University of Toronto) is Professor of Islamic Archaeology at the University of Bonn, the Director of the Research Unit of Islamic Archaeology at the same institution, and former Co-Speaker of the Annemarie Schimmel Kolleg of Mamluk Studies in Bonn. A historically trained archaeologist, Walker directs field projects in Israel and Jordan. Among her monographs, edited, and co-edited works are *Jordan in the Late Middle Ages: Transformation of the Mamluk Frontier* (Chicago 2011), *Reflections of Empire: Archaeological and Ethnographic Studies on the Pottery of the Ottoman Levant* (Boston 2009), *Living with Nature and Things: Contributions to a New Social History of the Middle Islamic Periods* (Bonn 2020), and the *Oxford Handbook of Islamic Archaeology* (Oxford 2020). She is the senior and founding editor of the Journal of Islamic Archaeology, and her research focuses on peasant society, land use and tenure, and landscape transformation.

Benyamin Storchan is a PhD candidate of the Department of Bible Studies, Archaeology and the Ancient Near East at Ben-Gurion University of the Negev. His dissertation, "The Metamorphosis of Holy Land Churches During the Early Islamic Period," examines the processes of change and continuity as seen from the archaeology of Byzantine churches in the Holy Land during and after the Islamic transition. Storchan completed his MA at Bar Ilan University and BA from Michigan State University. He has been employed at the Israeli Antiquities Authority for over 15 years as a research excavation archaeologist for the Jerusalem Region, and has directed a number of large-scale excavations, including the Glorious Martyr Excavation Project and the Early Bronze Age settlement at Eshta'ol. He has also published a number of final excavation reports and research articles. Storchan is currently a board member for The Israel Archaeological Association, research fellow at the Center for the Study of Conversion & Inter-Religious Encounters and has been recently awarded the David Amit Prize for Archaeological Research and a Humanities Faculty Scholarship from Ben-Gurion University of the Negev.

— 5 —

The Making of a Family Farmstead

Emergence and Demise of the Ottoman *'Ezbeh*

Bethany J. Walker, Benjamin J. Dolinka, Nicolò Pini, Roy Marom, and Benyamin Storchan

Introduction: It's a "fixer-upper!"

The Ottoman and post-Ottoman occupation of the site was framed by four events, the impact of which will be explored in more depth in the chapter that follows: the Ottoman conquest of Jerusalem in 1516; the Naqib al-Ashraf revolt of 1703; local implementation of the Tanzimat Land Law of 1858; and the earthquake of 1927. For at least a century, many of the architectural remains of Khirbet Beit Mazmīl continued to look much as they did in the Mamluk period, the new residents doing selective maintenance to keep it habitable; others, however, were slowly abandoned and covered with soil for cultivation. Only in the 19th century was the site dramatically transformed physically, though its function remained largely the same. This chapter follows the long, post-Mamluk history of occupation of Khirbet Beit Mazmīl, focusing on the site formation processes that created the ruins we saw when we began excavations in 2015, as well as the cultural practices that defined daily life there from the 16th through until the early 20th centuries.

From Mamluk *qaṣr* to a family farmstead (Stratum II: Early Ottoman, 16th to 17th century)

Historical background

Jerusalem, and its villages, fell under Ottoman control with the conquest of Bilād al-Shām in 1516. Besides replacing many, but not all, officials, most administrative and fiscal practices were initially kept as they were under the Mamluks. The land was subjected to a cadastral survey, Mamluk *Iqṭāʿat* were transformed into *timar*s, and endowed lands, while maintained as *awqāf*, were taxed for the first time (at the rate of 10 percent). The reign of Suleiman I (1538–1541) was pivotal in the resettlement of villages in the region, ensuring the safety of the roads and facilitating economic development. In the Holy City, he ordered the construction of the walls we see today, as well as the renovation of the Dome of the Rock. Pilgrimage and businesses thrived, as did local markets. The remainder of the century was a time of growing prosperity and security. Locals resettled many old ruins and brought new land under the plow. The dynamics of economic revival in the century following the Ottoman conquest of Palestine, however, differed from that of the late Ayyubid period described in the last chapter. Rather than the sale of state lands, it was the cooption of local elites (*al-aʿyān*) through the granting of offices, and the management of rural *awqāf* that created a new mid-ranking local/provincial elite. By the end of the century certain families came to political power, and would remain so for the next century, and Sufi sheikhs, the *ashrāf* (descendants of the

Prophet), and the leading clerics (the *ulamā'*) and merchants of Jerusalem acquired political position, wealth, and social capital. The growing availability of credit—made possible through a system called "cash *waqf*" and by borrowing money from the local tribal, religious, and mercantile elites—and the leasing of endowed lands made it possible for many rural peoples to rent orchards and for a diverse group of entrepreneurs—men and women, Muslims and Christians, locals and foreigners alike—to acquire property (Abdulfattah 2007; al-Sabagh 2014; Walker 2021b).

Khirbet Beit Mazmīl was abandoned for some decades after the Ottoman conquest, as the Mamluk military estate no longer served its original use, and the officers assigned to it were deposed, fled, or were killed. A visitor to the site in the late decades of the 16th century would have seen the partially roofless ruins of well-built structures on the top of the hill and at its base. Some vaults were still in place at the north end of the summit, and with a little effort at clearing out rooms and removing what architectural collapse was most in the way, the dilapidated buildings could be rehabilitated. Many of the terraced slopes below the hill and the terraced fields in the vicinity of the site were still being cultivated; the terraces that had fallen into disuse could have been repaired and put back into use.

At the end of the century, or perhaps at the turn of the 17th, the archaeological site was resettled in this manner as a peasant household. The names of families who lived in Beit Mazmīl during the 16th century are known from the detailed court records/registers. However, their origins are unknown. They could have been an extended family of peasants from neighboring villages, who rented land, or, alternatively, several households of sharecroppers, who worked the land for the absentee *timarcı* (*timar* owner). The granting of state lands to absentee military and civilian officials as land grants—or fiefs—was a common pattern by the end of the 16th century.

The division of the small site into multiple spaces for food preparation, laid out on different levels and accessible only through steps and single doorways, suggests independent living spaces for several households. Regardless, the household goods are distinctly different from that of the Mamluk period that preceded it, as we will describe below.

The excavations revealed traces of the process of resettlement of Khirbet Beit Mazmīl. The ruins at the bottom of the hill (the Southern Sector) were lived in again for only a short time, to be eventually converted into terrace walls. The buildings on the summit (the Northern Sector) were reoccupied and put entirely to domestic use. Rooms were cleared of their contents, which were deposited in a common midden. Some rubble was removed, but much was left in place between buildings, and soil was brought in to create a small, terraced space in the middle of a makeshift courtyard. This small, self-sufficient farmstead on the top of the hill continued to be occupied until the troubled times of the early 17th century.

Repurposing, reuse, and reoccupation

The built environment

The Ottoman occupation at the site has two markedly distinguished macro phases. The first one, towards the end of the 16th and beginning of the 17th century—and following a short period of abandonment after the Ottoman conquest of Jerusalem—has some elements of continuity with the Mamluk phase, since most structures appear to have been kept in place and limited transformations of earlier spaces occurred (see Appendix 5). However, it also saw a radical change in the function and spatial arrangements of the vast majority of the earlier buildings. During the second phase, taking place at the end of the Ottoman period after a longer period of abandonment, the site was entirely rebuilt and continued to be used as seasonal farmstead until the 1920s (see Appendix 6).

The first Ottoman phase is hard to recognize architectonically, since it appears that most of the structures were kept from the previous Late

Mamluk occupation. It is plausible that most building activities were limited to restorations and small rearrangements that did not alter consistently earlier structures. From a technical and architectonical perspective, the difference between new Early Ottoman structures and the earlier ones is not striking, as they both rely primarily on locally found and reused materials. The only evident difference is probably the degree of homogeneity in the building materials and the accuracy with which certain features are constructed. One good example is the remaking of the barrel vault in C3. Despite the fact that the technique and typology of the vault remains the same from the previous period, the building materials appear to be more irregularly arranged and diversified, suggesting a different and less skilled construction. Possibly, the original Mamluk vault required a complete restoration, that was completed by the new occupants of the site. In any case, the attempt to maintain not only the original typology of construction, but also the same aesthetic, is evident. As it will be illustrated later, no such attempt was made in the later phase, when the site was rebuilt with a completely different technique.

For certain contexts, and especially in Area C, it is impossible to distinguish the Early Ottoman structures from the Mamluk ones on the basis of the architecture alone. As noted in the previous chapter, changes to the built environment appear to have already taken place during the Late Mamluk period. The end of the 15th century is then to be seen as a truly transitional phase, at which point buildings were in general maintained and only marginally modified, but the function and organization of the site had started to change considerably.

The process that started in the Late Mamluk period was completed in the Early Ottoman period, with radical changes in the spatial conception of the site. Instead of having one single, articulated building extending over a large surface and multiple levels, a series of smaller, independent domestic units were created. It is interesting to see that in this case, the transformation of the built environment and its fragmentation did not imply a complete and extensive demolition of earlier structures, as seen for instance between the pre-Mamluk and the Mamluk periods. The existing spaces were simply rearranged in order to accommodate the new functions and needs required by rural households. One possible explanation is that the earlier structures were not in such a ruinous state to require demolition and building anew.

The Northern Sector is where the transformation of the Mamluk built environment is more appreciable, even though it is impossible to establish exactly the extent of such transformation. What is certain is that at least two smaller and independent dwellings were established. The first one occupies the North Building of the Northern Sector (Areas C and E1). It is extremely likely that this dwelling also had an upper floor: in an early phase of E1 a series of steps abutting against the former pillar (W124/1) were found in place, likely conducting to an upper floor or to a mezzanine. The mezzanine—or the continuation of the stairs—could have been supported by the arch or the demi-arch that covered the former press (L.245)—that went out of use towards the end of the 15th/beginning of the 16th century—and a new floor made of flagstones (L419 in E1 and L.229 in C6) covered the entire surface of C6 and E1 (Figure 1).[1] It is possible that contextually with the first transformation of the space, most notably the obliteration of what we believe to have been a press in C6, the door opening to the east was blocked and a new entrance was opened on the opposite side. Two steps abutted the outer side of the western wall of the press, suggesting that a new doorway was created here once the installation went out of use (Figure 2).

Above the flagstone floor in C6 (L.229), a cooking installation (L. 225) was excavated. In E1 the space appeared to have had a rather mixed function and to have had several modifications throughout the entire Early Ottoman occupation. In fact, after the first (L.419), two more flagstone floors (L414 and L409) were set on top of thick packs of alternating fill and plaster floors (Figure 3).

Figure 1 Remains of the Ottoman flagstone floor in Area E1 and C6 of the Northern Sector (photo: N. Pini).

Figure 2 Remains of the Ottoman flagstone floor obliterating the former installation L.245, and steps in correspondence with a possible new entrance in Area C6 of the Northern Sector (photo: N. Pini).

Interestingly, it appears that the southern half of E1 was elevated slightly higher than the northern half, where a short step can be identified on top of the former circular installation (L.258; Figure 4). Furthermore, in the southern part of the room, in association with the last flagstone floor, a step was identified, set on top of W136 and abutting against the eastern side of pillar W124/1, suggesting the existence of another staircase leading either to a mezzanine or to a second floor (Figure 4). The later reoccupation of the site prevents us from understanding exactly how the space was arranged and how a possible upper floor could be accessible and articulated. However, at least two possibilities can be hypothesized, based on examples of vernacular architecture from the region (for examples of extensive surveys see Hirschfeld 1995, 109–215; RIWAQ 2006).

The first solution would have a short stair leading to a mezzanine. In this case, comparisons in the region show that the stair is normally in a more-or-less central position, and at the top it opens both on its right side and to its left to a single, multifunctional space. Oftentimes, mezzanines were used as living quarters in otherwise single or two-room dwellings. The ground floor would then be used as an animal shelter, storage space, or for food processing and cooking. In other cases, especially in larger dwellings or when associated with productive installations, the mezzanine could be used as storage space.

Sometimes access is gained by a simple wooden ladder rather than a solid stairway (Fuchs 1998, 160–162). Nonetheless, the examples in stone are quite frequent in the region for Ottoman dwellings, being recognized as one of the traditional features of vernacular architecture (Fuchs 1998, 158). The mezzanine itself is normally supported by barrel

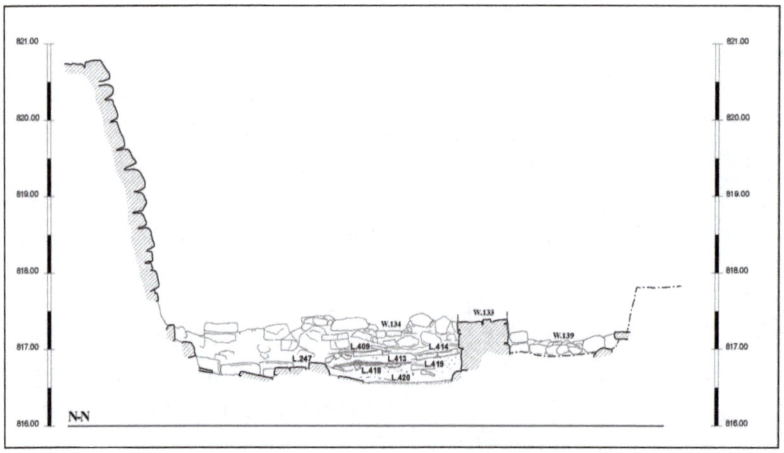

Figure 3 Section of E1 with the succession of flagstone floors, fills, beaten earth, and plaster floors (drawing: S. Pirsky).

Figure 4 The southern half of E1: in foreground on the right, the first short steps covering the circular installation L.258; in the background, to the right, the step abutting against pillar W124/1, associated with the last flagstone floor (photo: N. Pini).

vaults or demi-arches, springing from two lateral pillars or the external walls and abuts the pillar at the center that also supports the stairs—as seen, for instance, in Bani Naʿīm, east of Hebron, as surveyed by Hirschfeld (1995, fig. 142).

The presence of barrel vaults or demi-arches also characterizes the second solution of the space. In this case, though, instead of a mezzanine there is a landing, and the stairs turn either to the left or to the right to lead to a proper second floor. These structures are more often found in courtyards—so in an external space—but the construction technique is rather similar to the previous one. The first stairway leading to the landing abuts a pillar, the top of which often is the landing itself. The pillar is also abutted by the barrel vault or the demi-arch on the ground floor, which in turn supports the second portion of the stairway (RIWAQ 2006, 266, 324 and 330).

There is no conclusive evidence to tell which solution was really adopted in this first phase of Ottoman occupation, and to what extent this differed from the late Mamluk phase. The first solution, however, seems the most plausible. The dwelling in Area C6/E1 seems to closely recall the local vernacular typology of the single-room house. Even if not exclusive, the second arrangement—stairs leading to a landing and then to a second floor—is predominantly present in courtyards, and there is no indication that C6/E1 was converted into a courtyard dwelling after its functional transformation. Despite all changes in flooring and inner circulation, it seems to remain a covered space, a room closed on all four sides, eventually with two doorways on the western and eastern walls—the latter being walled in either in the Late Mamluk or Early Ottoman period.

Nonetheless, the dwelling was certainly associated with open-air spaces. As noted above, the new doorway created in C6 gave direct access to the space west of the dwelling, which appears to be a rather empty area. B5—the space that separates the dwelling in Area C from that of Area B—appears to have been used for refuse disposal, as suggested by materials coming from the former channel W127, the water system that went out of use after the Late Mamluk period.

It is likely that another open courtyard extended in C5. As said, it is unclear when the door to the east was blocked and if it gave direct access to this courtyard to the inhabitants of the dwelling during the Early Ottoman period. Evidence for a restoration of the northern portion of W134—closing the eastern side of E1—is quite evident, even if not clearly datable. It consists of more diversified and irregularly arranged building materials, differentiating itself from the rest of the walls creating the room C6/E1. The sack-masonry dates to the Mamluk phase of occupation, suggesting that the wall was restored in the Late Mamluk or Early Ottoman period. Whether the restoration came together with the blocking of the wall or predates it, is at present impossible to determine.

If the door was only blocked towards the end of the Early Ottoman period, it might still be the case that the courtyard was associated with this dwelling and possibly shared by other neighboring houses, namely the two vaulted chambers C3 and C4, and with Area D, which do not appear to undergo radical transformations until the drystone architecture of the 19th century was built. Yet, its spatial relationship with the rest of the buildings in Area C is still problematic. What can be documented, though, are the remains of a wall running north/west-south/east (W138 and W139), clearly abutting against the walls delimiting E1 (W133 and W134) and therefore later than them. It could have provided the southern limit and internal courtyard. Interestingly, in correspondence with the corner between W138/139 and W133, a circular installation, enclosed by a low wall made of a single row of field stones, was constructed (L.140). This might find direct comparison with some field walls found in the Southern Sector, abutting the main walls of the dwellings, and functioning as a partition wall for extra domestic areas (Figure 5).

One further wall line was identified just outside the door in E1 (W137), abutting the eastern face of the eastern wall (W134) and associated with a flagstone floor (Figure 6). It consists of piled flagstones and does not give the impression of having any structural function. It can be hypothesized that it was a partitioning wall in the courtyard.

It must be noted that it is impossible to determine where the entrance to this northern cluster was. It is possible that one of the vaulted chambers was used as entrance hall, thus opening to the north. There is no direct indication for an entrance to the east, even though the later drystone constructions might have obliterated all evidence. Also, it is possible that the associated extensive terrace (W105

Figure 5 Installation L.140 (photo: N. Pini).

Figure 6 W137 just outside C6 (photo: N. Pini).

and W114 in particular) substituted or altered a preexisting terrace wall or structure—possibly the continuation of W110—that could have compensated a possible gap in elevation.[2] What is certain is that no opening is documented to the south or to the west, suggesting a quite marked separation between this cluster and the second dwelling (South Building) developed in the southwestern corner of the hilltop.

In terms of our understanding of the built environment, the hypothesis of having a shared courtyard between different independent domestic units has solid foundations in the social interpretation of architectural arrangement. The shared courtyard is indeed one common feature in vernacular architecture of the region, also in more recent times: if the single dwellings are referrable to nuclear families, they are often aggregated into clusters belonging to the same extended family (Fuchs 1998, 160). This socially determined built environment has indeed a long tradition and might well be applicable to the Early Ottoman remains at Khirbet Beit Mazmīl (Pini 2020, 393–396).

The new arrangement creates a completely different living space at the site, and the construction of the walls connecting E1 to Area D (Walls 138 and 139) and creating an internal courtyard could be the direct consequence of the mutated occupation of the site and the co-residence here of different families. From a single complex occupying the entire summit and surrounding of the hill, where the different rooms and spaces were determined by the need to accommodate more specialized functions, we have now different, independent domestic units that need to accommodate different family units. The need arose, then, to assure privacy for each household and to clearly differentiate the space belonging to each one; this new requirement was not compatible with the earlier built environment, necessitating a reorganization of space.

The Northern Sector of Khirbet Beit Mazmīl was not occupied exclusively by the dwelling in C6/E1: a second dwelling in Area B, which was, likewise, physically defined by a preexisting structure, underwent a spatial reorganization. The architectural evidence is more limited in this part of the site, but a large single room, with internal separating walls was created (Figure 7). A doorway opening to the south was blocked, into which a small niche with a drain was inserted (Figure 8). At the same time, a new doorway was opened to the north, possibly opening to the open space between this dwelling and the northern, larger domestic unit. Despite the presence of Wall 124/2, it is impossible to verify if this new building also had an upper floor. The general impression is that this second dwelling was architectonically simpler and considerably smaller than the neighboring one: it would come as no surprise, then, if it consisted only of a ground floor. However, low walls in the single room suggest the existence of a partition of the inner space, possibly creating slightly elevated areas. A possible direct comparison would be the dwelling in Area B of the Southern Sector, where the earlier structures were completely obliterated by a two-room dwelling with an open, front courtyard.

The dwelling in Area B is associated with a wall running parallel to W138/W139 (W135 in E2), that

Figure 7 The dwelling in Area B of the Northern Sector (photo: N. Pini).

Figure 8 The blocked door in the dwelling in Area B of the Northern Sector (photo: N. Pini).

abuts—and partially destroys—the eastern side of the pillar W124/2. These two parallel walls appear to create a sort of alley winding between the two different residential clusters and leading to the Southern Sector. Wall W135 does not appear to support any superstructure, instead serving as an enclosure wall for a courtyard or as small retaining wall. We know from the small finds that this space was used, from the start, as a midden, possibly by both households.

The two dwellings of the Northern Sector, regardless of their exact spatial arrangement during the Early Ottoman period, present several other features that are considered typical of the "Traditional Palestinian house"[3]—and for this reason the comparison with the still extant regional vernacular architecture appears to be particularly fruitful. It is not the purpose of the present work to offer a renewed perspective or new elements for discussion on the historical development of such building typology. However, it is worthwhile to consider some aspects that are of interest for the understanding of our site and some of the developments taking place between the Late Mamluk and the Early Ottoman period.

The first striking similarity is the presence of a single, multi-functional room, which might have had internal partitions to eventually separate certain more specialized areas. One way of obtaining this spatial subdivision was to create elevated areas within the room, either mezzanines or *iwāns* (Fuchs 1998, 158–162; Ragette 2012, 54). The spatial arrangement can take different forms, but ethnographical comparisons show that normally the lower part of the room corresponded with the entrance. The elevated zones were then arranged on all or some of the other three sides of the dwelling: the different plans of the houses are ultimately determined by how many elevated areas are there, and the specific characteristics of these zones by the economic means or the specific needs of the inhabitants (Fuchs 1998, 164–165, fig. 7a).

While architectural surveys of vernacular houses offer potential parallels to the remains at our site, one factor in the way the site physically developed cannot be underestimated: the impact of preexisting ruins or ancient architecture. In many historical pictures (see Hirschfeld 1995, figs. 109 and 135), in fact, it is striking to see how frequently these houses embedded earlier structures and/or were surrounded by other ruins. Even if the dwellings do have recurring features that become part of a clearly distinguished vernacular building form, it must be taken into account that the exact shape of the single dwellings was highly affected by preexisting structures and cannot always follow standardized models, especially in a rural context.

The evidence from the rest of the site is less abundant and more fragmented, and there are clear indications for a progressive contraction of the site within the Northern Sector. The Southern Sector appears to have been either abandoned or converted into terraces and field plots. In both Areas A and B, the Mamluk structures are overlaid by new wall alignments. In Area B, a circular structure (W557), connected to walls (W553 and 551), was constructed as part of a newly established terrace system with cultivated plots. It is, unfortunately, impossible to determine from the architecture the exact chronology of this transformation.

Conversely, in the Lower Site, the entire area occupied by the previous massive Mamluk building appears to be rather intensively reoccupied, in this case without completely demolishing the earlier buildings, but instead heavily modifying them. Part of this construction activity involved the levelling of the natural irregularities of the bedrock with slabs. Also, the structures in S1 appear to have been heavily rebuilt. A tower-like building was constructed, with a flight of stairs connecting the Lower Site with the Northern Sector. As noted in Chapter 4, it is possible that a similar structure was first built during the Mamluk period, but the Ottoman activity cleared the area of most of materials dating to that period.

As in Area B of the Northern Sector, a *ṭābūn* was identified in S4. Another *ṭābūn* dating possibly to the same chronological horizon was installed in S1, suggesting the possible presence of multiple domestic units also in this part of the site. Unfortunately, there are not enough data to allow a more detail reconstruction of its built environment and no extensive new architectural remains have been documented.

To summarize the evidence for the Early Ottoman period, the 16th–17th century occupation sees a radical shift in the built environment and its function, as well as the beginning of a progressive contraction of the site. As discussed, such transformations probably started to take place already in the Late Mamluk period, toward the end of the 15th century, but were completed in the Early Ottoman phase, when the site was reoccupied after a short period of abandonment. It is possible that the blocked entrances found in several areas of the site are to be associated with this short abandonment: an attempt by the inhabitants to seal and protect their belongings while away.

In any case, the single complex with multiple specialized areas, and apparently fewer domestic functions, that was the Mamluk estate, was turned into a series of independent dwellings characterized by multifunctional spaces developed, all sharing similar spatial arrangement. At least three dwellings reoccupied the Mamluk remains in the Northern Sector, while at least one was set in the Lower Site. As best documented between Areas C-D and B of the Northern Sector, the dwellings appear to be totally independent—spatially and functionally—suggesting that different groups were inhabiting them. Nevertheless, the marked proximity of these dwellings also suggests the presence of different nuclear families belonging to the same extended social group. In this phase of occupation, the Mamluk estate was converted into a family home. What changes took place in other parts of the village cannot be determined from the limited exposures of our excavations.

The stratigraphic record (see Appendix 5 for full site plan)

The stratigraphic record confirms this phasing. There is considerable continuity in the late 16th–early 17th century (Early Ottoman, Stratum II). The Mamluk-era site was abandoned for most of the 16th century, but the largest part of its architecture remained standing and functional. At the turn of 17th century, these ruins were revitalized, with some repurposing and subdivision of space. The Northern Sector was spatially divided, creating two distinct domestic units out of reoccupied structures, modified largely through the laying of flagstone floors (Figure 9). They did not connect with one another, but stairs, comprised of spoliated flagstone, led from one unit to the other. The late Mamluk industrial and domestic units of Areas C and E1 were repurposed into the kitchen, storage, and living spaces of the North Building. The cooking and living spaces of late Mamluk Area B4 maintained the same purposes in the South Building. Areas B, C, and E1 produced many sealed contexts in domestic spaces, with well-stratified sequences of living and cooking surfaces taking the form of beaten earth surfaces, plastered floors, and flagstone floors.

The South Building (B3) is dominated in this period by a single-roomed dwelling (Figure 10) with heavily plastered walls and floor; it measures approximately 11 meters north–south by 6 meters east-west. It used W.118 (a construction of some antiquity) as its southern wall, W.117 in B3 (which joins W.102 along the eastern façade of Room B2) as its western wall and W.131 as its eastern wall. A new wall (W.122) was added in the north to completely enclose the structure as an independent unit. The B4 structure is aligned on a completely different axis than that of the other structures at the site in this period, as its layout was established in an earlier period. The B4 structure

Figure 9 Drone image of Northern Sector (courtesy of N. Pini).

Figure 10 Plan of B4 (courtesy of S. Pirsky).

was not apparently vaulted; as no rubble from the superstructure could be identified here, we do not know how the space was covered.

In the first season of excavations, an Early Ottoman coin hoard was discovered (see Figure 39). It was located in a stratified fill layer (L.119) along the south face of W.118 in Square B4, and the ten coins were covered by a large Handmade Geometric Painted Ware (hereafter HMGP) body sherd from a jar. The hoard consists of thick copper *mangīrs* that were minted in Cairo and date from the reign of Murad III (1574–1595). They appear to have been deliberately buried outside the house, away from the door, and hidden away under a large jar fragment and flat stones. They may have originally been bundled together in a small cloth bag that disintegrated in the acidic soil long ago.

Topsoil and sub-topsoil covered a layer of architectural collapse (L.112), which, in turn, sealed the contents of the interior of the building. The original doorway in Wall 118, to the south, was blocked, a new entrance opened in Wall 122 to the north, and the floors and walls plastered in this phase of occupation.[4] In the northeast corner of the room, and built against the south face of W.122, and below a cooking installation of the Late Ottoman period, was a well-stratified sequence of surfaces associated with cooking: the upper flagstone floor (L. 122); a beaten earth surface (L. 124) above a *ṭābūn* (L. 118); the plaster layer (L. 127) on which the *ṭābūn* was formed, sealing the southern face of W. 122; a second layer of plaster (L. 127a) below that; a thin, beaten earth surface (L. 127b); the middle flagstone floor (L. 128) that formed the surface of a second cooking installation; a thin plaster floor (L. 131) beneath that; the makeup of another plaster floor (L. 133); and the lower flagstone floor (L. 158) (Figures 11 and 12). The entire installation was sectioned in half, and all its contents were sifted and sampled for organic materials such as animal bones, plant remains, phytoliths, and ash and charcoal, and collected for laboratory analysis. The L. 127/127a sequence yielded a few small finds: a nearly complete Early Ottoman saucer lamp dating from the 16th to 17th century (from L. 127); the fragment of an iron horseshoe; a burnt glass bead (L. 127a); as well as metal slag from both loci. L. 127b was, by far, the richest of the lot, and produced a wealth of in situ objects, including two Early Ottoman copper *mangīrs* minted in Cairo during the 1590s; seven small and thin coins that were part of an ornamented veil; the bronze clasp from a book (possibly a Quran) or small box; a glass juglet rim; and a red agate bead. The makeup of the plaster floor (L. 133) below the cooking installation produced an Early Ottoman gray tobacco pipe dating from the late 16th/early 17th century.

Charcoal extracted from these layers, as well as ashy deposits and surfaces in other parts of the room, have yielded consistent radiocarbon dates of the Late Mamluk and Early Ottoman periods. The plaster (L. 127a) and beaten earth surfaces (L. 127b) below the *ṭābūn* were dated by radiocarbon analysis to 1446–1799 CE and 1446–1657 CE, respectively.

Figure 11 *Ṭābūn* in B4 cooking installation, view north (photo: B.J. Walker).

Ash (L. 130) from a pit dug against the south face of W. 122 (817.09), to the west of the cooking installation and *ṭābūn*, at a level below these features, produced a radiocarbon date of 1209–1399 CE. A beaten earth surface (L. 129) with embedded ash and charcoal, located in the southeast corner of this room and at an elevation lower than L. 127, was dated to 1428–1637 CE in the same fashion (Figure 13). Occupation in this room and associated with these food preparations activities should then be dated to the late 16th or early 17th centuries (Walker and Dolinka 2020, 201–202).

Excavations in C6 revealed numerous, contemporary activity areas related to cooking and dating to the Early Ottoman era (late 16th/early 17th centuries). We present here the two plastered cooking installations—the botanical analysis of the other cooking-related loci can be found in Chapter 7.2. An installation of construction identical to that in B4.127–128, with a fragmentary *ṭābūn*, was uncovered along the southern face of W.104, in the northwest corner of Square C6. It was constructed on flagstone floor L.229. This installation also yielded a well-stratified occupational sequence, which can be enumerated from top to bottom as follows: a beaten earth surface with widely dispersed flagstones (L.220); a plaster surface (L.221) below L.220, which yielded an Early Ottoman copper *mangīr*; another beaten earth surface (L.222) beneath L.221; and a layer of fill mixed with collapse (L.223), deposited above the main feature of the sector, which was a plastered cooking installation with *ṭābūn* fragments (L.224). After dismantling the cooking installation (L.225), along with the soil layer (L.228) surrounding it, a well-preserved flagstone floor (L.229, Figure 14)

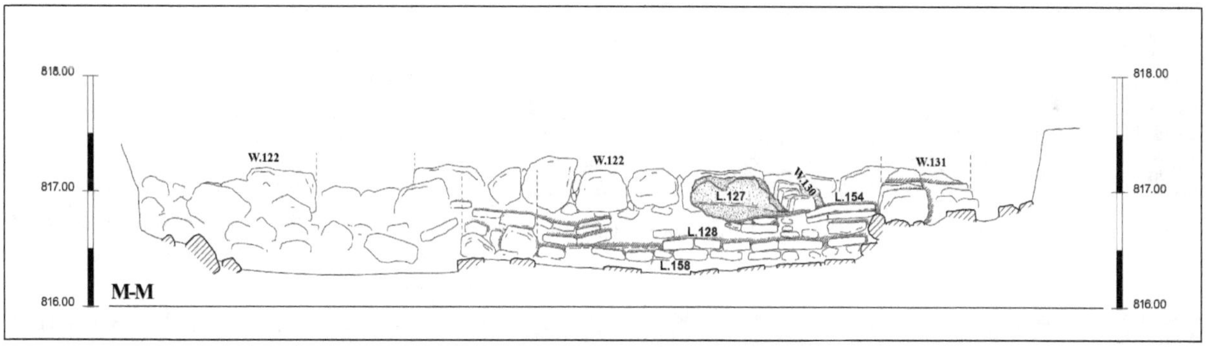

Figure 12 Section drawing of B4, north baulk (courtesy of S. Pirsky).

Figure 13 Section drawing of C6, north baulk (courtesy of S. Pirsky).

was revealed. The fill (L. 233) on the floor yielded a glass Mamluk sprinkler spout, arguably from the 15th century. Of interest from this installation was the thin double-plastered wall (W.129), which enclosed it on its western side and along W.104 (Figure 15). The cooking installation produced four late 16th century *mangīrs* and one gray tobacco pipe of the same period (Early Ottoman era).

Unlike the Early Ottoman cooking installation from B4, the entire C6 installation was completely dismantled, although a good stratigraphic section was left in situ for section drawing and photography. Like its counterpart in B4, all its contents were sifted and sampled for finds, organic materials such as animal bones, plant remains, phytoliths, and ash (for radiocarbon dating), which were collected for laboratory analysis. The L.225/L/226/L.228 stratigraphic sequence yielded a substantial amount well-dated in situ small finds. These included the following: four Early Ottoman *mangīrs* minted in Cairo during the 1590s; an Early Ottoman gray pipe; the metal pendant from a veil; an iron horseshoe fragment; a small piece from a glass bracelet; and a dark carnelian stone bead.

Figure 14 Flagstone floor (L. 229) in kitchen of C6 (photo: B.J. Walker).

Figure 15 Double-plastered walls (Wall 129) defining C6 cooking space (photo: B.J. Walker).

A second cooking area was identified in the southwest corner of C6. In this part of the square yet another cooking installation, this one oval-shaped (L.238), was revealed. It consisted of burnt beaten earth layers alternating with thin plaster layers and was cut into the L.229 flagstone floor. The stratigraphic profile for this cooking installation can be summarized as follows: L.238, the in situ installation itself; L.240, the layer below L.238, which consisted of a reddish-brown burnt beaten earth fill with ashy lenses; L.241, a thin plaster layer with ashy lenses below L.240; L.242, a beaten earth surface with whitish soft clay lenses, below L.241; and L.243, the thin plaster layer beneath L.242 at the bottom of the cooking installation. Unlike the other cooking installations from the site, this one was artifact-poor, and yielded no coins. The finds were mostly earlier in date and typical of the Mamluk II period (15th century) at the site. These included an almost complete small glass bottle, a glass sprinkler spout, and a green steatite whetstone. The installation, however, is Early Ottoman in date. It is built on and cuts through an Early Ottoman flagstone floor (L. 229). The installation, we believe, was constructed out of recycled soil, taken from close by. The two cooking installations of C6, then, represent an important stratigraphic transition at the site. Here the press room of the Mamluk period was repurposed as a kitchen at the turn of the 17th century. A similar development can be traced in B4.

Outside of kitchen spaces, the North Building also produced evidence for storage. E1, located in the courtyard to the south of C6, was a furnace room in the last Mamluk phase of occupation (15th century), and was repurposed in the early Ottoman period as a storage room with the laying of a series of flagstone floors. It is delimited by: W.136 on the west, which was built abutting the W.124/1 platform; W.133 on the south; and W.134 on the east. The room was entered from the north. A complete stratified sequence dating from the Early Ottoman period was revealed.

The stratigraphic profile for the Early Ottoman room in Square E1 consisted of the following (Figure 16): the patchy remains of a flagstone floor (L.409), which was only preserved in the westernmost portion of the square; a layer of fill (L.405) beneath the flagstone floor, which rested upon a plaster floor (L.406); below this were two loci, the first (L.414) a flagstone floor that was only preserved in the east sector of the square, and an orange hard-packed clayey surface (L.413) in the western part of the room, which had in situ mendable pottery vessels embedded into it; an artifact-rich hard-packed surface with plaster (L.418) below that; another flagstone floor (L.419) beneath that; and a final plaster floor (L.420), the first of a series of Late Mamluk era surfaces (Figure 17).

The stratified sequence from this room yielded a wealth of in situ finds, which provide important insights into the activities that took place there. L.412 produced the large fragment of an iron horseshoe and a small lens of glass slag, which would seem to indicate possible glass production at the site. Objects from L.413 included a nearly complete handmade basin and wheel-made redware jar embedded into its surface, as well as a glass jug base and basalt grinder. L.418 was the richest of the lot and yielded the following: an Early Ottoman *mangīr* minted in Cairo during the 1590s; a

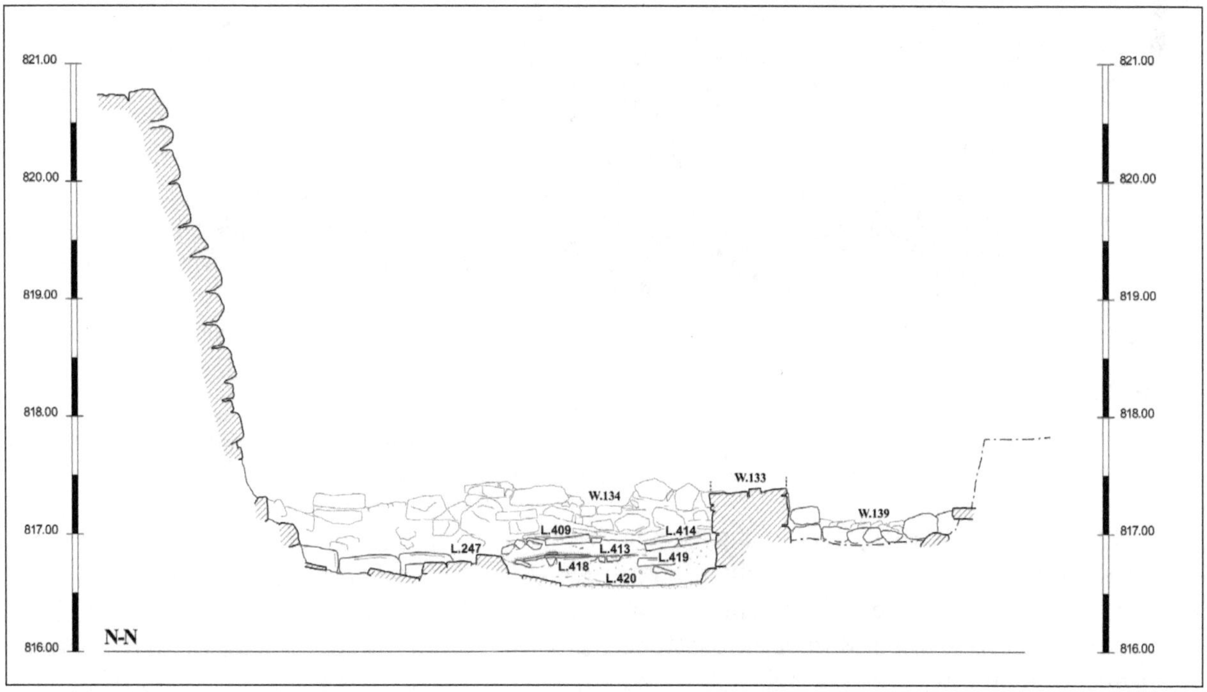

Figure 16 Section drawing of E1, view south (courtesy of S. Pirsky).

Figure 17 Broken flagstone pavements in E1 (photo: B.J. Walker).

Figure 18 Drone photo of Field C (courtesy of N. Pini).

complete iron horseshoe and several large fragments of others; an iron bridle ring or cavesson; an iron knife; iron functional rings; a glass bead and even more glass slag.

Loci 409–419 are collectively a series of plastered surfaces and flagstone floors that occupy a vertical depth of only 13 centimeters and represent, at best, a century of occupation. Most of the flagstone surfaces appear only in patches, and do not extend across the room. They may constitute, in fact, two damaged flagstone floors (L. 409 and 419), with intermittent repairs. The flagstones across the site show evidence of cracking and pounding, as would happen if heavy building blocks fell on the floors from an upper story. On the basis of the general stratigraphic profile across this part of the site, and the pottery from associated fills, the lower flagstone floor (C6.409) has been assigned to the Late Mamluk period, and the upper one (C6.419) to the Early Ottoman.

A final spatial reorganization of this part of the site occurred in the doorway of Wall 263 (C6.263) (visible in Figure 18, far right). As was noted in Chapter 4, this doorway connected the plastered room of E1 and the reused pressroom of C6 to a room in C5 with a flagstone floor, and of unknown dimensions and function, to the east (Figure 19). In the initial reoccupation of the site, this doorway was blocked with spoliated building stones, cutting off access between the rooms of C6-E and that of C5.[5] As there was no further work conducted in this part of the site in 2019, the reasons for this spatial restructuring remain unknown.

The Southern Sector of the site continued to be occupied in this period, but only for a short while. Square A12 (excavated within by the IAA in 2013) revealed the northeastern corner of a large plaster-lined installation (L201), enclosed by a corner formed by two poorly preserved walls (W402, W403) (Figures 20a and 20b). The interior faces of the walls were lined with plaster that rounded the corner adjoining them, in the same fashion as B4.119 and C6.224, both of which were dated ceramically to the Early Ottoman period. Within the installation were three superimposed white plaster floors. All the pottery from this square was Early Ottoman in date (see Figure 34 below).[6]

Figure 19 View east across rooms of North Building (C6 and E1) (photo: B.J. Walker).

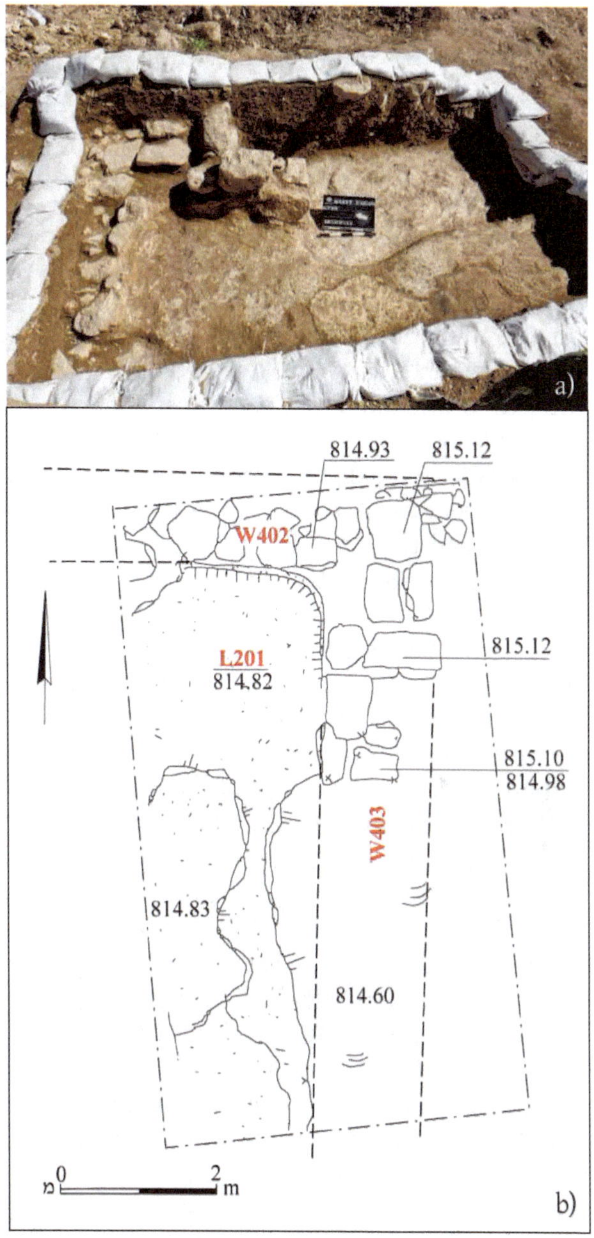

Figure 20 a) Plaster-lined installation (L.102) in A12, Southern Sector, 2013 season (courtesy of the IAA); b) Plan of installation L.102.

Further evidence of Early Ottoman occupation comes from the Western Building of the Lower Site (excavated by the IAA in 2017). The squares in the center of the excavation area (S2, S6) contained the wide foundations of three walls (W2–W4, width approximately three meters; Figure 21) enclosing a rectangular or square room (L205) paved with stone slabs (L3). The room's southern wall was not preserved, having apparently been damaged by development work prior to the excavation. Walls 2–4 probably supported a vaulted ceiling and possibly a second floor. Meager pottery finds from the Early Ottoman period (17th century) and a coin from the beginning of the Ottoman period were found beneath the flooring in Room 205 (L206). These included the shank end of an Early Ottoman light gray tobacco pipe, as well as an Ottoman *mangīr* from the reign of Murad III (1574–1595): as sealed deposits, they provide a *terminus post quem* for the construction of this floor and date when this room was occupied. Finds from the same period were recovered from the room's soil fills (L204). To the north of the corner of Walls 2 and 3 was another wall (W6; Figure 14) comprising two rows of stones (W405, W407), which form a corner. Wall 6 was partially excavated and is probably part of a large building that extends eastward. A wall (W5) to the west of the building's northwest corner similarly comprises two rows of fieldstones (W501, W604) preserved to a height of three courses. A coarse stone paving (L4) between the corner of the building and W5 abuts W5 and the west face of W4 (Figure 22). Wall 5 was probably built to separate the building from the land to the west, beyond the excavation area.

Figure 21 View north of West Building, Southern Sector, 2017 season (courtesy of the IAA).

Figure 22 Plan of West Building, Southern Sector, 2017 season (courtesy of the IAA).

This part of the site was not occupied for very long, however. A 17th century Ottoman *dirham* (silver coin) of the reign of Uthman II (1618–1622) recovered from upper fill (L. 200) provides a date for when the room, or the site, went out of use.[7]

The evidence from the Northern and Southern Sectors and the Lower Site suggests that at the time of the reoccupation of the site, sometime at the end of the 16th or beginning of the 17th centuries, the buildings of the Mamluk site were structurally sound, many of the walls and some vaults still standing, and the paved floors of several rooms still visible. The press room of C6 was converted into a kitchen, with the laying down of a flagstone floor and, with time, a long sequence of plaster and beaten earth floors. Much of the space of Area C located outside of the vaulted rooms was used for food preparation. The house in B4 served as a separate domestic unit, with its own areas for food preparation and stairs constructed with spoliated flagstones removed from Mamluk levels giving access to the Area C vaulted houses at the highest point of the hill. The structures located downslope in the Southern Sector (in A12) and the Lower Site (in the Western Building) appear to have served the same function. The site was thus transformed into something like a farmstead, housing several nuclear families. While the buildings of the Southern Sector and Lower Site eventually went out of use, those of the Northern Sector continued in occupation until the early 18th century.

The contours of family life

Social life of the kitchen

The household is the smallest unit of socialization. It is in the household that individuals first become socialized in their communities by learning (and accepting) social roles, community-relevant skills, and communal ethics. In this way the family is integrated into the community, and the communities into larger social and economic networks. At a rural site like Khirbet Beit Mazmīl, no spaces serve these functions better than kitchens and courtyards. They are often one and the same space, as some food preparation takes place in these out-of-door spaces. While storerooms and middens yield information about daily life—and these are explored further in this chapter—kitchens play a direct role in socialization and are a locus of socializing. It is where members of the household met to share food, watched over children, warm up on a cold day, shared daily chatter, and lit their pipes. It was far from only the women's space, as family members flowed in and out for a variety of purposes.

Archaeologically we define a kitchen as an activity area with evidence of repeated food-preparation: the presence of built installations (*ṭābūns*, *tannurs*, or other such clay ovens); residue from the cooking process (concentrations of ash, charcoal, and botanical remains); and fragments of the cooking vessels themselves, recovered from a primary context. At Khirbet Beit Mazmīl, kitchens, defined this way, are distributed throughout the site, and directly associated with individual domestic units. They take a variety of forms and, together with the cooking wares, reveal a diversity of food preparation practices.[8]

Early Ottoman kitchens have been identified in B4 (L.118, 122–124, 127) and C6 (L. 220–224 and 238). These cooking installations take the same form, comprised of alternating layers of plaster and hard-packed clay fill: the remains of regularly repaired floor surfaces (Figures 23a and 23b). Those layers contain ash and charcoal and are rich in botanical remains. They are spatially restricted to corners of the room, and do not extend across the room as surfaces. In B4, the installation is located in the northeast corner of the room, inside the door. In C6 the two installations were found in the northwest and southwest corners of a room, which was likely entered from a doorway in the east side and delimited by two thin walls of plaster (L.129); the function of those plastered walls, outside of marking a space, is unknown. *Ṭābūn* fragments were recovered from beaten earth surfaces and plastered floors, as well as in the vicinity of the installations. Wear and tear on these surfaces, through repeated walking, sweeping, cooking and food preparation, would have necessitated frequent floor maintenance and renovation through replastering or laying new floors. *Ṭābūns* can be in use for a generation or longer, thus the number of floor resurfacings (identifiable as alternating layers of compact earth and plaster) can add up over time.[9]

We have not identified places of food production for the last phases of occupation at the site. The sooting of the ceilings in the standing *manāṭīr* of B1, C3, and C4 are the result of open fires inside the vaulted rooms. These would have been hearths to keep the house warm during cold days, but also for cooking. Normally, however, the courtyard should have served this function. The continued cultivation of the courtyard areas and terraces of the site, along with the transport of new soils as needed, has eliminated any trace of this potential activity.

Figure 23 a) Cooking installations in B4; b) Cooking installation in C6 (photo: B.J. Walker).

The remains of the cooking vessels themselves do not shed much light on how kitchens—as places of food preparation—functioned, because the vast majority of cookpot sherds at the site come from secondary contexts: middens, collapse debris, abandonment levels, and courtyard and terrace fill.[10] Nonetheless, as a corpus they suggest a diversity of cooking practices and specialization of cooking wares. The Early Ottoman corpus is dominated by coarse, slipped globular cookpots (turned glossy black through repeated exposure to heat), and varieties of equally coarse (fabric with large quartz inclusions) handmade stewpots with horseshoe handles, incised handles or shoulders, or thumb-impressed rims. Wheelmade, thin-walled pots with upright and rounded rims, as well as Early Gaza Ware casseroles (of light grey fabric) were used in the households of Khirbet Beit Mazmīl in the late 16th through to the early 18th centuries. Although the state of preservation of the site is much better in the Late Ottoman era, there is a paucity of 19th and early 20th century cooking wares at the site.[11] One hole-mouthed cookpot from collapse debris, recovered in the 2015 season, is an exception (Walker and Dolinka 2010, 2, fig. 3:13). While some of the Early Ottoman wares simply continue a Mamluk-era tradition—handmade cooking pots and bowls, some with a burnished red slip, and the classic elephant-ear handled stew pot—there are some notable developments in the choices people made in how to equip their kitchens. The Mamluk cooking wares identified at the site include more wheelmade vessels and imports, such as globular glazed cookpots and Brittle Ware casseroles. By the 17th century, the occupants of Khirbet Beit Mazmīl were cooking primarily with handmade stewpots and wheelmade casseroles produced in the Gaza region. The burnished black slip seems to be an innovation of the time, replacing the red slip. While the reasons for the changes in overall cooking assemblages over time elude us at this point, it is apparent that peasant households here either produced what they needed on their own, or bought them at local rural markets, supplementing them with specialized wares for particular cooking needs from professional workshops in Gaza.[12]

Although cooking vessels were not recovered from kitchen spaces at Khirbet Beit Mazmīl, a variety of small finds were, revealing social activities, unrelated to food production, that may have taken place there in the Early Ottoman period. The dismantling of the cooking installations in B4 and C6 produced many of these objects; the rest were recovered near ashy deposits and *ṭābūns*. The objects found with the most frequency in these installations were coins (*mangīrs*). Seven *mangīrs*, discussed in detail below, were recovered from cooking installations; one more was found in an ash deposit and another beside a *ṭābūn*. One lamp (Obj. #6, B4.127/127a) and one tobacco pipe (Obj. #32, C6.225) were also retrieved from the contents of dismantled cooking installations. Another pipe was recovered in an in situ location in the kitchen space, sitting on a plastered floor near the cooking installation and below a *ṭābūn* (Obj. #22, B4.133). The installation in B4 was particularly rich in small finds, producing, in addition to the lamp and coins above, seven face veil coins, a clasp from a book or small box, a glass juglet rim, and even a fragment of a horseshoe. The association of such a wide range of everyday objects with kitchen space, and specifically with the heat-producing features, is intriguing. The ovens may have been places one hid money.[13] The lamps could have been lit with the embers from the oven; B4 and C6 were indoor spaces in dark houses that had few windows.[14] The myriad other small finds recovered from the floor layers that constituted the cooking installation were personal items and household goods (clothing, jewelry, glassware), used in a room that probably served multiple functions as a center of activity in the house. As for the pipes, like the lamp, they may have been lit with the embers from the oven.[15] This, too, was a custom widespread in Europe, where Ottoman and Ottoman-style *chibouks* were smoked. The kitchens, along with middens, are the richest repository of the traces of daily life at the site in the Early Ottoman period.

Tobacco consumption

Befordúltam a konyhára,	*I turned to the kitchen,*
Rágyujtottam a pipára . . .	*I lit the pipe . . . That*
Azaz rágyujtottam volna,	*is I would have lit it,*
Hogyha már nem égett volna.[16]	*if it hadn't been burning.*

Pipes and coffee cups are the most readily recognizable forms of Ottoman material culture on the local level (François 2005). Tobacco-smoking penetrated all classes of Ottoman society; was the main form of communal consumption in towns, villages, and among the Bedouin; and was enjoyed by men and women alike. Tobacco-smoking, the use of *chibouks*, the drinking of coffee, and the penetration of the *finçans* (small handless coffee cup) into most households are consumption patterns that resulted from the entanglement of the Ottoman Empire with global markets (Baram 1996). Because of the short life span (and high breakage) of the pipe bowls (which fractured with regular pounding of the bowl during cleaning); their mass-production at hundreds of sites across the Mediterranean and Central Europe; and their rapidly changing forms and surface decoration, *chibouk* bowls are often found at late historical archaeological sites throughout the Mediterranean and provide invaluable chronological anchors.

The spread of tobacco use and cultivation was a direct outgrowth of the European "discovery of the New World" and the conquest of the Mediterranean coasts by the Ottomans in the 16th century. Tobacco was introduced into the Middle East through the port sites from Africa or Europe or both (Baram 1996, 144) sometime between 1599 and 1606, and spread rapidly to the interior (Simpson 1998). When exactly tobacco was used in Palestine, and how early it was cultivated locally, remains a topic of debate. Tobacco was smoked in public spaces there as early as the first decades of the 17th century, and travelers' accounts of the period claim that it was a well-established pastime from at least 1599 (Bent 1893, 49, note 1). The earliest tobacco smoked in Palestine was a Lebanese product (Palestine Economic Policy Research Institute 2016); it was only later cultivated locally, from at least the late 17th century and perhaps earlier (Simpson 2000, 171). In Palestine, as throughout the Ottoman domains, smokers chose between locally grown and imported tobaccos: both were available in the local markets, as were locally (or regionally) produced and imported pipes.[17] Generally associated with middens and fills, and as surface finds on surveys, most pipe bowls in published reports date from the 17th to the 19th centuries (de Vincenz 2016; de Vincenz 2021; Shapiro 2022 for a series of pipes spanning the early 17th through early 20th centuries). Physical evidence of tobacco smoking from the 16th century is very rare (Walker 2023).

One fortunate exception may be the small, light gray *chibouk* bowls from Khirbet Beit Mazmīl, which can be dated from the 1590s. Nine pipes and pipe fragments from largely stratified contexts, in association with very early Ottoman handmade pottery; late 16th century *mangīrs* (ranging in date from the 1590s to 1617, the majority from Murad III's reign); and charcoal dated by radiocarbon analysis to the Early Ottoman period. These are among the earliest tobacco pipe fragments found to date in Palestine. As one pipe bowl (Obj. #133, E2.439) still contained the tobacco smoked in it, it was subjected to residue and isotope analysis to determine where the tobacco was cultivated. The pipe was also subjected to petrographic thin-section analysis, XRD, and XRF. As expected from written sources, the tobacco smoked this early at this rural site was likely grown in Lebanon; the pipe appears to have been produced in southern Palestine (possibly Gaza or Faluja).[18] The recovery of pipe fragments #133 and #134 on bedrock in the E2 midden provide a *terminus post quem* for the use of this space for refuse disposal.

The pipes and pipe fragments were recovered from contexts distributed throughout the site: two from the C6 kitchen (Objs. #22 and #32, loci B4.133 and C6.225, respectively), four from the E2 midden (Objs. #81, #131, #133, and #134, loci E2.402, E2.435, and E2.438, respectively), two from the B5 courtyard subtopsoil (Obj. #53, L. B5.142), one from the E1 storeroom (Obj. #85, L.E1.405), and a final one from the fill of the covered water channel of the B5 cistern (Obj. #31, L.B5.147). The recovery of pipes in a kitchen was discussed above. Finding pipes in middens and terrace fills (which comprised household garbage) is expected, as pipes had a high breakage frequency, and the bowls in particular needed to be replaced on a regular basis. The single pipe found in the E1 storeroom is more likely an object dropped in this room accidentally than one stored here with water jars and store jars.

In terms of tobacco consumption, one further find of the period, discussed in detail below, should be mentioned here. Fragments from two copper *narghilas* (water pipes) were discovered buried under an almond tree in C6. Although residue analysis was not done on the contents of the one bowl (head), macrobotanical analysis suggests that it was the tobacco smoked in the pipe. In what rooms the pipes were originally set up or used is unknown. We cannot reconstruct the physical setting of smoking in these households of any form (whether with a *chibouk* or a *narghila*), only the disposal patterns.

Meeting household needs: Ceramics

Most household needs were met through ceramic objects and vessels. The following presents the pottery of the Early Ottoman period as 12 separate and distinct assemblages. The assemblages are represented by 68 vessels, which include bowls, basins, cooking pots, store jars, jugs (*ibrīqs*), oil lamps, and tobacco pipes. An assemblage approach is particularly important when dealing with the Ottoman periods. Significant interpretative challenges, mainly chronological, have been duly noted in recent surveys of Ottoman pottery:

> "... when the archaeology of the Ottoman Levant is compared to that of nearly all other historical eras in the region, a short three-letter word says it all, and that is the word 'gap!' It is, indeed, quite shocking and even disturbing to have to admit that, where Levantine archaeology is concerned, we have less to show and tell about the four centuries that span 1516 to 1918 CE than is the case with any comparable span since the beginning of the Early Bronze Age." (de Vries, B., and Ø. LaBianca 2009, ix–x)

Research on Ottoman ceramics has progressed rather slowly as a discipline, with a very limited number of resources available for comparanda, especially within the Jerusalem region. The frustration echoed by scholars when dealing with these materials during the past decades is palpable, and this phenomenon was best summed up by Prag, when referring to the pottery uncovered from the fortifications of the Old City. In that report, she provided an *à propos* caveat regarding its study:

> ". . . so little has been published to date. The domestic unglazed repertoire is hardly known, and the domestic-quality glazed wares have little in common with the ceramics illustrated in fine colour art volumes which are mostly repetitive depictions of high-quality vessels in museum collections, originally produced for, and purchased by, elites." (Prag 2008, 254–255)

This sentiment was reiterated by Avissar, a few years later with regard to the publication of an Early Ottoman assemblage from the Jewish Quarter excavations, whereby she states:

> "Since few Ottoman assemblages have been published, parallels could not be provided for all the vessels . . . Plain Ottoman pottery is so far little known, and no comparable vessels can be cited." (Avissar 2012, 303–304)

Exacerbating the problems of dating and finding parallels are the archaeological contexts themselves at sites in the Jerusalem region. For example, Johns excavated several sherds dating from the 15th–18th centuries, but they came from ". . . filling which accompanied the Mamluke and Turkish building . . . from the cistern and . . . ditch" (Johns 1950, 188). He had to rely upon the expertise of Lane—Keeper of the Department of Ceramics at the Victoria and Albert Museum in London at the time—for the dates of what had been uncovered. Wightman also describes this scenario in reference to his pottery types from post-Crusader deposits uncovered at the Damascus Gate excavations:

> "About sixty percent of the pottery recovered . . . came from poorly stratified contexts . . . Almost none of this great volume of archaeological material could be excavated and recorded in a controlled manner. In the few instances where some measure of control was effected, the excavators found difficult and dubious stratification: levellings, massive pits and robbers' trenches . . ." (Wightman 1989, 61)

Finally, there has yet to be either a concise guide to, or typo-chronological schema for, the Ottoman ceramics of the region, unlike the monographs published for the earlier Roman and Byzantine periods (Magness 1993), and the Middle Islamic era (Avissar and Stern 2005). The first attempt-in-earnest to address this issue was the volume of articles produced and edited by Walker (2009), which offered the first collection of Ottoman pottery assemblages from across the southern Levant and opened the door for future research on the subject.

Fortunately, through the very recent work of de Vincenz (2021), the situation has improved somewhat. She created the first typo-chronology for the Ottoman ceramics utilizing an impressive and varied assemblage from the excavations at the Jaffa *Qishle*. That being said, she reiterates the sentiments of previous researchers at the beginning of her fundamental study, noting:

> "During my work on Islamic ceramics I noticed that [Ottoman pottery was] treated either as 'interesting souvenirs' or 'modern garbage.' Research on Ottoman period ceramics was mostly nonexistent from the archaeological excavation reports because [the period is too recent and] is not covered by the Antiquities Law . . . [and the study of this pottery] proved to be a difficult task since there was so little published comparanda." (de Vincenz 2021, 127).

Given the aforementioned statements, it is still clear at the time of this writing that there remain significant gaps in our knowledge about the ceramic repertoire of the Early Ottoman period in general, and the Jerusalem region in particular. Therein lies the importance of the well-stratified pottery from Khirbet Beit Mazmīl, the dating of which is supported by the presence of well-known types of tobacco pipes, numismatic finds, and radiocarbon dates.

Square B4

The domestic space of B4 produced two distinct assemblages. The first consists of bowls and lamps, and the corpus bears striking similarities to sealed Early Ottoman contexts from excavations at the Austrian Hospice and Cardo in the Old City of Jerusalem. One of the very few wheelmade bowls (Figure 24:1) from the site during this period was found directly on the beaten earth surface (L.124) above a *ṭābūn* (L.127) with several associated floors. It has a carinated body, a T-shaped rim, and the same form as an example from the Austrian Hospice (Dolinka 2017, fig. 2:5). Handmade incurved shallow bowls are commonplace among the Early Ottoman corpus at Khirbet Beit Mazmīl, and two specimens are presented here. The first (Figure 24:2) was found on the upper flagstone floor (L.122) near the north entrance to the B4 room. It has an upright rounded rim and is paralleled at Habonim-Kafr Lam in northern Israel (Avissar 2009b, 10, fig. 2.5:12), where it was found with an Early Ottoman gray tobacco pipe. The second example (Figure 24:3) was uncovered on the beaten earth surface (L.127b) associated with the B4 *ṭābūn*, which was just below a plaster floor (L.127) that yielded an Early Ottoman *mangīr* of Murad III (1574–1595). It differs slightly in its more pointed rim and is comparable with a bowl found in the excavations at the Austrian Hospice (Dolinka 2017, fig. 2:3). Two carinated handmade bowls with flat rims were also retrieved from the L.127b surface (Figure 24:4–5). These are much deeper than the incurved bowls and have a slight carination at the midpoint of the vessel walls, which is thickened on the interior; some have a vestigial vertical pressed-on handle, while others do not. The form is equal to an exemplar from the Austrian Hospice (Dolinka 2017, fig. 2:6). A final vessel found within L.127b is represented by a handmade deep bowl with a folded-over and rounded rim (Figure 24:6), which also has correlates with the Austrian Hospice assemblage (Dolinka 2017, fig. 1:9).

A pair of wheelmade saucer lamps with pinched nozzles were also recovered from the room in B4. The first (Figure 24:7) was found on the plaster floor (L.127) below the L.124 beaten earth surface and has parallels in Cardo Area X-3 in Jerusalem (Avissar 2012, pl. 10.4:3). The other lamp originated from a beaten earth surface (L.127b) that was sealed below a lower plaster floor (L.127a) that contained a *mangīr* of Murad III (1574–1595). The form is ubiquitous in Early Ottoman contexts from the site, acts as a great chronological indicator, and corresponds to the type of lamps uncovered in Area X-2 from the Cardo (Avissar 2012, pl. 10.1:19).

Another small assemblage from B4, represented by only three vessels, originated from contexts located along Wall 118, which ran roughly east–west. Two of these originated in L.155, the fill on lowest flagstone floor (L.158) of the structure. The first was a handmade bowl with flat rim and horizontal band of finger-indented appliqué just below the rim zone (Figure 25:1) that is comparable to one found at the Austrian Hospice (Dolinka 2017, fig. 2:8). The second was a wheelmade jug with everted rim that is triangular in section (Figure 25:2). While no parallels could be found for this vessel, the fabric is clearly local and the context is clearly Early Ottoman and this specimen can be utilized as a comparative tool for future research.

A very important find from the B4 was the body sherd of a HMGP store jar from L.119 (Figure 25:3), which was the first stratified fill layer along the south face of W.118. Its importance lies in the fact that this large broken piece of pottery was found covering an Early Ottoman coin hoard consisting of 10 thick copper *mangīrs* minted in Cairo (Kool 2017) during the reign of Murad III (1574–1595). Of course, it was found during the last hour of the last day of excavation for the 2015 field season.

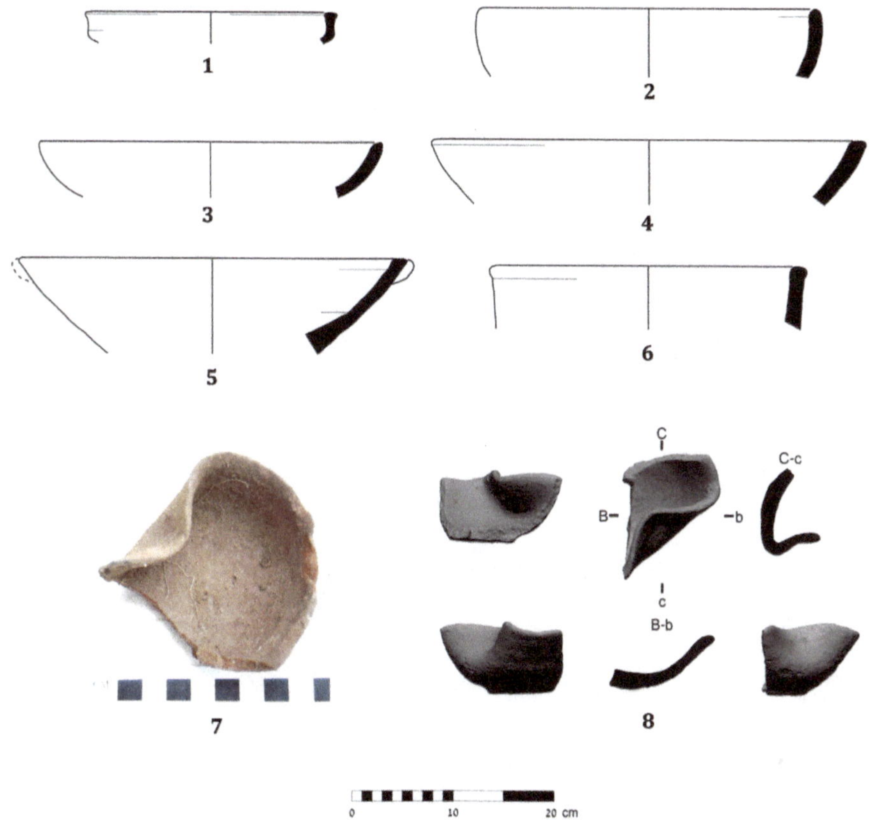

No.	Locus	Reg. No.	Form	Description
1	124	1034	BL	Wheelmade carinated bowl with t-shaped rim. Coarse reddish-yellow (7.5YR 6/6) fabric with chert and calcite inclusions.
2	122	1026	BL	Handmade incurved bowl with upright rounded rim. Coarse light brown (7.5YR 6/4) fabric with chert and calcite inclusions.
3	127b	1075/1	BL	Handmade incurved bowl with pointed rim. Coarse light reddish-brown (5YR 6/4) fabric with numerous tiny calcite inclusions.
4	127b	1039/2	BL	Handmade carinated bowl with flat rim. Coarse light reddish-brown (5YR 6/4) fabric with chert and calcite inclusions.
5	127b	1039/1	BL	Handmade carinated bowl with flat rim and vestigial vertical handle. Coarse light reddish-brown (5YR 6/4) fabric with chert and calcite inclusions.
6	127b	1075/2	BL	Handmade deep bowl with folded-over rounded rim. Coarse light reddish-brown (5YR 6/4) fabric with calcite inclusions.
7	127	1011	LP	Wheelmade saucer lamp with pinched nozzle. Yellowish-red (5YR 5/4) fabric with calcite inclusions and a reddish-brown (5YR 5/3) exterior.
8	127b	1039/4	LP	Wheelmade saucer lamp with pinched nozzle. Light reddish brown (5YR 6/4) fabric with calcite inclusions.

Figure 24 Early Ottoman pottery from the Domestic Space in B4.

No.	Locus	Reg. No.	Form	Description
1	155	1077/1	BL	Handmade bowl with flat rim and horizontal band of finger-indented appliqué just below the rim zone. Coarse reddish yellow (7.5YR 7/6) fabric with medium and large calcite inclusions.
2	155	1077/2	JG	Wheelmade jug with everted rim. Yellowish red (5YR 4/6) fabric with small and medium-sized calcite inclusions.
3	119	1021/14	SJ	Large Handmade Geometric Painted body sherd. Coarse light reddish brown (2.5YR 7/3) fabric, light gray (5YR 7/1) core, few medium chert inclusions, white (7.5YR 8.5/1) exterior slip, light reddish brown (5YR 6/4) painted decoration on exterior.
4	121	1025	LP	Wheelmade saucer lamp with pinched nozzle. Light reddish brown (5YR 6/4) fabric with calcite inclusions.
5	133	1046	TP	Shank-end of a gray tobacco pipe with asymmetrical wreath, stepped rim and termination, and simple rouletting. Well-levigated gray (2.5Y 5/1) fabric.

Figure 25 Early Ottoman pottery from the domestic space in B4.

A similar jar sherd was found at Belmont Castle (Grey 2000, fig. 6.3:64) in the Jerusalem hinterland. Because of the secure date of its find context, and although a body sherd, this fragment nevertheless provides important insights into the fabric and surface treatment for the Early Ottoman HMGP tradition for the Jerusalem region.

The penultimate vessel from this small corpus is the nearly complete example of a wheelmade saucer lamp with a pinched nozzle (Figure 25:4), that was revealed during the probe (L.121) which uncovered a *ṭābūn* in B4. A good parallel for this lamp was found in Cardo Area X-2 (Avissar 2012, pl. 10.1:19). The final piece from the B4 assemblage is the shank end of a tobacco pipe (Figure 25:5) that was found in L.133, the fill directly below the L.131 plaster floor. According to the Banias study, this type of shank end (Dekkel 2008, 126, fig. 4.2b:2) is from the early assemblage and includes both late 16th and 17th century versions. The Mazmīl specimen has an asymmetrical wreath with a stepped rim and termination. A band of clay appliqué was added to the upper part of the shank itself just before firing, giving the false impression that the pipe had an "extended" shank end. All of these elements are decorated with simple rouletting, and no parallel could be found for this unique specimen.

Square B5

Located to the north of B4 and south of Field C, Square B5 also yielded two small assemblages of Early Ottoman pottery. The first group consists of four vessels: two that were uncovered in L.152 (Figure 26:1–2), the fill on platform for arch springers (W.124/1); and two which were retrieved from L.147 (Figure 26:3–4), a loose dark-brown fill associated with the cistern. The first of these was a deep handmade hemispherical bowl with an out-turned rim (Figure 26:1), for which no parallel could be found. The second piece is the body sherd from a wheelmade strainer-jar neck with incised geometric decoration (Figure 26:2) in the so-called "Late Islamic Cream Ware," a term coined by Prag (2008, 272), which she dates to the 17th century. This moniker is somewhat misleading, however, as the examples recovered from the Jerusalem region (including the specimen presented here) are decidedly not "cream ware" by any means. The sherd from Khirbet Beit Mazmīl has a gray fabric and a pinkish white slip. A better description of this ware is offered with regard to an example from the Jewish Quarter, which explains that the "decoration has been scratched through the *light gray* . . . slip before firing, revealing the dark gray clay" (Avissar 2003, 438).

Two special ceramic finds were recovered from L.147. The first is a gray tobacco pipe (Figure 26:3) that is rouletted and incised on the upper portion of the bowl, and stamped decoration on the lower part of it that consists of two rows of bidirectional stamped arrowheads. On the bottom of the bowl is a production mark with an eight-petalled rosette. A very close parallel to this pipe is included in the Banias catalogue (Dekkel 2008, 136, fig. 4.6:24), where it is dated to the end of the 17th and beginning of the 18th century. That being said, the Khirbet Beit Mazmīl example differs in a few key respects: the fabric is finer and more delicate than the "heavy and massive" (Dekkel 2008, 136) one from Banias; the shape of the Khirbet Beit Mazmīl bowl is more oblong than Banias' "sack shaped" bowl (Dekkel 2008, 136); and, while that pipe has a six-petalled rosette as its production mark, the Khirbet Beit Mazmīl piece is eight-petalled. In addition, the rosette bears striking similarities to the stamp found on Early Ottoman *mangīrs* (Figure 26:3, far right) dating from the reign of Murad III (1574–1595), so it is suggested here that perhaps the eight-petalled rosette represents either a stamp, seal, or emblem of that ruler. The last vessel from the first assemblage in B5, also found in L.147, was another example of a wheelmade saucer lamp with pinched nozzle (Figure 26:4). A good parallel for this lamp was uncovered in Cardo Area X-3 (Avissar 2012: Pl. 10.4:3).

The second corpus of ceramics from B5 originates from a stratified sequence along Walls 124/1 and 124/2, which were the primary features from the square. This area had a paucity of diagnostic sherds,

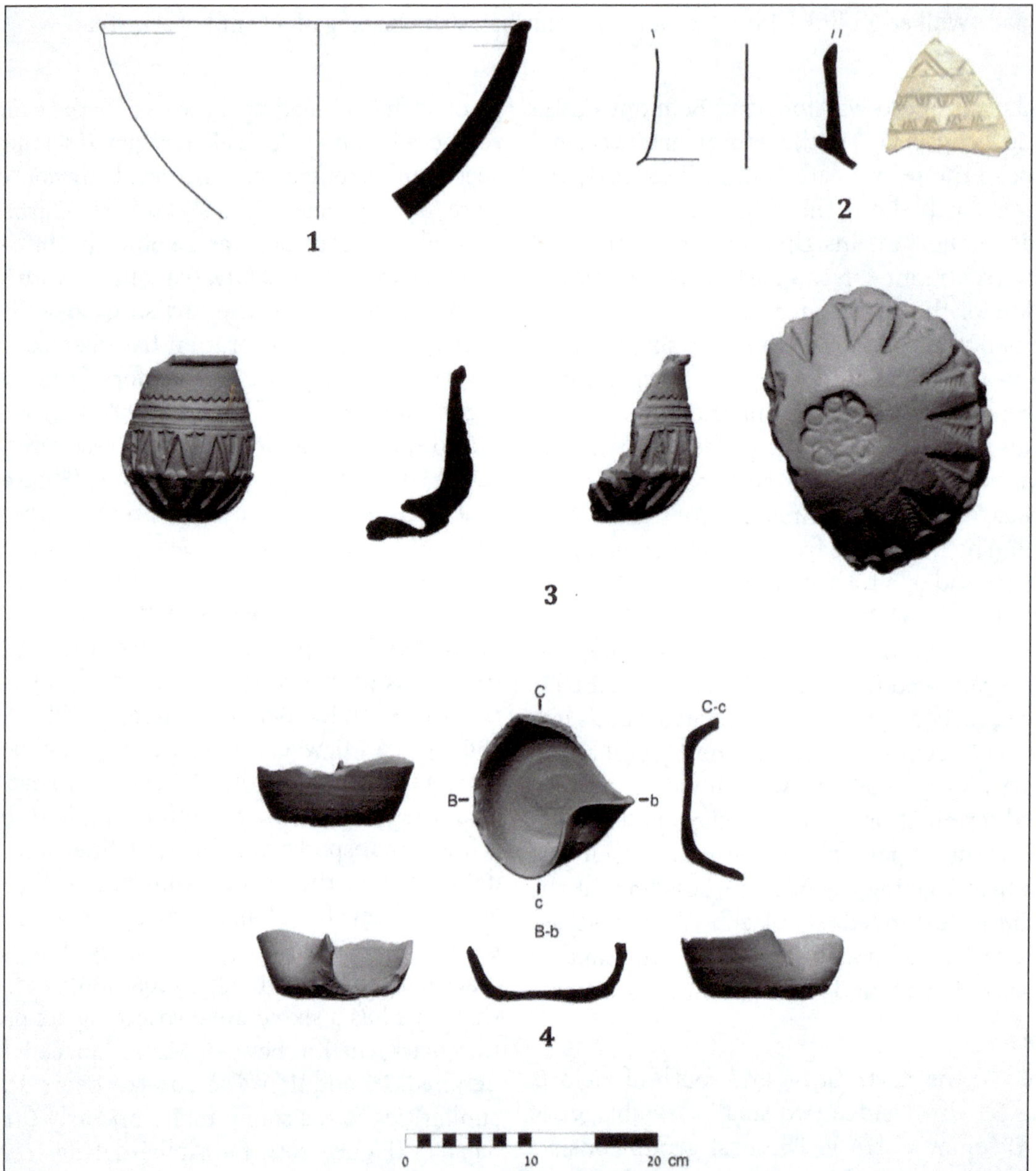

No.	Locus	Reg. No.	Form	Description
1	152	1072/1	BL	Deep handmade hemispherical bowl with an out-turned rim. Coarse light reddish-brown (5YR 6/4) fabric with calcite inclusions.
2	152	1072/2	SJ	Wheelmade Islamic Cream Ware strainer-jar neck with incised geometric decoration. Coarse gray (5Y 6/1) fabric with pinkish white (7.5YR 8/2) slip.
3	147	1066	TP	Gray tobacco pipe bowl with rouletted decoration, and production mark of a rosette on its base. Well-levigated gray (2.5Y 5/1) fabric.
4	147	1066/2	LP	Wheelmade saucer lamp with pinched nozzle. Light reddish brown (5YR 6/4) fabric with calcite inclusions.

Figure 26 Early Ottoman pottery from B5.

but the four examples presented here are very typical of the Early Ottoman period in the Jerusalem region. The first three vessels illustrated herein were recovered from L.160, which was located between the two walls. Included among them was a wheelmade Jerusalem Carinated Bowl (Figure 27:1) with an out-turned rim and a horizontal protruding ridge on the exterior at the carination point. Both glazed and unglazed versions of the form were produced locally in Jerusalem during the 16th century and later, and they are a hallmark of the Early Ottoman ceramic repertoire. Regarding this type, it was noted that "the form appears to imitate Italian imports such as the Italian Glazed Carinated Bowls dated to the 13th–15th centuries" (Avissar 2012, 305). Good parallels for this particular variation of this vessel type are known from stratified contexts in the Old City of Jerusalem (Dolinka 2017, fig. 2:4; Prag 2008, fig. 175:13).

Another vessel found within L.160 was a deep handmade bowl or basin (Figure 27:2) with an upright collared rim and crude finger indentations along the bottom of the exterior rim zone. It is much thicker and coarser than its Mamluk I-II

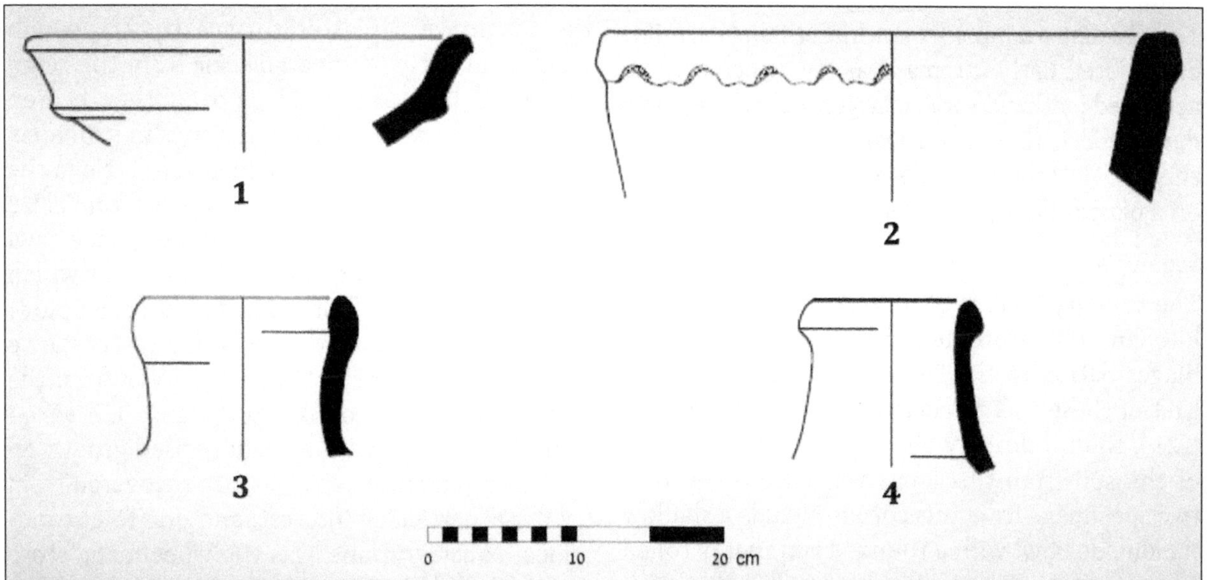

No.	Locus	Reg. No.	Form	Description
1	160	1090/1	BL	Wheelmade Jerusalem Carinated Bowl (Local Imitation of NIT) with flared rim and a horizontal protruding ridge on the vessel exterior at the carination point. Fairly well-levigated yellowish-red (5YR 5/6) fabric, a dark gray (5YR 4/1) core and small calcite inclusions.
2	160	1090/5	BL	Handmade bowl with an upright collared rim and crude finger indentations along the bottom of the exterior rim zone. Very thick and coarse light reddish-brown (5YR 6/3) fabric with a very dark gray (5YR 3/1) core and medium-to-large calcite inclusions
3	160	1088/2	SJ	Wheelmade store jar with incurved thickened rim that is oval in section. Coarse reddish-brown (2.5YR 6/4) fabric fired to a very pale brown (10YR 8/3), with small calcite inclusions.
4	159	1079/2	SJ	Wheelmade store jar with collared, everted rim that is triangular in section. Fairly well-levigated yellowish-red (5YR 5/6) fabric, a gray (5YR 6/1) core, with mostly small (but some medium) calcite inclusions.

Figure 27 Early Ottoman pottery from the Stratified Sequence along W.124/1–2 in B5.

forbears, and no parallel could be found. However, the context clearly dates to the Early Ottoman period. The final specimen from L.160 was a wheelmade jug with an incurved thickened rim that is oval in section (Figure 27:3) that is comparable with jug from an Early Ottoman context at Habonim-Kafr Lam (Avissar 2009b, 10, fig. 2.6:4). The final piece from this assemblage was uncovered in L.159, a stratified soil layer located between Walls 124/1 and 124/2. It is a wheelmade store jar with collared, everted rim that is triangular in section (Figure 27:4). These first appear during the Mamluk II period, where the vessels walls are much thinner and have a finer, more well-levigated clay. The present example from Khirbet Beit Mazmīl is of the later, Early Ottoman variety, which is characterized by thicker walls, larger inclusions, and a darker fabric than its forbears. A good parallel for this jar was found at the Austrian Hospice, laying on a plaster floor L.129 (Dolinka 2017, fig. 1:13).

Square C6

The primary feature of interest from C6 was the kitchen, which produced three ceramic assemblages dating to the Early Ottoman period. The first of these was associated with the removal of L.219, the fill directly above the kitchen, and all of the soil from this locus was sifted. The first two specimens from this corpus include a shallow handmade bowl with a rounded rim that is rolled and circular in section (Figure 28:1), and a small handmade bowl with a beveled rim and a vertical, vestigial lug handle (Figure 28:2). No parallels were discerned for either of these vessels. The third bowl from this context (Figure 28:3) was the type of handmade hemispherical bowl with a direct stance and rounded rim that is ubiquitous at the site and attested at the Austrian Hospice excavations (Dolinka 2017, fig. 2:6).

The next three vessels from L.219 consist of a type of cooking pot that essentially represents the standard form present during the Early Ottoman period at Khirbet Beit Mazmīl (Figure 28:4–6). All of them are handmade, high-necked, have an upright rounded rim, and have the same fabric. While this type first appears in the Mamluk II era, it continues essentially in the same form throughout the Early Ottoman period and is paralleled from the Damascus Gate excavations conducted by Bennett (Wightman 1989, 248, fig. 54:1). A variant of the form, with a slightly shorter neck that is offset from the shoulder and has two vertical loop handles (Figure 28:7) was also found in L.219. A similar cooking pot was found at Belmont Castle (Grey 2000, fig. 6.6:115).

The second ceramic assemblage from the kitchen was uncovered below the beaten earth and flagstone floor (L.220) associated with the cooking installation. The findspots for the pottery include the fill above a plaster surface (L.221), which yielded an Early Ottoman *mangīr* from the reign of Murad III (1574–1595), and another beaten earth surface with broken flagstones in its matrix (L.222). Both of these loci yielded very little in the way of diagnostic ceramics. The vessels from L.222 included a small, shallow incurved handmade bowl with an upturned flat rim (Figure 29:1), for which no parallel could be found, and another example of the ubiquitous handmade bowl with a direct stance and flat rim (Figure 29:2) that is known from the Austrian Hospice (Dolinka 2017, fig. 2:6). A variation of that form, with a slight incised groove on the rim top (Figure 29:3) was also recovered from L.222. Of particular interest, and due to chronological considerations, was the wheelmade store jar with beveled and internally thickened rim that is triangular in section (Figure 29:4) which was found in L.221. In the interim report for the project (Walker and Dolinka 2020, 208, fig. 4:4, ftn. 54–55), parallels that were cited for this form came from either Late Ottoman contexts (Boas 2000, 212, pl. I.1–3; de Vincenz 2021, fig. 8.5A:3 no. 3, in Black Gaza Ware) or were labelled simply as "Ottoman" (Grey 2000, fig. 6.2:33). Since this distinct vessel type at Khirbet Beit Mazmīl was found in association with an Early Ottoman *mangīr* in L.221, it makes sense that this form appeared much earlier than has previously been reported. The final piece from this corpus is another typical wheelmade

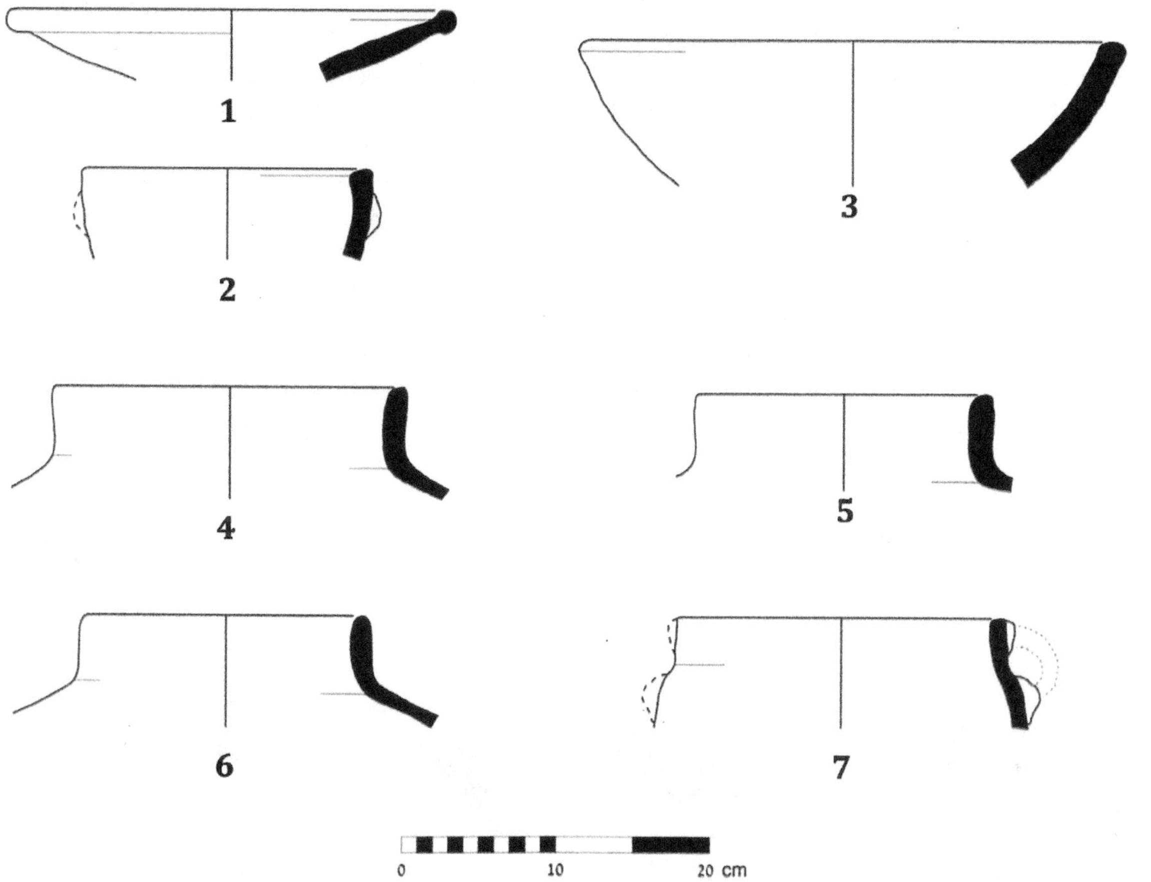

No.	Locus	Reg. No.	Form	Description
1	219	2022/4	BL	Shallow handmade bowl with a rounded rim that is rolled and circular in section. Coarse light reddish-brown (5YR 6/4) fabric with a dark reddish gray (2.5YR 4/1) core and large inclusions of chert and calcite.
2	219	2022/7	BL	Small handmade bowl with a beveled rim and vertical, vestigial lug handle. Coarse light reddish brown (5YR 6/3) fabric with small calcite and medium chert inclusions.
3	219	2022/5	BL	Handmade hemispherical bowl with a direct stance and rounded rim. Coarse light reddish-brown (5YR 6/4) fabric with small calcite and medium chert inclusions.
4 5 6	219	2022/1	CP	High-necked handmade cooking pot with an upright rounded rim. Coarse light reddish brown (5YR 6/3) fabric with burned out organic temper, crushed quartzite, calcite and chert inclusions.
7	219	2022/8	CP	High-necked handmade cooking pot with a flat rim, a neck that is slightly offset from the shoulder and two vertical loop handles.

Figure 28 Early Ottoman pottery from fill directly above the kitchen in C6.

saucer lamp with pinched nozzle (Figure 29:5), for which a good parallel is known from Cardo Area X-3 (Avissar 2012, pl. 10.4:5).

The last group of pottery retrieved from the kitchen was found within the contents of the L.225 plastered installation, which also produced an Early Ottoman *mangīr*. The first four vessels from this corpus are all handmade in the coarse local fabric. This locus yielded the following: a deep hemispherical bowl with an out-turned rim (Figure 30:1) that is the same type as illustrated in Figure 26:1 above, for which no parallel could be found; a bowl with a direct stance and rounded rim (Figure 30:2), which has correlates with an example from Habonim-Kafr Lam (Avissar 2009b, fig. 2.5:10); a handmade basin with folded-over rounded rim (Figure 30:3) and horizontal band of finger-indented appliqué just below the rim zone that is very worn, which is known from the Damascus Gate excavations (Wightman 1989, fig. 50:9); and another high-necked handmade cooking pot with an upright, flattened rim (Figure 30:4) like the one illustrated in Figure 30:5 above, which is also paralleled at the Damascus Gate (Wightman 1989. 248, fig. 54:1).

The final item from L.225 presented here is the complete profile of a gray tobacco pipe with

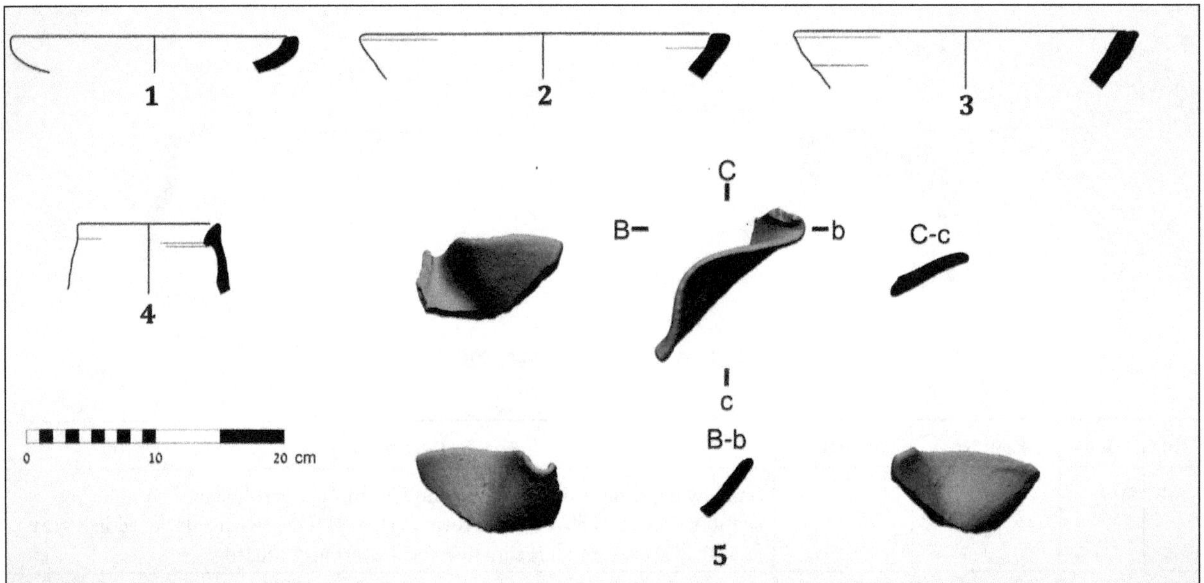

No.	Locus	Reg. No.	Form	Description
1	222	2025/1	BL	Small, shallow incurved handmade bowl with flat rim. Coarse very pale brown (10YR 7/3) fabric with small calcite inclusions.
2	222	2025/2	BL	Handmade bowl with a direct stance and flat rim. Coarse very pale brown (10YR 7/3) fabric with small calcite inclusions.
3	221	2024	BL	Handmade bowl with a direct stance and flat rim. Coarse very pale brown (10YR 7/3) fabric with small calcite inclusions.
4	221	2023	SJ	Wheelmade store jar with beveled and internally-thickened rim that is triangular in section. Pink (5YR 7/4) well-levigated fabric, gray (5YR 6/1) core.
5	222	2025/3	LP	Wheelmade saucer lamp with pinched nozzle. Well-levigated yellowish red (5YR 5/6) fabric with tiny calcite inclusions.

Figure 29 Early Ottoman pottery from below the L.220 Floor in the C6 Kitchen.

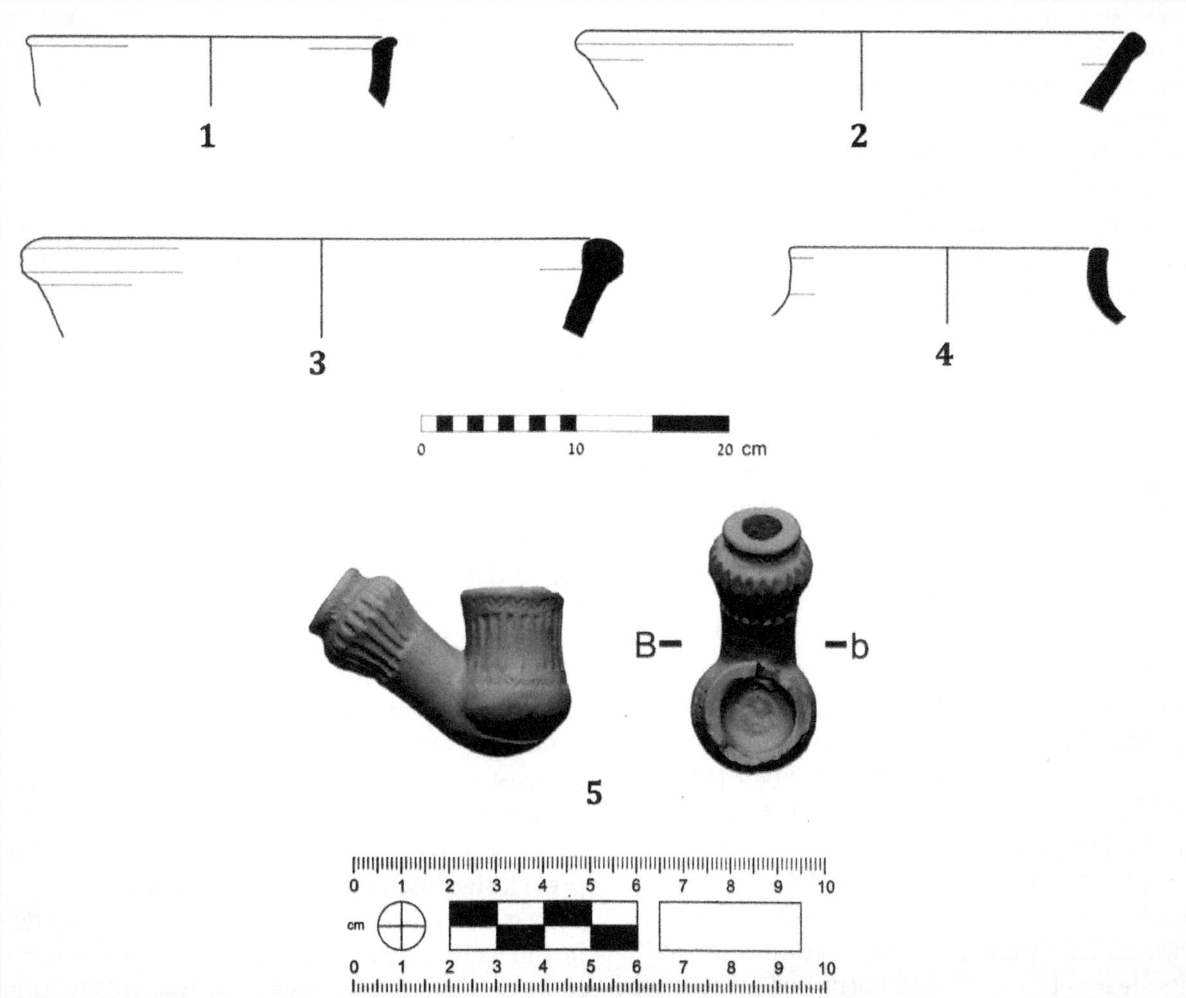

No.	Locus	Reg. No.	Form	Description
1	225	2028/2	BL	Deep handmade hemispherical bowl with an out-turned rim. Coarse reddish yellow (5YR 6/6) fabric with a gray (10YR 6/1) core, calcite and chert inclusions.
2	225	2042/2	BL	Handmade bowl with a direct stance and rounded rim. Coarse light reddish brown (5YR 6/4) fabric with a reddish gray (10R 6/1) core, calcite and chert inclusions.
3	225	2042/3	BS	Handmade basin with folded-over rounded rim and horizontal band of finger-indented appliqué just below the rim zone (very worn). Coarse pink (5YR /4) fabric with a light gray (5YR 7/1) core, calcite and chert inclusions.
4	225	2042/4	CP	High-necked handmade cooking pot with an upright, flattened rim. Coarse light reddish brown (5YR 6/3) fabric with burned out organic temper, crushed quartzite, calcite and chert inclusions.
5	225	2028	TP	Gray tobacco pipe bowl with vertical rouletting on both the bowl and the shank end; complete profile. Well-levigated light bluish gray (10B 7/1) fabric.

Figure 30 Early Ottoman pottery from the contents of the plastered installation in the C6 Kitchen.

vertical rouletting on both the bowl and shank end (Figure 30:5). In the typo-chronology for the Banias pipe assemblage, the bowl shape for this specimen appears in the earliest group that has a pronounced and distinct keel (Dekkel 2008, 4.2a:7a). This does not necessarily make the Khirbet Beit Mazmīl pipe later in date, however, because as noted in that study, ". . . the shank ends are of greater importance than the bowls for the purpose of dating and typological divisions into subtypes . . . [and] different bowls appear with identical shanks" (Dekkel 2008, 125). The rouletted shank end itself is typical of the assemblage dating from the 17th century (Dekkel 2008, 4.2b:2) and an exact parallel for this pipe from Banias is included in the catalogue (Dekkel 2008, 134, fig. 4:19).

Squares E1 and E2

Field E was located in the central portion of the walled enclosure in the upper/northern sector of the site. It yielded stratified sequences from two squares (E1 and E2) that provide the bookends for the chronology of the Early Ottoman period at Khirbet Beit Mazmīl. E1 functioned as a storeroom in this period, as it did in the late Mamluk era (see Chapter 4, this volume). The uppermost soil layers from E1 represent the last phase of that occupation. The lynchpin for this dating rests on the discovery of a red-slipped burnished tobacco pipe (Figure 31:1) that was recovered from plaster floor L.405. According to A. de Vincenz (*personal communication*), this pipe dates to the early 18th century and conforms to her Type J-18J from the Jaffa *Qishle* (de Vincenz 2021, 186). It has a plain rounded bowl that is decorated with horizontal rouletted bands on the upper half of the bowl, a plain undecorated shank, and a collared everted shank end typical of the 18th century pipes (de Vincenz 2021, fig. 8.6B: 8–14). Unlike the closest parallel from Jaffa (de Vincenz 2021, fig. 8.6B: 35), which has a pinkish ware and is unslipped but glazed (de Vincenz 2021, 187), the specimen from Khirbet Beit Mazmīl had a dark reddish-brown ware, is slipped and burnished, and has a dark purplish-brown exterior, reminiscent of the contemporaneous Akko pipe production (see Shapiro 2010, 78, fig. 107).

The majority of the corpus from the upper stratified sequence in the E1 storeroom was found on plaster floor L.406, which was directly below the L.405 floor. As this assemblage is dated to the 18th century, and very little has been published from this period, it is of little surprise that parallels could not be found for some of the vessels from this locus, which included: a handmade carinated bowl with an upright rounded rim (Figure 31:2), which has a parallel from Habonim-Kafr Lam (Avissar 2009b, 10, fig. 2.5:6); a handmade flat-bottomed bowl with rounded slightly out-turned rim (Figure 31:3), which has the same form as an example from the Damascus Gate (Wightman 1989, fig. 50:1); a larger and deeper handmade bowl with a T-shaped rim (Figure 31:4); and a wheelmade jug with a collared rounded rim (Figure 31:5). The final vessel from this corpus is a wheelmade saucer lamp with a pinched nozzle (Figure 31:6) that has a much darker and redder fabric than all the other aforementioned exemplars; it was retrieved from L.410, the stratified fill layer in the northeast corner of the storeroom to the east of W.134. Unlike the previously illustrated examples of lamps presented here, it is clearly a later version of the form, and differs in its rounded sides and disc base. It conforms to Gichon and Linden (1984, 159, fig. 1:E, Pl. 21:9) Type 2a, which was assigned a rather broad date range from the 18th to 20th centuries (Gichon and Linden 1984, 167). However, the typology from that study entirely lacks any forms dating from either 16th or 17th centuries, and given its context, this lamp more likely is contemporaneous with the red-slipped burnished pipe illustrated in Figure 31:1.

The lower stratified sequence from the storeroom consists of Early Ottoman pottery from contexts that were sealed below Floor L.413 = L.414. The first of two loci that were excavated include L.418, a hard-packed surface with chunks of plaster in its matrix. This was an artifact-rich context, that included an Early Ottoman *mangīr*; horseshoes and other equine accessories; and an iron knife.

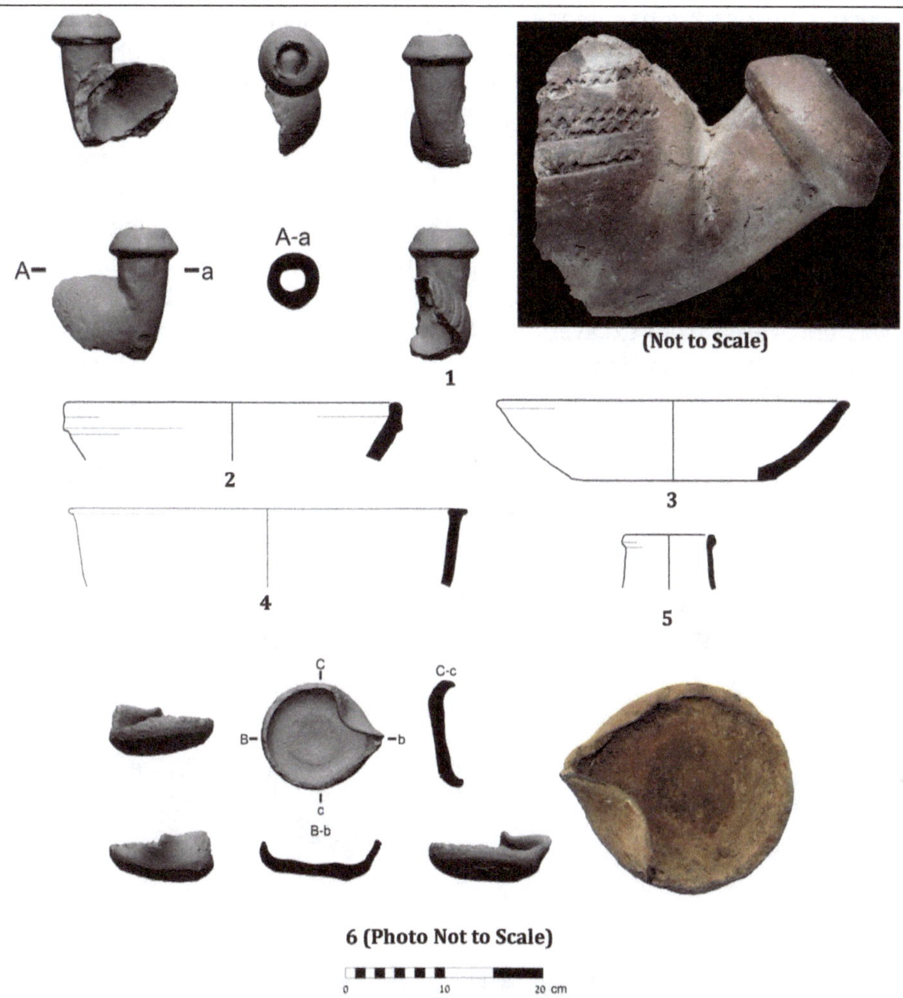

No.	Locus	Reg. No.	Form	Description
1	405	4005	TP	Bowl from red-slipped burnished tobacco pipe decorated with horizontal rouletted bands on the upper half of the bowl. Dark purplish-brown exterior.
2	406	4015/3	BL	Handmade carinated bowl with an upright rounded rim. Coarse light reddish brown (5YR 6/4) fabric with a gray (5YR 6/1) core, calcite and chert inclusions.
3	406	4015/2	BL	Handmade flat-bottomed bowl with rounded slightly out-turned rim. Coarse pink (5YR /4) fabric with a light gray (5YR 7/1) core, calcite and chert inclusions.
4	406	4015/1	BL	Handmade deep carinated bowl with a t-shaped rim. Pink (5YR 7/3) fabric with a gray (5YR 5/1) core.
5	406	4015/4	JG	Wheelmade jug with a collared rounded rim. Pink (5YR 7/4) well-levigated fabric, gray (5YR 6/1) core.
6	410	411	LP	Wheelmade saucer lamp with pinched nozzle. Red (10R 5/6) fabric with calcite inclusions, reddish-gray (10R 6/1) exterior.

Figure 31 Early Ottoman pottery from the upper stratified sequence in the E1 storeroom.

The ceramic corpus included: a handmade bowl with direct, slightly-rounded rim (Figure 32:1); a HMGP bowl with direct, slightly-rounded rim (Figure 32:2); a handmade bowl with upright flattened rim (Figure 32:3); another HMGP carinated bowl with rounded and folded-over rim (Figure 32:4); and a wheelmade *ibrīq* neck with everted rim that is triangular in section (Figure 32:5). A similar jug, albeit with a slightly different rim form but in the same fabric, was uncovered from excavations at the Diaspora Yeshiva on Mount Zion in the Old City of Jerusalem conducted by Tanami (Dolinka 2014a, L.314, B.040/4). The presence of the *ibrīq*, and two examples of Early Ottoman HMGP bowls, combined with the *mangīr* from this context, makes it possible to utilize these forms for future comparanda and as parallels for other Early Ottoman assemblages that will be studied in the future.

The upper stratified sequence from the midden in E2 yielded a corpus of Early Ottoman pottery that represented the typical handmade vessel forms found at the site during this period. These were recovered from L.408, the first stratified fill below the L.404 subsoil, and included the following: a hemispherical bowl (Figure 33:1) with a slightly rounded rim like the example found at the Austrian Hospice (Dolinka 2017, fig. 2:6); a bowl with a slightly upturned flattened rim (Figure 33:2) similar in form

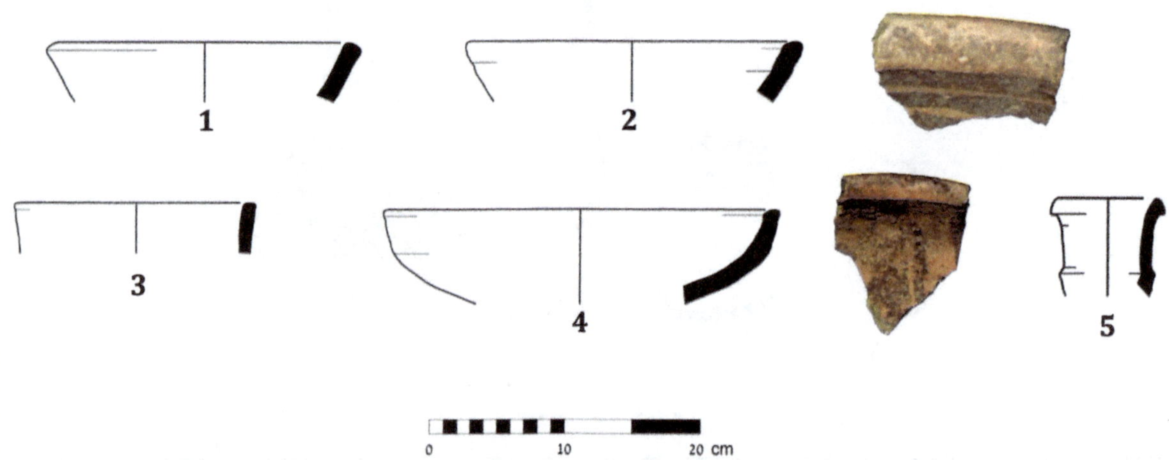

No.	Locus	Reg. No.	Form	Description
1	418	4023/1	BL	Handmade bowl with direct, slightly-rounded rim. Coarse light reddish brown (5YR 6/3) fabric with calcite and chert inclusions.
2	418	4023/3	BL	Handmade Geometric Painted bowl with direct, slightly-rounded rim. Coarse reddish yellow (7.5YR 7/6) fabric with calcite inclusions and a dark gray (5YR 4/1) painted decoration on the interior.
3	418	4023/4	BL	Handmade bowl with upright flattened rim. Coarse reddish yellow (7.5YR 7/6) fabric with medium and large calcite inclusions.
4	418	4023/2	BL	Handmade Geometric Painted carinated bowl with rounded and folded-over rim. Coarse light reddish brown (5YR 6/4) fabric with calcite and chert inclusions and a dark gray (5YR 4/1) painted decoration on the interior.
5	418	4023/5	JG	Wheelmade *ibrīq* neck with everted rim that is triangular in section. Red (2.5YR 4/6) fabric with calcite inclusions and a gray (5YR 6/1) self-slip.

Figure 32 Early Ottoman pottery from the lower stratified sequence in the E1 Storeroom.

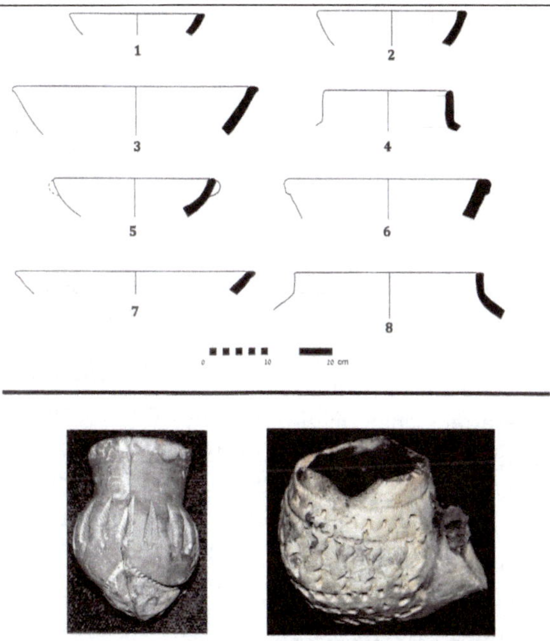

9 NB: Not to Scale 10

No.	Locus	Reg. No.	Form	Description
1	408	4009/3	BL	Handmade hemispherical bowl with a slightly rounded rim. Coarse reddish yellow (7.5YR 6/6) fabric with a gray (5YR 5/1) core, calcite and chert inclusions.
2	408	4009/4	BL	Handmade bowl with slightly upturned flattened rim. Coarse light reddish brown (5YR 6/4) fabric with a reddish gray (10R 6/1) core, calcite inclusions.
3	408	4009/2	BS	Handmade basin with direct and slightly folded-over rim. Coarse light reddish-brown (5YR 6/4) fabric with a dark reddish gray (2.5YR 4/1) core and large inclusions of chert and calcite.
4	408	4009/1	CP	High-necked handmade cooking pot with an upright rounded rim. Coarse light reddish brown (5YR 6/3) fabric with burned out organic temper, crushed quartzite, calcite and chert inclusions.
5	416	4020/2	BL	Handmade hemispherical bowl with a slightly rounded rim and a vestigial horizontal ledge handle. Coarse reddish yellow (7.5YR 6/6) fabric with a gray (5YR 5/1) core, calcite and chert inclusions.
6	416	4020/3	BL	Handmade bowl with rounded folded-over rim and horizontal band of finger-indented appliqué just below the rim zone. Coarse reddish-yellow (5YR 6/6) fabric with a dark gray (5YR 5/1) core and inclusions of chert and calcite.
7	416	4020/5	BS	Handmade basin with direct and slightly folded-over rim. Coarse reddish-yellow (5YR 6/6) fabric with a gray (5YR 6/1) core and inclusions of chert and calcite.
8	416	4020/4	CP	Handmade low-necked cooking pot with upright rounded rim that is slightly beveled on the interior. Coarse light reddish brown (5YR 6/3) fabric with burned out organic temper, crushed quartzite, calcite and chert inclusions.
9	439	4048	TP	Gray tobacco pipe with pronounced keel, decorated with deeply-stamped arrowheads in one register and direction; exterior blackened (2.5Y 2.5/1) from carbonization. Well-levigated fabric.
10	439	4048	TP	Gray tobacco pipe with small rounded bowl, rouletted/stamped decoration in low-relief. Well-levigated gray (2.5Y 6/1) fabric.

Figure 33 Early Ottoman pottery from the upper (Nos. 1–8) and lower (9–10) stratified sequence in E2.

to one recovered from the Ottoman defensive wall in the Old City of Jerusalem (Prag 2008, fig. 174:21); a basin (Figure 33:3) with direct and slightly folded-over rim, for which no parallel could be found; and the standard high-necked handmade cooking pot with an upright rounded rim (Figure 33:4), found throughout the site and known from the Damascus Gate excavations (Wightman 1989, 248, fig. 54:1).

The northwest sector of E2 produced a similar homogenous assemblage of ceramics that originated from the stratified layer in L.416, which was also below subsoil (L.411). It consisted of the following small corpus of handmade vessels: a hemispherical bowl (Figure 33:5) with slightly rounded rim and a vestigial horizontal ledge handle, paralleled at the Austrian Hospice (Dolinka 2017, fig. 2:6); a bowl with rounded folded-over rim (Figure 33:6) and horizontal band of finger-indented appliqué just below the rim zone (Dolinka 2017, fig. 2:8); a duplicate of the basin (Figure 33:7) recovered from L.408 and mentioned above (Figure 33:3); and a cooking pot (Figure 33:8) with a slightly lower neck than the typical form at the site, which has an upright rounded rim that is slightly beveled on the interior, and is known from the Ottoman defensive wall in the Old City (Prag 2008, fig. 173:8).

The lowest stratified context (L.439) in E2 yielded important finds with regard to the beginning of the Early Ottoman occupation at the site; these consist of two gray tobacco pipes that were found just above the bedrock. The first example (Figure 33:9) is a black-burnished pipe that has a pronounced keel and is decorated with deeply stamped arrowheads in only one register and one direction. The shape of the bowl has correlates from the Banias typo-chronology (Dekkel 2008, 134, fig. 4.5:15), where it is dated to the second half of the 17th century, although the decorative schema is different. A similar pipe was found near Ramla (de Vincenz 2020a), where it was dated from the 17th to early 18th centuries (de Vincenz 2020a, 388), but that example is much more elaborately decorated than the Khirbet Beit Mazmīl specimen, which has a rounder bowl and a larger keel. The second pipe (Figure 33:10) is a small, rounded bowl with rouletted and stamped decoration that is rendered in low relief. The general form is known from Banias (Dekkel 2008, figs. 4.2a:2, 4.4:3), where it is dated to the "seventeenth century or earlier" (Dekkel 2008, 131). A very close parallel was uncovered from the Old City of Jerusalem (Simpson 2008, 436, fig. 2:15).

Evidence from the Southern Sector (2013)

A small but significant corpus of Early Ottoman pottery was uncovered during the brief 2013 season of excavations at Khirbet Beit Mazmīl (License A-6719), conducted by Dolinka and Storchan for the IAA. The bulk of the evidence was uncovered in Square A12, where both the topsoil and subsoil had been previously removed by mechanical equipment during IAA-supervised inspection at the site. The first stratified fill layer to be revealed was L.200; below that was a plaster floor (L.201) that had Early Ottoman sherds above it.

The ceramic corpus from A12 consists of bowls, storage jars, and a tobacco pipe, all dating to the Early Ottoman period. The group of three HMGP bowls and basin (Figure 34:1–4) are sparsely painted with simple geometric patterns or horizontal linear bands that are rendered in a red, light red, or black paint on either the interior or exterior of the vessel, and occasionally the top of the rim top. A different group of these bowls was published in the project interim report (Walker and Dolinka 2020, 205, fig. 3:7–10), and a similar assemblage was uncovered from L.202 in Area B at the Cardo excavations in the Old City of Jerusalem (Avissar 2010, 334, pl. 18.1: 10–14), with individual findspots coming from Cardo Area X-5 (Avissar 2012, pl. 10.12:5) and Belmont Castle (Grey 2000, fig. 6.3:74). A bowl with the exact same rim form was found in the excavations at Misgav Ladach 26 Street in the Jewish Quarter of the Old City (Dolinka 2014b, L.116, B.1093/9), although this specimen had a smaller rim diameter. Rounding out the assemblage are the following: an HMGP store jar with upright rounded rim (Figure 34:5), which has close parallels from Belmont Castle in

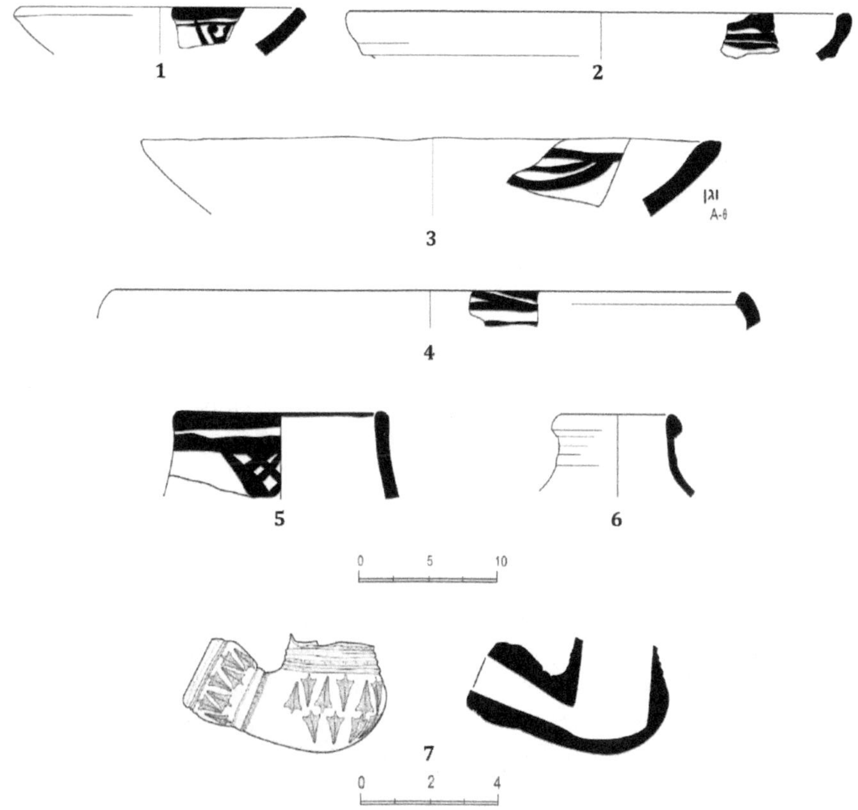

No.	Locus	Reg. No.	Form	Description
1	201	2002/3	BL	Handmade Geometric Painted bowl with direct, flat rim. Reddish brown (5YR 5/4) coarse fabric with large calcite inclusions and burned-out organic temper, very pale brown (10YR 7/3) interior slip, reddish black (2.5YR 2.5/1) painted geometric design on interior and rim.
2	201	2002/4	BL	Handmade Geometric Painted bowl with incurved rounded rim. Reddish brown (5YR 5/4) fabric, pale red (10R 6/4) interior slip, weak red (2.5YR 5/3) painted geometric design on interior.
3	201	2002/2	BL	Handmade Geometric Painted bowl with direct rim that is internally thickened. Reddish brown (5YR 5/3) fabric, pink (7.5YR 7/4) interior slip, weak red (2.5YR 5/3) painted geometric decoration on interior and rim top.
4	201	2001/3	BS	Handmade Geometric Painted basin with incurved and beveled rim. Light reddish brown (5YR 6/4) coarse fabric, pink (7.5YR 7/4) interior slip, dark grayish brown (10YR 4/2) painted geometric design on interior and rim top.
5	200	2000/2	SJ	Handmade Geometric Painted store jar with upright rounded rim. Light reddish brown (5YR 6/4) fabric, pink (7.5YR 7/4) interior/exterior slip, weak red (2.5YR 5/4) painted geometric design on exterior and rim top.
6	200	2000/1	SJ	Wheelmade store jar with collared rounded rim that is oval in section. Light reddish brown (5YR 6/4) fabric, very pale brown (10YR 7/3) interior and exterior slip.
7	200	2000/3	TP	Tobacco pipe decorated with rouletting and alternating rows of deeply stamped arrowheads on the bowl and shank, decorated with a six-petalled rosette production mark on the bottom of the bowl. Well-levigated gray (2.5Y 5/1) fabric.

Figure 34 Early Ottoman pottery from Square A12 in the Southern Sector.

the Jerusalem hinterland (Grey 2000, 6.5:99); and a wheelmade store jar with collared rounded rim that is oval in section (Figure 34:6), which is similar to an example from an Early Ottoman context from the Austrian Hospice (Dolinka 2017, fig. 1:11).

Perhaps the most interesting ceramic find from this corpus is the tobacco pipe (Figure 34:7), which was found in the fill above the plaster floor from Square A12. It is decorated with rouletting, double incised lines, and alternating rows of deeply stamped arrowheads on the bowl and shank. The bottom of the bowl bears a hexagonal star production mark (Dekkel 2008, fig. 4.3:1), and an almost exact parallel comes from the Banias assemblage (Dekkel 2008, 136, fig. 4.6:24), which has a six-petalled rosette on its base and is dated to the end of the 17th and beginning of the 18th century. Similar examples have also been found at Akko (Shapiro 2010, fig. 105, left), Banias (Simpson 2000, fig. 13.2:24–25), as well as from Jerusalem (Simpson 2008, fig. 268:7).

Evidence from the Lower Site (2017)

Two brief back-to-back field seasons at Khirbet Beit Mazmīl were conducted by ʿAdawī in 2017, with Stratum II providing evidence for the Early Ottoman period (ʿAdawī and Dolinka 2021). As was the case with the 2013 excavations conducted by the IAA at the site, the topsoil and subsoil had been previously removed by mechanical equipment during IAA-supervised inspection before the fieldwork commenced. The small pottery corpus presented here originated from Room 205 in the Western Building, which was located in the central sector of the area excavated and covered much of Squares S2 and S6. The room had a well-preserved flagstone floor (L.205), above which was a stratified soil layer (L.204) and below which (L.206) a thick copper *mangīr* from the reign of Murad III (1574–1595) was recovered.

The ceramic assemblage from Room 205 consists of the following vessels: the ubiquitous handmade bowl with direct, slightly-rounded rim (Figure 35:1) that is found throughout the site and has correlates from the Austrian Hospice (Dolinka 2017, fig. 2:6); a handmade bowl with upturned beveled rim and a thin pressed-on handle (Figure 35:2) of the same type that was found in L.127b from the Domestic Space in B4 (see Figure 24:5 and related discussion above), but has a slightly beveled rim instead of a flat one; a wheelmade store jar with an inverted rounded rim (Figure 35:3), which has the same form as an example from the Coptic Patriarchate at the Church of the Holy Sepulcher (Avni and Seligman 2003; Avissar 2009b, fig. 7:8); and a wheelmade store jar with a flat, thickened ledge rim that is offset from the rest of the neck (Figure 35:4). A good parallel for this vessel was recovered from an Early Ottoman context at the Diaspora Yeshiva excavated by Tanami in the Jewish Quarter of the Old City (Dolinka 2014a, fig. 4:7).

An interesting find of chronological importance is represented by the shank end of a gray tobacco pipe (Figure 35:5) that was uncovered in L.206—below the flagstone floor in Room 205—which also yielded the Early Ottoman *mangīr*. This shank end corresponds to an example in the Banias study (Dekkel 2008, fig. 4.2:3), where it was noted that:

> "In the earliest types and in the types that are clearly from the seventeenth century, these shank ends are typified by a relatively wide asymmetrical wreath usually decorated [by] stamping and a stepped ring protruding a considerable extent beyond the wreath." (Dekkel 2008, 126)

Thus, the combination of the *mangīr* and the tobacco pipe shank end confirm a date range for Room 205 between the late 16th and early 17th century.

Discussion and concluding remarks

Chronologically, the Early Ottoman pottery corpus from Khirbet Beit Mazmīl can be divided into two phases: the Early Phase, which dates from the late 16th to the mid-17th centuries; and the Late Phase, which covers the period from the mid-17th to the early 18th centuries. The Early Phase is represented by the following: the stratified series

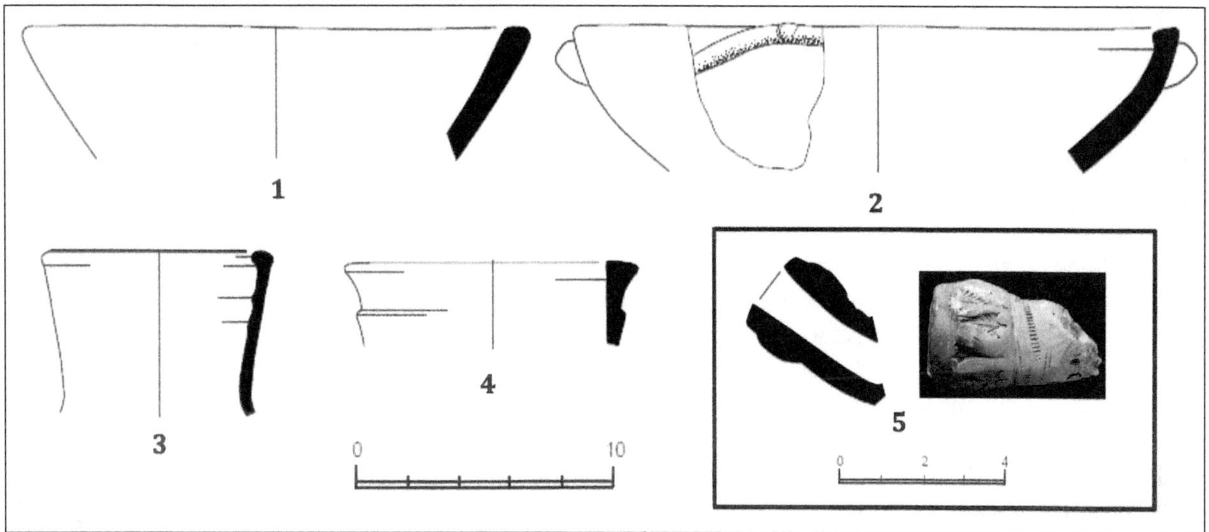

No.	Locus	Reg. No.	Form	Description
1	205	2011	BL	Handmade bowl with direct, slightly-rounded rim. Coarse very pale brown (10YR 7/3) fabric with small calcite inclusions.
2	204	2009/1	BL	Handmade bowl with upturned beveled rim and a vestigial horizontal handle. Coarse light reddish-brown (5YR 6/4) fabric with chert and calcite inclusions.
3	204	2009/2	SJ	Wheelmade store jar with an inverted rounded rim. Yellowish red (5YR 4/6) fabric with small and medium-sized calcite inclusions.
4	206	2013/3	SJ	Wheelmade store jar with a flat, thickened ledge rim that is offset from the rest of the neck. Fairly well-levigated yellowish-red (5YR 5/6) fabric, a gray (5YR 6/1) core, with mostly small (but some medium) calcite inclusions.
5	206	2013	TP	Shank end of a gray tobacco pipe with stamped decoration and rouletting. Light grayish-green (5GY 6/2) fabric.

Figure 35 Early Ottoman pottery from Room 205 in the Southern Sector.

of L.122, L.124 and L.127b (Figure 24), from the domestic space of B4, with its associated *ṭābūn*, both plaster and flagstone floors, and the presence of an Early Ottoman *mangīr*; the other loci from B4 (Figure 25), which included L.119, where the hoard of 10 *mangīrs* and a very early tobacco pipe were discovered; the sequence from the stratified fill layers (L.159–160) between the Walls 124/1 and 124/2 in B5 (Figure 27), which produced a Jerusalem Carinated bowl; the sifted fill (L.219) directly above the kitchen in C6 (Figure 28); L.221 and L.222, which were sealed below the L.220 beaten earth and flagstone floor (Figure 29), with the former producing a *mangīr*; the lower stratified fill (L.418) from the E1 storeroom (Figure 32) that was sealed under the L.413 = 414 floor and yielded a *mangīr* and some good examples of the HMGP from the Early Phase; both the upper (L.408, L.416) and lower (L.439) stratified sequences from the storeroom in E1 (Figure 33), the former of which consisted of an assemblage of bowls and cooking pots typical of the late 16th/early 17th centuries, and the latter of which produced two early types of tobacco pipes that were found directly above bedrock; and Room 205 from the Lower Site (Figure 35), where both an early tobacco pipe and a *mangīr* were found in a fill layer (L.106) sealed below the flagstone floor (L.205).

The Later Phase for the Early Ottoman ceramics from Khirbet Beit Mazmīl is a smaller corpus, with the evidence presented here consisting of four assemblages, all of which are dated from the second half of the 17th century to the beginning of the 18th century. The upper levels from B5 (Figure 26) assigned to this phase included the Late Islamic Cream Ware strainer jar from the fill directly on Wall 124/1 (L.152), and the gray tobacco pipe from the loose dark brown fill associated with the cistern (L.147). As the only clearly datable diagnostic sherds of the two contexts from which they originated, they both point to a date sometime in the mid-17th century, and there was no other evidence for anything that could be construed as being from the Early Phase found within these loci. Another assemblage attributable to the Later Phase originated from the contents of the plastered installation (L.225) in the C6 kitchen (Figure 30), which represents the latest era of its use for such a purpose. The primary dating evidence comes from a gray tobacco pipe that is a 17th century type (in the broadest sense). Although a *mangīr* was found within this context, it must be noted here that coins, especially very thick copper ones, typically have a longer "shelf life" than that of ceramics. The Southern Sector of the site also produced ceramic evidence from a solid 17th century context (Figure 34) that was recovered from the fill above and below a plaster floor from Square A12 (L.200–201), and consisted of HMGP bowls, a basin and a store jar of types known from the Old City of Jerusalem, as well as a gray tobacco pipe dating from the late 17th to early 18th centuries. The latest stratified evidence from the Early Ottoman period at Khirbet Beit Mazmīl originated from two successive plaster floors (L.405–406) in the E1 storeroom (Figure 31), which yielded a red-slipped burnished tobacco pipe with a clear 18th century date, and provides insights into the forms found along with it, from a little-known period in the Jerusalem region.

The pottery presented here is a representative sample that is mirrored in the overall Early Ottoman corpus recovered from the site. The pottery itself is rather homogenous with regard to the limited number of vessel forms (only seven) present in most contexts from the site. While bowls were the most common (34), types with lesser numbers included: store jars (9); tobacco pipes (8); cooking pots (7); basins (4); jugs/*ibrīqs* (3); and lamps (3). Glazed wares were practically nonexistent at Khirbet Beit Mazmīl during this period, and of the 68 vessels presented here, the vast majority of them (78 percent) were handmade; only 15 wheelmade ceramics (22 percent) were attested, and those included bowls (2), store jars (7), jugs (3) and lamps (3).

Household assemblages were recovered from the domestic space in B4 (Figures 24 and 25), as well as in the Southern Sector of the site from Square A12 (Figure 34) and Room 205 (Figure 35). These consisted of vessels needed for daily life, and included many bowls as well as basins, jugs, lamps, and tobacco pipes. Not surprisingly, the corpus from the kitchen in C6 (Figures 28 to 30) yielded a large homogenous quantity of high-necked cooking pots. Also included among the assemblage were serving vessels for after the food was prepared—that is, bowls and basins—as well as store jars and lamps. Finally, pottery from the storeroom in E1 was limited to bowls of varying sizes that could be stacked together in order to save space, as well as the occasional jugs and body sherds from store jars. Taken together, the Early Ottoman assemblages from Khirbet Beit Mazmīl provide valuable insights into life at a farming village from a ceramic viewpoint, with simplicity and homogeneity in its vessel forms, which mirrors the modest lifeways of those who inhabited the site.

Meeting household needs: Non-ceramic small finds

The well-stratified kitchen floors of B4 and C6 (constituting the "cooking installations") and the storeroom of E1 produced most of the objects of the Ottoman and Mandate eras in the Northern Sector. The many glass finds (including bracelets, a juglet, bowl, and bottle, but mostly fragments) beads, metal objects, and some non-descript, corroded copper fragments (including nails and hooks) and ground stone objects (saddle querns and hand stones)—the chronology of which could not be determined with any certainty—came from primarily secondary contexts.[19] The Southern Sector, on the other hand, produced small finds of the Ottoman era, from both stratified contexts and collapse and erosional fill in Fields A and B, expanding our knowledge of the ways the residents of the Northern Sector furnished their households. Only a single non-numismatic metal object was registered from Early Ottoman layers there. However, these contexts did produce bangles (glass bracelets) and coins.[20] Bangles, coins, and pipes are our most chronologically reliable small finds of the Ottoman period in Bilād al-Shām. Bangles are recovered in larger quantities from Ottoman-period sites than one would expect if they merely functioned as women's jewelry: Khirbet Beit Mazmīl is no exception in this regard. Bangles may have served other functions, as described below.

Forty-five plain and decorated bangle fragments of the Ottoman period were registered from the Southern Sector; 40 of these probably belonged to individual bracelets. The majority of the finds came from buildings in Area B; Area A yielded only two fragments. No bracelets could be restored to a complete form, but diameters of several could be measured from 6.2 to 8.7 centimeters. All the samples belong to well-known types of Ottoman-era bracelets, which were manufactured in different workshops throughout Bilād al-Shām and widely traded all around the Middle East (see detailed classification and chronology in Shindo 2001; Spaer 1992). They consist of four main groups. Type 1—plain twisted bracelets with circular cross-section (Figure 36: No. 1), (opaque or translucent glass); Type 2—twisted bracelets with circular cross-section (Nos. 2–4), decorated with applied monochrome or polychrome trails (translucent glass decorated with applied opaque trails); Type 3—bi-chrome bracelet with semicircular (No. 7) or pointed cross-section (No. 8) and monochrome trail decoration (translucent glass coated with opaque layer and decorated with applied trails and patches); Type 4—polychrome bracelets with pointed cross-sections (Nos. 5, 9) and marvered trail decoration (translucent glass coated with opaque layer and decorated with patches and trails). In addition, a small fragment with a flat section and enamel painted design was found (No. 6).

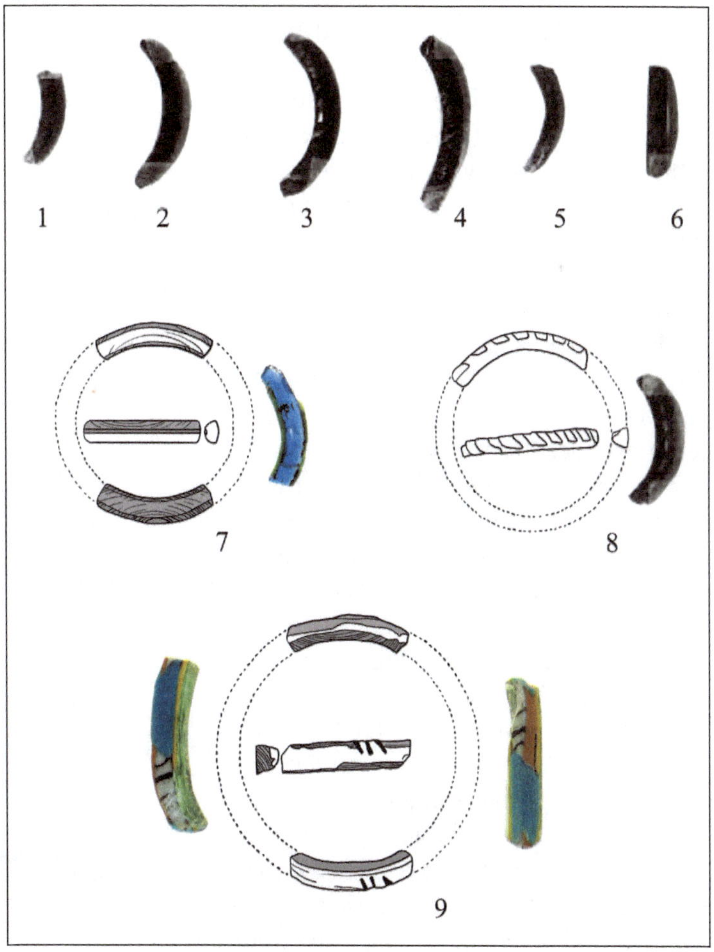

Figure 36 Glass bracelets from Southern Sector (courtesy of the IAA).

Types 1 and 2 comprise the largest amount of the excavated bracelets. Many of those represented on Figure 36 are made of good quality blue translucent glass, which is frequently decorated with bands of spiral trails in opaque white, red or yellow. No. 4 is olive-green. Most of the twisted bracelets have closely-spaced ridges; the trails are spirally wound around the exterior and fused together. Meanwhile, colorless bracelets differ by the slight twist and the trails are spaced with even wider intervals. An average glass thickness in the twisted bracelets' cross section varies from 6 to 7 millimeters. Some pieces show uneven tapering ends, which mark a seam at the joint. The type is likely to be produced in the same glass house, possibly in Jerusalem. Flattened bracelet No. 6, decorated with enamel paint, is exclusive. Its surface was incised with zigzag lines forming stock vector pattern, which was colored with translucent multicolored enamel (see discussion, Shindo 2001, 87, Type E).

The Lower Site also produced three metal objects with market exchange value: two coins—copper *mangīrs* minted in Ottoman Egypt dated 1595–1603 and 1595–1617, respectively (Figure 37:12–13)—and a brass jeton produced in Nuremberg during the 16th century (Figure 37:14). The latter were originally produced in large quantities as gaming counters for the Western market (Mitchener 1988, 17–18, 307). These counters reached the Ottoman Empire together with other brass products going eastwards like *Beckenschlägerschüssel* bowls from Nuremberg through commerce (Poyiadji-Richter, n.d.). In the Ottoman Empire they seemed to have been used as surrogate money up to the late 19th century (Kool 2005, 100; Kool and Ariel 2009, 109). From the area of Ottoman Palestine alone we have well over 120 documented finds of such brass tokens, many of them from excavations.[21]

Garbage disposal

A single midden located in square E2 of the Northern Sector of the site is probably our best context for reconstructing processes of site development from the end of the Mamluk period, as well as details about daily life in the Early Ottoman. It is a narrow space created originally by the spatial division of the Mamluk-era Northern Building from the Southern Building, located between Walls 138 and 135 and roughly a meter deep (Figure 38). It is the only context identified as devoted to refuse disposal at the site in this phase of occupation, and

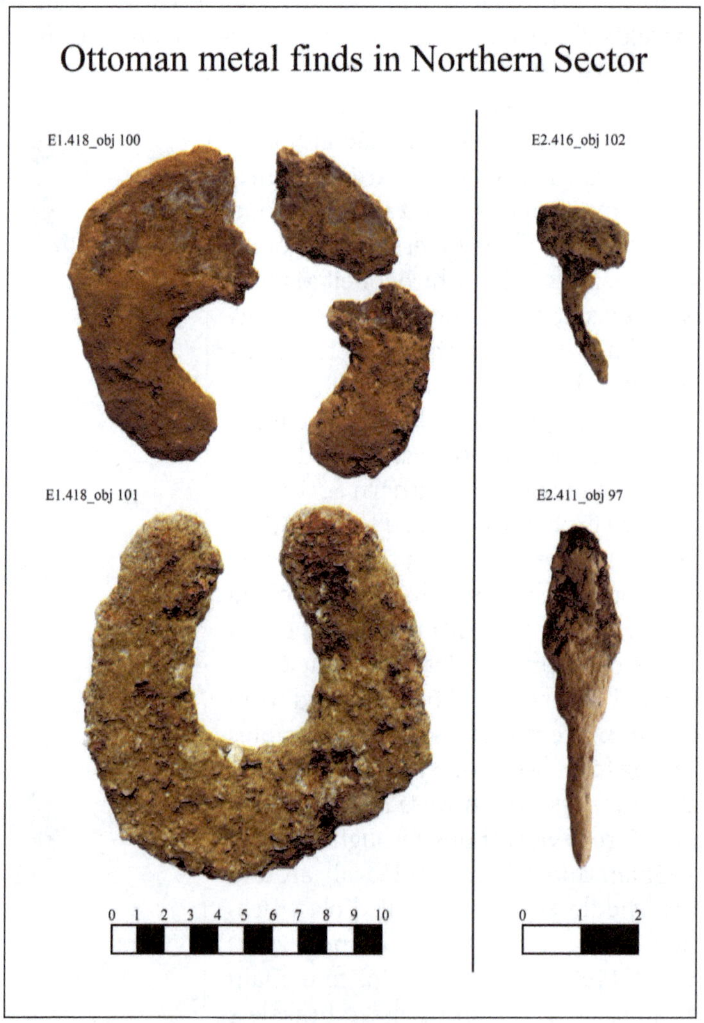

Figure 37 Metal finds from Northern Sector (courtesy of the IAA).

was used for a short period of time, roughly a century between the reoccupation of the site at the turn of the 17th century and the next stage of abandonment in the early 18th. The space-between-the-walls was used continuously and has created a stratified deposit of household refuse from the subtopsoil above the tops of the standing walls—only three courses high—to bedrock.[22] In this period, Wall 135 served as a terrace wall, demarcating the courtyard space to the south and separating garbage from cultivable soil. In addition to some tesserae and small finds, described below, most of the fill of this midden contained pottery sherds—some recently discarded, and some moved here during periods of clearance of old Mamluk-era buildings. While the soil was consistent in color and matrix, its contents can be divided into three phases (or episodes) of disposal, as follows.

This space was first used at the time of the reoccupation of the site in the late 16th or early 17th centuries. It seems to have begun with the clearance of the rooms and buildings in the nearest vicinity, as there were many sherds (some with joins) of late Mamluk vessels identified in the earliest levels of B4 and C6. Locus 439, which was the soil layer immediately covering the bedrock, included a nearly complete Mamluk pinched lamp, as well as modest-sized fragments (4–6 centimeters on average) of Mamluk-era creamware jars (combed store jars and tableware). The edges of the Middle Islamic sherds were worn, suggesting that they had been sitting somewhere exposed for some length of time, possibly decades. L.439 was a ceramically mixed deposit, with Late Mamluk and Early Ottoman sherds. Most notably, it was this locus that produced two early Gaza Ware *chibouks*, one of which still retained some tobacco residue. Additionally, burned olive pits from L.411 yielded a radiocarbon date of 300–600 CE.[23]

The contents of the middle layer of the midden were slightly different. The main locus (L.435) was also a mixed one ceramically, producing Mamluk and Early Ottoman sherds, and in rather large quantities. What distinguished L.435 from the loci above was in the composition of the Mamluk corpus. These sherds were rather large (an average of 6–12 centimeters) and included many sherd joins. The Mamluk wares included a wider range (and chronology) than in L.439: handmade cooking bowls (mendable), pinched lamps, combed creamware store jars, handmade stewpots with elephant-ear handles, and Egyptian sgraffitos of 14th century date (Walker 2004). The edges of these sherds are more worn than the Mamluk sherds of L.439. The locus, however, mostly produced Early Ottoman wares (of the 17th century). As for small finds, the locus produced a mix of objects both Mamluk—an iron knife, clay rattle, and soapstone bowl rim—and Early Ottoman—two coins and a face veil coin[24]—as well as metal fragments that could be assigned to either period—a metal ring and bronze buckle, for instance. The Mamluk sherds from this locus include material of an earlier date, and of more worn condition, than that of L.439. It appears that the midden at this point includes material cleared from some other part of the site, along with contemporary (17th century) household refuse. The midden, then,

Figure 38 E2 midden excavated to bedrock (photo: B.J. Walker).

went out of use not long after the discarding of a red burnished pipe of early 18th century date (Obj. #81, subtopsoil of E2.402).

Commercialization of rural life

Given its proximity to Jerusalem, the people of Khirbet Beit Mazmīl found ready access to urban markets and participated in the market agriculture that was quickly developing there from the 16th century. The commercialization of agriculture was made possible through purchases and leases of orchards and gardens and percentages of harvests, financed through access to new sources of credit, in a process discussed in detail in Chapter 8. The impact of urban markets and cuisine on kitchen culture and food systems at the site are explored in Chapter 7. The small finds at the site reveal ways that a rural community engaged in these changes, and how a greater incorporation of local society into the Ottoman orbit impacted daily life.

There is evidence of market activity throughout the site. It is not clear whether these markets were events that took place at the village itself, involved exchanges among villages in the region, or were conducted through the intermediary of Jerusalem. Nonetheless, the identification of weights and tokens, coins, and ceramic imports allow us to describe, in a general way, the economic worlds in which the people of this village lived, and how those networks changed over time.

A bronze weight (Obj. #139, B4.166) was recovered from a securely dated context in the Northern Sector of the site: sealed between two plaster floors, with a late 16th century *mangīr*, in the east half of the room in B4. Weights like these are normally associated with market activity, and the stratigraphic association with a coin used in rural market exchange may suggest that buying and selling of produce took place on site. As discussed above, the kind of brass jeton found in the Southern Sector was often used as surrogate money in this region in the 16th century. Similarly, the large numbers of glass bracelets, and a general deficit of other kinds of jewelry, may indicate that bracelets served another function than personal adornment. They have been found in abundance in the excavations of Damascus Citadel and Aqaba Castle, where they have been dated to the Ottoman period (Boulogne 2007; al-Shqour 2019). The presence of large numbers of glass bracelets at military sites and khans may be due to their role as mediators of market exchange, in lieu of cash (like the jetons), or as objects of exchange on their own (such as *Hajj* souvenirs).[25]

There was no shortage of coinage at the site, however. A total of 30 Ottoman coins were identified at the site over the course of the excavations from 2012 to 2019. Ten of the coins came from a hoard in B4, seven were sewn onto a veil, and one from under a floor in the Lower Site; most of the rest came from the cooking installations of B4 and C6, sitting on what were originally the plastered or beaten earth floors of kitchens. The regnal dates of the Ottoman sultans named on the coins range from the 1570s to 1617; the majority were minted in Egypt in the 1590s. All but one were copper *mangīrs*, the kind of small change used in everyday exchanges in the market.[26] The find spots are most relevant and suggest an attempt to hide away (or store in a secure place) small bags of change. They speak to a localized market activity, although we cannot identify the location or the nature of that marketplace.

The hoard of copper coins, discussed earlier, was discovered in a fill layer outside the building of B4 and buried against the exterior face of its south wall (B4.118). The pile of coins was partially covered by a large, curved body sherd of an Ottoman period HMGP jug, possibly dating to the 17th century (see Figures 39a and 39b). The entire assembly consists of the heavy copper *mangīr* denomination struck on small, thick round flans in Cairo by the new Ottoman rulers during the 16th–17th centuries (Figures 39a and 39b). Nine of the coins in the hoard date to the reign of the Ottoman sultan Murad III (1574–1595), while the remaining one could only be dated generally to the end of the 16th century. *Mangīr* copper coinage was introduced by the Ottomans in

Cat. No.	Weight (g)	Size (mm)	Obverse	Reverse	Date CE	Mint	Reference	IAA No.
			OTTOMANS Sultan Murad III (AH 982–1003 / 1546–1595 CE), *mangīr*					
1*	8.61	20	Center: [زع] نرصه/ رصم Tughra	Center: Arabesque	AH 982 (1574/75)	Miṣr al-Maḥrūsa	Kabaklari 1999:420, No.12-Msr-09	161119
2*	7.54	15	Center: مجرو[ةس]/ ةنس Tughra	Same	AH 982 (1574/75)	Same	Kabaklari 1999:420, No.12-Msr-09	161122
3*	7.54	15	Center: رصم/ المحروسة Tughra	Same		Same	Kabaklari 1999:420, No.12-Msr-10	161123
4*	9.06	18 × 21	Center: رصم/[المحروسة] Tughra	Same		Same	Kabaklari 1999:420, No.12-Msr-10	161115
5*	6.81	22	Same	Same		Same	Kabaklari 1999:420, No.12-Msr-10	161121
6*	8.32	15	Center: ضرب/ رصم Tughra	Same		Same	Kabaklari 1999:420, No.12-Msr-11 (small flan)	161116
7*	9.01	15	Same	Same		Same	Kabaklari 1999:420, No.12-Msr-11 (small flan)	161117
8*	8.27	18	Same	Same		Same	Kabaklari 1999:420, No.12-Msr-11 (small flan)	161118
9*	8.79	15	Center: رصم Tughra	Same		Same	Kabaklari 1999:420, No.12-Msr-11 (small flan)	161120
			Early Ottoman, *mangīr*					
10*	9.39	17	Center: رصم/ المحروسة Tughra	Center: Arabesque	End 16th c.	Miṣr al-Maḥrsa	(small flan)	161114

Figure 39 a) Early Ottoman *mangīrs* from Northern Sector; b) Catalogue of Early Ottoman *mangīrs* (courtesy of R. Kool).

the second half of the 14th century to be used for small change in daily transactions and for paying taxes, to be exchanged later for silver (Pamuk 2000, 38). Many of the *mangīrs* were produced in Anatolia, where copper mines were located. In theory, the coins circulated on a fixed nominal value (higher than their metallic value) set by the Ottoman state's centralized financial office (the *defterdarlık*), with their production and distribution farmed out to private entrepreneurs who leased land (*mültezim*) at each mint (Kabaklari 1998, 38, 46; Pamuk 2000, 39). In practice, complex economic, political, and monetary factors during Murad's III reign—among them the growth of the empire's population and massive influx of New World silver—put considerable pressure on its official silver *akçe* currency, leading to its debasement by approximately 50 percent (Ünal *et al.* 2020, 77–78). This in turn must have substantially reduced the buying power of the subsidiary copper *mangīr* coinage during this period (Kabaklari 1998, 40–42).

At present, virtually no published data exists about the circulation of Early Ottoman base coinage, except for the occasional reference to single coin finds in isolated excavation reports. Fortunately, more comprehensive data is available through unpublished excavation finds compiled in the IAA Coin Database over the past 80 years (Ariel 2016). These show that copper *mangīrs* must have circulated relatively abundantly in the area of Ottoman Palestine during the 16th to 17th centuries. In 2017, the database included more than 550 *mangīrs* from 130 excavations in 64 sites, from all over Ottoman-period Palestine. Among these the *mangīrs* from the Miṣr al-Maḥrūsah mint constituted by far the single largest group of coins from one mint, more than 35 percent of all coin finds. *Mangīrs* from Egyptian mints are particularly plentiful in excavation finds from Jerusalem. This hoard also provides further evidence for the use and circulation of Egyptian copper money in rural sites in Early Ottoman ruled Palestine.[27]

Finally, the consumption of imported pottery is an important indicator of Khirbet Beit Mazmīl's participation in market exchanges that go beyond Jerusalem. Imported wares represent only a small portion of the pottery used at the site.[28] They do, however, illustrate the degree to which Ottoman exchange networks had penetrated the Palestinian countryside. The quantities of sherds of imported wares and their find contexts are more numerous for the Early Ottoman period than the later centuries: numerous sherds of Ottoman glazed wares (mostly green-glazed bowls) have been found throughout the site, and Gray Gaza Ware jars, bowls, and pipes (equally distributed across the site) are consistently recovered.

There were far fewer ceramic imports recovered, in comparison, from Late Ottoman contexts at the site. Despite its proximity to the city of Jerusalem and its markets, only a few sherds of imported wares have been identified from the Late Ottoman period, and all of these from the 19th century. These include a sherd of Flow Blue porcelain (England, late 19th/early 20th century, from B4.111, a collapse layer); a sherd of English transfer print ware (late 19th century, from E2.422, a midden); a fragment of a Sarreguemines refined earthenware plate (France, England, or Scotland, late 19th/early 20th century, from C1.212, a construction fill for the resurfacing of Area C roofs); and a sherd of Rushayya al-Fukhkhar Ware (Lebanon, late 18th/early 20th centuries, from B4.178, the exterior of building B4). The site produced none of the eastern Mediterranean ceramic imports normally associated with Ottoman exchange networks, such as Çanakkale, Didymoteicho, and Kütahya wares. There were also no Chinese or Japanese porcelains identified. The number of imports for this period were minimal and reflect more the nature of the site than its location, the financial status of the residents, or the prices of such wares, which had become increasingly more affordable in the late 19th century. The site was a *khirbah* at this time—seasonally occupied by a family that lived full-time in Ein Karem. They may have brought smaller numbers of their European tablewares with them during the summers from that village, where they may have had many more.[29]

On the basis of the imports alone, one could argue that the site was part of a regional exchange network (typical of the Ottoman Empire) in the 16th and 17th centuries, but a more global, Europeanized network in the 19th century. A shift in trade from the Mamluk period is readily identifiable: Mamluk Beit Mamzil looked to Cairo economically, while Ottoman Beit Mazmil/Khirbet Beit Mazmīl was pulled into the orbit of Istanbul's trade networks, namely the eastern Mediterranean and Europe.

The *chibouks* present a different story about exchange and consumption. Seven of the nine pipes and pipe fragments recovered from the site belong to the earliest series in Palestine, with a distinctive light gray fabric. While these pipes appear to have been produced in the Gaza region (a mere 79 kilometers from Jerusalem), the tobacco was likely cultivated in Lebanon, although the results of the isotope analysis on the tobacco residues in the pipe from E2 cannot rule out Palestine. Smoking tobacco was an imported custom, one that was newly introduced into Palestine at this time. The people of Khirbet Beit Mazmīl in the late 16th/early 17th century would have been among the first to take up this foreign pastime.

Abatement and abandonment (Stratum Ic: Middle Ottoman, 18th to mid-19th centuries)

Historical background

The decentralization policies of the Ottoman state in the 17th century initially made economic growth and local political and administrative autonomy possible. The *timar* system, under which most land in Palestine was controlled by military officers, was slowly eroded by the leasing of the same lands to local elites: tribal sheikhs, Sufi sheikhs, and the urban *ashrāf* (descendants of the Prophet) and *ulamā*ʾ (clerics). Three local families became hereditary dynasties, controlling the governorships of the districts of Jerusalem, Gaza, Lajjun, and Nablus, acquiring great wealth and social capital in the process. The religious elites came to control the rapidly expanding rural and urban endowments, accumulating wealth in Jerusalem and its villages in the process. By the end of the century, much land in the region was owned by absentee landowners and leased to increasingly impoverished tenant farmers; many peasants became burdened by debt because of high taxes, leasing rates, and the inability to pay back loans which they took out to lease small plots of land (like orchards). In an attempt to reign in the increasingly independent notables and regain political and financial control of these districts (*sanjaks*), the Ottomans appointed a new governor in Jerusalem, Nablus, and Gaza: Mehmet Pasha Kurd Bayram, who promised a new phase of centralization. The new governor ruled with an iron fist, doubling taxes and leading punitive military expeditions against local Bedouin and peasants. This led to an armed revolt in 1703–1705 against the Ottomans, known as the Naqīb al-Ashrāf rebellion, led by the head of the *ashrāf* in Jerusalem. The rebellion was a turning point in the fortunes of Jerusalem and its villages. Nearly every sector of Jerusalem society, and the rural areas in its vicinity, took part in this revolt: the urban and rural religious elites, the tribal sheikhs, peasants, Bedouin, and even the Ottoman Janissaries stationed in Jerusalem. Although initially successful, divisions within the body of rebels led to a reoccupation of the city by Ottoman forces in 1705, and a brutal crackdown on rebels thereafter. Some villages were destroyed. Pilgrimage was disrupted, and the mercantile life of Jerusalem took some time to recover (Manna 1994, 52–58). Nothing had a greater impact economically on the Jerusalem countryside than this rebellion and its aftermath (Zeʿevi 1996, 84).

Beyond the reach of the state, tribal conflicts and migrations remolded the region demographically; they are among the most important factors behind settlement cycles of abandonment and resettlement in the Ottoman era. Following the wide-scale tribal migrations from the Hijaz in the 17th century, one group of people dislocated another, like balls on a pool table. In abandoning one village, the residents would move to a neighboring one, and sometimes eventually returned to their home settlements.[30] Tribal migration and the local fallout from the Qays-Yaman rivalry created rural instability throughout Palestine, as single villages could be serially abandoned, resettled, and abandoned once again (Marom 2022b). The anthropologist Susan Sutton refers to a similar phenomenon in Ottoman Greece, fueled by economic loss and opportunity, as "liquid landscapes" (Sutton 2000).

The abandonment of the farmstead at Khirbet Beit Mazmīl should be seen against the backdrop of the 1703 rebellion and the insecurity caused by frequent tribal confrontations. While there is no archaeological evidence for an armed conflict at the site, it is clear that the farmstead was eventually abandoned sometime in the early 18th century, when ceramic and numismatic evidence and radiocarbon dates come to an end. It was not to be resettled until nearly a century later.

Process of ruination

Subtle evidence of the process of abandonment from the site itself comes from the courtyard (in studies of vernacular architecture called the *manṭarah* yard). The trees planted in the almond orchard in the center of the complex, and which we encountered in 2015, are young; they are not contemporary with this phase of occupation and are associated with a later phase of courtyard development and soil transport. The smallest of the almond trees in the courtyard collapsed one workday, revealing below it, under 17 centimeters of topsoil and subtopsoil, a "flower bed." This raised bed (formed by Wall 125 in C6) was built for a single tree.[31] It is small and takes the form of a half-circle, constructed with a single course of well-cut (and spoliated) building stones, and demarcating a space of 60 centimeters radius. Wall 125 was built directly on top of the Early Ottoman C6.238 cooking installation, and against the plastered north face of Wall 124/2 (Figures 39a and 39b). The soil filling the bed (C6.218.2020) contained parts of two crushed, copper *narghilas*, replete with a charcoal basket. The pressed ash—which was still in one of the bowls—provided a radiocarbon date of 300–600 years before present.[32] The soil of the bed also produced an 18th century porcelain sherd, as well as a Black Gaza Ware jar handle and a fragment of late 19th or early 20th century European porcelain (see ceramic report below).[33]

The 18th century is an elusive period at the site, and it appears that occupation ends, for a time, early in the 1700s. Outside of the porcelain and the *dirham* from the Lower Site and possibly some of the Black Gaza Ware vessels from the Northern Sector (described below), the only reliable archaeological evidence for the 18th century are two burnished, red-slipped tobacco pipes (*chibouk*), the details of which are provided below.[34] The first pipe (Obj. #85) was found in fill (E1.405.4005) in a sealed context: sandwiched between a flagstone floor (E1.409) and a plaster floor (E1.406), in the storeroom adjacent to the space where the raised bed and *narghila* bowl were found. The soil in which the bed was laid was at the same level as that of adjacent L.405: the soil loci are contemporary and together constitute an 18th century stratum.[35] The second pipe (Obj. #81)

Figure 40 Raised bed (Wall 125) in C6—in center of photo (photo: B.J. Walker).

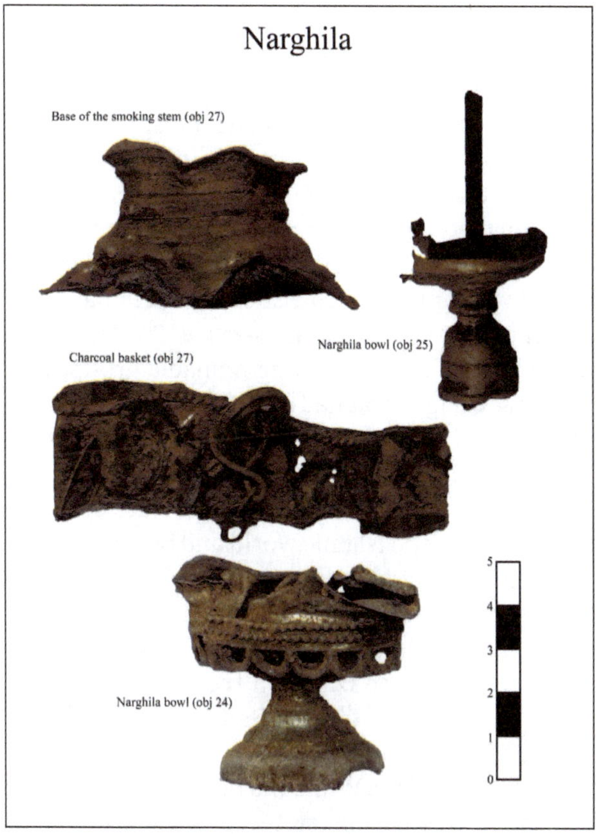

Figure 41 *Narghila* fragments buried in soil of C6. 125 (photo: B.J. Walker).

was recovered from subtopsoil in E2 (E2.402.4002) in the midden located between Walls 135 and 138. L.402 was the uppermost layer which produced finds in this midden; the thick topsoil above it was sterile. The last phase of the use of this midden was the early 18th century.

The inclusion of an entire water pipe set in the soil of this "tree bed" seems to have been quite deliberate (Figures 40 and 41). One plausible explanation for this phenomenon can be found in traditional peasant practice. Calcareous soil, such as that found in the Jerusalem region, is normally deficient in certain minerals (iron, copper, and zinc). Adding metal objects, of any form, to enrich the soil is still common practice in Jordan today, where there is lively social media debates over whether adding old metal utensils (after flattening them beforehand), wire, and metal shavings to garden soil will promote fruit tree growth.[36] The utility of such practices is also being debated in academic circles (Arif *et al.* 2016).

After this phase, the site was abandoned for a period of unknown duration. The clearest evidence of this is a 10 centimeters deep aeolian deposit (L.107) under the southeast corner of the B1/2 complex (Figure 42). From this fill was recovered a

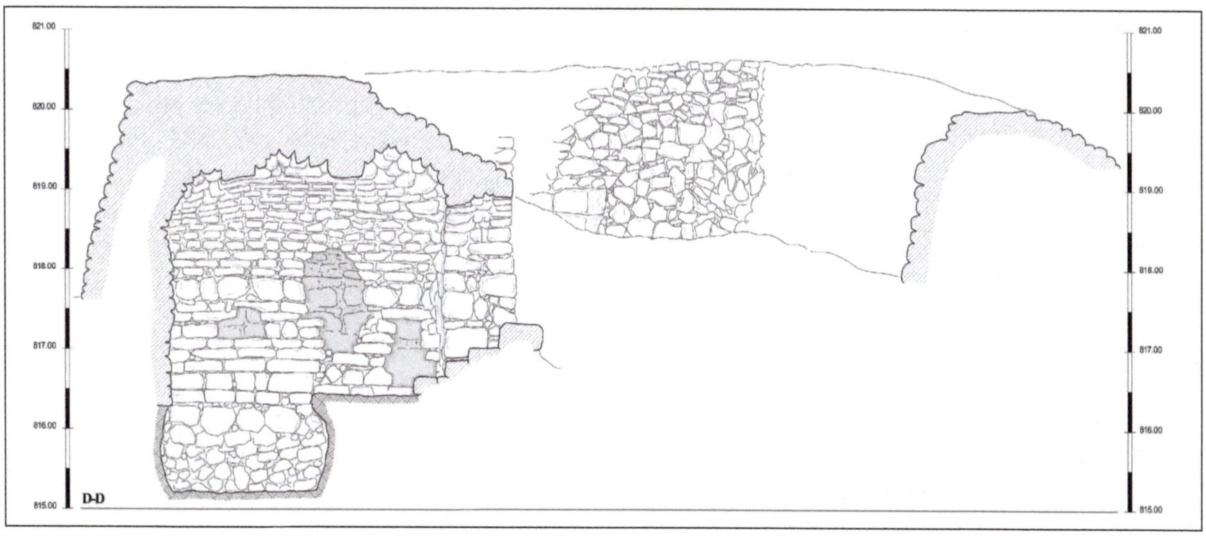

Figure 42 Section drawing of B1 and B2 *manṭarah* in Northern Sector (courtesy of S. Pirsky).

hole-mouthed cooking pot of a well-known type from the 19th century. The continuation of the L.107 abandonment fill in front of the building and at the level of the entrance's threshold was sealed by a layer of architectural collapse (L.105). Additionally, no pottery, coins, or small finds that could be associated with the late 18th or early 19th centuries could be identified. The *'ezbeh* described in the following section was, then, built after what must have been at least several decades of abandonment.

A new way of living: The seasonal farmstead (Stratum Ib: Late Ottoman, Tanzimat, and WWI, 19th century to the 1920s)

Historical background

The strong hand of the state and armed conflict form the backdrop of many of the changes experiences at the site from the 19th century. The 1800s witnessed a period of dizzying reforms, aggressively promoted by both Muhammad ʿAli's autonomous regime in Egypt (1805–1848) and by the Ottoman sultans across their empire. A range of political, economic, social, and educational reforms were aimed at strengthening the state domestically and internationally through modernization. Ibrahim Pasha, Muhammad ʿAli's son, led the conquest of Syria for his father. Jerusalem fell to his forces in 1831, and he served as the Governor until 1841. Many of his administrative reforms in Palestine—inspired by those promoted by his father in Egypt—were ardently opposed on the local level, particularly the granting of what was seen as Christian privileges, the implementation of a military draft, forced labor on imperial projects, government monopolies over agricultural production, secularization, and an increased presence of Europeans. A rebellion against Egyptian rule within the Muslim population began in Jerusalem in 1834. It was violently squashed, and villages outside Jerusalem, Bethlehem, and Hebron were destroyed and abandoned. The effects of this rebellion were to be felt in rural areas for years to come (Kark and Oren-Nordheim 2001, 27–29).

The Ottoman reforms—known as the Tanzimat ("Reorganization")—followed a similar trajectory from 1839 to 1876, resulting in the issuance of the empire's first constitution. Its most important legislation in Palestine was the implementation of the Ottoman Land Code of 1858/1859, which transformed land tenure. Those with customary use-right were now required to register land in their own names, which also made them liable for direct taxation and military service. The land code also played a role in settling nomadic groups and reviving villages that had been long abandoned. Its overall impact on peasants and rural communities, however, was mixed, as we will see below.

Dissatisfaction with the Ottoman state split the allegiances of the Islamic world and brought international conflict to Palestine. Armed conflict continued with the First World War: significant battles took place at Ein Karem and the hilltops nearby. It appears that the Ottoman trenches here were located due east of Khirbet Beit Mazmīl, and possibly quite near to it (Falls 1930, 243–245; Grainger 2006, 207–208, 244). While the heaviest fighting for Jerusalem on 9 December 1917, took place at Lifta and Ein Karem, there were skirmishes throughout the region; the major Ottoman offensive between 26–30 December was at Tall al-Ful (A. Peretz, IAA, *personal communication*). At one point in the war, Scottish soldiers positioned themselves from this hilltop (A. Peretz, IAA, *personal communication*; see also references to Scottish military presence in the region in Woodward 2006, 143, 145, 146). There is little physical trace of the war, however, at our site. One rifle cartridge and six bullets (Obj. #58, #59a, #59b, #60, #61, and #62) were recovered from topsoil covering the tops of walls (B4.126.1033, B4.141.1059, Wall 122), topsoil in front of B4 and B5 (B4.141.1059 and L.143.1063), and the roof of C1 (C1.121.2014 and 212).[37] The collapse of part of the C4 vaulted building superstructure may have been the result of this conflict, or that of 1929, or seasonal rains.

More important than armed conflict on the development of the site in this period, however, was the local implementation of Tanzimat legislation. The reforms of the 19th century, and those by the British Mandate authorities in the early 20th century, had an uneven impact on local communities. Rural populations did grow, with rural migration to the settlement of nomadic groups, incentives for registering and cultivating land, and a general improvement in security of the roads. Many villages could not sustain sudden population growth, however, and as populations grew, many households relocated to satellite settlements nearby or established new ones of their own, and cultivated the land there anew, reclaiming "dead lands," or *mawāt*, in the process (Kark and Oren-Nordheim 2001, 213). We would call these places homesteads, farmsteads, or hamlets. In historical geography they are known as offshoot settlements or "daughter villages." In Palestinian Arabic the same terms are used to describe such a settlement as for the hybrid manors described below—*qaṣr*, *'ezbeh*, and *khirbah*. Over time, these developed into fully functioning villages. This was a common settlement pattern in the 19th and early 20th centuries. At the same time, villages once occupied year-round could contract into small, seasonally occupied settlements of the same name (Grossman 1994; Kark and Nordheim 2001, 273–275; Marom 2022a, 159–163). Khirbet Beit Mazmīl in this period was a village of this latter form. The site illustrates well the fluidity of village size, function, and nomenclature in the last century of Ottoman rule.

The registration of land that was promoted by both the Ottoman Tanzimat and the British Mandate authorities accelerated trends begun a century before in Palestine: the rise of new rural dynasties and the commercialization of agriculture. Both had an unexpected impact on settlement and architectural form. Two forms of rural architecture typical of the central highlands become widespread in this period, though they have roots in the 18th century: the sheikhs' houses of the so-called "throne villages," and drystone hybrid manors: *'ezbehs* in local parlance. With the weakening of Ottoman rule in the late 17th century, and the delegation of many imperial prerogatives to local elites, local tribal leaders carved out an informal, autonomous rule in the Palestinian countryside. The physical manifestation of these sheikhdoms was the some 40 "throne villages" that occupied hilltops throughout the central Palestinian highlands. Of half-urban and half-rural form, these fortified complexes were both administrative centers and feudal mansions (Abdulfattah 2007; al-ʿĀmirī 2003).[38]

The hybrid manors are walled farmsteads used seasonally by peasants during the harvest of primarily olives, figs, and grapes. They are built of fieldstones cleared from the agricultural fields themselves and built without mortar, piling stones on top of one another to create defensive-like walls. At the center of the walled complex is a large, open-air courtyard, which is surrounded by living and storage quarters, stables, and installations for processing crops. A variety of vault forms and corbelled domes cover buildings used simultaneously as housing and agricultural watch towers (sg. *manṭarah*, pl. *manāṭīr*). There is considerable variety in the form, size, and plan of these complexes, which reflect very localized building traditions (al-ʿĀmirī 2018; al-ʿĀmirī and Rahhal 2003). Though they are concentrated in the central highlands, they can be found, as well, in the lowlands (the Israeli *Shephelah*), where they are associated with complex local and long-distance road systems, terraces used for the cultivation of orchards and irrigation of fields, and animal husbandry.[39] More than merely an architectural and settlement form, the *'ezbeh* played an important role in Palestinian agricultural development, and, like the terraces with which they formed an agricultural unit, are a rapidly disappearing component of Palestinian cultural heritage (Ghadban 2015).[40]

Architecture of this manner of construction does not survive long: the drystone construction of the central highlands rarely survives longer than two hundred years.[41] The heavy rains and frequent freezing events of Jerusalem's winters create cracks between the mortarless courses, leading to the collapse of walls and superstructure. The fragility of the standing architecture created numerous logistical challenges to our fieldwork from the start, as we tried to map the structures, clear them, and then plan for excavation. Only two (C3 and C4) of the four vaulted chambers (B1, C3, and C4, and the rooms of C6 and C7) were structurally stable enough to enter for study, and only one (C4) in a sufficient state of preservation to excavate. With each season, the complex suffered further damage, through the combined effects of winter weather and interseason vandalism.

The "hybrid manor" ('ezbeh*): An introduction*

Although largely destroyed by urban development, what stands above ground today, and which has been restored in the modern archaeological park (described at the end of this chapter), is one of these hybrid manors: a walled, multi-roomed complex of drystone masonry, perched at the summit of the hill, its vaulted rooms surrounding a central courtyard.[42] This was also the "ruin" (*khirbah*) that could be seen in the 19th century and was called locally Khirbet Beit Mazmīl. The ruins are imposing, appearing at first glance as a pile of randomly selected fieldstones, piled on top of each other to form an enclosure wall of irregular rectangular plan and dimensions, demarcating a complex of some 1300 square meters (roughly 36 × 36 meters). During site documentation by Danit Levy of the IAA in 2011, and at the start of our excavations in 2015, the site was heavily overgrown by trees, high grass, and weeds. Little was visible but the tops of the individual vaulted buildings; the wall that connected all of them could not be seen. It was not evident, then, that this was, indeed, a single complex. It is no wonder that people living in Bayit VaGan did not recognize this as an archaeological site. For them, it was a trash heap, where one threw away broken household goods among piles of stones.[43]

The complex once comprised as many as six well-built rooms (roofed in a variety of vaulting forms), some constituting towers (of a large *manṭarah* form) and making use of spolia of ancient buildings in their construction, and all connected to one another by a wall of irregular form and thickness and enclosing an open-air courtyard. The southeast corner of the walled space (Area D) was partially separated from the central courtyard by two walls (W. 106 and W. 115), enclosing a cistern complex and isolating it from the stabling area of Area C.[44]

The *'ezbeh* is remarkably well preserved, from its central courtyard to its vaulted ceilings and corbelled domes. External stairways attached to the walls lead to the roof, in a construction technique common to domestic, agricultural, and religious architecture in Late Ottoman Palestine and Transjordan.[45] A semi-circular tower (L. 110 in Field C8) is attached to the west side of the complex, recalling the historical *manāṭīr* (Arabic; in Hebrew *shomera*): agricultural watch towers/dry stone huts traditionally placed in vineyards and orchards to guard the harvest and provide seasonal housing for harvesters.[46] Larger vaulted rooms (only those in A1, B1, and C4 were extant) once stood at the corners and along the enclosure wall of the walled complex, serving a range of domestic and storage functions. That of C4 was particularly well preserved, heavily reusing elements of an Abbasid structure: a plaster floor (L.310) covered the foundational cobbled floor (L.307) of the original building, which was a modification of the bedrock, and the walls supporting the Phase Ib vault were built directly on six courses of an Early Islamic one, the foundation trench (L308), which was ceramically dated to the Abbasid period (see Figure 8 in Chapter 3). Unlike the cloister vaults of C6 and C7, the vaulting system of this room has been heavily repaired, converting what appears to have been originally a cross-vault (of possible Mamluk date) into a dome of corbelled construction. This form of corbelling is characteristic of the roofing systems of *manṭarahs* in the central highlands (al-ʿĀmirī and Rahhal 2003).

Radical rebuilding—construction of the ʻezbeh

Building a new life: Phases of construction

The physical structures of the site were continuously inhabited until sometime in the early 18th century, when they were abandoned. There are no traces for a violent end of this phase of occupation. Only in the 19th century, and exclusively in the Northern Sector, was the site reoccupied, with a complete rebuilding that obliterated most traces of earlier structures.

A series of drystone buildings were built, either completely replacing the preexisting structures or abutting against or entirely covering the few standing ruins. It is not clear in what state the previous buildings were when the choice was made to rebuild the site. It is possible that the collapse was quite extensive, ultimately leaving a huge amount of building materials immediately available for reuse. Indeed, the materials used to build the new complex entirely consist of field stones and reused blocks, the latter likely found on the spot and gathered while clearing the areas where the new structure would be built. They are normally extremely diversified blocks—therefore belonging to multiple earlier structures—laid without any kind of mortar and in an irregular way. The building technique is therefore extremely poor: no attempt to create regular courses or to select materials was made, and the frequency of small stones used as wedges is an eloquent proof that the rebuilding activity was done by just piling up unsorted materials at the place of their immediate disposal. For this reason, this dry stonework is very easily recognizable, and extremely fragile. During the excavation, it was possible to witness firsthand how heavy rains could badly damage large portions of the structures (Figures 43 and 44).

Figure 43 Drystone architecture before the collapse (photo: N. Pini).

Figure 44 Collapsed drystone architecture (photo: N. Pini).

The new Khirbet Beit Mazmīl consisted of two buildings organized around a central, enclosed courtyard (Figures 45 and 46). Only two portions of the earlier site were kept, although with several transformations: the series of vaulted chambers in Area A and the cistern house of Area D. The rest of the site was filled with earth—possibly brought from somewhere else, since it was sterile (with no archaeological remains)—and levelled, thus creating the central courtyard.

Figure 45 The northern building (three-dimensional model: N. Pini).

The rooms with the cloister vault and the two adjacent with the barrel vaults (one likely Mamluk and the other probably restored in the Early Ottoman period) seem to have undergone limited interventions. C4 was, on the contrary, almost completely rebuilt: the main difference is the roofing, namely a dome built through corbellings according to a completely different typology and technique (Figure 47). The same technique can be found the B1 building (*manṭarah*) on the south-western corner of the courtyard (Figure 48). Both complexes might have had an upper floor or an accessible, flat roof: in the case of the North Building, it was accessible thanks to a staircase made of flagstones inserted into the wall. In B1, a small, rather high window was documented, possibly associated with an upper floor or an elevated internal area (see Figure 46).

Figure 46 The dwelling in B1, before the beginning of the excavations (courtesy of D. Levi).

Figure 47 The interior room of C4 with 19th century dome (photo: N. Pini).

In the ethnographic surveys (see below) there are references to at least one large restoration of some of the complexes forming the new *'ezbeh*: unfortunately, the building technique is so irregular that it is extremely hard to clearly identify different architectural phases. The only clear interfaces are between the drystone architecture and the few surviving structures from the earlier period. It is the case for instance in Area C6, between the drystone wall W104 and the cloister vault (L.111). The other area where this difference and architectural phasing is particularly striking is in Area D, particularly in its northern part.

The eastern portion of the *manṭarah* in B1 seems to abut the structures in the southwest part of the Northern Sector, directly built on top of the corner of the delimiting wall. This is potentially the only architectonical subphase of the 19th century

Figure 48 The interior room of B1.

drystone buildings that can be documented. It is impossible, however, to determine whether this stratigraphic posteriority really relates to a reconstruction or an expansion of the original structure.

One of the most characteristic features of the site in its new configuration is its complete integration into the terraced landscape. To the east, the massive drystone wall W105, which encircles the site on its east side, serves as the retaining wall for the terrace located uphill of the *'ezbeh*. To the west, the former wall dividing the Upper and Lower sectors of the site was partially restored and three *manāṭīr* were added; the southern one ultimately erased the more ancient tower structure in S1, and its collapse created the gigantic stone heap described by the excavators (see Chapter 4). The northern one (A1) is ultimately one of the few documented structures allowing a direct passage from the lower part of the site to the top of the hill. These three *manāṭīr* (S1, A1, B1/2) are visible in historical photos of the 19th century (see Figure 8 in Chapter 1).

The different *manāṭīr* follow a well-documented typology of towers in the region. Of particular interest to this concern is the survey conducted in the Battir area, west of Bethlehem—one of the best-preserved examples of terraced landscapes in the region and inscribed since 2014 as a Cultural Landscape in the UNESCO World Heritage List (WHL). The survey was conducted in the preparation of the dossier for its inscription in the WHL and was included in the nomination document (UNESCO Nomination Text 2012). The survey identified three types of *manāṭīr*, differing in shape: "round-corbelled," "rectangular or squared quadrilateral, built of dressed ashlar stone," and "solid stone heaps with oval or rounded shapes" (UNESCO Nomination Text 2012, 56–59).

These different forms do not seem to represent a chronological difference, rather are determined by a different placement within the terraced landscape and/or by different uses. Indeed, precisely dating the original construction of such structures is often impossible. They are known even from late prehistory onwards and are frequently rebuilt and reoccupied. Given that all *manāṭīr*, regardless of form, were meant to guard the terraces and shelter those working the fields for short periods of time, in some examples a slightly more specialized function seems to be recognizable: it can be related to productive activities taking place there, or also to the need to shelter groups of people for longer periods of time.

The two preserved examples at Khirbet Beit Mazmīl appear to belong to the "round-corbelled type," built with reused, mostly unsorted, and undressed materials—the only exception being the lintels of the doorways. As mentioned in previous chapters, one of the unsolved questions is how much this new arrangement and these new structures changed the earlier built environment of the site and the circulation patterns between the different sectors.

Despite the widespread destruction affecting the structures and even more the surrounding terraced landscape, it is possible to have a clearer picture of how the site was organized in its latest phase of occupation. A series of historical sources offer us a glimpse into the past of the area, as well as a series of ethnographical investigations, during which the last inhabitants of the *'ezbeh* were interviewed. Furthermore, a series of well-known and preserved examples represent important comparisons.

First of all, it must be underscored that the site is literally embedded in the terraced landscape, as evident from a series of pictures taken between the end of the 19th century and at the beginning of the 1930s from Ein Karem. Despite the fact that Khirbet Beit Mazmīl is barely recognizable in the background, some interesting features are identifiable along the slopes: first of all, the entirety of the descent was terraced; among the terraces, there were at least two more *manāṭīr* (see Figures 7 and 8 in Chapter 1).

This finds further confirmation in another photo, taken few years later, this time looking from the top of the hill towards the valley (see Figure 9 in Chapter 1). Here, what appears to be the site of Khirbet Beit Mazmīl is in the foreground. The low resolution of the photo prevents us from grasping the details of the structures, but a few important elements can be described. The site had been abandoned for several years when the picture was taken, and it already appears to be in an extremely ruinous condition. This is a confirmation of the fragility of the drystone architecture, which would collapse quite fast unless constant maintenance was guaranteed. This is one of the reasons why multiple restorations are identifiable at different sections of the standing structures: they bear witness to a frequent, near-continuous act of maintenance during the last phase of occupation of the site.

The other element is that the structure did not have a second floor: the flat roof was accessible from the courtyard through the stairs we documented in front of vaulted room C3. The roof also enabled access to the higher level of the terrace system extending to the east of the site, as the *manṭarah* offered access to the lower one in the west.[47]

One final interesting document to consider is a 1950s map of the modern suburb (Figure 49). Here, the site is depicted in relation to the terrace system and the topography of the area: the limits of the 19th century site clearly coincide with the Northern Sector, which seems to occupy a rather level surface within a generally rather steep slope, especially towards the south. To the west, the wall dividing the Upper and Lower sites clearly coincides with a major elevation step along the slope. The slope seems particularly steep to the east of the site, in particular behind Area D and immediately northeast of the Southern Sector.

Unfortunately, the map lacks detail, and only the contour of the drystone buildings is clearly recognizable. The immediate surrounding of the *'ezbeh* appears to be free of ruins or other structures, with

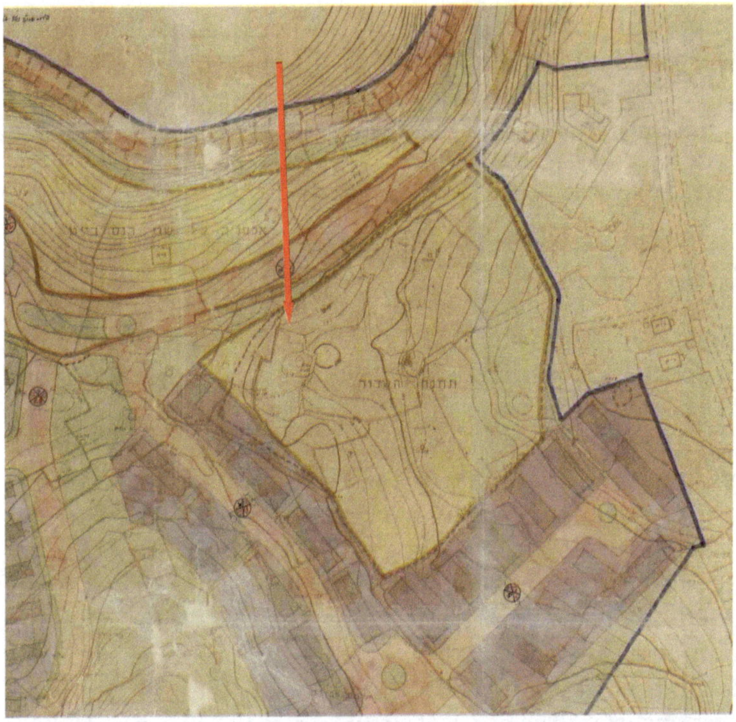

Figure 49 The 1950s plan of the modern suburb: the red arrow points at the site of Khirbet Beit Mazmīl.

the exception of possible surviving terrace and field walls. In any case, the map might be a further indication of how the topography determined and influenced the spatial arrangement of the site.

Certainly, the topography of the place underwent transformation over the course of the centuries, but it is possible that a similar morphology characterized the area also in the pre-Ottoman period. As mentioned in previous chapters, there might be pieces of evidence suggesting that the site had to cope with elevation differences also in its earliest phases, in particular the first phases of the wall dividing the Upper and Lower sites and some possible tower structures in Area S of the Lower Site. The map might give a further element in support of this hypothesis, even if we should be careful in projecting to earlier periods evidence from modern times.

To conclude, with its final occupation, the site makes a complete rupture with its architectonical history. The Early Ottoman village was now substituted by a single enclosed seasonal farmstead. Despite remaining tightly embedded into the terrace system, its function dramatically changed, turning from being a proper settlement—from which the terrace system ultimately depended—into a simple feature in the landscape itself.

A labor of love: Moving soil and stones

To revitalize the site and make it livable again, selective effort was put into site clearance. Much of the rubble from architectural collapse of the previous period was moved to the spaces between buildings (such as on the bedrock of C7). The collapse appears to have come primarily from the upper wall courses from the Phase II buildings, which were still in place in the Early Ottoman period. Sterile soil, likely collected from off-site, was brought to the courtyard area and dumped on top of the rubble and covering earlier Mamluk and Early Ottoman-period walls (see Figure 38).[48] Because of the uneven bedrock, which was deeper in the north and east of the site (50–60 centimeters deep in E2) than in west (10 centimeters under B1), significantly more soil was transported to Fields C6 and E2.

This soil deposit created a level building and use surface, covering what were then ruins of the North and South Buildings and creating a central courtyard.[49] Sections of the Mamluk-era vaults of the rooms in Field C appear to have been standing at the time, though in need of repair, and the ceiling above C4 repaired through the construction of a corbelled vault. On top of the ceilings of these vaulted rooms, a large volume of field stones—cleared from fields prior to cultivation—were piled on one another to create a massive, walled farmstead of irregular form. It was this massive wall that preserved remnants of the Mamluk and Early Ottoman complexes.

In an effort to calculate the labour involved in such a task in man-hours, we have adopted the labour models of the Kölner Tableau, developed by the Institute of Archaeology at the University of Cologne.[50] This tool was used to roughly determine the man-hours needed to do the soil and rock work described above, to fill these spaces, bury old walls, and create a courtyard of these dimensions:

1. E2: 64 meters square, with a combined soil and rock deposit of 55 centimeters deep—the space being filled was done so primarily with soil (two-thirds soil and one-third rock).
2. B4: 42 meters square, with a soil deposit of 10 centimeters deep.
3. C5: 35 meters square, with a soil deposit of 10 centimeters deep.

Labor calculations, of course, had to factor in the kind of tools used and the distance the soil and rock was carried. Loosening and carrying the soil (two-thirds of the total courtyard space thus filled) with a hoe or pickaxe—without the use of a shovel—and moving it 3 meters away; transporting the architectural collapse (one-third of the total courtyard space thus filled) 3.5 meters away; and refilling the space would have required the following man hours: E2–66 hours and 38 minutes in total; B4–7 hours and 56 minutes in total; and C5–6 hours and 36 minutes in total. This is the equivalent of 81 man-hours of total labor. If the work was done by a peasant household, and by the men of the family (as is noted in ethnographic accounts), then it would have taken five adult men, each working 16 hours, to create the central courtyard. In comparison to the rural labor patterns in the Mamluk period, peasant labor in the Ottoman era was more self-exploitative (Walker 2022). The difference was in the way the two sultanates managed land and important changes in land tenure from the 15th century, a point that will be explored in more detail in Chapter 8. The back-breaking work of creating cultivable courtyards, of building heavy drystone towers, and constructing and maintaining agricultural terraces was largely done under peasant initiative, and it was this labor that drove the commercialization of agriculture in the Jerusalem area in the Ottoman centuries.

A cultivable courtyard was thus created as a large open space of irregular form, encircled by a series of *manāṭīr* (serving as both agricultural watch towers and living space) and wide walls connecting them. A well house, the foundations and basic form constructed long before (as described in Chapters 3 and 4), occupied the east side of the complex. A number of tether stones, built into the southern faces of the C3 and C4 vaulted rooms, watering troughs, and grinding stones, attest to the use of the courtyard for stabling of animals and food preparation.

While all the trees one sees at the site today are young, no older than a few decades, the courtyard was first cultivated as a garden long ago. The raised bed described earlier was built in an 18th century level. The fill (C6.218) of the bed (Wall 125) was mixed, as noted before, producing both early 18th and late 19th/early 20th century porcelain, as well as Black Gaza Ware. The bed was built in the early 20th century, and the tree for which it was originally constructed planted then. The tree that occupied that space in 2017 was among the youngest almond trees in an orchard that was no more than 40 years old, with many trees as young as 10 years. It is possible that this was a wild tree, the seed of which fell into this place quite by accident, and never sunk deep enough roots to stand on its own for long.[51]

One further courtyard-related intervention should be mentioned in this regard. Prickly pear cactuses have surrounded the site since the IAA survey of the site in 2011 (Figures 51a and 51b). Before removing them in 2017, a mature hedge of cactuses also lined the south face of the North Building. There the cactuses hid from view a seam in the drystone masonry between C6 and C7 (Figure 50a) and obscured the springer of a vault in Wall 121, as well as the west side of a cloister vault in C6 (L.111) (Figure 50b). Prickly pear cactuses were imported from the New World during the

Figure 50 a) Cactuses of Field C hiding wall seam between C6 and C7 and vault springer (Wall 121); b) Wall seam and springer exposed (photo: B.J. Walker).

Ottoman period and used to delineate property lines and protect orchards and vineyards in the late 19th and early 20th centuries.[52] They were found in abundance in vineyards, and often eaten with bread (Schick 2010, 43, citing the mid-19th century account of the German missionary Friedrich Klein).

Their function inside and alongside standing architecture begs for an explanation. Outside the walled site, they would have functioned as property markers, perhaps at a time when the ruins were used only as an orchard. Inside the space outlined by the *'ezbeh* walls, however, they must have served another function. In this position, they appear to have been planted to hide repairs to and rebuilding of the drystone construction. It is not clear when they were planted. Prickly pear cactuses can live up to a century, though their average life span is 20–30 years. The most likely period of their cultivation at Khirbet Beit Mazmīl, is the first half of the 20th century (and before 1948), perhaps to hide repairs to the building and during a time when only the courtyard was used.

More labour-intensive than the transport of soil was the transport of stone. The construction of the *'ezbeh* required back-breaking work. As the *'ezbeh* was both residence and garden, further effort was put into making the most use of limited space for multiple functions. Terraces were built into the drystone construction from the beginning. Several walls (W. 109 in Field C; W. 107 and 113 behind Field D; and the terrace above D3) served this function, creating a two-tiered, orchard and freeing up courtyard space for other activities. These developments are most visible above D3, where the roof of the main cistern provides the base of a second garden level (Figure 51a). The medieval cistern house (C3) thus served the extra function as a terraced almond orchard, with Walls 110 (a rebuild of an older wall), 107, and 112 creating a terrace on the roof of the easternmost cistern's vault (Figure 51b). The upper courses of Wall 106, as well, formed its own terraces that encircled the *'ezbeh* on its east side. Walls 111 and 113 extend well into the space of what is now a modern parking lot, which may cover the northernmost parts of the Mamluk and Abbasid complex; their function is unknown.[53] Construction of this system of terracing required further soil transport and deposition.

Figure 51 Two-tier terracing in Field D: a) Above the D3 cistern; b) View west (photo: B.J. Walker).

around a complex roughly 25 meters square in dimension.[55] The house of B4 and the *manāṭīr*—which include the vaulted structures of A1, B1/B2, and C3/C4—comprise individual domestic units, with their own courtyards, cooking spaces, and activity areas. If our reconstructions of phases of soil transport are correct, there were four courtyards used simultaneously at the site, each likely corresponding to a multiple-household family unit that opened onto it: one in C5 (approximately 7 × 10 meters), servicing C3 and C4; a second in E2 (approximately 7.8 meters east-west; the dimensions for north–south unknown), separated from the former by Wall 135 (identified as a terrace wall built to hold back soil); a third in Field B, (between 12 and 13 meters east-west; the dimensions north–south unknown) for the use of the households of B1 and B2; and a fourth in C8, for the household in B4 (approximately 8.6 meters east-west; the dimensions north–south are unknown, as this area was not excavated in its entirety).[56] They all share the terraced orchard and cistern house of Field D. The fields surrounding the site and along the slopes of the hilltop on which it sits, are linked to the site physically, spatially, and functionally. The small *manṭarah* (in this case very literally an agricultural tower) labelled L110 on the site map and attached to exterior face of the west wall of the complex (W.102, see Appendix 1), overlooked the terraced fields on the west slope; Walls 107 and 112 created a terrace above the housing complex on the east side; and Walls 113 and 111 appear to have served field walls for cultivated land adjacent to the site on its east side (and now under concrete). The space defined by the encircling wall (Walls 100, 101, 102, 107, and 108), then, constitutes the residence of the people living in the village of Khirbet Beit Mazmīl from

The stratigraphic record (see Appendix 6 for full site plan)

After a period of abandonment from the mid-18th century, a new housing complex—a walled farmstead—was built on a completely different model and with a vastly different construction technique, sometime in the 19th century. The four extant tower-shaped domestic complexes (*manāṭīr*) were built on 10 centimeters of an aeolian deposit (the abandonment layer: B2.107),[54] and in some places on top of stubs of ancient walls that could not be removed. The heavy walls between those towers created a defense-like wall of irregular form the mid-19th to the mid-20th century. This drystone complex was built in a single phase of construction sometime in the 19th century.

The individual single-room houses betray a surprising diversity. Those in Field C (and described above), are repurposings of earlier medieval buildings, and their layout and size are determined by a plan that was, by then, hundreds of years old. Excavation of the vaulted buildings in this field were limited to C4, the best-preserved building of the site. C4 is a single vaulted room with an interior space of 4 × 2 meters. A single niche provides space for a candle in the east wall. After removal of the deep subsoil layer (L.301) in the building, the entrance and threshold (L.302) were cleared, revealing a three-step stairway (L.305) leading from the threshold in the entrance of C4 down to the floor of the room. The occupational levels inside the room comprised of the following stratigraphic sequence: a beaten earth floor (L.306), which was directly under the L.301 subsoil; below this was a very thin white plaster layer (L.310) which probably represented another floor; and a floor that had been hewn into and modified the bedrock (L.307). In the northeast corner of C4, a robber's trench (L.308) was cut directly into the modified bedrock floor, of a form repeated elsewhere on the site—C6. 229 and the modified bedrock of E1 (no locus number assigned) of the Northern Sector, and Cisterns I and II in A12 of the Southern Sector—and attributed in all cases to the Abbasid period (see Figure 8 in Chapter 3); this was excavated and cleared. Of interest was the pottery that was retrieved from the pit contents (L.309): all the diagnostic ceramics dated from the Abbasid period (9th–10th centuries), and included Early Islamic fine-ware bowls, a fine-ware jar with an everted rim, and an Abbasid lamp. The east wall of the room (W.119) was built on top of several courses of an earlier wall, visible in the robber's trench. Outside the room, to the right of the doorway, three courses of a wall (W.120) of building stones of rather monumental size and well-cut form project from the building's façade and appear to disappear into the drystone construction of the east side of C4, behind Wall 119. During his survey of the site in the 1940s, Bagatti identified these as Crusader construction (Bagatti 1948, 16). Clearly, C4 is built on top of and incorporates much older building components, of the Abbasid and arguably Crusader periods.

The B1/B2 complex is in many ways a self-sufficient unit, with a clearly defined, small courtyard (defined by Walls 116 and 117 and extending to the A1 *manṭarah*) that expands into a larger space south of B4, and a platform or bench (*masṭabah*), L. 100. A spur wall (Wall 116) separates the bench from the rest of the small courtyard. The subdivision of courtyards in this manner is well known in Ottoman Jordan, where walls like this are used to separate different animals from one another (goats from chicken, for example), or to create spaces for special activities (cooking, sleeping, etc.).[57] The B2 room/house has windows placed at each of two stories (see Figure 42). This unit, while not excavated for structural reasons, was surveyed and documented. It was built anew on the abandonment level (L.107), without repurposing of any earlier structures; only a single massive block of basalt was reused as the lintel in the doorway of B1 (also visible in Figure 42). It represents the purest form of *manṭarah* architecture at the site, with corbelled vaults supporting a dome.

The building in B4 is a bit of an enigma. There is no evidence of occupation here in this period. The architectural collapse that we are associating with the 1927 earthquake (L.112) fell directly on the top of the extant wall that defines the building in the Early Ottoman period. In other words, three courses of the walls of the house were still standing and visible when the site was reoccupied in the 19th century, however it remained in ruins. It was not covered by soil brought there to create a courtyard, as was done elsewhere at the site, nor was it put to use as a midden. In the Mandate era, the space behind (south of) Wall 118 was used for garbage disposal, as suggested by the butchered remains of a cow.[58]

There was only one water system in use for all the housing units. The Field D water system was a reuse of the two cisterns located at the south-east end of the site. The complex was likely covered at this point. All the drystone terraces above it (Walls 106, 107, and 115) and the wall separating it from the rest of the complex (W.115) date to this occupational phase. The stratigraphic sequence in front of the C3 cistern was shallow and limited: a shallow deposition of topsoil (L.500) and fills (L.501 and L.502) of only 8–10 centimeters in depth. They, in turn, covered a thin plaster floor (L.506), which was very patchy, poorly preserved and ranged in thickness between 1.5–2 centimeters. Below the plaster floor was a flagstone floor (L.504), which covered most of the eastern sector of the area and has been dated to the Mamluk period by OSL. Most of the pottery recovered from the work Area D was Late Ottoman, with a fair amount of Black Gaza Ware *ibrīqs* present in the assemblage. These represent the last phase of use of the water system, and the end of full-time occupation at the site.

The roof of the drystone complex in C1 was also investigated, with the hope of exploring the vaulted chambers hidden by the drystone architecture in C7. The roof was resurfaced at least once in this phase of occupation, with a layer of small stones, reused tether stones, and pottery sherds of the late 19th century. The sherds included a fragment of a Sarreguemines Ware plate, of the kind produced in France, England and Scotland in the late 19th and early 20th centuries and illustrated in Figure 52.11 below (A. de Vincenz *personal communication*).

ʻEzbeh *as a cultural form: Daily life on the walled farmstead*

Experiencing the built environment

While rapidly disappearing today, the *ʻezbeh* was, in the recent past, the most characteristic man-made landscape feature of the central highlands. These walled farmsteads were inhabited by extended families during the harvest season, built through family and communal labor, and primarily connected to olive, grape, and fig cultivation, but also, more generally, orchards. Because they were seasonally occupied (for only a few months each year) and built largely of fieldstones without mortar, foreign visitors considered them ruins and abandoned places, describing them as such in their travel accounts. That locals also referred to them as *khirbahs* (ruins) simply reinforced the confusion over the function of this very special settlement type. The Arabic term carries many meanings, but in this place and time it was used to describe a seasonally occupied farmstead (Walker 2021a). Never located far from villages, which were occupied year-round, *ʻezbehs* were part of a larger system of terraced agriculture that also included agricultural watch towers (*manāṭīr*), cisterns, and the cultivated fields and orchards (physically marked with terraces and plot walls). Among those documented during our surveys in the Khirbet Beit Mazmīl-Ein Karem region were systems also equipped with elaborate networks of stairs and paths, which led the peasant farmers from one level of terraced fields to another, and from one set of fields to another. *ʻEzbehs* played an important role in the local economy, growing out of the endowment, privatization, and leasing of land and the rapid development of market agriculture during the centuries of Ottoman rule (Walker 2022). In social terms, they were the architectural expression of the extended family unit and remained part of family memory and communal lore until long after they fall out of use.

There are considerable sub-regional differences in the general layout and size of these complexes, though the construction techniques and function are the same (al-ʿĀmirī 2018; al-ʿĀmirī and Rahhal 2003). The *ʻezbeh* at Khirbet Beit Mazmīl has no close published parallels, as the plan and building material are heavily impacted by the reuse of preexisting architecture at the top of the hill. The drystone walls and roofing are laid on top of and utilize stubs of preexisting walls, a practice that has created the irregular plan of the enclosure wall. It does share, however, many attributes identified in hybrid manors across the central highlands, namely drystone construction, a walled complex centered on an open courtyard, corbelled domes over living spaces and *manāṭīr*, and proximity to terraced fields. While many rooms could have served multiple functions, at least one was clearly domestic (C4), as it was fitted with wall niches (for lamps), the interior of the corbelled dome was covered with soot (from cooking or burning fuel for heating), and a well-laid plaster floor (L.310) served as the main living surface. Sleeping, however, was done on the roof: mats were often laid there under a makeshift hut for this purpose (Kark and Oren-Nordheim 2001, 270).[59] The extant staircase attached to the courtyard side of room C3 is the only extant access to the flat roof at this site. Fields and roads could have also been watched from this vantage point. It was the courtyard, however, that would have been the center of activity for the household. The tether stones built into the façades of the vaulted rooms of Field C attest to the use of the courtyard for keeping livestock. Many basic household chores, such as food preparation and child rearing, took place in the same space.

Structures supporting agriculture and post-harvest processing were also found within the complex walls. The placement of large *manāṭīr* at the corners of the complex enclosure wall (at A1, B1, L.110, and possibly some of the vaulted rooms of Field C) served two purposes: to provide a view from all directions of the family's fields, and to store harvested and processed crops. Unfortunately, these buildings were not structurally sound enough to excavate and take soil samples from them for botanical study. We therefore miss an opportunity to document cropping practices from the farmstead itself.

Moreover, the domed and vaulted rooms of Fields B and C, where they were accessible for investigation, produced few in situ objects. Unlike the Phase II small finds, most artifacts from this phase of occupation (pottery, glass, and metal) were retrieved from middens and fills in the courtyard. These rooms were apparently cleaned out at some point in the past, and perhaps subsequent to the site's abandonment. The courtyard produced no surfaces where we could identify activity areas. A second period of soil transportation and the construction of a terrace above Field D, to create the almond orchard one sees today, along with subsequent episodes of planting (described below), erased most evidence of on-site cultivation, animal husbandry, and outdoor domestic activities from the later Ottoman decades. As a result, we could not identify any kitchen spaces nor conduct a botanical study from courtyard soils. A picture of daily life in the 19th century, however, can be pieced together through the pottery and small finds recovered from secondary deposits, discussed below.

Ceramics

The final ceramic phase at Khirbet Beit Mazmīl is represented by a small assemblage dating from the 18th to early 20th centuries, most of which was found in topsoil or subsoil contexts that were heavily disturbed by modern construction activities. Nevertheless, they provide insights into the final phase of occupation at the site before it was largely damaged and gradually abandoned following the earthquake of 11 July 1927 (Amiran *et al.* 1994, 277), which measured approximately 6.2 on the Richter Scale. The first corpus consists of a selection of Black Gaza Ware vessels, a hallmark of the ceramic repertoire for the region during the Late Ottoman and Mandate periods that was produced mainly in Gaza, Khan Yunis, and Faluja (Israel 2006, 39–47). The chronology for the emergence of this unique ware is contested, with some scholars downdating its appearance to as early as the 16th or 17th centuries (de Vincenz 2021, 168). However, it has been traditionally dated by most scholars (Israel 2006) from between the 18th to 20th centuries. That being said, recent research, based upon archaeological finds as well as the historical record, have demonstrated that the apogee of this production was in the 1940s, during the Mandate period (Erickson-Gini *et al.* 2022, 65, 86; Israel and Saidel 2021, table 1; Ustinova and Nahshoni 1994, 176). The manufacturing center for the production of Black Gaza Ware that was closest to the Jerusalem region was Faluja (Israel and Saidel 2021), which was located approximately 35 kilometers southwest of Jerusalem, near Hebron, literally half the distance to the production center in Gaza, and this seems to be the likely candidate from where this ware found its way to Khirbet Beit Mazmīl.

Most of the corpus of Black Gaza Ware presented here originated in the Northern Sector, with two vessels coming from the Lower Site, and only a few of these were from stratified contexts. The first example is a bowl (Figure 52:1) with oblique inverted rim that is triangular in section that was found in L.436, the fill from the stone-lined space in the northeast corner of E1; no parallel could be found for this bowl. A store jar with everted ledge rim (Figure 52:2) and vestige of one of the two handles, which ran from below the rim to the upper shoulder, was recovered from a subsoil layer in Area D (L.501). The form is known from al-Qubab near Latrun, where it was referred to as a "cooking pot" (Avissar 2009a, 12, fig. 2.11:13), but it is actually a *burniye*, which was used for the storage of *leben* cheese (Israel 2006, 234, subtype D1). These vessels correspond to Type J-GAZA-JR-1A from the Jaffa *Qishle* (de Vincenz 2021, 170, fig. 8.5A:3 no. 1), where they are dated to the middle of the 19th century. Another store jar (Figure 52:3) with an outturned and pointed rim was found in L.109, the subsoil layer in B4. It conforms to Water Jar Type 5 in the typo-chronology provided by Israel, where it is dated to the 18th century (Israel 2006, 242, fig. 259).

Perhaps the most ubiquitous form in Black Gaza Ware found at sites with Ottoman occupation levels was the *ibrīq*, a pouring vessel for liquids that has an ovoid body and two handles, one that runs from the bottom of the neck to the upper shoulder and another that attaches to the long spout (de Vincenz 2021, 172). Five examples are presented here. The first (Figure 52:4) is a neck fragment with horizontal inverted and collared rim, vestigial remnants of a handle on the lower neck/body join, and incised diagonal ridges on the neck. It was found in a topsoil layer (L.125) from the Lower Site, and the form is attested at Yoqne'am (Avissar 2009a, 9, fig. 2.3:8). The next specimen, from the same L.125 topsoil, is a neck fragment with an oblique inverted rim and incised horizontal ridges on the neck exterior (Figure 52:5). This vessel closely resembles Type C from the Black Gaza Ware typo-chronology provided by Israel (2006, 156, fig. 136), and an exact parallel was uncovered from excavations at the Gloria Hotel in the Christian Quarter of the Old City of Jerusalem (Dolinka 2021, 212, fig. 6:9). Another *ibrīq* neck fragment with oblique inverted rim (Figure 52:6) and undecorated neck was recovered from the midden fill between W.135 and W.138 in E2 I the Northern Sector; no parallel could be found for this vessel. The penultimate specimen (Figure 52:7), which was uncovered from the earthquake collapse layer (L.112) in the northeast sector of B4, is also a neck fragment, but it has an upright flattened rim that is triangular in section. The form is related by Israel (2006, fig. 134) to Ibrīq Type 4, which can date from the 18th to the 20th century, and also corresponds to Type J-GAZA-JUG-7A from Jaffa (de Vincenz 2021, 172, fig. 8.5A:4 no. 10), with a suggested date of the 19th to 20th century. Its presence in the 1927 earthquake collapse therefore fits well within the chronological range provided by both typo-chronologies for that form. The last Black Gaza Ware sherd consists of an *ibrīq* spout (Figure 52:8) with the vestigial remains of a broken handle and traces of worn pale red paint on its exterior, which was found during the clearing of the water reservoir in D1 (L.507). The spout conforms to those found on Ibrīq Type 3 (Israel 2006, 153–154, fig. 133 on the far left), and the painted decoration on it can be ascribed to the second half of the 19th century (A. de Vincenz *personal communication*).

Of importance, due to the secure dating of its context, are the Flow Blue porcelain sherds (Figure 52:9) found in L.111 from Area B, which was situated below a large collapse layer attributed to the earthquake of 1927. This blue-and-white Transfer Ware—from Staffordshire, England, and produced by the manufacturers Wedgwood, Minton, Royal

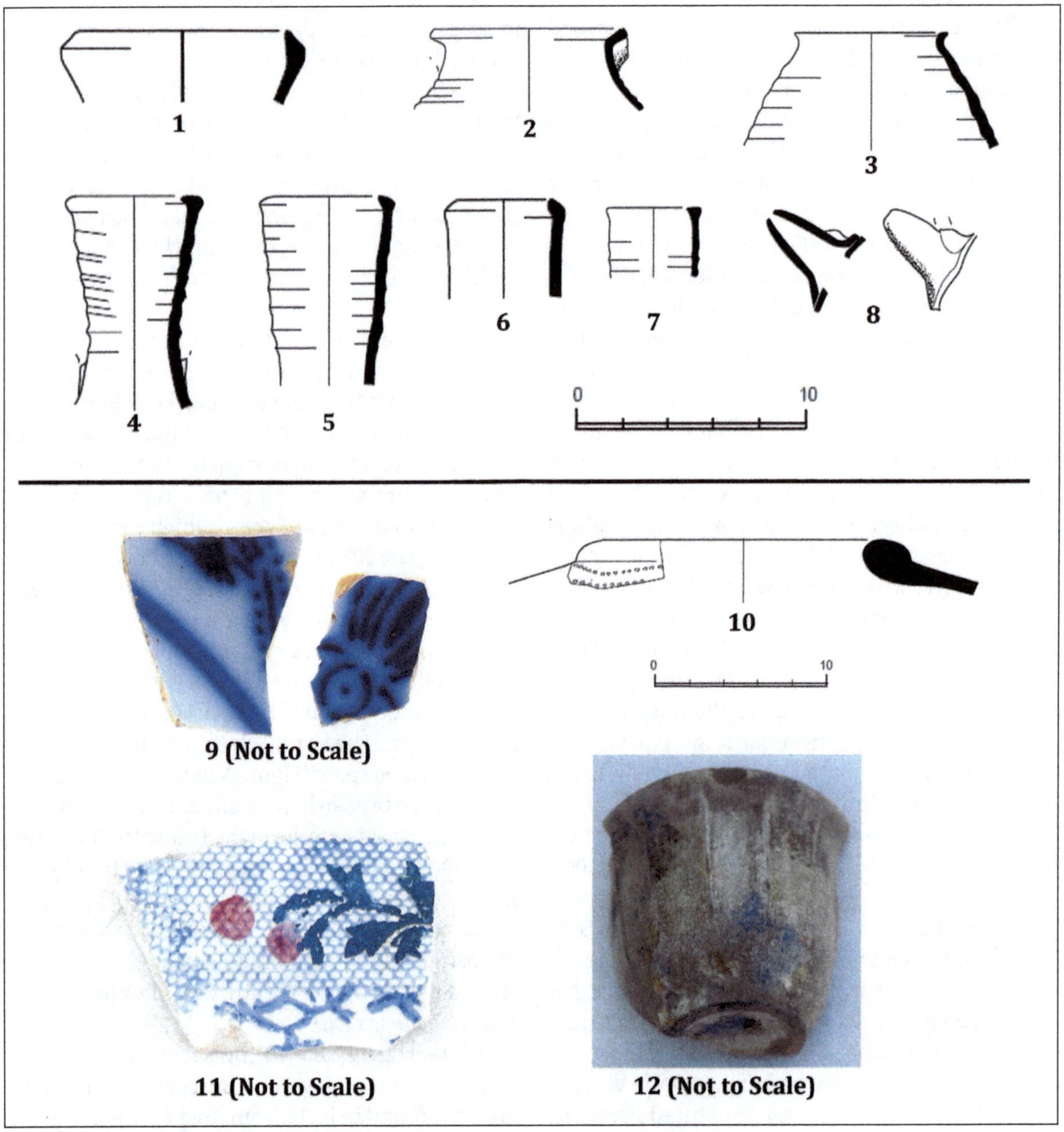
1
2
3
4
5
6
7
8
0
10
10
0
10
9 (Not to Scale)
11 (Not to Scale)
12 (Not to Scale)

No.	Locus	Reg. No.	Form	Description
1	436	4047/8	BL	Bowl with oblique inverted rim that is triangular in section. Coarse gray (2.5Y 5/1) fabric with small calcite, mica and quartz inclusions with dark gray (10YR 4/1) exterior.
2	501	5001/1	SJ	Store jar with everted ledge rim and vestige of one of the two handles which ran from below the rim to the upper shoulder. Coarse grayish brown (10YR 5/2) fabric with calcite, mica and quartz inclusions, gray (10YR 5/1) exterior.
3	109	1008/1	SJ	Store jar with out-turned and pointed rim. Coarse gray (2.5Y 6/1) well-fired fabric with calcite, mica and quartz inclusions.
4	125	1054	JG	*Ibrīq* neck fragment with horizontal inverted and collared rim, vestigial remnants of a handle on the lower neck/body join, and incised diagonal ridges on the neck exterior. Coarse grayish brown (10YR 5/2) fabric with calcite, mica and quartz inclusions and gray (10YR 5/1) exterior.
5	125	1056	JG	*Ibrīq* neck fragment with oblique inverted rim and incised horizontal ridges on the neck exterior. Coarse grayish brown (10YR 5/2) fabric with calcite, mica and quartz inclusions and gray (10YR 5/1) exterior.
6	439	4053/2	JG	*Ibrīq* neck fragment with oblique inverted rim and undecorated neck. Coarse grayish brown (10YR 5/2) fabric with calcite, mica and quartz inclusions and gray (10YR 5/1) exterior.
7	112	1013/6	JG	*Ibrīq* neck fragment with an upright flattened rim that is triangular in section and incised horizontal ridges on the neck exterior. Coarse grayish brown (10YR 5/2) fabric with calcite, mica and quartz inclusions, gray (10YR 5/1) exterior.
8	507	5010/2	JG	*Ibrīq* spout with vestigial remains of a broken handle. Coarse gray (2.5Y 6/1) well-fired fabric with calcite, mica and quartz inclusions, traces of worn pale red (10 7/4) paint on exterior.
9	111	1019/5–6	BL	Body sherds from a Flow Blue imported porcelain bowl. Pinkish white (7.5YR 8.5/2) fabric, thick white glaze, cobalt blue freehand linear, geometric and brush-stroke decoration.
10	107	1005/2	CP	Hole-mouth cooking pot with a thickened rolled rim and horizontal rouletted decoration below the rim zone. Coarse light reddish brown (5YR 6/3) fabric and a very dark gray (5YR 3/1) exterior with chert and crushed quartzite inclusions.
11	212	2014	BL	Rim sherd of a hard-paste bowl rim with stencil printed with blue net pattern and decorated with polychrome (blue, green and red) sponged decoration. Pinkish white (7.5 YR 8.5/2) fabric with a thick white slip and covered with a transparent glaze.
12	208	2009	CC	Glazed earthenware coffee cup with slightly out-turned rim, low ring-base and vertical fluting. Olive-brown (2.5Y 4/3) fabric decorated with mottled green and brown underglaze painting along the vessel exterior.

Figure 52 Late Ottoman and Early Mandate pottery from the Last Occupational Phase.

Doulton, and Swansea—reached its apogee during the late Victorian period, between the 1880s to early 1900s (Snyder 2000, 11). This porcelain conforms to Type J-TW-FB in the recent typo-chronology developed for the Ottoman period in Jaffa by de Vincenz (2021, 134), where most examples date from the end of the 19th century.

A good example of a Late Ottoman hole-mouth cooking pot (Figure 52:10) with a thickened rolled rim and horizontal rouletted decoration below the rim zone was also found in a fill below the earthquake tumble (L.107) in the southwest sector of B3, immediately to the south of the B1/2 *manṭarah*. This cooking pot is a well-known form, and the specimen from our site has parallels from al-Qula (Avissar 2008, fig. 2.11:11), Belmont Castle (Grey 2000, fig. 6.6:112), and the Jaffa *Qishle* (de Vincenz 2021, fig. 8.5B:1 no. 2, Type J-CP-2B), with a suggested date of the late 19th to early 20th century.

The final two vessels presented here were found during the clearing and/or cleaning of architectural elements in Area C, and both were contemporaneous in date to the hole-mouth cooking pot mentioned above. The first (Figure 52:11) is an interesting hard-paste glazed bowl sherd that was found in the fill on the rooftop of the C1 Vaulted Structure (L.212). It has an unusual hybrid decorative schema that combines both the Transfer Ware (de Vincenz 2021, 129) and Polychrome Sponged Decoration (de Vincenz 2020a, 502, fig. 4:6) traditions. In a fruitless attempt to find a parallel for this seemingly unique sherd, an exhaustive search was conducted, but alas none could be found with the comparative materials available to the writer. An expert on the subject consequently provided the following synopsis of it:

> "... the ware seems to be unpublished, but I have seen it before . . . this is a piece of hard-paste (refined earthenware). The blue net pattern was probably stencil-printed (you can actually see the seam of the stencil sheet!), while the other blue and green patterns were sponge-printed. The red dots are either sponge-printed or, more likely, just blobs of paint added with a round brush. These are rather cheap plates, produced mainly in France (e.g. Sarreguemines), England and Scotland (Portobello potteries and similar) in the late 19th and early 20th centuries." (A. de Vincenz *personal communication*)[60]

Given the paucity of glazed sherds that were recovered in post-Mamluk contexts, this sherd provides a rare glimpse of evidence for long-distance international trade at the rural site of Khirbet Beit Mazmīl. The final item from the Late Ottoman/Early Mandate assemblage retrieved from the excavations at the site is a glazed earthenware coffee cup with vertical fluting, a slightly out-turned rim and a low ring-base (Figure 52:12) that was found during the cleaning of W.104 in the northeast corner of C6 (L.208). It roughly corresponds to Type JER-CC-6B (de Vincenz 2021, 225) from the Jaffa *Qishle*, although no clear parallels for this coffee cup could be found. The vertical fluting on this vessel "is the most common modern way of decorating [these] vessels" (Savage and Newman 2008, 180), which date from the early 20th century (de Vincenz 2020b, 305). Thus ends the final chapter in the long biography of Khirbet Beit Mazmīl, ceramically speaking. It is imperative to tell the complete history of a site to better understand its changes over time; therein lies the importance of the Late Ottoman and Early Mandate pottery assemblages from the site.

Non-ceramic small finds

Outside of the glass bracelets discussed earlier in this chapter, the dates of which are generally assigned to the Ottoman period, the Late Ottoman phase of occupation at the site (Stratum Ib) was poor in non-ceramic small finds in comparison to previous ones. There were many fragments of heavily corroded iron implements and a few ground stone objects, as in the Early Ottoman Stratum 1a; these recovered primarily from courtyard fill. The assemblage as a whole would have serviced a rural peasant household engaged in agriculture, on-site processing of cereals, and some degree of animal husbandry. In combination with the ceramic assemblage of this period, the objects used in daily life at this site are not so different from those available to the resident of a manor house of a throne village, such as at Majdal Yābā; only the number and diversity of porcelain and fine earthenware imports distinguish them (Taxel 2017; Tsuk *et al.* 2016).

The people of the place—the Abū Hadhūd family

The lives and identity of the people who lived in Khirbet Beit Mazmīl in this period can potentially be traced through two sources, outside of the archaeological record. The series of 400 Ottoman *sijillāt* continue to 1917, but the latest files have yet to be transcribed and published in Turkey.[61] The most reliable sources, therefore, are a series of interviews concerning local history published by Palestinian residents or recorded by oral history initiatives like the Palestinian Rural History Project. The former residents of this village have long passed away, but their descendants, now of advanced age, carry the communal memory of the place and personal experiences of playing as children in the "ruins" and spending summers in the cultivated orchards. Fuad Abū Hadhūd, one such descendant, published on 11 January 2011, a rich account replete in information about the site in the 19th and 20th centuries. Abū Hadhūd's account is provided in translation below.

The origins of the Abū Hadhūd family[62]

> In the name of God, the most gracious, the most merciful—
>
> May the peace, blessings, and mercy of God be upon you.
>
> And may God's prayers and peace be upon His Nobel Messenger, our master Muhammad, his family and companions, and whoever follows his guidance until the Day of Judgment.

Now, I, the author of this clarification, Fuʿād Ismāʾīl Sulaymān Abū Hadhūd, was born in my beloved village of Ein Karem, seven kilometers west of Jerusalem, into which it was later absorbed/integrated. By the grace of God, I was probably seventy-six years old by the time I completed writing this report. Its purpose is to correct misconceptions by providing a detailed account of the link of the family Abū Hadhūd to our beloved Ein Karem, which often brings tears to my eyes when mentioned, and love, and longing.

I beg your permission to start with the following incident:

In the early fifties of the last [20th] century, one of my relative on my mother's side—may God have mercy on them both—died in a traffic accident. He was ʾAwnī Ḥussayn Rābiʾa, and he was a young, handsome, and popular young man; the surroundings of the cemetery, where he was being laid to rest, God's mercy on him, were filled with his lovers of different ages. I was then about fifteen or sixteen years old.

Among the mourners was an elderly sheikh, and God knows that he may have been over ninety years old. He was sitting on the edge of a nearby grave, turning his face to the lovely spring sun, and his body was shaking due to his old age. He is an elder of the al-Ḥamza Family, and I believe that he was Ḥajj Ismāʾīl al-Ḥamza. I fell under his gaze by chance, so he asked those standing by, "Whose boy is this?" He was told, "Son of ʾAbdullah Abū Hadhūd." He looked at me and said in the rural dialect, "Oh, Uncle, your origins are from Najd." I said to him in astonishment, "from Najd." He said, "Yes, your family hailed from Najd, my Uncle."

As for the second part:

We, the elders of the family, know—from what we heard from those who came before us, and on my part, from what I heard from my father, mother, and my father's grandmother, and the stories of some elderly people from Ein Karem—that our family, Abū Hadhūd, was based in Beit Mazmil, located directly on the top of the mountain to the east of Ein Karem (that is, between Ein Karem and al-Quds al-Sharīf [Jerusalem]).

The family lands extended from the end of the area of Jaljīl—it is said that it is a grave of one of the Righteous—which forms the eastern end of the territory of Ein Karem, to al-Qaṭamūn in southwest of Jerusalem, and to the borders of the village of al-Malḥa, southeast of Beit Mazmil. It included all the lands of Wādī al-Badawīyyah and the lands of Umm al-Niʿāj, northeast of Beit Mazmil, up to the main Jerusalem-Ein Karem road, between the settlements of Beit HaKarem and Bayit VaGan.

This was before my family suffered their greatest calamity, which will be mentioned below.

I know the area very well, and I know what was left of my family's lands in Beit Mazmil, which my grandfather divided among some of his children before his death. Part of my father's share was cultivated (*ma'mūr*): a vineyard of grapes and figs, and part of it was uncultivated. Al-Khirbeh was where my family's former dwellings existed. It remained in my grandfather's and grandmother's possession, and we used to dwell there in the summers during the school holidays until the end of the season.

However, I do not know how it happened that a small portion [of the land] was removed from it, on which the Jewish settlement of Bayit VaGan was established to the northeast of Beit Mazmil.

[The] third part is also the most important:

My people were, as I said before, perhaps a single family—Abū Hadhūd—and they dwelled in one place surrounded by a simple wall with a door, and it is the same place where we used to seasonally spend our summer holidays (*'izbat al-ṣayf*), after it became known as "al-Khirbah."

It happened one day that a child from the family was killed—the reasons for the killing are unknown—but this dead child was an orphan. And his mother was a stranger; she was from the village of Siʿir in the district of Hebron. So, this woman, the mother of the murdered child, called for the help of her kin. A group of young men from her village came out with their weapons in support of their daughter, and they went to the homes of my family in Beit Mazmil. My family welcomed them well and slaughtered animals for them.[63]

Here it must be said that the society of that time had a well-known custom, that in such cases, if the food served was salty, that meant that the other party was ready for reconciliation and understanding, and if it was without salt, then it meant the opposite.

When the people of Siʿir tasted the food and found it unsalted—whether that was intentional or by mistake—fighting broke out between the two parties in that narrow space, which, according to my estimation, does not exceed six dunams. Some members of my family were killed, while others managed to escape. All residences/dwellings but two rooms were destroyed. Thus the *'ezbeh* became a *khirbah*.

Meanwhile, the sounds of the voices reached some of the people of Ein Karem. Among them was Mr. al-Mishnī, the notable of Ḥāret al-Dhawār [quarter]. He mobilized his men to support his Abū Hadhūd neighbors, but by the time he arrived, everything was over. However, he continued to pursue the aggressors in order to inflict harm on them, until the leader of the Siʿir party stopped and said to him, "Oh, Mishnī, there is no blood or revenge between us and you.[64] It is enough for you to loot these lands that no longer have an owner." So al-Mishnī stopped.

As for me, as a grandson who came into being later, I do not know and have not heard how the lands were occupied and distributed.

As for the fourth part:

So how did those who survived among the Abū Hudhūd end up in Ein Karem?

It was said that there was a woman with her child, who raised sheep outside the village of al-Khirbeh. She survived, but had nowhere to go, so she sought shelter in one of the homes of Ein Karem near her. There she planted the first seed of the family of Abū Hudhūd. I think that her child was my father's grandfather. I think so, because I never heard a word about my grandfather Sulaymān from my father's grandmother or my father or anyone else. I remember the features of my grandfather: he was tall, board-shouldered, a strong farmer, who loved the land, and he was very old—I imagine that he may have been over ninety when he died. He had two daughters, Ḥalīma and Fāṭima, and four sons; they were uncles: ʾUthmān,

Mūsā, Maḥmūd, and my father, ʻAbdallah, who was the youngest. I was about sixteen years old when he died, God rest his soul.

However, there is some mystery to me—how did this happen? While my father also had cousins and uncles in Ein Karem, all of them were older than him. They had homes and children: Uncle Saʾīd, who has four sons (including Muḥammad and Yiḥyā) and some daughters; Uncle ʾĀlī, who has two sons (Muḥammad, whose nickname is ʿĪd, and Musṭafa) and some daughters; and Uncle Ḥajj Sulaymān, who has two sons (Yūsūf, who is known as Abū Sulaymān and Aḥmad, who is known as Abū Bakr) and some daughters. Ibrahīm, and Khalīl joined later, but I do not know where they came from, and they did not leave any offspring behind them.[65]

Thus, the researcher or interested person can distinguish from the small number of those who belonged to the Abū Hudhūd family in Ein Karem before the migration from Palestine, that the Abū Hudhūd family are not from the roots of Ein Karem's people, but from Beit Mazmil, and that they sought refuge there due to the circumstances of their ordeal. I do not know how they joined the Dar ʿĀlī clan (*ḥāmūlah*).

May God have mercy on our dead, your dead, and all those mentioned in this clarification. God is Forgiving, Merciful.

I have written honestly what I know to correct some confusion and inaccuracies.

I conclude the topic, and I have put down my pen. I ask anyone who has more information, clarification, or objections to please provide what he has.

—Fuʿād Ismāʾīl Sulaymān Abū Hadhūd,
6 January, 2011.

Important information about Beit Mazmil and the people of this village can be gleaned from this account. The author of the account, Fuʿād Abū Hadhūd, was born in 1935 in Ein Karem. He describes the initial abandonment of the village by his family, during a blood feud surrounding a killed dependent with another village. The property was apparently looted. From this point on, the village would be properly called "Khirbet Beit Mazmīl." This event happened four generations earlier, during the childhood of his great-grandfather. After the destruction of Beit Mazmil in the 19th century, one member of a branch of the family survived. The restored Abū Hadhūd clan retained possession of the abandoned ruin and one of the sons of the next generation took advantage of the improved security situation after the quenching of the Qays-Yaman rivalries in the 1850s to reestablish a presence in the ruin; this was the farmhouse which Fu'ad described—and not the original village in which his progenitors lived originally. This attempt failed in the long-term: while the grandparents continued to live in Khirbet Beit Mazmīl, and the grandchildren spent summers there, the children (Fuʿād's father and siblings) moved to Ein Karem as adults. The land, which was divided among the sons after the grandfather's (Sulaymān) death, continued to be cultivated.

The narrator also describes the physical form of the homestead (surrounded by a simple wall). It was one large walled complex, with a single entrance, and the entire space, in his estimate, was no larger than six dunams. Further interviews with members of the Abū Hadhūd family, and other residents in Ein Karem, have provided additional information about the family and the village. The Abū Hadhūd clan may have come to Palestine from Najd, perhaps through southern Jordan; exactly when and under what circumstances is unknown. The Abū Hadhūds maintained extensive social connections and marital alliances with villages as far as Si'ir. After leaving Beit Mazmil, surviving family members moved to Ein Karem, where they already had relatives, and the rest moved to other nearby villages. We can also deduce that the residents of Beit Mazmil were allied with a local strongman of Ein Karem's Eastern Quarter.

Like other family histories of lineages displaced from abandoned villages, the Abū Hadhūd's narration telescopes the history of the Beit Mazmil to that of one family, like the case of the Abū Ḥāmeds of Mulabbis.[66] It is not clear from interviews when the Abū Ḥāmeds left Beit Mazmil for Ein Karem, and it is possible that they lived at Beit Mazmil after 1948. The many people interviewed as part of the Palestine Rural History Project do not describe a "closeness" between the different clans that once lived there.[67] These family structures should be reflected in the architecture of the site, with each walled complex (each individual *'ezbeh*) accommodating one *ḥāmūlah*.

The orchards of Beit Mazmil were remembered by the people of Ein Karem as among the best gardens east of their village.[68] Beit Mazmil was one of many ruins surrounding Ein Karem; these included Ḥamāmah (which appears in the 19th century woodcut in Figure 7, Chapter 1).[69] In discussing weather and climate, informants often explained that the eastern wind comes down from Beit Mazmil, where its walls slows it down.[70] While this oral history project is ongoing, the close relationship between the villages of Beit Mazmil and Ein Karem, and their people, is palpable in these narratives. This is a theme to which we will return in Chapter 8.

Khirbeh in a suburb (Stratum 1a: British Mandate and Contemporary, 1920s to 2021)

Historical background

> "Wherever almond and fig trees, olive groves or clusters of cactuses are found, there once stood a Palestinian village: still blossoming afresh each year, those trees are all that remain." (Pappe 2006, 228)

While the oral history narratives describe occupation of the site until Fuʿād Abū Hudhūd's grandparents' generation, and cultivation of its lands until 1948, the archaeological record documents a gradual abatement of domestic use of the built environment by the late 1920s. While the First World War does not seem to have had a direct or immediate impact on the site's occupational history, a collusion of natural events and local social conflicts did. The earthquake of 11 July 1927, measured 6.2 on the Richter Scale and was felt all over Palestine and Jordan, causing 250–500 casualties across the region. The earthquake was so powerful that the flow of the Jordan River was interrupted, and there were landslides on the road connecting Jerusalem and Jericho. Jerusalem and its environs were heavily hit, causing extensive damage to buildings (Blackenhorn 1927).[71] Aftershocks struck Jerusalem for a year (Amiran *et al.* 1994). Drystone architecture, such as used in *'ezbehs* like Khirbet Beit Mazmīl, would have sustained considerable structural damage under these conditions.

The increasing urbanization of Jerusalem from the 1920s onwards also physically impacted the site. The Jewish neighborhood of Bayit VaGan was established in 1928 as a garden suburb of Jerusalem (Kark and Oren-Nordheim 2001, 161).[72] Two years after the earthquake, the villages of Ein Karem, al-Malḥa, and the newly established Jewish neighborhood of Bayit VaGan were caught up in the riots of 1929.

Bayit VaGan has experienced significant growth in recent years, and many of its high buildings now cover what was the Southern Sector and the eastern border of the Northern Sector, as well as the farmland to the east of the site.[73] In 1948 Ein Karem was depopulated, and with it seasonal occupation of Khirbet Beit Mazmīl ceased, and the site (that is, the walled housing complex) was abandoned completely. The agricultural land that belonged to it, however, appears to have continued to have been cultivated, at least for a few years more, as is apparent in Figure 8 in Chapter 1. This land eventually disappeared, with the establishment of the neighborhood of Kiryat HaYovel in 1952.[74]

The site abandoned

At what point is a site considered abandoned? According to the Abū Hadhūd interview, and the archaeological evidence, this was a gradual process of decreasing population size and reduced human activity on site. This process was precipitated by events of conflict, as well as more steady demographic changes inside the family. The family moved away from the village by the mid-19th century, but continued to call it home and live there, at least in the summers, apparently until 1948.

The land was cultivated continuously through that time, and its orchards were the pride of the Ein Karem region, as also noted by their neighbors from al-Malḥa. In 2018, our project investigated the relic agricultural terraces and rock piles of the 1950s Jewish neighborhood of Beit Mazmil (now part of Kiryat HaYovel), located 200 meters south of the archaeological site. None of the OSL dates confirmed an Islamic-period construction or maintenance; these terraces were constructed in the Early Roman period.[75] Much of the rest of the cultivated land that belonged to the Palestinian Beit Mazmil—to the west, east, and southeast of the site—is now under lost to construction, and with it a rare opportunity to study the cultivation practices of the Late Ottoman village.

Archaeological evidence for the gradual abandonment of the housing complex, on the other hand, comes from a British Mandate copper mill of 1927, sealed by architectural collapse (C5.304) under topsoil, directly in front of the well-preserved vaulted room of C4 (Figure 53). Similar collapse layers were identified throughout the site (B3.105, B4.111, and B4.112)—all covered by subtopsoil—and represent this phase of structural damage and partial abandonment. Two sherds of Flow Blue porcelain, likely from the same vessel and recovered from B4.111 provide a *terminus post quem* of the late 19th century for this collapse. Despite the structural damage to parts of the drystone complex, the courtyard and terraces continued to be used, as well as some of the domestic spaces in the rest of the complex, until 1948. The sub-topsoil, that covered the collapse layers, produced fragments of metal tools and nails; glass bracelets, bottles, and jugs; and *finçans* (Turkish coffee cups);[76] as well as coins (Obj. #2 in C6.203 and those sewn to veils) from the Mandate era.

The secret garden: Rebirth of ruins

The site laid in ruins until the 1970s, when it was repurposed as a garden and used by local neighborhood children as an informal playground. The almond orchard that one sees at the site today was planted in the last 15–40 years. In 2015, the orchard also included pomegranates and a wild apple tree or two; in the 1970s roses were planted there, though they are long gone.[77] The combined topsoil and subtopsoil—which was 42 centimeters thick in Field C and 33 centimeters in Field D—represents the last phase of development of the courtyard space as garden space in the last five decades. This thick topsoil consisted of some soil added for the modern garden, along with garbage disposed by the two neighborhoods that have grown up to the east and west of the site.[78]

Succumbing to urban sprawl

Today all the agricultural land, the smaller *manāṭīr* and field walls, and the lower half of the housing complex—nearly everything that constituted the village of Beit Mazmil—is covered by modern construction. What remains is the northernmost part of the housing complex at the top of the hill, which is today the Khirbet Beit Mazmīl Archaeological

Figure 53 Collapse in front of C4 doorway (photo: B.J. Walker).

Park (discussed below). Cellcom, one of the first telecom providers in Israel, expanded into the 3G wireless service market in 2006, and at that time it built a tower behind the *'ezbeh*, at one of the highest locations in the city. The parking lot was built at the same time. Although that tower was gone when the IAA conducted fieldwork there in 2011, this site has been known as "Giv'at Antennot" (Hill of the [Cell] Tower) ever since. After this, the land surrounding the site became prime real estate, and there has been non-stop construction ever since. The site of the Northern Sector (the buildings and the young almond trees in its courtyard) were set on fire (by kerosene) twice during the IAA's excavations there in 2013. We have witnessed the collapse of parts of the drystone construction since 2015—the combined result of hard winter rains and thaws and regular vandalism.

The vaulted room of C4 and the cistern complex in D1 were repurposed in the 2010s by local children, teenagers, and young men, in admittingly creative ways: the vaulted room at different times as a "bachelor's pad" (replete with sofa and dartboard) and prayer room, and the cisterns as swimming pools (in spite of our efforts to close them off for safety reasons). The site today is an archaeological park, developed by the IAA and preserving the *'ezbeh* of the last phase of occupation.

The Khirbet Beit Mazmīl Archaeological Park today (2022–2023)

The abandonment of Khirbet Beit Mazmīl in the 20th century ushered in a new era for the site marked by drastic changes. Without routine maintenance, the site began transforming from an active living space to that of a ruin. Slowly the drystone walls and constructions weakened, shifted, and collapsed. As a dormant site, it developed a new life, beginning to catch the attention and imagination of archaeologists. In the late 1800s Khirbet Beit Mazmīl was apparently a site well known to local guides, and the buildings were visited by Guérin during his extensive geographical, historical, and archaeological survey of Judaea (1869, 82–83). Confused with the identification of the site, he called the site "Khirbet 'Aïn Karim," suggesting that the ruin was a subordinate settlement related to the much larger village, although his Arab guide told him the site was known locally as "Khirbet Beit Mizmîr." According to Guérin's description, the site contained architectural remains of several houses built of large irregular stones. Impressed by the ruins, he associated the site with the biblical residence of Obed-Edom, which served as temporary resting place for the Ark of the Covenant (2 Samuel 6 and 1 Chronicles 13). Testament to the archaeological interest in the site can be seen by the inclusion of the site in the map drafted for the Survey of Western Palestine conducted by the Palestine Exploration Fund (Conder and Kitchener 1883, 108, Sheet XVII).

On 6 August 1926 the Hebrew newspaper *Dvar* reported the groundbreaking for the first house of the new Bayit VaGan neighborhood.[79] The construction of Bayit VaGan increased throughout the next two decades, and in the 1950s construction of the Qiryat Ha-Yovel neighborhood was begun with Khirbet Beit Mazmīl at its northeastern extremity. The neighborhood was built and named to celebrate the 50th anniversary of the Keren Kayemeth LeIsrael-Jewish National Fund.[80] Access to the neighborhood from the north was via a winding road, today known as Arther Hantke Street, which rides the ridge east of Ein Karem and veers around Khirbet Beit Mazmīl to its north. Moreover, the summit of the Khirbet Beit Mazmīl hill was commandeered by the large antenna towers erected by the Israel Broadcasting Authority. The antennas became the new dominant feature, casting shadow on the ruins below, and locals began to refer to the site as "Hill of the [Cell] Tower" (as mentioned above), or "Antenna Hill." While these constructions could have been potentially fatal for the archaeological site, the *khirbah* became an integral part of urban life and was taken into consideration without any formal mandate from the government.

It was only in 1996 that Khirbet Beit Mazmīl became officially recognized as an antiquities site

(Site Nos. 17613/0 and 27050/0) and thus protected under Israeli law. The legal name of the site appears as "Jerusalem, Rehov Qiryat Ha-Yovel" with other names listed as "Beit Mamil," "Beit Mizmir," and "Sharafa," and is described as a medieval village with structural remains, including watch towers built upon ancient structures, and agricultural terrace walls covering 300 × 300 meters. The registration of the site granted the IAA legal authority over the site and allowed it to be involved in all future building projects both around and at the site.

As urban sprawl in modern western Jerusalem accelerated, the ruins at Khirbet Beit Mazmīl were threatened. After the inauguration of the Jerusalem light rail system, which runs along Arther Hantke Street, Khirbet Beit Mazmīl became a high traffic "short cut" for local residents cutting through the site to access the station. The increased informal activity at the site caused potions of walls and vaults to collapse and crumble at an advanced rate. In reaction to the issuance of development plans for construction in the land plots near the site, the IAA and the Jerusalem Municipality zoned the site as a "green area" for public usage. However, some of the plots had already been slated for schools. The development works of the education campus led to the first of three rescue excavations at the site. The results of the excavations led to an agreement between the IAA, Jerusalem Municipality, and private developers of the nearby housing projects to draft a plan for conservation of the site as an open archaeological park.

Monitoring the progress and results of the continued excavations at the site, plans for the Khirbet Beit Mazmīl Archaeological Park were drafted by Yonatan Tzachor, conservation architect on behalf of the IAA. The plans included numerous variations which could be implemented depending on a range of budget constraints. After the 2019 excavation season, the site was subject to vandalism and some of the stone masonry was irreversibly affected. As a sort of blessing in disguise, the destructive acts at the site emphasized the urgency for the creation of the park. In 2022 the final version of the park plan was approved, and restoration of the site began at the start of 2023 by a team of IAA conservators led by Yossi Vaknin and Mohammed Hamdan, with the help of manual laborers.

The plan of the Khirbet Beit Mazmīl Archaeological Park was chosen based on public safety and upkeep cost. For example, the cisterns and vaulted structures were deemed unsafe to keep open if the site were to be open access to the public. Furthermore, the park would need routine maintenance, and in order to keep costs to a minimum, most of the built remains from the Mamluk phase occupation were to be covered over. In short, most of the architectural features uncovered in the center of the site were to be covered (Areas B, C, and E), which would create an open, grassy yard enclosed by stone masonry. The conservation work would then focus on the peripheral walls, which were preserved to a considerable height, and restoration of the collapsed facade of the *manṭarah* in Area B and water facility in Area D. The plan also was designed to limit access to the site to a single path near the *manṭarah*, with paths from the main road and light rail train line as well as the Bayit VaGan and Qiryat Ha-Yovel neighborhoods.

After filling in the central area of the excavation and the vaulted rooms in the north, the restoration work focused on the B1/2 *manṭarah* in the southwest corner of the site which had been damaged after fieldwork ended. The reconstruction of the tower was accomplished using the excavation documentation and photographs taken by IAA inspector reports. The collapsed stones were put back into place, preserving and maintaining a uniform architectural style. The stones were placed using a simple mortar made from lime and sifted earth from the site, in order to strengthen the walls and prevent future collapse. In a similar manner, the mortar was added to the tops of all the peripheral walls of the site (Figure 54).

The most significant feature and focal point of the Khirbet Beit Mazmīl Archaeological Park is the water facility uncovered along the eastern edge of the site (Area D). In this area the remains of a hewn ritual bath were designated to be covered up. To close off the bath from public access, the conservators chose to rebuild a collapsed arch with stone masonry and mortar (Figure 55). The cistern and drainage system in the east were to be kept open, however, with an iron grate installed above the water shaft to prevent visitors from falling in. While this area was once arched over, the restoration plan decided not to rebuild the arch to prevent future vandalism. Between the water cistern and ritual bath, the excavation uncovered the remains of a flagstone floor. The conservation plan for the site identified this as an important feature to be preserved, and many flagstones were consequently added to create a completely paved floor area. The various stages of conservation were documented by the IAA Conservation Department. The culmination of the work, in the form of an internal report, was submitted discussing the specifics of modern intervention. With the restoration and conservation works complete, the Beit Khirbet Mazmīl Archaeological Park awaits its inauguration, planned for 2024. In the last 50 years the site has undergone remarkable transformations, beginning from a ruin at the edge of a new neighborhood, to that of an active archaeological excavation, and as a park that resembles, in many ways, the site as it stood 50 years ago. The Khirbet Beit Mazmīl Archaeological Park is now ready for its final transformation, and we await to see the site filled with residents and tourists alike, who can enjoy this wonderful piece of Jerusalem's cultural heritage.

Figure 54 View of the eastern wall of the *khirbah*, looking southeast (photo: M. Hamdan, courtesy of the IAA).

Figure 55 View of the water installation after the conservation work (photo: M. Hamdan, courtesy of the IAA).

Notes

1. As mentioned in the previous chapter, the stratigraphy of this area is extremely complicated and to a certain extent blurry. With the chronological information we have, we cannot exclude the eventuality that the transformation of the built environment took place at the end of the Late Mamluk occupation. What is certain, though, is that during the Ottoman period the entire room had a flagstone floor covering the entire surface, and a series of beaten earth surfaces and plaster floors were regularly laid on top of it. This finds direct parallels with all the other structures dating to the Ottoman period at the site.
2. The average ground level of the supposed courtyard is around 817 meters ASL. The ground on top Wall 110 and W111 has an elevation ranging between around 818.76 and 819.21 meters ASL. The top of the later Late Ottoman period W114 is slightly over 820 meters ASL.
3. The literature on the subject is vast but written largely for the purposes of documenting cultural heritage; few are archaeological studies. Among the most important works are al-ʿĀmirī (2012, 2018) and al-ʿĀmirī and Tamari (1989). For an overview on the debate on the subject and an extended bibliography, see also Fuchs (1998).
4. The charcoal in the wall plaster of B4 was subjected to radiocarbon analysis, which gave a date of 300–600 YBP. A date in the Early Ottoman era is, thus, reasonable.
5. Doorways in Mamluk-era structures in A6 of the Southern Sector were blocked in the same manner during this phase of occupation.
6. This pottery was not included in Storchan and Dolinka (2014).
7. The object number for the coin in the IAA collections is 171722. The find context may be related to erosion from the Northern Sector part of the site.
8. The study of kitchen culture at this site, and others of the Mamluk period in Jordan and Syria, are the focus of the ongoing PhD thesis (University of Bonn) by Shurouq Munzer, entitled "Cuisine Traditions and Cooking Vessels During the Mamluk and Early Ottoman Periods in Southern Bilād al-Shām." It will be published as a separate monograph.
9. A.M. Hansen *personal communication* (2 August 2023).
10. As for the find spots of Mamluk sherds, while most have been recovered from fill in abandoned spaces (as refuse in the C6 furnace room and C6 settling basin), we were fortunate to identify an in situ kitchen in B4 in the final season of excavation. This context was discussed in the previous chapter. The cooking wares here came from hearths and, in combination with *ṭābūns*, were made up mainly of small, poorly fired (friable), handmade cooking pots and bowls.
11. Metal pots and pans may have gradually replaced clay pots for cooking.
12. We know from medieval Arabic cookbooks that different vessel forms and materials (clay, metal, stone) were used to prepare different dishes (see Gascoigne [2013] for Early Islamic Egypt and Walker *et al.* [2024] for Mamluk Transjordan). Recent research on cookpots from a late Ayyubid kitchen at Tall Hisban (Jordan) has combined organic residue and isotope analysis of the vessels and archaeobotanical analysis of the surrounding soils to document the relationship among cooking vessels and ovens, cuisine, and food preparation practices (Walker *et al.* 2024).
13. This was a common practice in villages across Europe only a generation ago, when large clay ovens were built into the corners of kitchens.
14. The plastered floors and walls suggest this, as well as vault springers in C6.
15. They could have also been dropped in the kitchen, if the women were smoking there. Smoking tobacco was, from the moment it was introduced into Palestine, a pastime of both women and men.
16. Sándor Petőfi—*Befordúltam a Konyhára*...—a Hungarian poem of 1843 (I am grateful to Prof. Béla Bodó, University of Bonn, for bringing my attention to this poem). The renowned Hungarian pipe tradition of the 19th and early 20th centuries was inspired by the importation of Ottoman *chibouks*.
17. There is historical evidence that in the 19th century, the towns of Jerusalem, Jaffa, and Nazareth had their own pipe-makers, who also supplied

the nearby villages (Simpson 2002, 169). The evidence of Khirbet Beit Mazmīl, however, suggests that there was a local industry much earlier than this.

18. At the time of completing the writing of this monograph, the results were being prepared for publication. The residue analysis was done by Silvia Polla, of Freie Universitat in Berlin, and the isotope analysis by Luana Bontempo at the Fondazione Edmun Mach of the Istituto Agrario San Michele All'Adige, Italy. Salam Kassem, PhD student in Islamic Archaeology at the University of Bonn, conducted the materials analysis of the pipes at the Steinmann Institute of the same institution.
19. The beads and glass bracelets (often called bangles in archaeological literature) are typical small finds in rural Syria in this period. For a discussion of them in a larger burial assemblage from late 19th-century Jordan, see Walker (2001).
20. The following glass and numismatic reports are based on the unpublished project reports of Katsnelson (2014) and Kool (n.d.), respectively.
21. Data from the IAA menorah coin database (26/01/2015).
22. The midden loci are as follows: 402, 404, 408, 411, 416, 421, 422, 426, 427, 435, and 439.
23. There is, of course, a problem in using this date to understand the phasing of the midden fill. The olive pits are in a secondary context, thrown away like the pottery in the same soil level. The sampling for radiocarbon dating and botanical analysis was originally done because we initially defined this space as an agricultural terrace.
24. At the time of writing, the three coins had not yet been formally read. However, in the field we gave a preliminary reading of "Early Ottoman," as the three coins had the same dimensions and weight (in the hand) as late 16th-century *mangīr*.
25. The site of Tall Hisban in Jordan provides an illuminating parallel from an earlier period. Numerous glass bracelets have been recovered from Mamluk contexts throughout the site. In addition, a hoard of over 60 silver *dirhams*, hidden in a lamp; a copper scale pan (for weighing coins); and metal and glass weights (for the same process) were found in a series of vaulted buildings on the upper, west slope of the *tall*, just below the citadel. Written sources attest to the importance of the Hisban market (*sūq*) in the 14th century, which was said to have served 300 villages (Walker 2011). Here, these objects have been collectively interpreted as material evidence for the market activity documented in period chronicles.
26. For an analysis of the centrality of copper coins in the economy of Early Islamic Syria, see Heidemann (2011); for Middle Islamic (Mamluk) Jordan, see Schultz (2022).
27. Reproduced from Kool in Walker and Dolinka (2020, 217–220).
28. Because quantification of the pottery was conducted only during the last season of excavation (2019), we could not do statistical analysis across the site. We have done so, however, for those areas that were excavated only, or primarily, in 2019, such as the midden and courtyard of E2.
29. For Jerusalemites, *quṣūr* (fortress-like villas) functioned a bit like summer homes, where people vacationed in the countryside (Kark and Nordheim 2001, 119). Khirbet Beit Mazmīl has a different history than these, but the relationship between permanent and summer residences is informative.
30. The villages of northern Jordan have a similar history, as economic opportunities (and failures), population transfers by the state, and social conflicts have moved people back and forth between villages (Walker 2016). On the destabilizing impact of tribal conflicts on villages in Transjordan in this period, see van der Steen (2014).
31. For parallels of this practice in other Palestinian *'ezbeh*s, see Shadi (2012, 22, fig. 8a).
32. The ashy substance in the charcoal tray was sampled for macrobotanical analysis, as well as for phytoliths and radiocarbon dating. The results of the macrobotanical study concluded that "it appears to be very finely pressed ash that was mixed with a substance; it eventually hardened and crystallized over time into a large mass, while resting inside the container. The fragments are fairly brittle, and the fine substance looks like a dark brown glossy, crystallized substance. The mixture was very firmly pressed into

the container, since the larger fragments have impressions from the inside of the object. In terms of the contents that made up the ash, they were already pulverized prior to being put into the container, as is attested by the absence of any recognizable morphological features of plant tissues (seeds/fruits, leaves, flowers, wood, etc.)." This was, in short, the remnants of the tobacco.

33. There is fast-growing scholarship now on this ware, which is beyond the scope of this study to address. There is, nonetheless, growing evidence that the dark black Gaza Ware, as opposed to the earlier light gray ware, was a late Ottoman product that continued to be produced into the 20th century.
34. De Bruyn's (1715) drawing of the village in the early 18th century (see Figure 8 in Chapter 1) provides further visual support for an early 18th-century occupation.
35. The *narghila* set was recovered from the raised bed fill (C6.218.2020) at an elevation of approximately 817.25 meters ASL. The red slipped, burnished pipe came from fill E1.405.4005, at an elevation of 817.26 meters ASL.
36. I am grateful to University of Bonn PhD students, Ahmad al-Ghizawat and Muaffaq Hazza, for pointing me to these rich online discussions.
37. At the time of writing, these objects had not been cleaned or studied by a specialist, so it is not certain that they are from the First World War or the 1929 riots. However, the First World War is more likely. Excavations from around Jerusalem have also yielded First World War rifle cartridges at Khirbet ʿAdassa, Qalunia, Shuafāt, Beit Ḥanīna, and Nabi Samwīl (A. Peretz, IAA Specialist on First World War military equipment, *personal communication*). For a recent study of First World War rifle cartridges at Majdal Yābā in the early 20th century, see Glick in Taxel (2017, 74–76).
38. For an archaeological report on the excavations of the manor house of one of these "throne villages" (Majdal Yābā), see Tsuk *et al.* (2016) and Taxel (2017).
39. The ruins of the expansive walled farmstead of Khirbet Ezra, for example, consists of a farmhouse, terraced orchards, a *manṭarah* and mill, cisterns, and a threshing floor. Today, as likely in the past, cows are being grazed in its fields.
40. For ancient, local parallels of the association of watch towers, terraced fields, and drystone housing, see Edelstein *et al.* (1998).
41. *Manṭarah* complexes are a rapidly disappearing component of the landscape of historical Palestine. They have been the focus of cultural heritage efforts to conserve and reconstruct them, along with old agricultural terraces.
42. On traditional drystone architecture in the Palestinian highlands, see al-ʿĀmirī (2018), al-ʿĀmirī and Rahhal (2003), al-Awad and Bshara (2012), Fuchs (1998), and Jäger (1911).
43. Informal interviews with people from the neighborhood, and with former residents, produced the same narrative.
44. This hygienic practice would have kept animals away from the household water supply.
45. The British Mandate-era mosque in the Jordanian village of Ḥubrāṣ, excavated in 2006, carries such a staircase (Walker and Kenney 2006; Walker *et al.* 2007). For other contemporary examples serving *ʿezbehs* in the central Palestinian highlands, see al-ʿĀmirī and Rahhal (2003, 48–49).
46. See Ronn (1977); and Schick (2010, 42) (based on Paul Klein's reports of the mid-19th century). For a modern architectural study, see al-Houdalieh and Ghadban (2013).
47. During fieldwork we often walked across these roofs to conduct daily photography of Areas C and D.
48. The soil is of dark color and rich in humus. It has the consistency of potting soil and could have been collected from the fields immediately outside the residential area of the site.
49. The 18th-century date was provided by a late 18th-century coin recovered from the fill on which was built one of the corner towers (*manāṭīr*), providing a *terminus post quem* for the construction of the entire drystone complex, which was built in a single phase. The date is supported by what we know about the survival rates of this kind of vernacular architecture. Drystone (construction with field stones and without mortar) rarely survives longer than a century or two, as

it is highly vulnerable to winter rains and freezing and thawing.

50. The tool is publicly available and can be found online at: https://www.iak.uni-bonn.de/de/institut/abteilungen/klassische-archaeologie/forschung-1/grk-1878.
51. I am grateful to Yehoshua Frenkel (Professor Emeritus, Haifa University) for this suggestion about almond tree propagation.
52. On the use of large cactuses to separate orchard plots in this period, see Guérin (1869, 129). On the use of large cactuses to protect vineyards from animals and humans, see Klein's 19th-century account in Schick (2010, 42).
53. As is apparent from Figure 8 in Chapter 1, the northern side of the Late Ottoman *'ezbeh* ended where the parking lot now begins. That lot did not seem to interfere with the physical remains of this structure.
54. The same layer was identified in section throughout the site, but was not assigned locus numbers, as they were not excavated. Further excavation against the drystone walls presented a structural challenge and could have caused building collapse.
55. The south end of the complex was destroyed by bulldozers in the same construction work that removed part of the A1 tower.
56. A1, which was surveyed but not excavated in 2013, was largely destroyed by bulldozers during the construction work done at the edge of the site before 2015. While it is probable that this *manṭarah* also had its own courtyard, we cannot be certain.
57. For archaeological parallels from Tall Hisban, Jordan, see Area P in Walker *et al.* (2018).
58. For the full faunal report, see Chapter 7.3.
59. A light-weight pergola (*'arīsh*), likewise provided with mats, was also built at ground level behind the house for sleeping in the hot summer months (Dalman 2013, vol. I/2, 488).
60. Exact parallels for the stenciled and sponge-painted sherd have subsequently been identified at Caesarea, recovered from a Bosnian house excavated in the summer of 2022 (personal communication, Asa Eger). Unpublished at the time of writing this volume, they will appear in a forthcoming article by Asa Eger and Anna de Vincenz.
61. Microfilm copies are available in libraries in Israel, Palestine, and Jordan.
62. The Arabic text can be found in its entirety at the following internet blog: http://ainkarem.blogspot.com/2011/01/blog-post.html?m=1.
63. The slaughter at the feast by the aggrieved party is a common trope, ahistorical, that is used to explain the destruction of villages during inter-village warfare. This theme is documented in dozens of villages in the Palestine Rural History Project.
64. This is a warning against initiating a new blood feud, like the one which ended in the destruction of Beit Mazmil.
65. That is, they are outsiders who did not belong to the family.
66. The PRHP interview of the Abu Ḥāmed family took place on 16 May 2011. The Arabic transcription can be found online at: https://waqqad.7olm.org/t38-topic (last accessed 21 August 2023). For a detailed study of another such abandoned village—Jindās—see Marom (2022b).
67. PRHP interview conducted on 14 April 2023.
68. M.S. Huzayen *personal communication*, 4 June 2011.
69. M.S. Huzayen *personal communication*, 4 June 2011.
70. M.S. Huzayen *personal communication*, 4 June 2011. This is a common narrative among interviewees and comes from a local song that is often used at weddings. The transcription literally reads:

    أما الرياح الشرقية فتنحدر من بيت مزميل بعد ان تقوم
    شُرفات بيت مزميل بتصديعها!

71. The rapid replacement and repairs to flagstone floors across the site, the poor preservation of some of these floors across room spaces, and the general buckling of the uppermost floors is likely the result of the collapse of superstructure caused by this earthquake.
72. The land purchased for this purpose was that of al-Karrama (Kark and Oren-Nordheim 2001, 325).
73. At the time of writing, these include high-rise apartments, a community center, and a mall.
74. Kiryat HaYovel was a neighborhood initially created to settle Jewish immigrants from numerous

Arab countries. The neighborhood originally carried the name of the Arab village: Beit Mazmil (Wikipedia—"Kiryat HaYovel").

75. O. Ze'evi *personal communication*, 27 July 2023. These terraces and rock piles were Area A of the Tel Aviv University hinterland study of our larger project.
76. Surprisingly few *finçans* were recovered from the site from any of the seasons of excavation.
77. The planting history is based on visual assessment of the age of trees and personal communication from Gideon Avni (IAA), who grew up in Bayit VaGan in the 1970s.
78. The modern refuse is made up of mostly corroded metal fragments: a chisel, a metal can, a mechanical piece that looked like part of a firearm, and so on.
79. The newspaper is available online at: https://www.nli.org.il/en/journals/NNL-Journals 990018701260205171/NLI.
80. *Yovel* is a biblical measurement of a 50-year cycle.

References

Abdulfattah, K.

2007. "Throne villages of the highlands: Local nobility and their mansions in Ottoman Palestine." *Near Eastern Archaeology* 70.1: 43–50.

'Adawī, Z., and B.J. Dolinka

2021. "Jerusalem, Qiryat Ha-Yovel: Preliminary report." *Hadashot Arkheologiyot: Excavations and Surveys in Israel* 133 (04.05.2021). Online Edition: https://hadashot.iaa.org.il/Report_Detail_Eng.aspx?id=25962&mag_id=133.

Amiran, D.A.K., E. Ariel, and T. Turcotte.

1994. "Earthquakes in Israel and adjacent areas: Macroseismic observations since 100 B.C.E." *Israel Exploration Journal* 43.3–4: 260–303.

al-ʿĀmirī, S.

2003. *Throne Village Architecture*. Ramallah: RIWAQ-Centre for Architectural Conservation.

al-ʿĀmirī, S.

2018. *The Peasant Architecture in Palestine*. Ramallah: RIWAQ-Centre for Architectural Conservation.

al-ʿĀmirī, S., and V. Tamari

1989. *The Palestinian Village Home*. London: British Museum Publications.

al-ʿĀmirī, S., and F. Rahhal

2003. *Manāṭīr: Quṣūr al-Mazāriʿ fī Rīf Filisṭīn*. Ramallah: RIWAQ-Centre for Architectural Conservation.

Ariel, D.

2016. "Coins from a small country: How excavated coins are managed in Israel, from the dig to the bookshelf." In *Les monnaies de fouille du monde grec (VIe-Ier s. A.C.). Apports, émoires et méthodes*, edited by F. Duyrat and C. Grandjean, 99–111. Bordeaux: Ecole française d'Athènes.

Arif, N., V. Yadav, S. Singh, S. Singh, P. Ahmad, R.K. Mishra, S. Sharma, D.K. Tripathi, N.K. Dubey, and D.K. Chauhan

2016. "Influence of high and low levels of plant-beneficial heavy metal ions on plant growth and development." *Frontiers in Environmental Sciences* 4:69. Internet Edition: https://www.frontiersin.org/articles/10.3389/fenvs.2016.00069/full.

Avissar, M.

2003. "Early Islamic through Mamluk pottery." In *Jewish Quarter Excavations in the Old City of Jerusalem, Volume II: The Finds from Areas A, W and X-2—Final Report*, edited by H. Geva, 433–446. Jerusalem: Israel Exploration Society.

Avissar, M.

2009a. "Ottoman pottery assemblages from excavations in Israel." In *Reflections of Empire: Archaeological and Ethnographic Studies on the Pottery of the Ottoman Levant*, edited by B.J. Walker, 7–14, Figs. 2.1–12. Boston: American Society of Overseas Research.

Avissar, M.

2009b. *The Pottery from the Excavations at the Coptic Patriarchate in the Church of the Holy Sepulcher*. Jerusalem: Unpublished manuscript on file at the Israel Antiquities Authority.

Avissar, M.

2010. "Early Ottoman pottery." In *Jewish Quarter Excavations in the Old City of Jerusalem, Volume IV: The Burnt House of Area B and Other Studies Cardo—Final Report*, edited by H. Geva, 334–336. Jerusalem: Israel Exploration Society.

Avissar, M.

2012. "Pottery from the Early Islamic to the Ottoman Period from the Cardo and Nea Church." In *Jewish Quarter Excavations in the Old City of Jerusalem,*

Volume V: The Cardo (Area X) and the Nea Church (Areas D and T)—Final Report, edited by H. Geva, 301–345. Jerusalem: Israel Exploration Society.

Avissar, M., and E.J. Stern

2005. *Pottery of the Crusader, Ayyubid, and Mamluk Periods in Israel. IAA Reports* 26. Jerusalem: Israel Antiquities Authority.

Avni, G., and J. Seligman.

2003. "New excavations at the Church of the Holy Sepulchre compound." In *One Land—Many Cultures: Archaeological Studies in Honour of Stanislao Loffreda OFM*, edited by G.C. Bottini, L.D. Segni, and L.D. Chrupcala, 154–162 (Studium Biblicum Franciscanum, Collectio Maior 41). Jerusalem: Franciscan Printing Press.

al-Awad, J. and Kh. Bshara

2012. *Buyūt al-Rīf fi Filisṭīn*. Ramallah: RIWAQ-Centre for Architectural Conservation.

Bagatti, B.

1948. *Il Santuario della Visitazione ad ʿAin Karim (Montana Judaeae). Esplorazione archeologica e ripristino.* Jerusalem: Gerusalemme Tipografia dei Francescani (Studium Biblicum Franciscanum 5).

Baram, U.

1996. "Material culture, commodities, and consumption in Palestine, 1500–1900." Amherst: Unpublished PhD dissertation, Department of Anthropology, University of Massachusetts, Amherst.

Bent, J.T.

1893. *Early Voyages and Travels in the Levant, vol. I: The Diary of Master Thomas Dallam, 1599–1600*. London: The Hakluyt Society.

Blackenhorn, M.

1927. "Das Erdbeben im Juli 1927 in Palästina." *Zeitschrift des Deutschen Palästina-Vereins* 50.4: 288–296.

Boas, A.

2000. "Pottery and small finds from the Late Ottoman village and the early Zionist settlement." In *Ramat Hanadiv Excavations. Final Report of the 1984–1998 Seasons*, edited by Y. Hirschfeld, 547–590. Jerusalem: Israel Exploration Society.

Boulogne, S.

2007. "Reflet d'un art mémoires: les bracelets de verre coloré du Bilad al-Sham mémoires et ottoman." Paris: PhD dissertation, Université de Paris-IV (Panthéon-Sorbonne).

de Bruyn, C.

1715. *Voyage au Levant, c'est-à-dire, dans les principaux endroits de l'Asie Mineure, dans les isles de Chio, Rhodes, et Chypre etc.* Paris: G. Cavelier.

Conder, C.R., and H.H. Kitchener

1883. *The Survey of Western Palestine. Memoirs of the Topography, Orography, Hydrography, and Archaeology. Volume III: Sheets XVII-XXVI. Judaea*, edited by E.H. Palmer and W. Besant. London: Palestine Exploration Fund.

Dalman, G.

2013. *Work and Customs in Palestine. Volume I/1 and 2*, translated by N. Abdulhadi-Sukhtian. Ramallah: Dar Al Nasher.

Dekkel, A.

2008. "The Ottoman clay pipes." In *Paneas Volume II: Small Finds and Other Studies, IAA Reports* 38, edited by V. Tzaferis and S. Israeli, 113–164. Jerusalem: Israel Antiquities Authority.

Dolinka, B.

2014a. *Notes on the Ceramic Assemblage from the Excavations at the Diaspora Yeshiva on Mount Zion, Old City of Jerusalem*. Unpublished manuscript.

Dolinka, B.

2014b. *Medieval Pottery from Misgav Ladach 26, Jewish Quarter, Old City of Jerusalem*. Unpublished manuscript.

Dolinka, B.

2017. *The Middle and Late Islamic Pottery from the Excavations at the Austrian Hospice, 2011-2012*. Internet Edition: https://www.academia.edu/10347780/THE_MIDDLE_AND_LATE_ISLAMIC_POTTERY_FROM_THE_EXCAVATIONS_AT_THE_AUSTRIAN_HOSPICE_2011_2012.

Dolinka, B.

2021. "The Islamic pottery from Gloria Hotel, Jerusalem." In *New Studies in the Archaeology of Jerusalem and Its Region: Collected Papers*, Volume 14, edited by Y. Zelinger, O. Peleg-Barkat, J. Uziel, and Y. Gadot, 199–217. Jerusalem: Israel Antiquities Authority.

Edelstein, G., I. Milevski, S. Aurant, H. Gitler, L.K. Horwitz, M.E. Kislev, R. Reich, A.M. Rosen, S.A. Rosen, S. Siegal, and P. Smith

1998. *Villages, Terraces and Stone Mounds: Excavations at Manahat, Jerusalem, 1987–1989*. Jerusalem: Israel Antiquities Authority.

Erickson-Gini, T., Y. Israel, D. Varga, and V. Lifshits
2022. Historical research and archaeology in Amaẓya/Duwāyima. In *Archaeological Excavations and Research Studies in Southern Israel, Collected Papers Volume 5 (18th Annual Southern Conference)*, edited by A. Golani, D. Varga, Y. Tchekhanovets, and M. Birkenfeld, 65–90. Jerusalem: Israel Antiquities Authority.

Falls, C.
1930. *Military Operations: Egypt and Palestine from June 1917 to the End of the War, Part I*. London: Macmillan and Co.

François, V.
2005. "Tabak, ibrik, fincan et autres pots d'époque mémoire au Bilâd al-Châm." *Turcica* 37: 281–308.

Fuchs, R.
1998. "The Palestinian Arab House and the Islamic 'Primitive Hut.'" *Muqarnas* 15: 157–177.

Gascoigne, A.L.
2013. "Cooking pots and choices in the medieval Middle East." In *Pottery and Social Dynamics in the Mediterranean and Beyond in Medieval and Post-Medieval Times*, edited by J. Bintliff and M. Caroscio, 11–24. London: British Archaeological Reports.

Ghadban, S.
2015. "Palestinian dry stone structures—an endogenous expression of cultural landscape." *Sustainable Development, Culture, Traditions Journal* 1b: 34–47.

Gichon, M., and R. Linden.
1984. "Muslim oil lamps from Emmaus." *Israel Exploration Journal* 34, 2–3: 156–169.

Grainger, J.D.
2006. *The Battle for Palestine 1917*. Rochester: Boydell & Brewer Ltd.

Grey, A.D.
2000. "The unglazed pottery." In *Belmont Castle: The Excavation of a Crusader Stronghold in the Kingdom of Jerusalem*, edited by R.P. Harper and D. Pringle, 87–100. Oxford: Oxford University Press.

Grossman, D.
1994. *Expansion and Desertion. The Arab Village and its Offshoots in Ottoman Palestine*. Jerusalem: Yad Ben Tzvi.

Guérin, V.
1869. *Description géographique, historique et archéologique de la Palestine II: Judée*. Paris: E. Leroux.

Heidemann, S.
2011. "The agricultural hinterland of Baghdad, Al-Raqqa and Samarra: Settlement patterns in the Diyar Mudar." In *Le Proche-Orient de Justinien aux Abbassides. Peuplement et dynamiques spatiales*, edited by A. Borrut, M. Debié, A. Papaconstatinou, D. Pieri, and J.P. Sodini, 43–58. Turnhout: Brepols.

Hirschfeld, Y.
1995. *The Palestinian dwelling in the Roman-Byzantine period*. Jerusalem: Franciscan Printing Press.

al-Houdalieh, A.H., and Sh.S. Ghadban
2013. "Agricultural watchtowers in al-Tireh Quarter and ʿAin Qinia Village, Ramallah, Palestine." *International Journal of Architectural Heritage* 7: 509–535.

Israel, Y.
2006. "The Black Gaza Ware from the Ottoman period." Beersheba: Unpublished PhD dissertation, Ben-Gurion University of the Negev.

Israel, Y., and B.A. Saidel
2021. "Archaeological evidence for the production of Black Gaza Ware pottery at the former Arab village of Faluja, Israel." *Journal of Eastern Mediterranean Archaeology and Heritage Studies* 9:1–2: 116–134.

Jäger, K.
1911. "Das Bauernhaus in Palaestina: mit Rucksicht aus dem Biblische Wohnhaus Untersucht und Dargestellt." Tübingen: PhD dissertation, University of Tübingen.

Johns, C.N.
1950. "The Citadel, Jerusalem. A summary of work since 1934." *QDAP* 14: 121–190.

Kabaklari, N.
1998. *Mangır: Copper Coins of the Ottoman Empire 1299–1808. (Osmanli imperatorluğu bakir paralari)*. Istanbul: Uşaklılar eğitim ve kültür vakfı yayınlari.

Katsnelson, N.
2014. "The glass finds from Bet Sheʿan (Youth Hostel)." *Atiqot* 77: 23–57.

Kark, R., and M. Oren-Nordheim
2001. *Jerusalem and Its Environs: Quarters, Neighborhoods, Villages 1800–1948*. Jerusalem: Magnes Press.

Kool, R.
n.d. *The Coins and Lead Seals from Khirbet Beit Mazmīl*. Unpublished project report of 2014.

Kool, R.
2005. "The coins." In *Tel Yoqne'am: Excavations on the Acropolis, IAA Reports* 25, edited by M. Avissar, 99–102. Jerusalem: Israel Antiquities Authority.

Kool, R.
2017. *An Early Ottoman Hoard from Khirbat Beit Mazmīl (late 16th c.).* Jerusalem: Unpublished report on file at the Israel Antiquities Authority.

Kool, R., and D.T. Ariel.
2002. "Coins from the salvage excavations at Meron." *'Atiqot* 43: 109–114.

Manna, A.
1994. "Eighteenth and Nineteenth-century rebellions in Palestine." *Journal of Palestine Studies* 24.1: 51–66.

Magness, J.
1993. *Jerusalem Ceramic Chronology, circa 200–800 CE. JSOT / ASOR Monograph Series 9.* Sheffield: JSOT Press.

Marom, R.
2022a. "Dispelling desolation: the expansion of Arab settlement in the Sharon Plain and the western part of Jabal Nablus, 1700–1948." Haifa: PhD dissertation, Department of Middle East and Islamic Studies, University of Haifa.

Marom, R.
2022b. "*Jindās*: A history of *Lydda*'s rural hinterland in the 15th to the 20th Centuries CE." *Lod, Lydda, Diospolis* 1: 2–31.

Mitchiner, M.
1988. *Jetons, Medalets & Tokens I: The Medieval Period and Nuremberg.* London: Hawkins Publications.

Palestine Economic Policy Research Institute.
2016. *The Economic and Social Impact of the Expansion of Tobacco Cultivation in Palestine.* Background Paper, Roundtable 6. Online Edition: http://www.mas.ps/files/server/20160510092303-1.pdf (page no longer extant).

Pamuk, S.
2000. *A Monetary History of the Ottoman Empire.* Cambridge: Cambridge University Press.

Pappe, I.
2006. *The Ethnic Cleansing of Palestine.* Oxford: One World.

Pini, N.
2020. "Walls of identities: Built environment as social marker." In *Living with Nature and Things: Contributions to a New Social History of the Middle Islamic Periods,* Mamluk Studies—Band 23, edited by B.J. Walker and A. al-Ghouz, 375–400. Göttingen: V&R Unipress.

Poyiadji-Richter, E.
n.d. "Metalwork products destined for Cyprus? The fifteenth-sixteenth century dishes in the Leventis Municipal Museum of Nicosia." Unpublished conference paper presented at *The Art and Archaeology of Lusignan and Venetian Cyprus (1192–1571): Recent Research and New Discoveries,* 12.12–14.2014. Cyprus: Archaeological Research Unit, University of Cyprus.

Prag, K.
2008. "Site SI, III-VI: The defences of the Ottoman city." In *Excavations by K.M. Kenyon in Jerusalem 1961–1967 Volume V: Discoveries in Hellenistic to Ottoman Jerusalem,* Levant Supplementary Series 7, edited by K. Prag, 243–378. Oxford: Oxbow Books.

Ragette, F.
2012. *Traditional Domestic Architecture of the Arab Region.* Zlin: Edition Axel Menges / American University of Sharjah.

RIWAQ
2006. *Riwaq's Registry of Historical Buildings in Palestine (Riwaq's Monograph Series on Architecture History of Palestine 9.1).* Ramallah: RIWAQ-Centre for Architectural Conservation.

Ronn, Z.Y.D.
1977. "Stone huts as an expression of terrace agriculture in the Judean and Samarian Hills." Tel Aviv: PhD dissertation, Department of Geography, Tel Aviv University.

al-Sabagh, M.H.
2014. "Before banks: Credit, society, and law in sixteenth-century Palestine and Syria." Santa Barbara CA: PhD dissertation, Department of History, University of California-Santa Barbara.

Savage, G., and H. Newman.
2008. *An Illustrated Dictionary of Ceramics.* London: Thames and Hudson.

Schick, R.
2010. *Palestinian Life, Customs and Practices: German Articles from the Late 19th and Early 20th Centuries.* Amman: University of Jordan Press.

Schultz, W.
2022. "Coins where there were no mints: Mamluk coins from Jordanian archaeological sites." In *The*

Mamluk Sultanate and Its Periphery, Mémoires de l'Association pour la Promotion de l'Histoire et de l'Archéologie Orientales, Université de Liège, edited by F. Bauden, 141–162. Louvain: Peeters.

Shadi, Gh.
2012. "The architecture of the Palestinian dry-stone hut: *al-mantarah*." *Journal of Comparative Cultural Studies in Architecture* 6: 16–25.

Shapiro, A.
2010. "Ottoman-Period Smoking pipes and nargile heads from excavations in the Old City of Akko." In *One Thousand Nights and Days: Akko through the Ages*, edited by A. Killebrew, 75–80. Haifa: Hecht Museum.

Shapiro, A.
2022. "Ottoman tobacco pipes from Khan et-Tujjar." *ʿAtiqot* 107: 263–270.

Shindo, Y.
2001. "The classification and chronology of the Islamic glass bracelets from al-Tür, Sinai." In *Cultural Change in the Arab World*, Senri Ethnological Studies, edited by T. Nishio, 73–99. Osaka: National Museum of Ethnology.

al-Shqour, R.
2019. *The Arab Khans and the Origin of Khans in Jordan: An Archaeological Approach*. Piscataway: Gorgias Press.

Simpson, St. J.
1998. "Where there's Smoke there's Fire . . .': Pipe-smoking in the Ottoman Empire." *British Museum Magazine* 31: 15–17.

Simpson, St. J.
2000. "The clay pipes." In *Belmont Castle: The Excavation of a Crusader Stronghold in the Kingdom of Jerusalem*, edited by R.P. Harper and D. Pringle, 147–171. Oxford: Oxford University Press.

Simpson, St. J.
2008. "Late Ottoman pipes from Jerusalem." In *Excavations by K.M. Kenyon in Jerusalem, 1961–1967, V: Discoveries in Hellenistic to Ottoman Times*, edited by K. Prag, 433–446. Oxford: Oxbow Books.

Snyder, J.B.
2000. *A Pocket Guide to Flow Blue (2nd ed.). Schiffer Book for Collectors*. Atglen: Schiffer Publishing.

Spaer, M.
1992. "The Islamic glass bracelets of Palestine." *Journal of Glass Studies* 34: 44–62.

van der Steen, E.
2014. *Near Eastern Tribal Societies During the Nineteenth Century: Economy, Society and Politics Between Tent and Town*. London: Routledge.

Storchan, B., and B. Dolinka
2014. "Jerusalem, Bayit Va-Gan." *Hadashot Arkheologiyot: Excavations and Surveys in Israel* 126 (09.10.2014).

Sutton, S.B.
2000. "Liquid landscapes: Demographic transitions in the Ermionidha." In *Contingent Countryside: Settlement, Economy, and Land Use in the Southern Argolid Since 1700*, edited by S. Sutton, 84–106. Stanford: Stanford University Press.

Taxel, I.
2017. "Majdal Yābā: More insights on the site in Medieval and Late Ottoman to Mandatory Times." *Journal of Islamic Archaeology* 4.1: 49–86.

Tsuk, T., I. Bordowicz, and I. Taxel
2016. "Majdal Yābā: The history and material culture of a fortified village in Late Ottoman- and British Mandate Palestine." *Journal of Islamic Archaeology* 3.1: 37–88.

Ünal, R.M., F. Kringzinger, M. Alram, and S. Pfeiffer-Tas
2010. *Die Münzschatz von Beçin*. Vienna: Österreichischen Akademie der Wissenschaften.

UNESCO
2012. *Palestine, Land of Olives and Vines. Cultural Landscape of Southern Jerusalem, Battir* [1492–2012-Nomination Text]. Internet Edition: https://whc.unesco.org/en/list/1492/documents/.

Ustinova, Y., and P. Nahshoni
1994. "Salvage excavations at Ramot Nof, Be'er Sheva." *ʿAtiqot* 25: 157–177.

de Vincenz, A.
2016. "Chibouk smoking pipes—secrets and riddles of the Ottoman Past." In *Arise, Walk through the Land*, edited by J. Patrich, O. Peleg-Barkat, and E. Ben-Yosef, 111–120. Jerusalem: Israel Exploration Society.

de Vincenz, A.
2020a. "Ceramic and porcelain vessels of the Ottoman period from the French School on Yefet Street, Yafo (Jaffa)." *ʿAtiqot* 100: 497–512.

de Vincenz, A.
2020b. "A clay tobacco pipe from Ramla." In *Ex Oriente Lux: Studies in Honour of Jolanta Młynarczyk*, edited

by K. Jakubiak and A. Lajtar, 85–80. Krakow: Archeobooks.

de Vincenz, A.

2021. "Porcelain and ceramic vessels of the Ottoman period." In *Excavations at the Ottoman Military Compound (Qishle) in Jaffa, 2007, 2009*, The Jaffa Cultural Heritage Project Series, Vol. 4, edited by Y. Arbel, 127–304 (Pipes: 174–215). Münster: Zaphon.

de Vries, B., and Ø. LaBianca

2009. "Preface." In *Reflections of Empire: Archaeological and Ethnographic Studies on the Pottery of the Ottoman Levant*, edited by B.J. Walker, ix-xiii. Boston: American Schools of Oriental Research.

Walker, B.J.

2001. "The Late Ottoman cemetery in Field L, Tall Hisban." *Bulletin of the American Schools of Oriental Research* 322: 47–65.

Walker, B.J.

2004. "Ceramic evidence for political transformations in Early Mamluk Egypt." *Mamluk Studies Review* 8.1: 1–114.

Walker, B.J.

2009. *Reflections of Empire: Archaeological and Ethnographic Studies on the Pottery of the Ottoman Levant*, edited by B.J. Walker. Boston: American Schools of Oriental Research.

Walker, B.J.

2011. *Jordan in the Late Middle Ages: Transformation of the Mamluk Frontier*. Chicago: Middle East Documentation Center, University of Chicago.

Walker, B.J.

2016. "The Northern Jordan Project and the 'liquid landscapes' of Late Islamic Bilad al-Sham." In *Landscapes of the Islamic World: Archaeology, History, and Ethnography*, edited by S. McPhillips and P. Wordsworth, 184–199. Philadelphia: University of Pennsylvania Press.

Walker, B.J.

2021a. "Khirba." In *Encyclopaedia of Islam* III, Part 2021.1: 73–75. Internet Edition: https://referenceworks.brillonline.com/entries/encyclopaedia-of-islam-3/khirba-COM_35538?s.num=32&s.start=20.

Walker, B.J.

2021b. "Peasants, rural economy and cash crops in medieval Islam." In *Markets and Exchanges in Pre-Modern and Traditional Societies*, edited by J.C.M. García, 91–113. Oxford: Oxbow Books.

Walker, B.J.

2022. "Peasant dependencies in Medieval Islam: Whose agency in food production and migration?" In *BCDSS Working Paper*. Internet Edition: https://www.dependency.uni-bonn.de/images/pdf-files/working-papers/wp-walker.pdf.

Walker, B.J.

2023. "Imperial interventions in daily life: The eastern Mediterranean under early Ottoman rule." *Historical Archaeology* 58.4: in print. Special issue entitled "Mediterranean historical archaeology in the twenty-first century: a critical assessment," edited by R. Palmer and M. Given.

Walker, B.J., and E. Kenney

2006. "Rural Islam in Late Medieval Jordan: NJP 2006—the Mosques Project." *Newsletter of the American Center of Oriental Research* 18.2: 1–4.

Walker, B.J., E. Kenney, L. Carroll, L. Holzweg, S. Boulogne, and B. Lucke

2007. "The Northern Jordan Project 2006: Village life in Mamluk and Ottoman Hubras and Sahm: A preliminary report." *Annual of the Department of Antiquities of Jordan* 51: 429–470.

Walker, B.J., T. Greer, R. al-Shqour, A. LaBianca, R.D. Bates, J.P. Hudon, W. Schultz, J. Henderson, C. Corbino, S. Laparidou, A.M. Hansen, and Ø.S. LaBianca

2018. "Tall Ḥisbān 2016 excavation season: Household archaeology in the Medieval village." *Annual of the Department of Antiquities of Jordan* 59 (2018): 557–595.

Walker, B.J., and B. Dolinka.

2020. "*Ḫirbet Bēt Mazmīl.* Investigations of Medieval Jerusalem's Hinterland: Interim report on the 2015 and 2017 Field Seasons." *Zeitschrift des Deutschen Palästina-Vereins* 136.2: 191–234.

Walker, B.J., A. Farahani, S. Laparidou, S. Polla, and L. Bontempo

2024. "Kitchen culture in 13th-century Syria: A cookpot in context at Tall Hisban." *Journal of Islamic Archaeology* 11.1: forthcoming.

Wightman, G.J.

1989. *The Damascus Gate, Jerusalem: Excavations by C.M. Bennett and J.B. Hennessy at the Damascus Gate, Jerusalem, 1964–66. BAR International Series No. 519.* Oxford: British Archaeological Reports Oxford Ltd.

Wikipedia

2023. Kiryat HaYovel, 05.30.2023.

Woodward, D.R.
2006. *Hell in the Holy Land: World War I in the Middle East.* Lexington: University of Kentucky Press.

Zeʿevi, D.
1996. *An Ottoman Century: The District of Jerusalem in the 1600s.* Albany: State University of New York Press.

About the authors

Prof. Dr. Bethany J. Walker (PhD 1998, University of Toronto) is Professor of Islamic Archaeology at the University of Bonn, the Director of the Research Unit of Islamic Archaeology at the same institution, and former Co-Speaker of the Annemarie Schimmel Kolleg of Mamluk Studies in Bonn. A historically trained archaeologist, Walker directs field projects in Israel and Jordan. Among her monographs, edited, and co-edited works are *Jordan in the Late Middle Ages: Transformation of the Mamluk Frontier* (Chicago 2011), *Reflections of Empire: Archaeological and Ethnographic Studies on the Pottery of the Ottoman Levant* (Boston 2009), *Living with Nature and Things: Contributions to a New Social History of the Middle Islamic Periods* (Bonn 2020), and the *Oxford Handbook of Islamic Archaeology* (Oxford 2020). She is the senior and founding editor of the Journal of Islamic Archaeology, and her research focuses on peasant society, land use and tenure, and landscape transformation.

Dr. Benjamin J. Dolinka is an archaeologist and ceramics specialist who has conducted research and fieldwork in Greece, Jordan, and Israel during the past 31 years. He received his PhD in Archaeology from the University of Liverpool in 2007 with a dissertation on Nabataean pottery and worked for the Israel Antiquities Authority from 2010 to 2018. Mentored on the Islamic pottery by the late Dr. Miriam Avissar (IAA), he has come to be one of the leading specialists for the Jerusalem area. His forthcoming *The Pottery of Medieval Jerusalem, ca. 1050–1550 CE* will be published by Equinox as part of its *Monographs in Islamic Archaeology* series.

Dr. Nicolò Pini (PhD in Classical Archaeology at the University of Cologne, Germany) is an archaeologist, specialized in ancient urbanism, architecture, and social structures in the Roman and Byzantine periods. Over the last years he has ventured into the Early, Middle, and even Late Islamic periods. He currently conducts a postdoctoral research project at the CReA-Patrimoine, Université libre de Bruxelles (Belgium), funded by the Belgian Fonds de la Recherche Scientifique—FNRS (CR—Chargé de recherches). He was previously awarded a COFUND Marie Skłodowska-Curie Individual Fellowship at the same institution. He collaborates on several projects in the Near East, in particular with the Islamic Archaeology Research Unit at the University of Bonn and is also external research associate of the Panorama Platform (based at the Universitè libre de Bruxelles), specialized in acquisition and digitization of objects and architectural surveys.

Dr. Roy Marom is a Dan David Fellow in Tel Aviv University. Previously he served as a Fulbright Postdoctoral Fellow in University of California, Berkeley (2022–2023). Marom earned his PhD at the Department of Middle Eastern and islamic Studies in the University of Haifa (2022). Marom's research focuses on the history of rural Palestine during the Late Ottoman and British Mandate periods. So far, his Palestinian Rural History Project (PRHP) documented oral histories and information about 700 Palestinian communities. His broader academic work concerns Palestine's historical geography from the Umayyad to the British Mandate periods.

Benyamin Storchan is a PhD candidate of the Department of Bible Studies, Archaeology and the Ancient Near East at Ben-Gurion University of the Negev. His dissertation, "The Metamorphosis of Holy Land

Churches During the Early Islamic Period," examines the processes of change and continuity as seen from the archaeology of Byzantine churches in the Holy Land during and after the Islamic transition. Storchan completed his MA at Bar Ilan University and BA from Michigan State University. He has been employed at the Israeli Antiquities Authority for over 15 years as a research excavation archaeologist for the Jerusalem Region, and has directed a number of large-scale excavations, including the Glorious Martyr Excavation Project and the Early Bronze Age settlement at Eshta'ol. He has also published a number of final excavation reports and research articles. Storchan is currently a board member for The Israel Archaeological Association, research fellow at the Center for the Study of Conversion & Inter-Religious Encounters and has been recently awarded the David Amit Prize for Archaeological Research and a Humanities Faculty Scholarship from Ben-Gurion University of the Negev.

— 6 —

Jerusalem's Farmland Transformed

The Expansion of Agricultural Terracing

Omer Ze'evi-Berger and Nitsan Ben-Melech

The Jerusalem highlands to the west of the city are characterized by moderate slopes and winding wadis. The land beyond the city limits was always used for agriculture, and the Medieval Islamic period is no exception. Throughout its entire history of occupation, Khirbet Beit Mazmīl was a site intimately engaged with agriculture, its fields likely located nearby. In this chapter, we explore "life on the farm" through the lens of its farmland, and through the transformation of its fields over time.

Terraced landscapes are a hallmark of the Judean hill country, the Bilād al-Shām highlands in general, and, indeed, the entire Mediterranean basin (Asins Velis 2006; Gadot *et al.* 2016; Hamarneh 2022; Kinnaird *et al.* 2017; Porat *et al.* 2018; Vetter and Rieger 2019). These terraces are directly linked to issues of land use and land tenure, human-landscape interactions, and societal interactions. Each terrace represents a highly labor-intensive activity that required a high level of social organization and coordination to construct and maintain. Terraced landscapes are very difficult to date however, mainly because they are palimpsests: built and rebuilt constantly. Additionally, because of their explicit connection to "off-site" activities, one rarely finds any datable material in their fill. If one does, for the most part it is in a secondary or tertiary context. For that reason, many attempts to date terrace complexes have relied on tenuous spatial relationships (Davidovich *et al.* 2012; Elgart-Sharon *et al.* 2020; Gibson 2015).

To better understand the terraced landscape around Jerusalem and to better date terrace complexes, the Formation of Terraced Landscapes in the Judean Highlands, Israel Project was launched in 2013. A key component of this initiative was the application of Optically Stimulated Luminescence (OSL) to date landscape developments (Gadot *et al.* 2016; Porat *et al.* 2018). OSL has been introduced in recent years as a method of dating sediment movements based on its most recent exposure to light. The project employed OSL dating as well as archaeological and geomorphological analyses to date sediment fills coming from terraces and contexts associated with other agricultural installations. Extensive work in five selected locations surrounding the modern city of Jerusalem documented terrace-construction in the Hellenistic, Late Roman, and Early Islamic periods. The terraced landscape that one sees today, however, was constructed during the Late Islamic period: in the past 800 years.

Whether demographic growth drove this terrace-building has been a topic of scholarly debate. Elinson (2016) has shown that there were sufficient flat land reserves in the highlands around Jerusalem to meet its needs during its periods of greatest population growth in pre-modern times. In their seminal work, Elgart-Sharon (*et al.* 2020) postulated that, in contrast to previous notions, the construction of terraces in the Jerusalem region was not the result of population growth and the need to feed greater Jerusalem at any given time. Instead, large-scale terrace construction was connected to governmental land-use systems that encouraged extensive land exploitation for the growing of cash crops.

Khirbet Beit Mazmīl is an ideal site to test hypotheses about the factors behind agricultural terracing. The site was constructed at the top of a hard dolomite hill (the Weradim Formation). The area around it consists of harder dolomite hills and ridges. At the same time, the wadis between them have outcrops of mixed dolomite and in some softer limestone (the Aminadav Formation) and even softer marl and chalk in others (the Kefar Shaul Formation) (Sneh and Avni 2011). It is a geology suited to terrace construction. Concerning water sources sufficient for agriculture, while freshwater springs are abundant in the Jerusalem highlands, only one such spring is in the vicinity of the site, towards the village of Ein Karem. The area west of the Jerusalem watershed line enjoys a semi-arid Mediterranean climate. With wet, cool winters, dry, hot summers, and an average annual rainfall of about 500–600 mm (Arad *et al.* 2016, 30–31), the environment is sufficient to support at least some forms of dry farming, or a combination of dry and watered farming.

Methodology

With this information in mind, a landscape component was incorporated into the Medieval Jerusalem Hinterland Project in 2017, to explore the terraced areas around Khirbet Beit Mazmīl. Fieldwork was preceded by inspection of aerial photography from the early 20th century onwards, together with different maps from different time periods detailing cultivated lands, plots, and land-use percentages. Based on information drawn from these tools, the team conducted a pedestrian survey, recording different agricultural features, rather than a surface-collection of sherds. Each survey group consisted of three to ten people. All features identified were recorded in detail on a dedicated FileMaker-based database. The spatial data was recorded with GPS-RTK systems (Leica GS05 and Trimble R2) and analyzed with ESRI ArcMap 10.1 and ArcGIS Pro 2.8.

Several locations were selected for further investigation of the areas and features recorded during the survey. This investigation was carried out based on the framework detailed in Davidovich (*et al.* 2012) and shown to be effective in subsequent research (Gadot *et al.* 2016). Probes, between 1 × 1 and 2 × 2 meters wide, were excavated down to bedrock in several places. While most probes were excavated into terrace fills, some were sunk into other features, such as watch towers and stone piles. Most probes in terrace fills were excavated at the intersection with other features, mostly plot fences. The terrace walls' construction were rather similar across the slope: all were constructed of one row of medium-large field stones and preserved to various course levels. This type of construction is Davidovich's "Type I" (Davidovich *et al.* 2012, 200).

After the excavation, sections from each probe were sampled for OSL dating. All OSL ages are presented in this chapter in approximate Gregorian calendar years, and a full report on each sample is provided in Table 1. Sampling locations were based on sedimental and feature stratigraphy. Generally, terrace fills were sampled at the top and bottom of the sections, representing the layers of fill we could discern in the field. For the most part, we started by dating the bottom sample from each terrace and proceeded to date additional samples in selected places only. Alongside the samples for absolute OSL ages, Portable OSL (POSL) samples were collected with an aim of studying the sedimentation processes of the fills. Portable OSL readers are a relatively new tool that provides quick assessments of the luminescence signal but do not provide a full OSL date (Munyikwa *et al.* 2021). Its implementation in archaeological studies, and specifically studies related to archaeological installations such as terraces, have enabled scholars to gain more information regarding the rates, homogeneity, and development of the sediment fills (Munyikwa *et al.* 2021; Porat *et al.* 2019). For this study, the POSL samples were collected from both the vertical and horizontal axes of the sediment fills of terraces and the accumulations around other built installations.

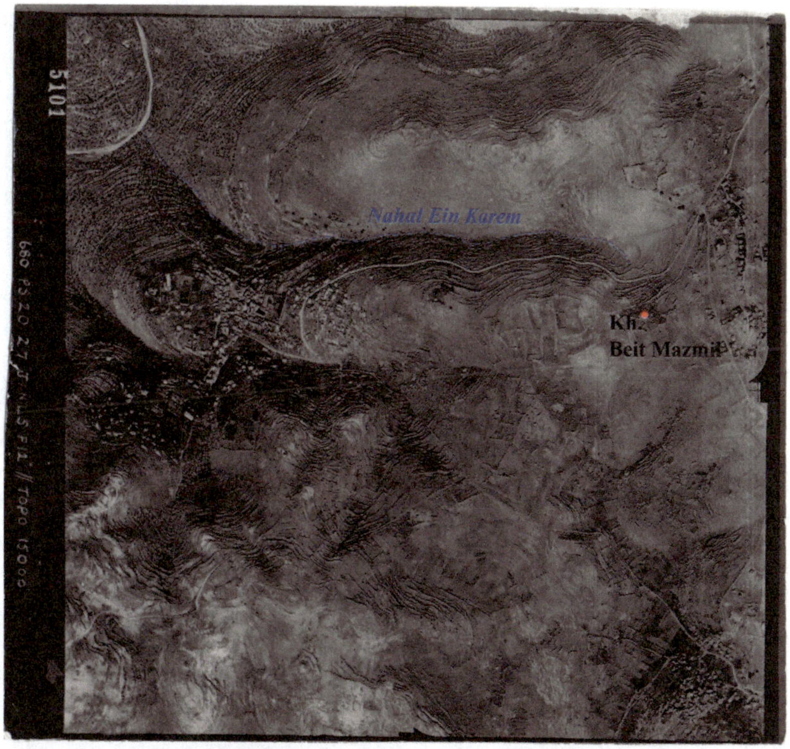

Figure 1 RAF aerial photo of Khirbet Beit Mazmīl and its surroundings.

Survey

When one inspects aerial photographs from the first half of the 20th century, it is evident that the landscape surrounding the site has been heavily terraced (Figure 1). With the expansion of modern-day Jerusalem, however, only a handful of open areas around Khirbet Beit Mazmīl are not covered in concrete and remain available for study. Three pockets that remain undeveloped were selected as being possibly related to Khirbet Beit Mazmīl. These were two wadis in the vicinity of the site—Nahal HaYovel and Nahal Ein Karem—and an additional wadi slightly farther away to the west—Nahal Karmit. These were surveyed to map agricultural features and systems in every available area (Figure 2).

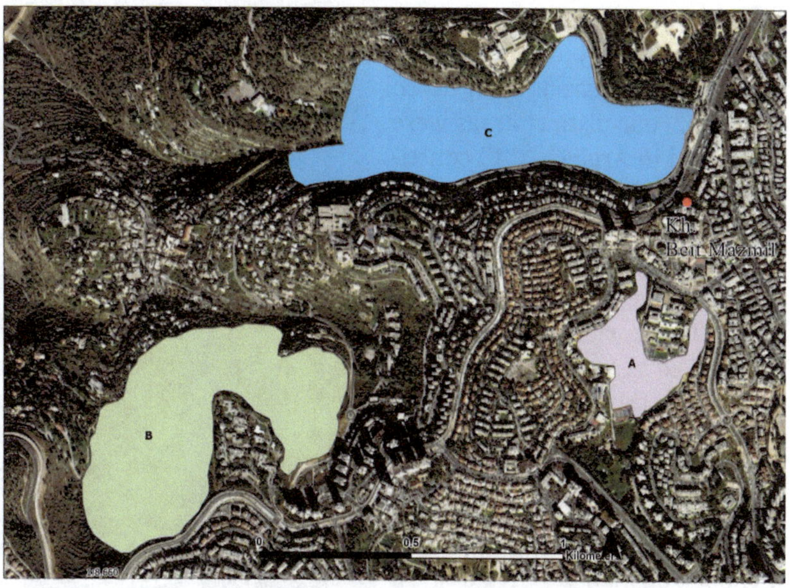

Figure 2 Surveyed areas around Khirbet Beit Mazmīl, Nahal HaYovel, Nahal Ein Karem, and Nahal Karmit.

Nahal HaYovel, designated "Area A," is a tributary of the Refaim Wadi, leading west from Khirbet Beit Mazmīl. Though it is the closest of all survey areas to the site, it is also the least preserved of the three. A total of 32 features were documented, including terraces, stone piles, and rock carvings. Unfortunately, no plot fences were identified in the area. It was therefore challenging to identify the overall agricultural system and land tenure mechanisms in Area A (Figure 3). Nahal Karmit, designated "Area B," as a wadi is a tributary of the Nahal Ein Karem stream. It is the farthest area from Khirbet Beit Mazmīl but has the most complete terrace systems. These include clear, discernable plots, plot fences, stone huts (*manāṭīr*), stone piles, and lime kilns. Its proximity to the village of Ein Karem suggests that at least the latest archaeological phases detected are related to this settlement. Out of all the examined areas, Area B contains the highest number of features. Nahal Ein Karem, designated "Area C" (Figure 4), as a wadi is a tributary of the Nahal Soreq River, leading west from Khirbet Beit Mazmīl towards Ein Karem. This wadi included terraces, plot fences, *manāṭīr*, and stone piles of different preservation conditions, including a huge plot fence bisecting Nahal Ein Karem from north to south. Four distinct areas were discerned in Area C. "C I" constitutes the eastern part of Nahal Ein Karem, east of the large plot fence, on both banks of the stream. This area holds well-preserved terraces and other agricultural installations. "C II" constitutes the northern bank of the stream, west of the large plot fence. It is characterized by a distinct lack of agricultural installations, except for a press and what seems to be a road with a small built support leading up from the wadi. "C III" constitutes the southern bank, west of the large plot fence. While terrace walls are prevalent in this area, they are poorly preserved when compared to areas C II and C IV. "C IV" constitutes the western edge of Nahal Ein Karem, on the southern bank of the stream. This area holds two roughly equal-sized plots with terrace walls and stone piles of proper preservation.

Both Nahal HaYovel (Area A) and Nahal Ein Karem (Area C) were selected for excavation. Despite its

Figure 3 Surveyed features in Area A.

Figure 4 Surveyed features in Area C.

high state of preservation, it was decided to not excavate the agricultural systems in Nahal Karmit because of its distance from Khirbet Beit Mazmīl and its proximity to Ein Karem and its monasteries. Altogether, 18 spots within the two wadis were selected for excavation: three in Area A and 15 in Area C. The excavation process is presented in the following subchapter by area and according to agricultural units.

Excavation

Area A

Area A (Figure 3), or Nahal HaYovel, was understood as a singular unit of agricultural activity. The area is roughly 200 meters, as the crow flies, from Khirbet Beit Mazmīl. Three probes were excavated in Area A, one into a stone pile on the southern bank of the wadi and two into terrace fills: one on each side. Stone pile A/F203 is roughly 9 × 3.5 meters and comprises small to large field stones. The section cut into it uncovered a sediment layer at the bottom of the pile, and an OSL sample obtained from it dated to 450–170 BCE. It is still unclear whether this layer represents the sediment present when the pile was created, or whether it is washed sediment accumulated over time.

The two terraces probed in Area A appeared to have similar characteristics. Both were very poorly preserved and consisting of a dark-brown fill behind a small terrace wall. Both walls—A/F201 and A/F204—were built of a single row of small field stones. The low state of preservation of the terrace walls could be due to excessive erosion of the walls and their associated fills over time. The small stones comprising the wall suggest that it was always a low construction, however. One sample from the fill of terrace A/F201 was dated to between 40 BCE and 240 CE. As the OSL ages from Area A did not coincide with activity at Khirbet Beit Mazmīl, we therefore chose to abandon further investigations in this area.

Figure 5 Two bronze coins found in probe C2.

Area C

Area C (Figure 4), or Nahal Ein Karem, spans the wadi between the hill of Khirbet Beit Mazmīl and the village of Ein Karem. Area C I is an area of well-preserved terraces on the eastern end of the wadi. Five probes were excavated in this area, on the southern bank of the wadi. Three abutted large plot fence C/F201 from the east, one abutted plot fence C/F225 from the west, and one abutted *manṭarah* C/F206. Plot fence C/F201, the most distinct manmade feature in the area, consists of two rows of medium-large field stones at a width of 1–1.2 meters, preserved to a height of up to 1 meter. Area C I holds within it several other man-made elements, including at least two *manāṭīr*, two water cisterns, stone piles, and several concrete structures.

Probes C1, C4, and C5 were excavated at the intersection of plot fence C/F201 and three different terraces (C/F213, C/F203, and C/F211). At the bottom of probe C1 the base of the plot fence was uncovered, sitting directly on bedrock. Terrace wall C/F203 came up to and abutted the plot fence from the east. In probe C4, plot fence C/F201 was built on top of the terrace wall, which continued in a poorly preserved manner to the west. In probe C5 the remains of a cornered terrace were discerned. This corner was later incorporated into the line of plot fence C/F201. In all cases, despite partial collapse, the terrace wall itself and the terrace fill inside of it was preserved to a significant depth. A similar architectural strategy was observed in all probes: a terrace wall had been built at the edge of the natural bedrock step; the terrace had then been filled with a layer of medium to large field stones laid on the bedrock; and a deep layer of dark brown sediment had then filled the remaining space to the top of the terrace wall. Following the excavation, the fills of all terrace probes were sampled from top to bottom for Portable OSL (POSL) measurements, and samples for full OSL ages were collected from the bottom and the uppermost parts of the sections, representing each layer of the fill. This same methodological approach was taken in all subsequent probes discussed in this chapter, unless otherwise stated. The earliest OSL sample was collected from the bottom of probe C5, with an age range of 580–740 CE. The bottom of the fill in probe C4 was dated to between 840–1000 CE. The top of the fill meanwhile dated to between 1710–1770 CE. The fill in this probe was further sampled underneath the stones of plot fence C/F201. This sample was dated to 1380–1460 CE. This sets the timeframe for the construction of the plot fence as between the 14th and 15th centuries. The bottom of probe C1 was dated to 1470–1530 CE. Together with the evidence of the stratigraphic occurrence of the wall, it is clear that this terrace was added after the construction of the plot fence.

Probe C15 was excavated around the northeastern corner of *manṭarah* C/F206, in the vicinity of plot fence C/F201. C/F206 was built on a large clearing of bedrock within plot C/F207, abutting

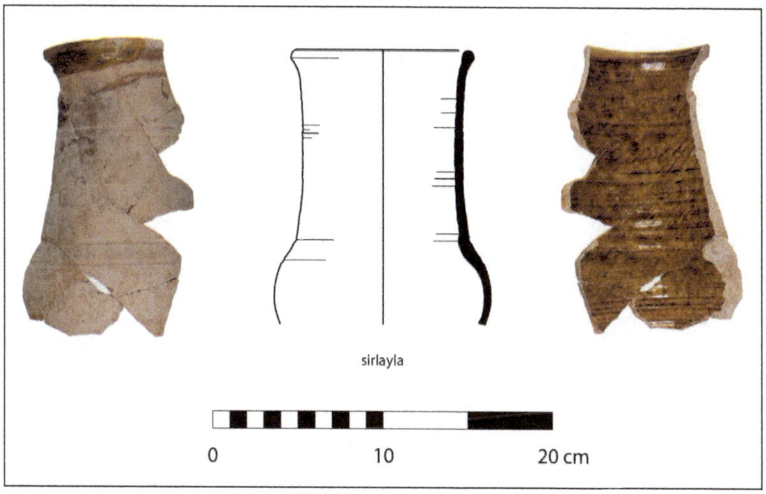

Figure 6 Pottery vessel found in square C9.

terrace wall C/F211 from its outer face, near water cistern C/F228. The structure, consisting of two rows of large, roughly hewn stones with a "dibs core"—a mixture of small stones, sediment, and plaster—was built directly on bedrock. The structure has an inner circular construction, presumably culminating in a dome, and an outer square façade built around it. Sediment accumulation around it was very shallow and included only dark brown topsoil, sloping northward. The aim of the probe was to reach the point of interface between the base of the structure and the bedrock and sample any sediment that was possibly trapped there at the time of construction. The OSL sample was ultimately collected from sediment trapped between the stones of the walls and the bedrock after removing one of the stones at the base of the *manṭarah*. The sample was dated to 1640–1720 CE.

Probe C16 was an elongated probe excavated at the intersection of terrace wall C/F226 and plot fence/steps C/F225, at the eastern edge of plot C/F207. Fence C/F225 functioned as both a method to delineate one plot from another and as a stairway. The probe extends from bedrock down to terrace wall C/F226. Sediment accumulation in this probe was relatively shallow and seemed to have been only one layer of fill between topsoil and bedrock, without a layer of stones. This type of fill continued under the stones of the steps, where it had been presumably sealed at the time of construction of the steps. The fill itself was dated to 490–710 CE, while the steps were dated to 1150–1290 CE. It seems that the fill and the terrace itself had been in place since the earlier stages of the agricultural activity on the slope, and the steps were added during a later phase between the 12th and 13th centuries.

Area C II, on the northern bank of Nahal Ein Karem, west of plot fence C/F201, was devoid of any terracing activity. Two important elements were identified in the area, however. The first was a press complex (C/F227) that included a pressing floor, collection vat, and a rock-cut niche, most likely for wine production. Rock-cut presses are very difficult to date, though niched presses are dated stratigraphically to the Iron Age (Amit and Yezerski 2001). The second recorded feature was a single row of small to medium stones at an angle to the slope, identified as a road (C/F228).

Area C III, bounded by plot fences C/F201 and C/F213, is a long expanse of a poorly preserved terraced slope stretching circa 600 meters. The area was designated as plot C/F229. Three probes were excavated in Area C III: one abutting plot fence C/F201; one abutting plot fence C/F213; and one in a partially preserved terrace wall in the middle of the area.

Probe C2 was excavated at the intersection of plot fence C/F201 and terrace wall C/F203, opposite probe C4. The top layer of the probe was comprised of a mix of small pebbles and loose dark brown sediment, most probably related to stone-clearing activities from the terraces on either side of the fence. As seen elsewhere, probe C2 yielded a layer of small to medium-sized stones with a light brown matrix over the bedrock. In this layer two coins (B1009.01 and B1010.01, Figure 5) were found, dating to the 13th century.[1] Through OSL the

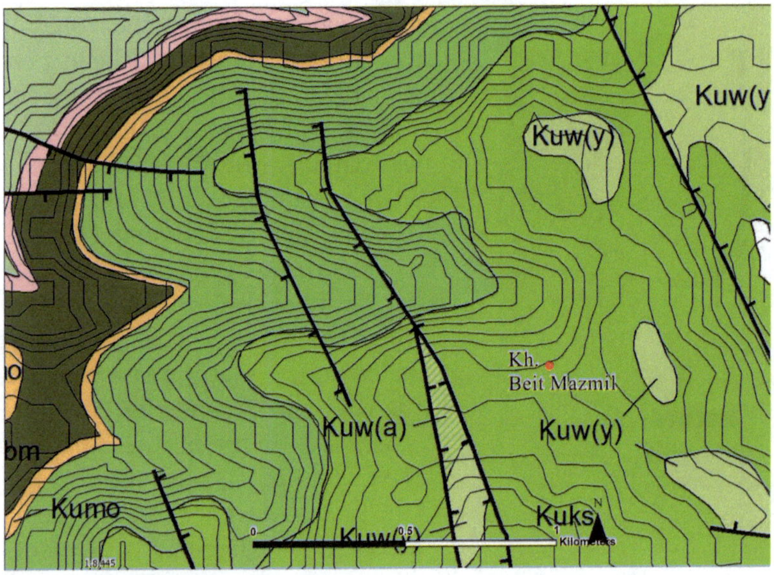

Figure 7 Geological map of Nahal Ein Karem (after Sneh and Avni 2011).

bottom of the fill was dated to 1120–1220 CE and this date range coincides with the dating of the coins, giving a *terminus post quem* for the construction of the terrace transected by fence C/F201. The upper part of the fill was sampled from below the stones of plot fence C/F201, similar to probe C4, and dated to 1290–1410 CE. This age largely corroborates, and slightly expands, the possible timeframe for the construction of the plot fence. As the methodology for the dating of the plot fence relies on the fills of the terraces that it covered, we can suggest the following scenario: the terrace was used and maintained since at least the 13th century. Agricultural activity such as plowing, watering, etc., led to the sediments of the fill being exposed to light, which explains the resetting of the OSL signal. The construction of the plot fence sometime during—or around—the 14th century then covered a part of the fill, resulting in similar dates coming from probes C2 and C4.

Probe C3 was excavated in the middle of plot C/F229, into a partially preserved terrace wall, where here too, a layer of stones was uncovered above bedrock. Both terrace walls generally seemed to be of the same construction as all nearby terraces: one row of field stones, or Davidovich's Type I (2012). One sample alone was collected from the bottom of this fill, which was neither processed nor dated.

Probe C7 was excavated where a free-standing step of soil was noticed, abutting plot fence C/F213. It initially seemed to be the remains of a collapsed terrace. Following the excavation however, no clear evidence of an intentional fill or a construction of a terrace was noticed but a thin layer of orange sediment and above it a layer of orange-brown loamy fill was uncovered above the bedrock. The first course of plot fence C/F213 was uncovered at the bottom of this probe, built directly on the bedrock. The POSL results from the section did not indicate a deliberate, one-time fill. The step of soil was therefore not sampled for a full OSL age. A sample was instead collected from the bottom of the plot fence, above the bedrock, behind a stone removed from the first course of the wall. This sediment was theoretically trapped underneath the stones during the time of the plot fence's construction. This sample was dated to 1240–1460 CE, agreeing excellently with the date for plot fence C/F201.

Area C IV, bound between plot fences C/F213 and C/F215, is the best preserved of Area C's sub-areas. It is made up of two, similar-sized rectangular plots—C/F216 and C/F217—that seem to have been built with a plan in mind, rather than developing organically. Some of the terraces in said plots were built together with the fences while others abut them, indicating either a later phase of construction or simply a technical phase. However, there seems to have been at least one earlier phase of terraced agriculture on this side of the slope. Some terrace remains are of a slightly different orientation than the general plot lines, forming corners and possibly canceling older terraces. Six probes were excavated in this area, three in each plot.

Figure 8 Earliest terracing activity spanning the southern bank of the wadi.

Figure 9 Plot fence C/F201 separating areas C I and C III.

Probe C6 was cut into stone pile C/F215, which consisted of small to medium field stones, representing clearing endeavors in the area. The stones were piled on top of a high bedrock ledge, with a small amount of sediment between them, and a thin layer of sediment at the interface between the bedrock and the pile. It is unclear exactly what this layer represents—whether this is the original sediment before the pile, or a gradual accumulation after the piling of the stones. We chose not to pursue the investigation of this feature further for this reason.

Probes C8, C10, C11, and C13 were excavated in various terraces in Area C IV. In all probes the excavation uncovered the same pattern seen throughout Area C: a stony fill under an orange-brown loamy fill and above bedrock. These probes were significantly deeper than the ones excavated in other parts of Area C due to a higher state of terrace preservation: possibly stemming from a later date of abandonment. Of note are the high concentrations of phytoliths in the loamy fill in probes C10 and C13, which are probably indicative of a time where this level was exposed. Probes C8 and C10 were excavated at opposite ends of plot C/F217, at the interface between terraces and plot fences. The wall of the plot fence was only uncovered in the excavation of probe C10, however, while in probe

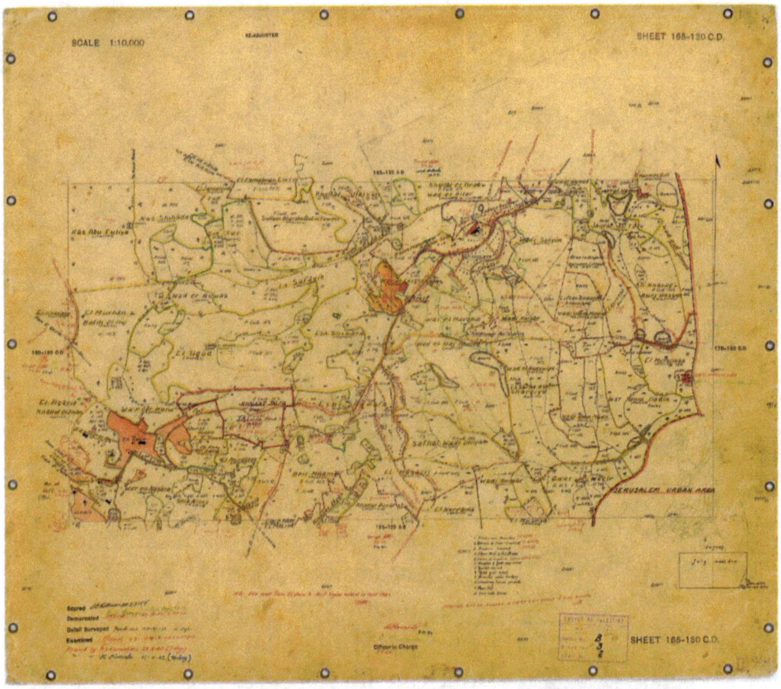

Figure 10 Cadastral map from the British Mandate period marking Area C III Waqf Khallat Jaljīl (courtesy of Survey of Israel).

C8 the stones of the fence seemed to have been sitting above the fill. This indicates that in this case the terrace predated the fence. The sample collected from the bottom of probe C8 was dated to 1130–1250 CE, while the sample from probe C10, collected from underneath the fence itself (as in probe C7), dated to 1490–1570 CE. Probes C11 and C13 were excavated in plot C/F216, with C11 in the northeastern corner of the plot, while C13 was situated in the middle of terrace C/F224. Probe C11 was excavated at the intersection of plot fence C/F213 and terrace wall C/F211. The fill of the terrace was highly eroded, situated between the poorly preserved terrace wall and the high-outcropping bedrock ledge on which plot fence C/F213 was at the point constructed and due to its poor preservation, the probe was abandoned. Probe C13 was excavated into the fill of terrace C/F224. Like terrace C/F211, this terrace wall was also poorly preserved and partially eroded. The probe was excavated at a point where the stones of the wall had partially collapsed but the sediment fill was freestanding almost to the top. A sample from the bottom of this probe was dated to 260–400 CE, and the top of the fill was dated to 1610–1670 CE.

Probe C9 was excavated in a self-contained terrace corner designated C/F219, where the terrace was preserved to a height of 0.9 meters. A layer of very compact brown sediment was found at the bottom of the fill, contained between a natural step in the bedrock and wall C/F219. Atop the bedrock step a one-course wall was uncovered (C/F222), likely an earlier terrace wall that had fallen out of use. Between wall C/F222 and wall C/F219, and above the first layer of compacted sediment, was a layer of medium field stones and brown fill. Above this, a layer of orange-brown sediment had covered the entire probe, covering the top of wall C/F222. A complete profile of a yellow monoglazed vessel dating to the 15th to 16th century was recovered from the bottom of the fill near wall C/F219 (Figure 6). See Chapter 8 for a discussion of its further function in agriculture.

The bottom layer of compacted soil was dated to 100–280 CE. The fill in the uppermost layer of the terrace dated to 1300–1400 CE. The stoney layer in the middle dated to 1830–1850 CE. This presented a particular challenge, with the middle layer being dated to a considerably younger age than the sample above it. Layers of stoney fill are notoriously difficult to date using OSL. This is due to the unstable nature of the fill, which could lead to sediment mix and the contamination of the layer during excavation or sampling, resulting in the assigning of an unreliable age. This complexity is also seen in terrace wall C/F219 itself. C/F219 was built above an

Figure 11 Cistern and *manṭarah* in Nahal Ein Karem.

Figure 12 The earliest iteration of terracing at Nahal Ein Karem (C I, C III, and C IV).

earlier terrace wall, C/F218, though its OSL dates seem to suggest that the base of terrace C/F219 is actually older than the lower C/F218. Finally, the glazed vessel found near the bottom of the terrace typologically dates to an earlier period than the top layer that covers the entire upper part of the probe, which is to be expected stratigraphically though it conflicts with the OSL readings. Based on the numerous methodological issues and contradictions, it is evident that the dates from probe C9 should be considered carefully. The archaeological data points to an earlier terrace being covered by a later one, and the unusual find of the glazed vessel offers a provisional date from sometime during the Late Mamluk to Early Ottoman period. The POSL results also point to a two-stage fill. It is possible that the early stages of the terrace can be dated to as early as the 2nd or 3rd centuries (CE), temporally close to other dated samples collected from across the slope. Later use of the terrace can also be attributed to the Medieval period.

Discussion

The decision to divide Area C into sub-units C I to C IV was the result of natural and anthropogenic causes. First and foremost, the lack of terrace activity and the presence of a wine press in the same area of Area C II is possibly the result of a geological difference. In broad strokes, Nahal Ein Karem's stepped, rocky slopes are part of the Aminadav Formation, while the softer limestone of the Kefar Shaul Formation can be found at the top of

Figure 13 Terracing in Nahal Ein Karem: second phase of activity and expansion of parcelization.

the hill (Figure 7, Sneh and Avni 2011) although local variation within the wadi is possible. As seen in several instances, several rock formations were utilized for different purposes. On many occasions, stepped landscapes were used for agricultural terracing, while large stone surfaces, like in Area C II, were utilized mostly for agricultural production including wine and olive presses. It is possible that there was some terracing activity on this slope in earlier times, although it seems unlikely. It is likewise possible that to a limited degree some cultivation took place on the slope in earthen pockets, although this would likely be a low investment/ low yield activity. An alternative explanation for the differences in land exploitation between the northern and southern slopes of the wadi could be connected to a difference between south-facing and north-facing slopes. It is possible that due to higher levels of sun exposure, farmers refrained from cultivating the south-facing slope.

Areas C I, C III, and C IV seem to have originally been part of a larger terrace system spanning the southern bank, and probably the eastern flank of the northern bank of Nahal Ein Karem (Figure 8) as is evident through the continuation of terrace wall lines under and beyond different plot fences. It is possible that the southern bank was parcelized with plot fences that were later removed when the area was re-parcelized. This option is however unlikely, since one would expect to find some

Figure 14 The third iteration of terracing activity at Nahal Ein Karem, after the construction of plot fences C/F201 (C I) and C/F213 (C IV).

evidence of such activity in the landscape. It is more probable that the area was cultivated as a single unit as part of a larger estate, possibly that of Khirbet Beit Mazmīl. Alternatively, the area was potentially parcelized using "soft" boundaries: visible landmarks such as large trees, boulders, or terrace corner endings. The earliest iteration of terraced cultivation at Nahal Ein Karem is dated to between the late 2nd and mid-8th centuries: the late Roman and Early Islamic period, respectively (see Appendix 8 for a full OSL age report). Additional OSL samples were dated to the 10th and 12th centuries, and together with two Ayyubid coins found at the base of a terrace in probe C2 show that the terrace system in the wadi continued to be maintained and developed over the centuries.

The subsequent construction of plot fence C/F201, bisecting the wadi, is dated between 1290–1460 CE. It is possible that the division of the landscape with this plot fence resulted in the abandonment of the area west of it (Area C III). It is also possible that the majority of Nahal Ein Karem was already abandoned at this stage, and the construction of C/F201 represents a partial revival of terrace cultivation in the wadi (Figure 9). This could be the result of the wholesale privatization of land and introduction of land grants from the mid-14th century onwards (Igarashi 2014). Another possibility is that both sides of the plot fence continued to be cultivated and that Area C III was abandoned at a later date, while Area C I continued to be cultivated into the early 20th century.

Figure 15 Terracing in Nahal Ein Karem: organized plots and further expansion of parcelization.

The Abu Madyan endowment (see Chapter 8) of lands around the village of Ein Karem is bordered by Sataf and Qalunya to the north, al-Shaqaq to the west, Maliha to the south, and Beit Mazmīl to the east. Aside from these markers, the endowment provides no additional spatial data for the location and nature of the land grant. The endowed lands were additionally likely fragmentary and did not constitute a single, continuous territory. A British Mandate period topocadastral map from 1942 marks Area C III—between plot fences C/F201 and C/F213, the hill above it, and down the wadi to the stream bed—as "Waqf Khallat Jaljīl" (Figure 10). Within our area of study there are therefore potentially two endowments: one with a secure date but an unclear location and another clearly marked but with an unknown date. The date and location of plot fence C/F201 in relation to the two aforementioned endowments can imply that there is a link between the three. It is possible, and indeed even likely, that the fence delineated the border between the two endowed lands. Alternatively, the fence marked the end of a Waqf Abu Madyan fragment to the east, and an area that would be later endowed under Khallat Jaljīl.

The majority of cultivated areas in Nahal Ein Karem were terraced lands, and while the terraces themselves seemingly share a uniform construction method (Type I), they are part of a more elaborate agricultural system that is in no way uniform. The western plots included terraces, cornered terraces, and stone piles, while the eastern plots included stairways, several *manāṭīr*, and at least two water cisterns (Figure 11). The existence of such features in some tentatively earlier plots, and their absence from other, later plots, might testify to a complex decision-making process and different cultivation methods. Possibly, plots that included water cisterns could be connected to the hand-watering of fruit trees, while *manāṭīr* may point toward high-value products that required safekeeping.

While the majority of terraced lands in general—and specifically in Nahal Ein Karem—were terraces constructed strictly for dry farming, a handful of terraces appear to have been intended for watered crops. The modern average rainfall in this part of the Jerusalem highlands is circa 500 mm per year. The soil of the area is characterized by terra rossa with a high water absorption rate, and brown rendzina sediments with a lower absorption rate compared to the terra rosa soils (Arad *et al.* 2016). Since terraces do not rely solely on naturally forming soils but rather on intentional fills, it is likely that there was a mixing of different soils, perhaps with different goals in mind.

An additional means of solving the issue of water management is seen in the layers of medium to large field stones observed at the base of most terraces. In some cases these are topped by an additional layer of smaller pebbles, gradually transitioning to relatively clear sediment. It has been suggested that these stone layers were used as a water management tool in order to expel excess water in the terraces, prevent root decay, and prevent terrace erosion and collapse (Gibson 2001; Gibson and Edelstein 1985; Hamarneh 2022, 115). In instances where pebbles were used, it may indicate the planting of young trees, which need an easier perch for their roots.

Finally, both OSL and POSL have proven to be invaluable tools when trying to unravel spatial stratigraphy over large areas. OSL and POSL have aided us in the grueling task of peeling back and distinguishing layers of human activity in the landscape when nearly no other datable materials could be found. It has been shown that in most cases ceramic typology and radiocarbon dating are unreliable dating methods when inspecting terracing activity: more often than not sherds and organic material are found without a clear context (Davidovich *et al.* 2012, Gadot *et al.* 2016). Furthermore, in a case such as the one at hand, with three distinct areas of activity and a uniform terrace typology, it would be virtually impossible to connect activity between separate areas without the use of OSL and POSL.

Figure 16 Terraced plots in Area C I and the construction of a *manṭarah.*

Conclusions

Based on stratigraphy, material finds, OSL dates, and portable OSL readings, we would like to suggest a spatial stratigraphy for Nahal Ein Karem divided into five stages.

Stage One: The earliest phase of agricultural terracing activity on the slopes spans the entire southern bank of Nahal Ein Karem. This phase is evidenced by poorly preserved terrace walls in Area C III, and in several deeper spots in preserved terraces in Areas C I and C IV (Figure 12). The oldest OSL age (Sample 135) may indicate an even earlier activity phase, which requires additional inspection. OSL results in probes C13, C15 and C16 represent agricultural activity during the Late Roman to Early Islamic periods. A younger OSL age from probe C4 may indicate a continuous use and reconstruction of the terraces.

Stage Two: Later terracing activities, mostly (re)construction and renovation works, are evident in probes C2, C8, and C16, and have been dated through both OSL and two recovered coins to the Crusader-Ayyubid periods. In addition to terrace (re)construction and renovation, it is in this period that we see the first evidence for landscape parcelization at Nahal Ein Karem. The most distinct example is a series of steps, functioning both as a delineator between and pathway connecting the two separated areas (Figure 13).

Stage Three: This period, dated to the 14th century, sees the construction of two large plot fences demarcating Area C II. Contemporary with the construction of the fences, Areas C I and C IV clearly continue to be cultivated, with the construction and reconstruction of terraces. Area C III, meanwhile, seems to have been abandoned. These activities are evident in probes C2, C4, C7, and possibly C9. The construction of the fence might be related to the Abu Madyan endowment and its delineation. This wall also coincides spatially with the Khallat Jaljīl endowment. It is possible that this fence represents the border between the two endowments, or between Waqf Abu Madyan and the lands that will later become Waqf Khallat Jaljīl (Figure 14).

Stage Four: An expansion of land use and parcelization of the slope is evident on both sides of the wadi, represented in probes C1 and C10. This stage dates to the Late Mamluk to Early Ottoman periods. OSL ages from probe C10 were taken from under the westernmost plot fence and indicate that an expansion of parcelized lands occurs to the west at this time (Figure 15).

Stage Five: This stage, represented in probes C4, C15, and C16, is dated to the Late Ottoman period. The terraces undergo further renovation and maintenance work, and there is simultaneously the construction of a large *manṭarah* (Figure 16).

Peasant decision-making was a complex process influenced by a variety of actors and factors, including, but not limited to geology, climate, local and regional interactions, and larger political trends. The lands surrounding Khirbet Beit Mazmīl played an important part and were subjected to changes in land use and parcelization. The five phases we have identified at Nahal Ein Karem reflect five different factor assemblages that affected farmers' choices and transformed the rural landscape.

Note

1. These coins were Damascene *fulūs* of the reign of the Ayyubid sultan al-Malik al-ʿĀdil. We thank Prof. Warren Schultz, De Paul University in Chicago, for this reading.

References

Amit, D., and I. Yezerski

2001. "An Iron Age II cemetery and wine presses at An-Nabi." *Israel Exploration Journal* 51 (2): 17.

Arad, U., U. Oren, E. Romem, R. Sinai, A. Perlberg, A. Mendleson, M. Ron, L. Admor, and U. Ramon

2016. *Northern Jerusalem Highlands: Survey and Analysis of Nature, Landscape and Human Heritage*. Tel Aviv: Deshe Institute.

Asins Velis, S.

2006. "Linking historical Mediterranean terraces with water catchment, harvesting and distribution structures." In *Proceedings of the 1st Conference on the Archaeology of Crop Fields and Gardens*, edited by J.P. Morel, J. Tresseras, and J.C. Matamala, 21–40. Bari: Edipuglia.

Davidovich, U., N. Porat, Y. Gadot, Y. Avni, and O. Lipschits

2012. "Archaeological investigations and OSL dating of terraces at Ramat Rahel, Israel." *Journal of Field Archaeology* 37 (3): 192–208. Internet Edition: https://doi.org/10.1179/0093469012Z.00000000019.

Elgart-Sharon, Y., N. Porat, and Y. Gadot

2020. "Land management and the construction of terraces for dry farming: The case of the Soreq catchment, Israel." *Oxford Journal of Archaeology* 39 (3): 274–89. Internet Edition: https://doi.org/10.1111/ojoa.12201.

Elinson, R.

2016. "The physical potential and the subsistence economy in the Jerusalem Hills." Tel Aviv University.

Gadot, Y., U. Davidovich, G. Avni, Y. Avni, M. Piasetzky, G. Faershtein, D. Golan, and N. Porat

2016. "The formation of a Mediterranean terraced landscape: Mount Eitan, Judean Highlands, Israel." *Journal of Archaeological Science: Reports* 6 (April): 397–417. Internet Edition: https://doi.org/10.1016/j.jasrep.2016.02.028.

Gibson, S.

2001. "Agricultural terraces and settlement expansion in the highlands of Early Iron Age Palestine: Is there a correlation between the two?" In *Studies in the Archaeology of the Iron Age in Israel and Jordan (Journal for the Study of the Old Testament Supplementary Series 331)*, edited by Amihai Mazar, 113–146. Sheffield: Sheffield University Press.

Gibson, S.

2015. "The archaeology of agricultural terraces in the Mediterranean Zone of the Southern Levant and the use of the Optically Stimulated Luminescence (OSL) dating method." In *Soils and Sediments as Archives of Environmental Change. Geoarchaeology and Landscape Change in the Subtropics and Tropics (Erlanger Geographische Arbeiten Band 42)*, edited by B. Lucke, R. Bäumler, and M. Schmidt, 295–314. Erlangen: Erlanger.

Gibson, S., and G. Edelstein

1985. "Investigating Jerusalem's rural landscape." *Levant* 17 (1): 139–55. Internet Edition: https://doi.org/10.1179/lev.1985.17.1.139.

Hamarneh, C.

2022. "Investigating ancient man-made terraces of Petra–Jordan." Berlin: PhD dissertation, Humboldt-Universität zu Berlin. Internet Edition: https://edoc.hu-berlin.de/handle/18452/25400.

Igarashi, D.

2014. *Land Tenure and Mamluk Waqfs (Ulrich Haarmann Memorial Lecture 7)*. Berlin: EB-Verlag.

Kinnaird, T., J. Bolòs, A. Turner, and S. Turner

2017. "Optically-Stimulated Luminescence profiling and dating of historic agricultural terraces in Catalonia (Spain)." *Journal of Archaeological Science* 78: 66–77. Internet Edition: https://doi.org/10.1016/j.jas.2016.11.003.

Porat, N., U. Davidovich, Y. Avni, G. Avni, and Y. Gadot

2018. "Using OSL measurements to decipher soil history in archaeological terraces, Judean Highlands, Israel." *Land Degradation and Development* 29 (3): 643–50. Internet Edition: https://doi.org/10.1002/ldr.2729.

Munyikwa, K., T.C. Kinnaird, and D.C.W. Sanderson

2021. "The potential of portable luminescence readers in geomorphological investigations: a review." *Earth Surface Processes and Landforms* 46(1): 131–150.

Porat, N., G.I. López, N. Lensky, R. Elinson, Y. Avni, Y. Elgart-Sharon, G. Faershtein, and Y. Gadot

2019. "Using portable OSL reader to obtain a time scale for soil accumulation and erosion in archaeological terraces, the Judean Highlands, Israel." *Quaternary Geochronology* 49: 65–70.

Sneh, A, and Y. Avni

2011. "Jerusalem: Sheet 11-II." In *Geological Map of Israel 1:50,000*. Jerusalem: Geological Survey of Israel. (Partly revised, 2016.)

Vetter, T., and A.K. Rieger

2019. "Water harvesting as a key for understanding adapted ancient livelihoods in an arid environment—approaches to dryland archaeology in the eastern Marmarica (NW-Egypt)." *Journal of Arid Environments* 171: 103940. Internet Edition: https://doi.org/10.1016/j.jaridenv.2019.02.009.

About the authors

Omer Zeʿevi-Berger is a PhD candidate in the University of Bonn's Islamic Archaeology Research Unit. Currently, his research centers on landscape archaeology and human-environment interactions, with a focus on terraced agriculture. His dissertation deals with the meshwork of relationships among cities, villages, and landscapes around Jerusalem and the Shephelah during the Middle and Late Islamic periods. In addition to the Medieval Jerusalem Hinterland Project, Mr. Ze'evi-Berger is affiliated with the Tel Hadid excavation project, and the "TERRSOC—Reading Ancient Landscapes" project, funded by the German Research Foundation.

Nitsan Ben-Melech is a PhD candidate in the school of Jewish studies and archaeology at Tel Aviv University. Nitsan's primary focus is on landscape archaeology, specializing in OSL dating of archaeological materials. Her dissertation deals with chronological aspects of the landscape in the Jerusalem highlands.

— 7 —

Land Use and Foodways

7.1: Evidence from Phytoliths

Sofia Laparidou and Georgia Kasapidou

To reconstruct the medieval history of land-use in Southern Greater Syria, it is necessary to study local-scale agricultural and pastoral practices, and state-related industrialization and intensification of agriculture. The region of Jordan and Israel is a diverse landscape in ecological and climatic terms (Cordova 2007) and has a climate that ranges from Mediterranean to arid (Mithen and Black 2011). The Ayyubid/Mamluk periods are marked with increased demographic and agricultural trends, investment in large-scale agricultural production (King 1992; MacDonald 2007; Jum'a Mahmoud 2000; Walker 2011), as well as changes in settlement patterns (Walker 2004, 2005, 2011). To better understand the local medieval economies of this period and the place of people within them, we could look at the archaeobotanical remains of medieval villages such as Khirbet Beit Mazmīl (Walker 2008, 2011, 2022), using historical sources and the archaeological record relevant to the region. Historical texts from the medieval period provide information on the degree to which rural communities depended on regional and interregional markets and if peasants were self-sufficient based on access to water, land ownership rules, selection of crops, and soil enrichment through manuring (Walker 2021). For the region of Transjordan in particular, Mamluk chronicles show that rural communities were dependent on each other for the exchange of specialized goods and that villages were united within a market network and financial exchange. This is further indicated by the copper coins found in abundance in medieval villages across region, including Khirbet Beit Mazmīl (Walker 2008, 2011, 2021, 2022).

During the medieval period of state investment in Transjordan, annual and interannual fluctuations of environmental systems must have affected imperial agricultural regimes and impacted local peasant communities. The success of both imperial and peasant agricultural regimes relies on local environmental parameters, such as the quality of agricultural land, soil properties, local rainfall variations and local land-use changes (Butzer 1982). In Southern Greater Syria, agriculture heavily depends on local environmental systems that are related to seasonality and the rotation of winter rainfall and summer droughts, making farmers susceptible to a variety of natural and cultural hazards (Butzer 1996; Palmer 2002; Halstead 1990).

Khirbet Beit Mazmīl is located in southwest Jerusalem, at 819 meters above sea level and seven kilometers southwest of the New Gate of Jerusalem's Old City (Walker and Dolinka 2020). During the medieval period the site was found within Jerusalem's gardens and was surrounded by terraced fields. It was a satellite settlement of the Arab village of Beit Mazmīl and becomes textually visible for the first time during the 14th century. According to the tax registers and Jerusalem's Sharia court registers of the 16th and 17th centuries, the land surrounding the village grew cereals including wheat and barley. There is no mention of "summer crops," such as fruits and vegetables, among the taxed products. During the 16th century tax-farming became commonplace, while by the 17th century many vineyards, olive groves, and orchards of figs, apples, and quinces were documented in the surroundings of the site (Walker and Dolinka 2020). According to Walker and Dolinka (2020), the site was an "amiral estate"—a *qasr*—in the Mamluk period (14th and 15th centuries). It served as the residence of an estate manager and as a center for rural administration, situated in the middle of a landscape that during this period was beginning to be developed for orchard cultivation through terracing. According to a 14th century endowment, local peasants invested in terracing as an economic strategy (Walker *et al.* 2020). In the 16th century this walled complex was transformed into a family estate for commercial agriculture.

The relationships between land and peasants

Walker, based on narrative sources, shows that peasants could make decisions for crop selection and cultivation independently of state decision-making across the diverse landscapes of Southern Greater Syria (Walker 2022). Some land, primarily orchards and gardens, passed onto peasant ownership after the collapse of the *iqtā* system in the post 14th century period (Walker 2022).

Imperial water systems and traditional, localized water systems make up the complex hydrological systems of Transjordan. While the largescale irrigation systems established by the Mamluk state in Southern Syria, particularly in the Jordan Valley, are well-documented; indigenous, traditional systems comprising of springs, cisterns, check dams, and small canals that collected water from various sources are not mentioned in historical texts (Walker 2022). Understanding their usage therefore depends largely on archaeological research. Botanical analysis, through the phytolith record, could provide proxy data for the use of such localized water supplies and complete the historical picture of peasant water culture and resilience mechanisms in times of drought. Water installations were abundant around Khirbet Beit Mazmīl, including several Early Islamic large-scale water systems comprising of large and interconnected cisterns (Walker and Dolinka 2020). The combined results of the phytolith analysis (Walker 2022) and the botanical, faunal, and textual studies paint a picture of intensive and diversified cereal and orchard-based agriculture at a market level of production (Walker and Dolinka 2020), facilitated by privatization of land, semi-legal leasing of endowed lands, further developments in terrace construction, and the empire-wide commercialization of agriculture.

State-related intensification and common winter and summer crops cultivated in medieval Syria

Walker (2022) argues that changes in land tenure reflect changes in land-use, and for day-to-day peasant decision-making those changes were market-driven. Using information taken from 14th to 16th centuries written sources—including endowment documents, tax registers, and court documents—Walker describes the local development of orchards in the vicinity of Khirbet Beit Mazmīl and establishes that these changes were related to the market-orientation of agricultural production in the Jerusalem area between the 14th and 17th centuries. According to Walker, terrace construction in the site's vicinity is indicative of the development of farmland and agricultural intensification for market production that followed the largescale privatization and endowment of lands in the Late Mamluk period (Walker 2022). On the Madaba Plains, Jordan, Tall Hisban is an archaeological site that was part of the same farmers' market network. Botanical evidence shows that in the hinterland of Tall Hisban, diversification of cereal production and intensification of production of wheat through irrigation occurred in the 14th century (Laparidou and Rosen 2015; Walker and LaBianca 2003). What was cultivated in those terraces and in the orchards of Khirbet Beit Mazmīl is studied through both the historical texts and botanical record.

Phytolith evidence for the agricultural activities of the medieval community of Khirbet Beit Mazmīl

Previous evidence for agriculture in the medieval Khirbet Beit Mazmīl community derives from architectural installations and evidence of different varieties of grain: primarily wheat and barley, but also including the grains of date palms, fruits, spices, and a variety of summer crops (Walker 2022). Botanical, faunal, and textual studies paint a picture of intensive and diversified cereal and orchard-based agriculture at a market-level of production (Walker and Dolinka 2020). The phytolith data derived from the Khirbet Beit Mazmīl excavations in 2017 and 2019 are presented in this chapter. The dataset consists of 20 sediment samples, and locus and context information are seen in Figure 1.

Methods

Sediment samples were processed in Greece at the Wiener Lab in the American School of Classical Studies in Athens, and at the Precision Agriculture Lab in Perrotis College at the American Farm School in Thessaloniki. Samples were analyzed by Miss Georgia Kasapidou (MSc) and Dr. Sofia Laparidou using the following protocol as adapted from Rosen (2005). 800 mg of sediment per sample were processed. Samples were sieved using a 0.5 mm sieve, and a 10 percent HCl solution was used to remove pedogenic carbonates while any remains of HCl were removed using a centrifuge at 2000 rpm for five minutes. A 0.5 mm sieve—as opposed to the usual 0.25 mm sieve—was used to retrieve large multi-cell silica skeletons that are useful in the identification of the economic crops utilized in the period studied. These include wheat and barley husk phytoliths. Clays were removed using a Calgon solution (sodium hexametaphosphate). Calgon, filtered water, and the samples were added to beakers and stirred. The phytoliths were then left to settle, while the clays were poured off. Air dried samples were burned at 500 degrees in a furnace for two hours to remove organic matter. Finally, phytoliths were separated by adding a 3 ml sodium polytungstate solution on all samples—including centrifuge samples—to extract phytolith content. Distilled water was used during the phytolith separation processing stage. Phytolith remains were left to dry and 2 mg of phytoliths per sample were used to mount microscope slides using Entellan solution. Microscope slides were left in the fume cupboard for two weeks before counting under the microscope. When possible, a minimum of 200 single-cell phytoliths and 100 multi-cell phytoliths were counted for every sample. All slides were counted at ×40 magnification under a light transmitting microscope. Absolute counts of phytoliths per gram of sediment (n/gm) are used to make comparisons between stratigraphic layers possible and to graph the data.

Introduction to the dataset

Ten sediment samples were processed for phytoliths derived from the 2017 excavation season, and 10 more were processed that derived from the 2019 excavation season.

Regarding the samples processed and analyzed from the 2017 season (Figure 1), four samples deriving from Trench B were selected for analysis. One sample deriving from Locus 124 (B.1028) was used as a control sample and was collected from the uppermost beaten earth surface in Trench B, located immediately below the plastic cover of the trench. Two samples deriving from Locus 127b, (B.1039) were also analyzed. These samples came from a beaten earth surface below Locus 127 and Locus 127a and were adjacent to a burnt concentration in the southeastern corner. Additionally,

Sample number		Context	C14	OSL	Date
1	B4.124.1028	Control sample collected from the top beaten earth surface of trench B	X		16/17th century
2	B4.127b.1039.5	Beaten earth surface below Locus 127 and Locus 127a, and adjacent to a burnt concentration in SE corner			16/17th century
3	B4.127b.1039.28				16/17th century
4	B4.130.1044	Breaks within the beaten earth surface/floor that included ashy material	X	x	16/17th century
5	C6.221.2024.23	Plaster surface below Locus 220 (a beaten earth surface)			16/17th century
6	C6.221.2024.36				16/17th century
7	C6.222.2025.4	Beaten earth surface underneath Locus 221	X		16/17th century
8	C6.222.2025.7		X		16/17th century
9	C6.226.2031.14	Fill of a basket excavated in trench C6			16/17th century
10	C6.226.2031.4		X		16/17th century

Figure 1 Context information: Sediment samples from the 2017 excavation season processed for phytoliths (1–10).

one sample originating in breaks within this beaten earth surface/floor and including ashy material was also analyzed for phytoliths (Locus 130.B.1044).

Six samples from trench C6 were analyzed for phytoliths. C6 is a domestic space with both plaster and beaten earth surfaces. The samples analyzed included two samples from Locus 221 (2024), which was a plaster surface below Locus 220 (itself a beaten earth surface); two samples from Locus 222 (2025), which was a beaten earth surface below Locus 221; and two samples from Locus 226 (2031), which were collected from the fill of a basket taken from trench C6. All samples date from the 16th to 17th centuries.

Regarding the samples taken for phytolith analysis from the Khirbet Beit Mazmīl 2019 excavations (Figure 2), one sample came from trench C6, a domestic space with plaster and beaten earth surfaces. Sample C6.259.2068 came from the fill of a plastered settling basin designated Basin #2. Three samples came from "storeroom" E1: sample E1.428.4035, from the soil inside stone-lined pit #140; sample E1.431.4037, which could be bitumen and came from the fill of installation 432, a stone-lined space inside a firing room; and sample E1.433.4038, from the fill below "storeroom" E1's flagstone floor. Sample E2.438.4058 came from the fill of a possible agricultural terrace located on-site. Four samples came from context B4: sample B4.128.1084, from the final plaster floor; sample B4.129.1085, from the soil below Locus 128; sample B4.167.1094, from an ash concentration in a plaster surface in the eastern half of the room, designated B4.166; and sample B4.176.2008, from the fill of a hearth designated Locus 180. Sample D1.510.5015 came from a thin soil layer under the flagstone floor in the cistern complex, immediately above bedrock. The majority of samples date from the 15th to 16th centuries (for more information see Figure 2).

Phytolith data: Preliminary results, Khirbet Beit Mazmīl 2017

Single-cell grass phytoliths were found in all 10 samples. Four categories of phytoliths—rondels, bilobes, saddles, and quadralobates—are presented

Sample number		Context	C14	OSL	Date
MAZ1	E1.428.4035	Soil inside stone-lined pit #140	X		14/15th century
MAZ2	E1.431.4037	Fill of installation 432. Stone-lined space inside firing room (cf. bitumen)			14/15th century
MAZ3	E2.438.4058	Fill of a possible agricultural terrace on-site			Later Ottoman?
MAZ4	C6.259.2068	Fill of Basin #2, plastered settling basin	X	x	14/15th century
MAZ5	B4.128.1084	Final plaster floor			16th century?
MAZ6	B4.129.1085	Soil below Locus 128			16th century?
MAZ7	B4.167.1094	Concentration of ash in plaster surface B4.166 (eastern half of room)	X		16th century?
MAZ8	E1.433.4038	Fill below flagstone floor in "storeroom" E1	X		15th or 16th century
MAZ9	B4.176.2008	Fill of hearth (Locus 180)			?
MAZ10	D1.510.5015	Thin soil under flagstone floor in cistern complex, just above bedrock	X		Pottery date TBA

Figure 2 Context information: Sediment samples from the 2019 excavation season processed for phytoliths (MAZ1–10).

in Figure 3. Those morphotypes can represent three different Poaceae grass subfamilies: the Pooid, Chloridoid, and Pancoid grasses. In general, Pooid grasses, which are C_3 plants, are indicative of cooler climatic conditions and produce rondel-shaped short cells. This could be, in turn, indicative of temperate plants like wheat and barley. Chloroideae, which create saddle-shaped short cells and indicate dry land grasses and warm, dry habitats, were found in three samples. Rondel-shaped short cells could also be produced by Panicoid grasses. Panicoid grasses grow in warm, humid environments and produce bilobes and cross-shaped phytoliths. Some Panicoids live in cooler, drier habitats as well. It is important to bear in mind that Phragmites, which is a grass that grows in marshy areas, also produce saddles.

Rondels were found in all samples analyzed from Khirbet Beit Mazmīl and the results could be influenced by the presence of economic plants, such as wheat and barley, that produce rondel-shaped single-cell phytoliths. Chloridoid short-cells were found in three samples from trench C6. Higher counts of saddle-shaped phytoliths were counted in samples from Loci 221 and 222, while lower counts of saddle-shaped phytoliths were counted in samples derived from the fill of the basket. Bilobes were found in abundance in sample B4. 130.1044. It is likely their presence indicates the presence of reed grasses, something that is evident in Figure 7.

The main economic crop identified in the samples was wheat. It was identified by the presence of wheat husk multi-cell phytoliths. The wheat husk phytoliths were present in three samples and were found in higher counts in sample B4.130.1044. This context contains much wheat chaff as well as reed phytoliths. Palm phytoliths were also found, but only in sample B4.130.1044 (Figure 6). Wheat chaff was overall rare in the samples. The highest counts of wheat and unidentified cereal husks came from sample B4.130.1044, while unidentified cereal husks were present in samples from contexts B4.127b.1039, C6.221.2024, and C6.222.2025. Figure 4 shows the numbers per gram of sediment of multi-cell wheat husk phytoliths. No phytoliths

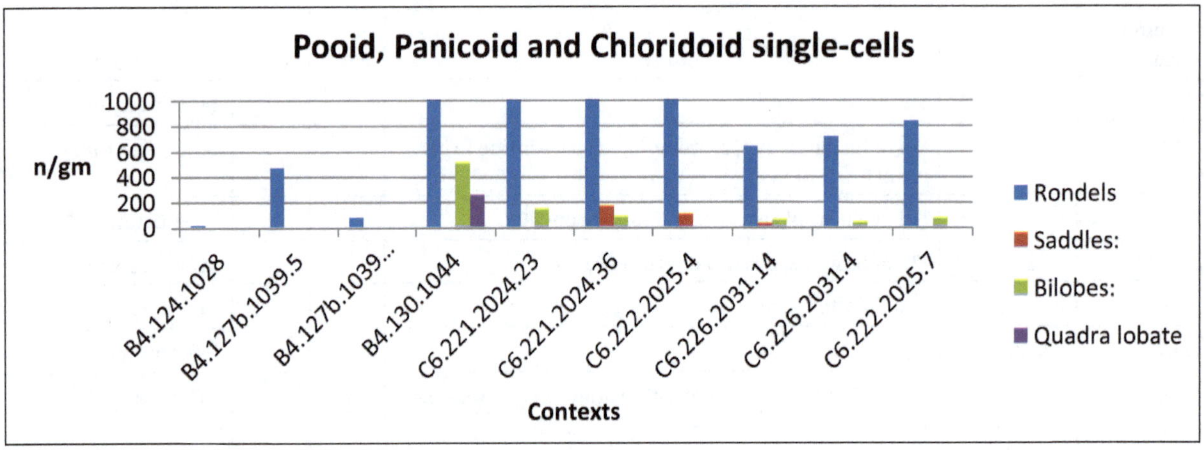

Figure 3 Pooid, Panicoid, and Chloridoid single-cell grass phytoliths.

associated with wheat processing, such as cereal, straw, and weed phytoliths, were present in those samples. The absence of such early-stage crop-processing by-products suggests that the room in Trench B4 was not a space for crop-processing and grain and fodder storage.

Dicot plants, such as wood and bark, were found in low numbers in five samples (Figure 5). The highest numbers were recorded in samples from trench C6.

Reed phytoliths were present in all contexts from Khirbet Beit Mazmīl and appear to form a basic component of the local landscape and play a major role in the site's economy (Figures 5 and 6). Reeds entered the site either as fodder or construction material and were present in their highest number in sample B4.130.1044. Reed phytoliths were also present in all contexts from trench C6 and appear to be present in all the beaten earth and plaster surfaces sampled (Figures 5 and 6).

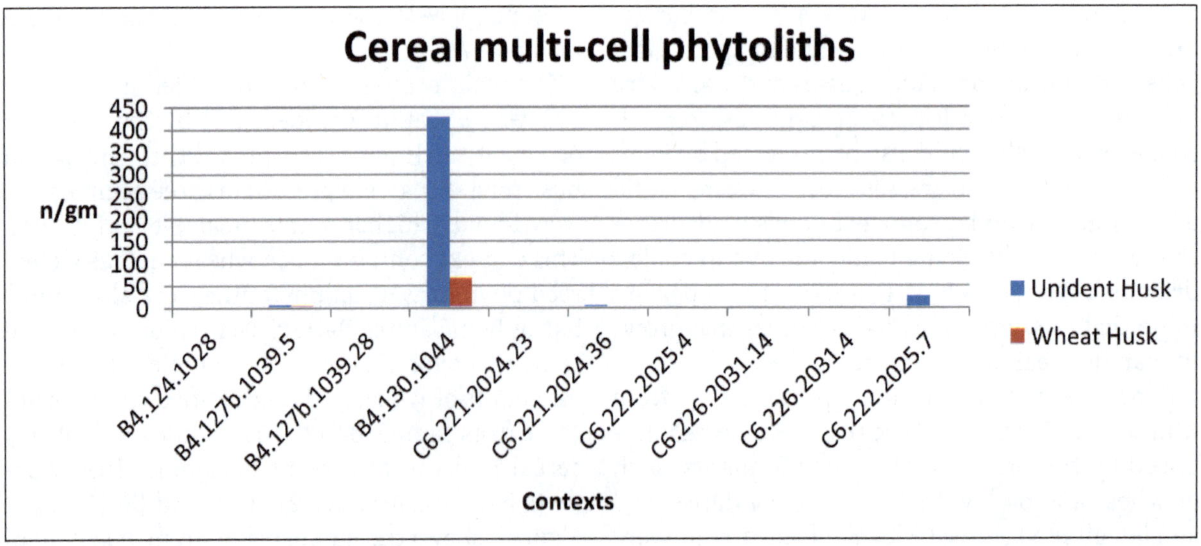

Figure 4 Cereal multi-cell phytoliths.

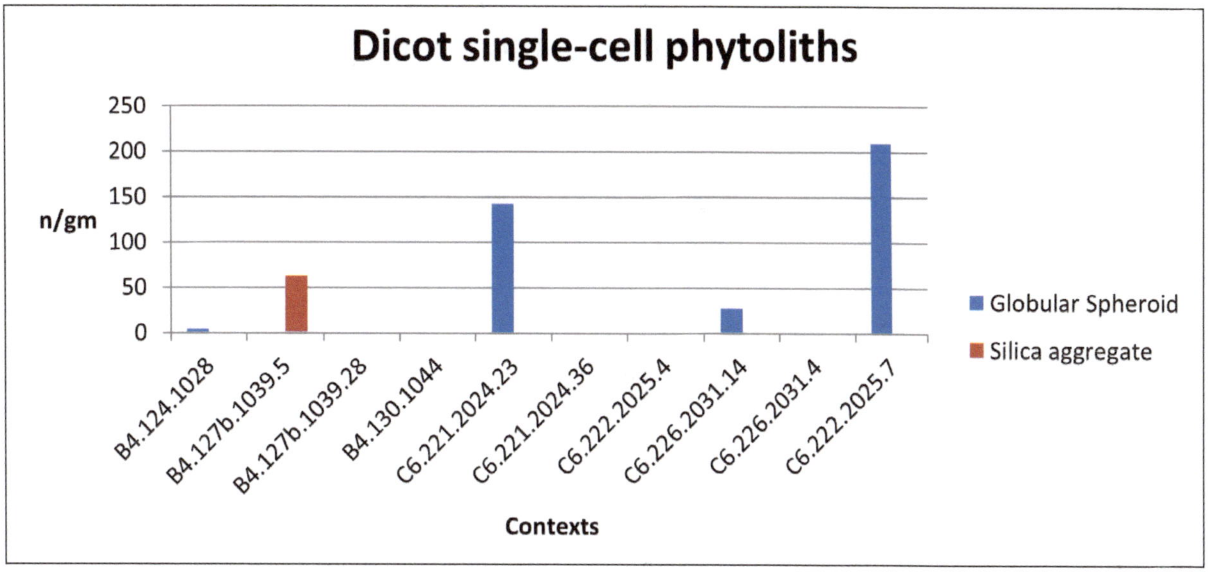

Figure 5 Dicot single-cell phytoliths.

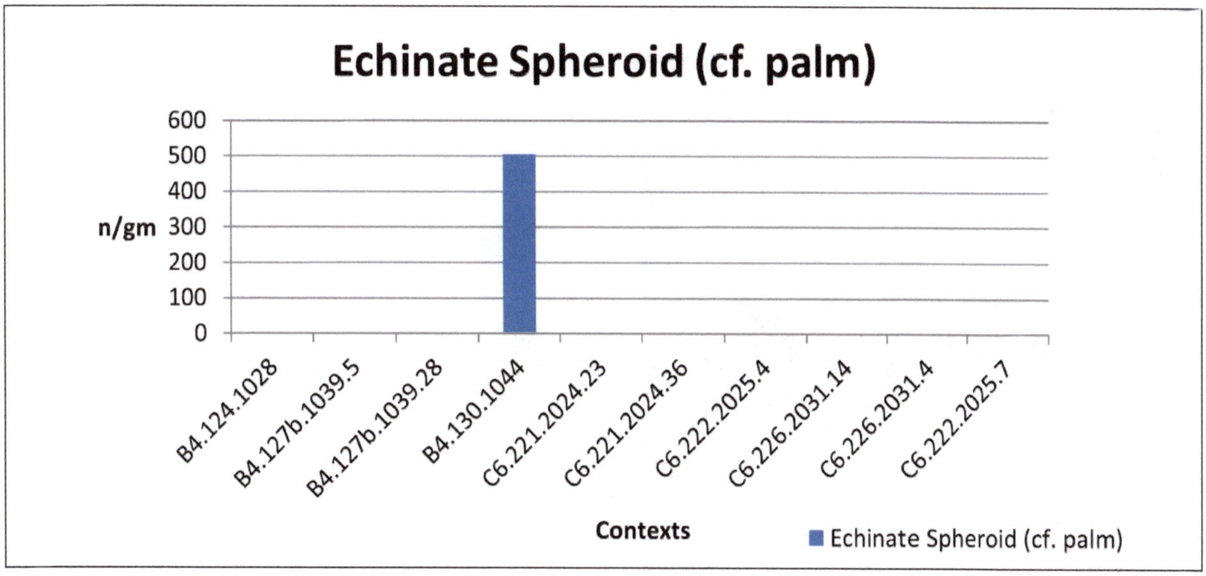

Figure 6 Echinate spheroid phytoliths (cf. palm phytoliths).

Phytolith data: Preliminary results, Khirbet Beit Mazmīl 2019

Single-cell grass phytoliths were found in all 10 samples. Three categories of phytoliths—rondels, bilobes, and saddles—are presented in Figure 3.

Rondels were found in all samples, while Chloridoid short-cells (saddles) and Panicoid short-cells (bilobes) were found in only one sample (MAZ8), from the fill below the flagstone floor in "storeroom" E1. The incidence of bilobes in this sample is likely an indication of the presence of reed

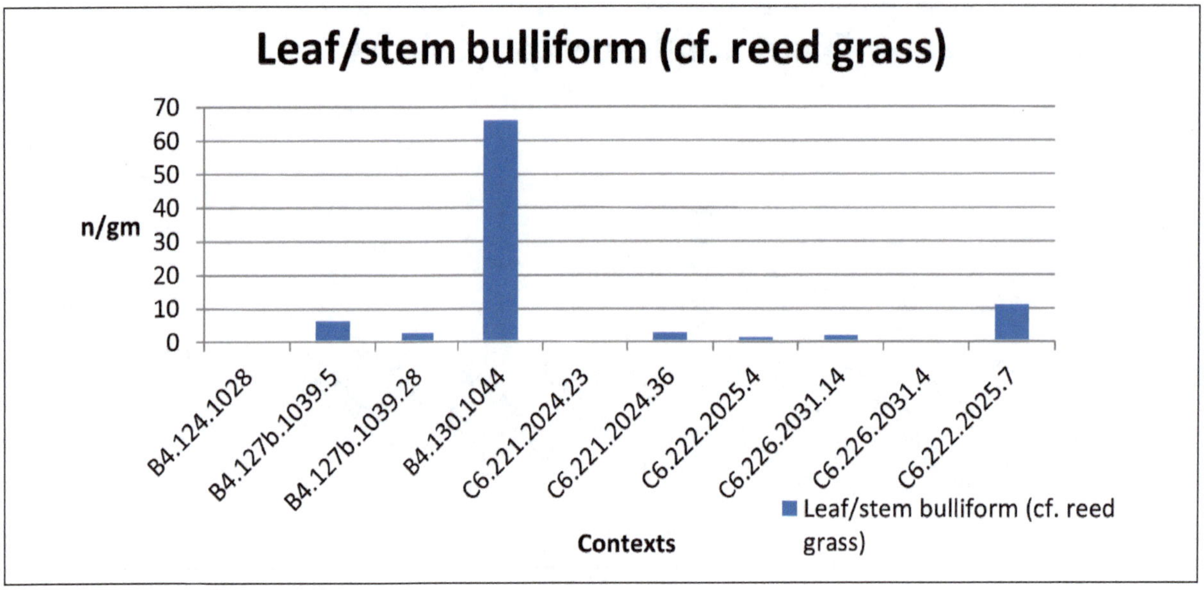

Figure 7 Leaf/stem buliform phytoliths (cf. reed grass).

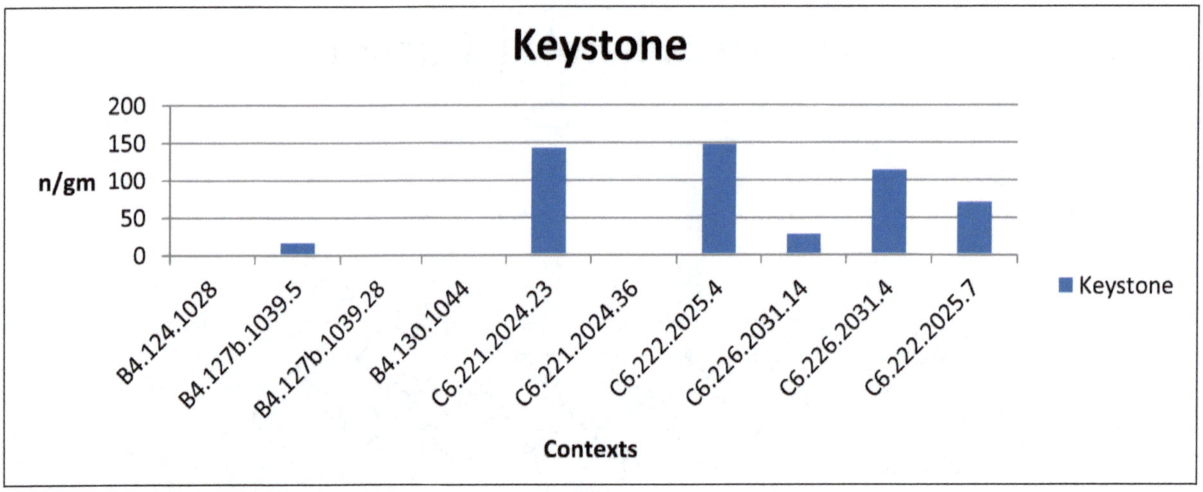

Figure 8 Single-cell reed phytoliths.

grasses, something that is illustrated in Figure 6. Due to the lack of multi-cell husk phytoliths and the overall low level of preservation of economic crops, such as cereals, in all contexts, there may be another explanation for the noted presence of rondels: possibly the existence of Festucoideae grasses.

Figures 2 and 3 illustrate the presence of cereal phytoliths stemming from husks that were present in all samples analyzed. Besides the presence of one wheat husk in sample MAZ6, all other silica skeletons remain unidentified and cannot provide any evidence of economic cereal crop cultivation or processing. The question of storage for husked

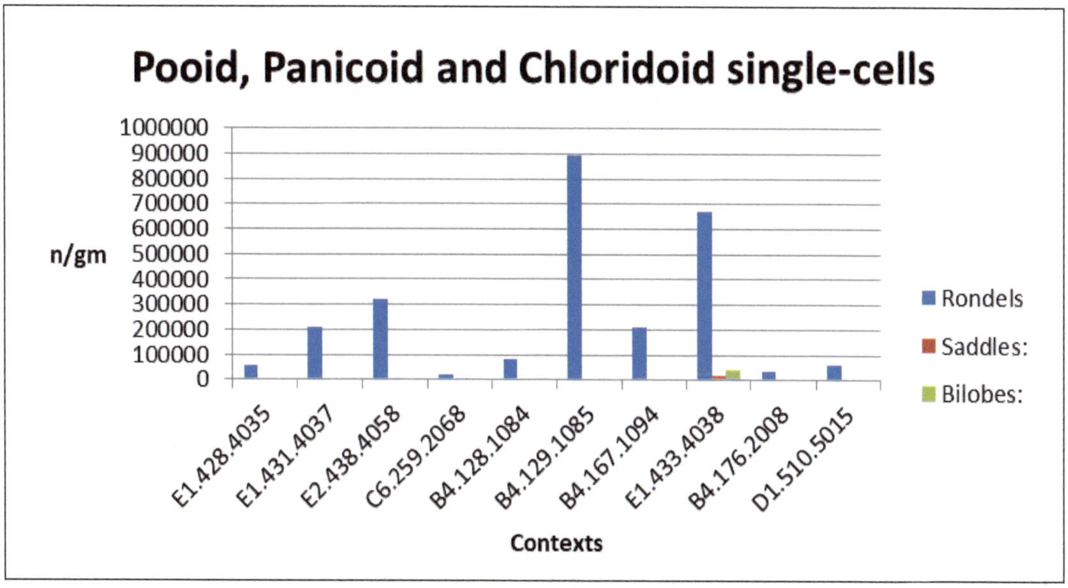

Figure 9 Pooid, Panicoid, and Chloridoid single-cells.

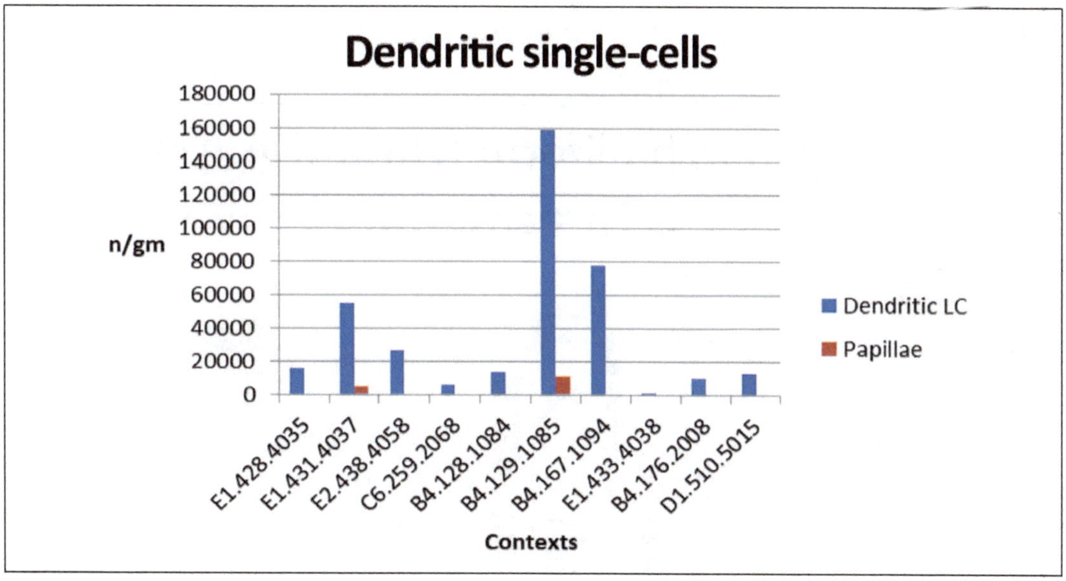

Figure 10 Dendritic single-cells.

cereals is also rather uncertain based on the evidence as provided above.

The presence of reed grass phytoliths in three samples from MAZ6, 8, and 10—all of which are related to floor fills, or the soil covers of room floors—most probably stem from reed matting, or from other construction materials used in the walls or ceilings where reeds were widely used.

Pooid grasses were overall dominant and recorded in higher concentrations in all phases of occupation

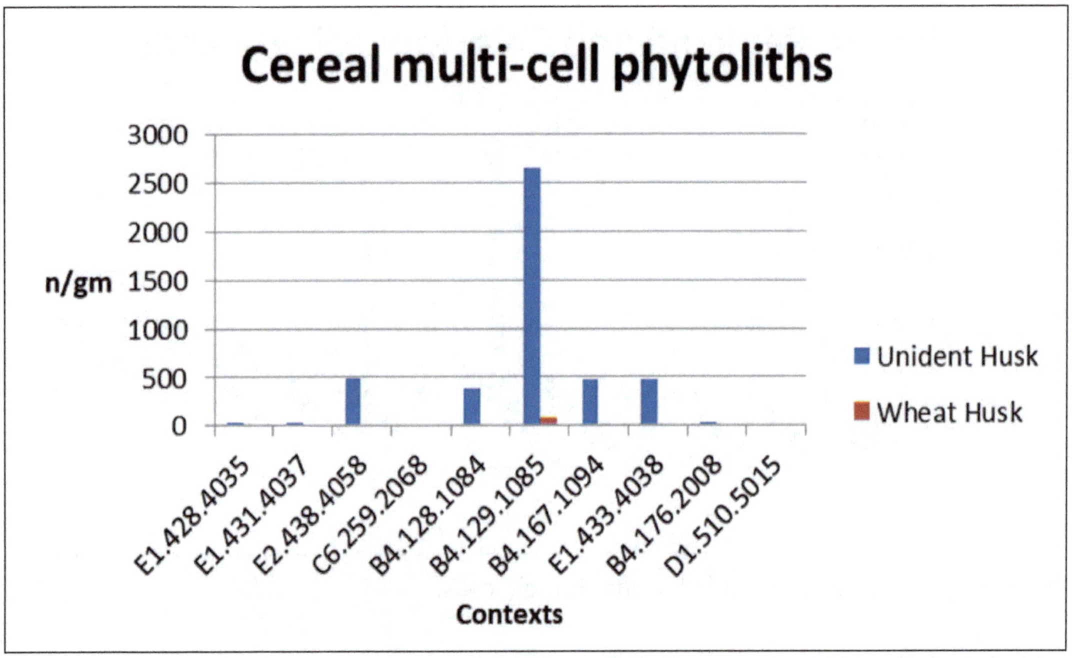

Figure 11 Cereal multi-cell phytoliths.

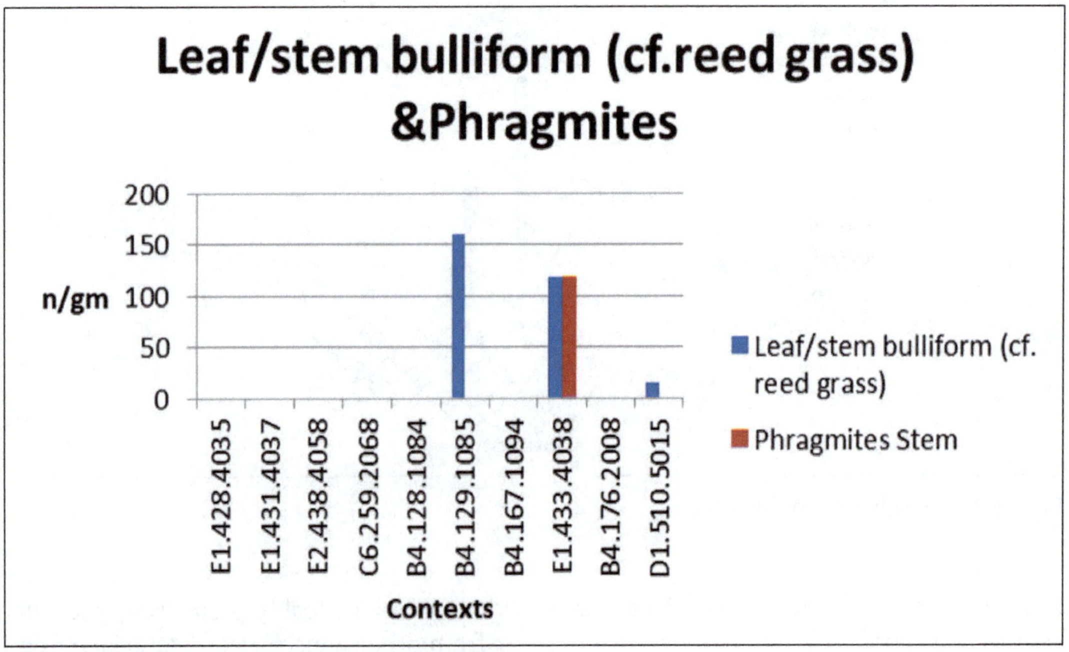

Figure 12 Reed phytoliths.

and in a variety of contexts. Rondels were found in all samples, while Chloridoid short-cells (saddles) and Panicoid short-cells (bilobes) were found in only one sample (MAZ8). This sample came from the fill below the flagstone floor in "storeroom" E1. Panicoid grasses—a grass subfamily that includes weed grasses that thrive in warm, dry environments—were not dominant in the assemblage (Figure 3). Cereal phytoliths were identified according to methods outlined by Rosen (1992). Aside from one wheat husk in sample MAZ6, all other silica skeletons remain unidentified and do not provide any evidence of economic cereal crop cultivation or processing, while the picture of storage for husked cereals remains rather unclear. Although in generally low concentrations, wheat (*Triticum* sp.) was at its highest in 16th century contexts (Figure 4).

Discussion

The level of phytolith preservation was generally poor in samples analyzed from Khirbet Beit Mazmīl, and phytoliths consequently had limited potential for archaeobotanical and environmental analysis. Given the preservation of the single-cell and multi-cell morphotypes presented above, however, and combined with the macrobotanical data derived from the same contexts, we can nonetheless propose a scenario for Khirbet Beit Mazmīl's agricultural economy according to the phytolith data.

The main economic crops identified through phytolith analysis at this stage are cereals (2017 and 2019 excavation seasons) and date palms (2017 excavation season). The dates are only found in one sample, but their importance for the site's economy should be investigated further. Cereal husks were found in most samples and in all contexts. It was only possible to identify the husks from wheat. Barley husks have not been identified at this stage in the analysis, and it will be interesting to see where barley chaff was deposited or stored on the site, either as fodder or as a stock for other uses, such as plaster temper. Cereal chaff is present in larger percentages at the expense of weed chaff of wild grasses. This possibly indicates that the economic crops were entering the site in clean crop form—of hulled barley or wheat—rather than as processed material or fodder. The absence of cereal straws further strengthens this hypothesis. Phytoliths associated with crop-processing by-products—straw and wild grass husks—were not present in the samples analyzed.

The phytolith record reflects a subsistence farming economy: both production for local markets and the economic activities of a family estate geared towards commercial agriculture. This assumption is based on the absence of evidence for heavy investment in and intensification of grain production. Diagnostic multi-cell phytoliths of wheat derived only from one sample from Trench B4, which was a domestic space. On the other hand, no barley husks were found in the phytolith record to suggest that the storage, processing, or management of such a crop occurred on-site.

The phytolith record indicates that the inhabitants of Khirbet Beit Mazmīl produced cereal grains for consumption at a subsistence level and for local markets. According to the phytolith evidence, the site produced grain during the 14th to 15th centuries (Figures 2 and 9), possibly increasing the production of wheat grain in the 16th to 17th centuries. There is no evidence to suggest that cereal processing occurred on-site, specifically in Trench B4 and C6; both domestic spaces. According to the macrobotanical evidence, cereal processing waste such as straw and chaff were used as a temper.

Interestingly, reed grasses were found in abundance in all samples analyzed and indicate wet, marshy areas surrounding the site. It also appears that reeds were used as a construction material in either the roofs, walls, or posts. For example, Phragmites, which belongs to the Poaceae plant family and Panicoid plant subfamily, is a grass that grows in marshy areas and produces saddles (Ollendorf 1992). Both Phragmites phytoliths and saddles were found in all samples from both domestic spaces (Trenches B4 and C6). The reed phytoliths were in higher densities in the ash lenses found in Trench B4, interrupting the flagstone floor (Locus 128). Just as with palms, reed grasses could be a source of fuel. The presence of reed phytoliths within the ashy context of Trench B4 L.130 indicates that reeds were either used as fuel, or the phytoliths come from animal dung that was itself burned for fuel. Reed phytoliths, also found in the samples taken from the floor fills, could indicate that reeds were used as construction material (Figures 5 and 6). The abundance of reed phytoliths is indicative of the presence, as well as the use, of plants from wetland areas during the 14th/15th centuries through to the late 16th/early 17th centuries. Their presence in the samples suggests a wet and moist microenvironment around the site that undeniably favored cereal cultivation during the periods under investigation. Such results could also indicate marshy environments close to the site where animals could graze. High densities of Phragmites phytoliths from "storeroom" E1 (Figure 7) could be the remains of roofing material or basketry used within the storage area. Phragmites, aside from being a reed used in construction, could also have entered the assemblage as a forage plant, assuming that the inhabitants were agropastoralists and kept their livestock within the settlement. One can observe in the analyzed ashy contexts (MAZ2 and MAZ7) a low concentration of reed phytoliths. This low number suggests, however, that the reeds were not used primarily as fuel, nor that they could have come exclusively from animal dung deposits used in firing. Reed phytoliths were found in higher concentrations in the samples deriving from the floor fills (MAZ6 and MAZ8), which suggests that reeds were used primarily as construction material (Figure 6).

Regarding the cultivation of cereals, a general trend in the phytolith data suggest a preponderance of Pooid grasses in all occupation periods (Figure 3). This grass subfamily includes cultivated cereals such as wheat and barley. The cereal component of samples taken from Khirbet Beit Mazmīl includes wheat (*Triticum* sp.) but not barley (*Hordeum* sp.). The dominance of rondels (Figure 3) observed during analysis could therefore suggest that both crops had a presence on-site, though barley chaff was deposited or stored elsewhere. Barley could have entered Khirbet Beit Mazmīl as fodder or as a stock for other uses—such as plaster temper—while wheat was treated as an economic and food crop.

Trench B4 revealed macrobotanical remains of barley, while Trench B7 exhibited the remains of hulled barley. According to the macrobotanical remains, barley husks were used at Khirbet Beit Mazmīl for fuel along with other early crop-processing remains, while hulled barley was cultivated along with free-threshing wheat. If free-threshing wheat was the dominant crop and cultivated throughout the periods under investigation, the husks would be considered an early-stage by-product from threshing. The general absence of such remains from the areas sampled suggests that early processing was not taking place on-site and that surfaces were kept clean of such remains. If hulled barley was a dominant species and was cultivated at Khirbet Beit Mazmīl, the husk would have not necessarily been removed from the grain, especially if it was used for fodder.

An emphasis on wheat cultivation as opposed to barley could indicate a prosperous climatic and environmental setting for cultivation, or the ability of Khirbet Beit Mazmīl's economy to sustain irrigated cultivation. Environmental uncertainty and drought conditions would demand barley to be present in higher concentrations, as it is a more drought-resistant crop than wheat. As discussed by Hansen in Chapter 7.2 of this volume, however, the remains of cereal crops present on-site include free-threshing wheat—likely hard wheat and bread wheat—and hulled barley. If hulled barley was the dominant species, the hulled grain would be used for fodder and phytolith analysis would not reliably pick up any husk on-site, as the husk is not easily removed from the grain.

Phytolith data point towards an intensive and diversified cereal and orchard-based agriculture at a market level of production, favored by prosperous environmental conditions in the site's vicinity, and that major investment in agricultural infrastructure for cereal production, of both free-threshing wheat and hulled barley, could occur throughout the periods under investigation. Water installations would support large-scale wheat cultivation and local peasants could depend on a diversified local-scale agriculture based around cereal and orchard cultivation over the 14th to 16th centuries. This system, related to seasonality and the rotation of winter rainfall and summer droughts, makes farmers susceptible to a variety of natural and cultural hazards. A number of agricultural risk-buffering strategies could have been employed by the peasants of Khirbet Beit Mazmīl and could have included diversification, storage, trade, and exchange. Cereal chaff, according to the phytolith record, entered the site and was present in all samples. The recovery of the dendritic single-cell phytoliths (Figure 10) that form in cereal husks from different sampling areas of a site can indicate areas of cereal storage or cereal processing as well (Harvey and Fuller 2005; Rosen 1992). The production of a surplus of cereal processing by-products such as chaff and straw can also be of primary importance for pottery making, as seen with coarse ware pottery. According to the phytolith record, during droughts the farmers could have depended upon the production, storage, and control of agricultural surplus cereals, and crop by-products of economic value, such as chaff.

The macrobotanical assemblage collected from Khirbet Beit Mazmīl shows a diverse range of cereal crops, though free-threshing cereals predominated (see Hansen, Chapter 7.2). We can infer from this that free-threshing cereals, hulled barley, and hard wheat played a significant role in the diets of both the site's human inhabitants and their domesticated animals and were also important economic crops. Bread wheat and emmer wheat—a glume wheat—would have meanwhile played a more minor role in the diet and local economy. There were also a wide variety of fruits and pulses at the site (see Chapter 7.2). Combined with the phytolith record, the macrobotanical remains show that the inhabitants of Khirbet Beit Mazmīl practiced small-scale cereal and garden cultivation, providing a diversified economy for the inhabitants of the semi-arid region.

Conclusive comments on phytolith evidence

Even when both the 2017 and 2019 Khirbet Beit Mazmīl excavation season data sets were taken into account, phytolith preservation in those samples analyzed was overall poor. It was nonetheless possible to identify the main economic crops, including cereals—namely wheat—and date palm. The dates were only found in one sample (B4.130.1044), deriving from the 2017 excavation season, indicating that it became an important crop during the 16th to 17th centuries. Cereal husks were found in most samples analyzed and in all the contexts sampled. It was possible to identify the husks as likely coming from a free-threshing wheat species. According to Hansen, free-threshing wheat species were identified on-site (see Chapter 7.2). Depending on the cereal species cultivated, cereal husks can be considered as early or late-stage by-products (Harvey and Fuller 2005). If free-threshing wheat was cultivated, the husks and straw are considered as early-stage by-products from threshing. It is likely that hulled barley species entered the site as economic crops, while wheat chaff could indicate that some kind of processing took place on-site. Barley husks have not been identified at this point in the analysis of the samples, and it will be interesting to see where barley chaff was deposited or stored on the site, as either fodder or as a stock for other uses, such as the production of plaster temper. Interestingly, reed grasses were found in an abundance in all samples analyzed and are suggestive of a wet, marshy area surrounding the site, and also suggest that reeds were used as a material in the construction of roofs, walls, or posts.

Phytolith analysis of sediment samples derived from Khirbet Beit Mazmīl showed some evidence of peasant agricultural activities. It appears that the agricultural society around medieval Jerusalem practiced small-scale agropastoral activities. Phytolith analyses indicated that during periods of prosperous environmental conditions agricultural production sustained an autonomous community that practiced a mixed agropastoral economy. This community depended upon both the distribution of agricultural surplus to local markets, and small-scale production and storage of cereals, other crops, and agricultural surplus. Using micro- and macrobotanical analysis it can be surmised that peasants at Khirbet Beit Mazmīl invested in agricultural strategies of risk minimization and practiced small-scale cereal and garden cultivation, which provided for a diversified economy. This created a sustainable and resilient agricultural society throughout the 14th to 16th centuries. Phases of declining water availability would have impacted the practice of agriculture and the production of free-threshing wheat. Peasants could, however, employ various low-level mechanisms to protect against crop failure and food shortage, including diversification of production. For example, an emphasis on barley cultivation as opposed to wheat could offer security during periods of uncertainty, as barley is a more drought-resistant crop than wheat. Also, peasants at Khirbet Beit Mazmīl depended on a wide variety of crops, including date palms, and on livestock and other domestic animals.

The phytolith analysis of samples taken from the medieval village of Khirbet Beit Mazmīl, in southwest Jerusalem, shows that the local peasant community sustained a resilient subsistence agricultural economy, with an emphasis on cereal production.

Acknowledgments

Sofia Laparidou would like to thank the project directors, Professor Bethany Walker and Dr. Benjamin Dolinka, for involving her in the project.

References

Butzer, K.W.
1982. *Archaeology as Human Ecology: Method and Theory for a Contextual Approach*. Cambridge: Cambridge University Press.

Butzer, K.W.
1996. "Ecology in the long view: settlement histories, agrosystemic strategies, and ecological performance." *Journal of Field Archaeology* 23: 141–150. Internet Edition: https://doi.org/10.2307/530500.

Cordova, C.E.
2007. *Millennial Landscape Change in Jordan: Geoarchaeology and Cultural Ecology*. Tucson: University of Arizona Press.

Halstead, P.
1990. "Waste not, want not: traditional responses to crop failure in Greece." *Rural History* 1(2): 147–164. Internet Edition: doi:10.1017/S0956793300003290.

Harvey, E.L., and D.Q. Fuller
2005. "Investigating crop processing using phytolith analysis: the example of rice and millets." *Journal of Archaeological Science* 32: 739–752.

Jum'a Mahmoud, H.K.
2000. *The Settlement Patterns in the Jordan Valley in the Mid-to Late Islamic Period*. Oxford: Archaeopress.

King, G.
1992. "Settlement patterns in Islamic Jordan: The Umayyads and their Use of the land." *Studies in the History and Archaeology of Jordan* 4: 369–375.

Laparidou, S., and A.M. Rosen
2015. "Intensification of production in Medieval Islamic Jordan and its ecological impact: Towns of the Anthropocene." *The Holocene* 25(10), 1685–1697. Internet Edition: https://doi.org/10.1177/0959683615594470.

Macdonald, B.
2007. "Four archaeological surveys in southern Jordan." In *Crossing Jordan: North American Contributions to the Archaeology of Jordan*, edited by T.E. Levy, P.M. Michele Daviau, and R.W. Younker, 161. London: Routledge.

Mithen, S., and E. Black
2011. *Water, Life And Civilisation: Climate, Environment and Society in the Jordan Valley*. Cambridge: Cambridge University Press. Internet Edition: https://doi.org/10.1017/CBO9780511975219.

Palmer, C.
2002. "Milk and cereals: identifying food and food identity among Fallāhīn and Bedouin in Jordan." *Levant* 34 (1):173–195. Internet Edition: https://doi.org/10.1179/lev.2002.34.1.173.

Ollendorf, A.L.
1992. "Toward a classification scheme of sedge (Cyperaceae) phytoliths." In *Phytolith Systematics: Emerging Issues (Advances in Archaeological and Museum Science*, 1), edited by G. Rapp and S.C. Mulholland, 91–111. Boston: Springer.

Rosen, A.M.
1992. "Preliminary identification of silica skeletons from Near Eastern archaeological sites: an anatomical approach." In *Phytolith Systematics: Emerging Issues (Advances in Archaeological and Museum Science*, 1), edited by G. Rapp and S.C. Mullholand, 129–147. Boston: Springer.

Rosen, A.M.
2005. "Phytolith indicators of plant and land use at Çatalhöyük." In *Çatalhöyük Project Volume IV: Inhabiting Çatalhöyük*, edited by I. Hodder, 203–212. Cambridge: McDonald Institute for Archaeological Research, University of Cambridge.

Walker, B.J., and O. LaBianca
2003. "The Islamic Qusur of Tall Hisban: Preliminary report on the 1998 and 2001 seasons." *Annual of the Department of Antiquities of Jordan* 47: 443–471.

Walker, B.J.
2004. "Mamluk investment in Transjordan: A "boom and bust" economy." *Mamluk Studies Review* 8: 119–139. Internet Edition: https://doi.org/10.6082/M1222RX3.

Walker, B.J.
2005. "The Northern Jordan Survey 2003—agriculture in Late Islamic Malka and Hubras villages: A preliminary report of the first season." *Bulletin of the American Schools of Oriental Research* 339: 67–111. Internet Edition: https://doi.org/10.1086/BASOR25066903.

Walker, B.J.
2008. "The role of agriculture in Mamluk-Jordanian power relations." *Bulletin d'études orientales* 57: 79–99.

Walker, B.J.
2011. *Jordan in the Late Middle Ages: Transformation of the Mamluk Frontier*. Chicago: Middle East Documentation Center, on behalf of the Center for Middle Eastern Studies.

Walker, B.J., and B. Dolinka
2020. "*Ḫirbet Bēt Mazmīl*. Investigations of Medieval Jerusalem's Hinterland: Interim report on the 2015 and 2017 Field Seasons" *Zeitschrift des Deutschen Palästina-Vereins* 136.2: 191–234.

Walker, B.J., Y. Gadot, Y. Elgart-Sharon, and O. Ze'evi
2020. "Agricultural terracing and rural revival in Late Medieval Palestine." In *Proceedings of the 11th International Conference on the Archaeology of the Ancient Near East* 2, edited by L. Korn and A. Heidenreich, 651–662. Wiesbaden: Harrossowitz Verlag.

Walker, B.J.
2021. "Peasants, rural economy and cash crops in medieval Islam." In *Markets and Exchanges in Pre-Modern and Traditional Societies*, edited by J.C.M. García, 91–113. Oxford: Oxbow Books.

Walker. B.J.
2022. "The changing face of agricultural 'estates' in 15th and 16th-Century Palestine: The commercialization of Khirbet Beit Mazmīl." In *The Mamluk-Ottoman Transition: Continuity and Change in Egypt and Bilad al-Sham in the Sixteenth Century*, Vol. II, edited by S. Conermann and G. Şen, 325-357. Bonn: V&R Press.

About the authors

Dr. Sofia Laparidou is a secondary teacher working at the Ministry of Education, within the Ministry of Culture, Greece. She is an environmental archaeologist interested in the changing nature of human-environment relationships as shaped by climate, cash cropping, and the expansion of intensive agricultural production into semiarid regions and has used microbotanical analysis to investigate ancient agricultural and pastoral economies. Dr. Laparidou obtained a PhD in Anthropology at the University of Texas; an MSc in GIS at the University of Greenwich; and an MSc in Environmental Archaeology at the University of Sheffield. She has published her research on the political ecology and microbotanical methods of medieval Jordan in the *Proceedings of the Eighth International Conference on the Archaeology of the Ancient Near East*. Her paper, entitled "Intensification of production in Medieval Islamic Jordan and its ecological impact: Towns of the Anthropocene," has been published in *The Holocene* (2015).

Georgia Kasapidou (MSc) is a researcher with a focus to archaeobotany. Her dissertation was based on the analysis of plant micro-remains, such as phytoliths and starch grains, from grinding stone tools. She has taken part in various archaeological projects analyzing microbotanical remains from artifacts in prehistoric contexts.

7.2: *The Macrobotanical Record*

Annette M. Hansen

In this chapter the results of the analysis of the archaeobotanical assemblage from the rural village of Khirbet Beit Mazmīl, located on the western edge of Jerusalem, will be presented.[1] The site is located on a hill, now enclosed by a walled apartment complex and next to a community center. It had been occupied, reused, and rebuilt almost continuously since the Late Byzantine period and was a part of a network of villages that dotted the countryside surrounding the city (Walker and Dolinka 2020). The site is of import and interest as it is considered "the best-preserved farmstead in the immediate outskirts of Jerusalem" and is an ideal case study to investigate Jerusalem's dynamic rural hinterland and agricultural areas, especially during the Late Medieval period, that has not previously been studied in detail (Walker and Dolinka 2020, 191). The name of the farmstead site Khirbet Beit Mazmīl ("the ruins of the House of the Flute") is mentioned in 19th and early 20th century written sources, though its historical name and background are difficult to trace. It has been hypothesized that the site is associated with the village Beit Mazmīl ("House of the Flute") that is located close by. Beit Mazmīl is documented in a 14th century Mamluk endowment document (*waqf*), and in the 16th century it and the surrounding land are mentioned in an Ottoman tax register (*defter*) and a court document (Walker and Dolinka 2020). These latter sources mention that the land in the village's immediate vicinity were devoted to cereal and fruit tree cultivation (Walker and Dolinka 2020); see Walker (this volume) for a detailed study of the written sources on the site and surrounding area during the Mamluk and Ottoman periods in Chapters 4 and 5 respectively. As this chapter will argue, such observations are also reflected in the archaeobotanical assemblage from the village.

Interest in Khirbet Beit Mazmīl grew during a rescue excavation of the southern part of the site between 2012 and 2013, when, in addition to Byzantine (5th to 6th century) and Crusader/Ayyubid (mid-12th to mid-13th century) occupational phases, two additional occupational phases from the Mamluk period (14th and 15th centuries, respectively) were uncovered. These excavations attested to a well preserved Mamluk period structure with various features, including stables, kilns, and shallow-domed bread ovens (*ṭābūns*). Some samples rich in macrobotanical materials were taken from within this structure (see Section Trench B7). This rescue excavation confirmed the site's potential for rigorous study and suggested the presence of an unexcavated complex located uphill immediately to the north (Walker and Dolinka 2020). Therefore, a follow-up study was set into motion in 2015 by the project's site directors Prof. Dr. Bethany Walker (University of Bonn) and Dr. Benjamin Dolinka (formerly Israel Antiquities Authority and Gerda Henkel Stiftung Research Fellow) within the Medieval Jerusalem Hinterland Project, funded by the German Israeli Foundation for Scientific Research and Development (GIF). Archaeobotanical sampling and analyses were of key importance to the planning of the three field seasons between 2017 and 2019 to aid in the reconstruction of the agricultural and food economy as well as the local environment of Khirbet Beit Mazmīl during the Late

Medieval period, particularly during the Mamluk and Ottoman occupational phases of the site. This chapter integrates the results of the analysis of the archaeobotanical samples taken in 2013 and the 2017–2019 field seasons.

This chapter will begin with a description of the applied archaeobotanical methodology with respect to sample selection, excavation, and processing and the botanical identification and quantification of the remains. Next, I will make a brief note on the selection of samples for radiocarbon dating that were carried out within the project. In the subsequent section, I will both describe the different archaeological loci per trench and present the archaeobotanical results from the trench within this context in order to interpret the activities and processes that took place within those spaces. Here I will also discuss the significance of certain notable or rare finds within their respective archaeological context. In the final section, I will summarize the main findings of the study and what the archaeobotanical evidence may contribute to the understanding of economic activities at Khirbet Beit Mazmīl and those at similar medieval rural sites.

Methodology: Sample selection, processing, and botanical identification and quantification

For this study a total of 56 samples taken from seven main trenches during the 2017 through 2019 field seasons at Khirbet Beit Mazmīl were assessed in terms of archaeobotanical remains, plant inclusions, and plant impressions. The number of samples studied per type are: 30 soil samples (totalling 1057.5 liters); 13 plaster floor samples; 8 samples of fragments of *ṭābūn* ovens or earthenware vessels; 4 charcoal samples; and 1 plant fiber sample. The studied trenches are: B4 (Domestic Space); B7 (Extramural Structure); C6 (Domestic Space); C6–7 (Transitional Area); D1 (Water Reservoir); E1 (Potential storeroom with installations); and E2 (Potential agricultural terrace). The most common preservation state of the archaeobotanical remains was a carbonized or charred state, although some mineralized plant remains were present within the samples. Relevant observations on taphonomy will be made within the respective discussions on the plant remains per trench in relation to their archaeological context in the sections below. Some specimens for which a certain identification to the species or subspecies level could not be made were identified with the notation "cf." to indicate the closest genus/species to the archaeological specimen(s). Some archaeobotanical specimens could not be identified to a lower taxonomic level due to damage to their morphological features during intensive charring or to post-depositional breakage.

Specific archaeological contexts were chosen for sampling to help understand the various economic activities occurring within the spaces at the site, especially those related to construction—including beaten earth and plaster floors—and food processing, preparation, and consumption—including installations such as *ṭābūn* ovens. The assemblage of archaeobotanical remains, inclusions, and impressions will be interpreted in relation to the excavated trench and locus to help discern the potential use and function of the plant remains at the site, the broader agricultural activities taking place in the area, and to obtain potential signals reflecting the local environment. All archaeobotanical contexts and samples from the 2017 and 2018 seasons were excavated by the author, while the excavation team conducted sampling in the 2019 field season.

In addition, as a part of the methodological protocol set up for the site, phytolith samples were systematically taken from each of the contexts sampled for macrobotanical remains following the protocol by Laparidou in 2017 and carried out in the 2017–2019 field seasons. For the complete results of the phytolith analysis, see Chapter 7.1 by Laparidou and Kasapidou. It was also deemed methodologically relevant and interesting to set aside all zooarchaeological remains from the macro-archaeobotanical soil samples in order to study smaller animal taxa. These taxa are not usually found with traditional sieving on archaeological sites, since the 4 millimeter sieves (and larger) that are typically used are not suitable for collecting the bones of these taxa. A protocol was established by Hansen and Corbino and put in effect during the 2017 to 2019 field seasons. For the full results of the zooarchaeological analysis, see Chapter 7.3 by Corbino.

All soil samples were processed with a floatation machine in the garden of the W.F. Albright Institute of Archaeological Research (AIAR) in Jerusalem. This is a process in which soil samples are filtered in a large water tank and archaeobotanical material floats to the surface and is collected in a fine-meshed cloth bag. This fraction of the sample is referred to as the "flot." Heavier archaeobotanical remains, such as larger pieces of charcoal, and materials such as rocks, ceramics, glass, metal slags, coins, and bones sink and are collected in a second sieve of 1 millimeter. This material is later laid out to dry and then dry-sieved to collect, separate, and label such finds for later study by the material group's respective specialist. This part of the sample is referred to as "heavy residue" (HR) or as the "heavy fraction." Silt and soil are filtered to the bottom of the tank. Sub-samples of plaster surfaces that had impressions of seeds and cereal crop-processing waste were also taken to explore cereal cultivation that occurred at or nearby the site and to study the use of such materials in plaster flooring. Similarly, fragments from *ṭābūn* ovens and coarse, unidentifiable earthen ware vessels were collected for the analysis of their plant impressions, typically originating from cereal crop-processing waste. Some loose charcoal fragments that were separated during dry-sieving were collected for further analysis alongside charcoal from the floatation samples. One potential plant-based textile fragment was sampled to determine the taxon of which it was made. Context descriptions and metrics—soil volume, flot volume, and volume of the archaeobotanical remains—for the archaeobotanical samples were recorded in tables per trench and are presented in their respective sections.

After initial processing, the archaeobotanical samples were shipped to the laboratory of the author, initially at the University of Groningen, and later at the Interdisciplinary Historical Food Studies (FOST) and Industrial Microbiology and Food Biotechnology (IMDO) research groups at the Vrije Universiteit Brussel in Brussels. The samples were first sieved, screened, and sorted for richness. Subsequently, the archaeobotanical remains, including seeds/fruits, rachis, culm, chaff, and wood were sorted, quantified, and identified using a stereo light microscope and consulting a botanical reference collection and archaeobotanical identification manuals. For the identification of archaeobotanical specimens, *Flora Palaestina* Volumes 1–4 (Feinbrun-Dothan 1978, 1986; Zohary 1966, 1972), *Digital Atlas of Economic Plants* Volume 1, 2a, and 2b (Cappers *et al.* 2009), *Digital Atlas of Economic Plants in Archaeology* (Neef *et al.* 2012), *Maris Nostri Novus Atlas Project: Seeds and Fruits of the Mediterranean Basin* (Sabato and Peña-Chocarro 2021), and *The Identification Guide for Near Eastern Grass Seeds* (Nesbitt 2009) were the main flora and archaeobotanical seed atlases consulted. Plant remains were identified and quantified using a stereo microscope. For reasons of epistemology, microscopic pictures of specimens of each taxon identified at the site are presented in the figures throughout this chapter. The archaeobotanical remains were photographed using an iDu Optics LabCam Pro by the author, and images were later edited using Adobe Photoshop by A.J. Bronkhorst.

Plant remains were quantified according to the protocol presented in Heinrich and Hansen 2020, which will be summarized here. Whole or complete specimens and the embryos of partial specimens of seeds and fruits were counted as "1." For rachis and culm (i.e., the quantifiable unit for cereal straw) each node and its internode were counted as "1." Though internodes are not formally quantified in standard archaeobotanical practice, these fragments, as a by-product of cereal processing, were also documented in the table to attest their presence within the sample(s). Broken specimens of seeds and endocarps, for which no quantifiable features were present (i.e., no whole specimens, apical, embryo, basal ends, or nodes) were quantified as "+" for 1–10 fragments; "++" for 11–50 fragments; "+++" for 51–100 fragments; "++++" for 101–150 fragments; and "+++++" for 151–200 fragments. While not as precise as absolute quantification in providing the minimum number of individuals (MNI) of a specific taxon, such data may still help to glean information about food consumption and the types of plants/plant parts potentially used for construction, fuel, and fodder on the site. The volume (in ml) of wood charcoal present within the samples was measured to assess the potential for a future study on the identification of wood taxa, which is beyond the scope of the current chapter.

Radiocarbon sampling

The objective for collecting samples for radiocarbon dating was to help improve the dating of the site's stratigraphy. Samples were therefore selected from within sealed contexts, such as plaster layers or beaten earth surfaces, ideally extracting single seeds such as grain kernels or other plant parts such as (olive) endocarps. Grain kernels and endocarps, prior to selection for radiocarbon dating, were identified to the lowest possible taxonomic level and photographed. These specimens have been taken up into the final quantification of other macrobotanical material from their context of origin. An overview of samples selected for radiocarbon dating may be found in Figure 1.

Single seeds were selected from surfaces to narrow the date range and avoid non-systematic offsets in age that can arise from sampling heartwood, branch wood, or mixed charcoals from archaeological contexts (Brock *et al.* 2010). A few wood charcoal samples were taken from specific contexts, such as the *narghila* from Trench C6–7, for an indication of the date range of charcoal used in a specific function. Radiocarbon dating was carried out at the University of Cologne Centre for Accelerator Mass Spectrometry. For a detailed discussion on the site chronology and both relative dating and absolute dating, see Chapters 4 and 5 for the Mamluk and Ottoman occupational phases respectively.

Season	Square	Locus	Basket	Material	Identification
KBM17	B4	L.124	B.1028	Seed/Fruit	*Hordeum vulgare* (hulled)
KBM17	B4	L.127	B.1037	Seed/Fruit	*Hordeum vulgare* (hulled)
KBM17	B4	L.127a	B.1038	Seed/Fruit	*Triticum turgidum* ssp. *durum*/*Triticum aestivum* ssp. *aestivum*
KBM17	B4	L.127b	B.1039	Seed/Fruit	*Hordeum vulgare* (hulled)
KBM17	B4	L.129	B.1043	Seed/Fruit	*Hordeum vulgare* (hulled)
KBM17	B4	L.130	B.1044	Seed/Fruit	*Hordeum vulgare* (hulled)
KBM17	B4	L.130	B.1048	Seed/Fruit	*Triticum turgidum* ssp. *durum*/*Triticum aestivum* ssp. *aestivum*
KBM13	B7	L.537 (East)	B.5240	Fruit	*Olea europaea* (endocarp)
KBM13	B7	L.537 (West)	B.5236	Fruit	*Olea europaea* (endocarp)
KBM17	C6–7	L.218	B.2020	Charcoal	Pressed ash / Charcoal from *narghila*
KBM17	C6	L.221	B.2024	Seed/Fruit	*Hordeum vulgare* (hulled)
KBM17	C6	L.222	B.2025	Fruit	*Olea europaea* (endocarp)
KBM17	C6	L.226	B.2031	Seed/Fruit	*Hordeum vulgare* (hulled)

Figure 1 Seed/Fruit Identification for Radiocarbon Dating.

Context descriptions and archaeobotanical evidence at Khirbet Beit Mazmīl

Trench B4—domestic space

Summary of the contexts

Trench B4 is a domestic space that was partly excavated during the 2015 field season. A total of 22 samples were taken from this trench during the 2017 and 2019 field seasons to better understand the use and chronology of the space as well as the hypothesized cooking activities taking place within it. A brief description of the sampled contexts can be found in the following section and a summary of the contexts sampled and sample metrics are presented in Figure 2.

A series of surfaces were uncovered in the northeast corner of the trench that were excavated and systematically sampled. Two ash lenses (L.130 B.1044 and L.130 B.1048) interrupted the flagstone floor (Locus 128). Locus 124 was a beaten earth surface that was excavated in B4 along the northern and eastern balks. It lay directly underneath the plastic covering from the exploratory field season in 2015 and corresponds to the previous, now outdated, Locus 114. The layers in the section were recorded in 2015, but these did not correspond to the surfaces that were uncovered in the 2017 season. These outdated Loci (113–117) were therefore not followed during the 2017 field season and new loci were designated.

Below the beaten earth surface (L.124) two plaster surfaces (L.127 and L.127a) were encountered. These surfaces were distinct, yet fairly inconsistent, making it difficult to determine where one surface ended and the other began. Both surfaces were sampled. Due to the presence of Ottoman *mangīr* coins in both L.127 and L.127a, a date of the 16th to early 17th century was suggested. These coins were minted in Cairo in the 1590s (B. Dolinka *personal communication*). *Mangīr* coinage in general was minted in great quantities and widely distributed until the second half of the 16th century (Pamuk 2000). The "multiple-plastered" surfaces as encountered in this context were common in domestic spaces at the site and imply a regular, perhaps seasonal (or more frequent) replastering of surfaces. Seasonal settlement patterns in this area have been attested ethnographically (B.J. Walker *personal communication*). The (re-)plastering of surfaces involves the use of cereal crop by-products as temper. Though this material in itself does not provide evidence for the mobility of inhabitants, it may reflect regular, seasonal building activities at the site. These likely occurred when the relevant resources were more readily available and could be allocated straightforwardly, such as immediately after the harvest and when threshing took place. Cereal chaff impressions in the plaster floors attest the use of this material as temper (Hansen *et al.* 2017). Sometimes it was not possible to excavate these plasters separately from each other as they were thin, friable layers, especially in places where they had already been partially damaged over time. These plaster surfaces were also tempered with cereal processing waste, mainly straw. They were also often separated by beaten earth surfaces. These beaten surfaces were rich in archaeobotanical remains and often contained archaeozoological remains as well. A rim of a cooking installation (L.118), perhaps a *ṭābūn*, was discovered at the end of the 2015 season, part of which was exposed on L.124. However, it was not removed and could not be recovered in the 2017 season. One *ṭābūn* fragment was also recovered in L.127a, the second plaster surface below L.124. This may suggest that after the *ṭābūn* fell out of use, it was then plastered over, albeit poorly, leaving part of it exposed in the beaten earth surface above the plaster levels.

A beaten earth surface (L127b B.1039), consisting of a reddish, *terra rossa* soil was discovered below plaster surfaces L.127 and L.127a, with a burnt concentration located in the southeast corner of the locus. This concentration was sampled and assigned a separate locus and basket number (L.129 B.1043). This surface was rich in archaeobotanical remains and contained many objects, including a cut bead, metal slag, glass, pottery, and a small, coin-like decorative metal disk that was attached to a piece of

Sample ID	Context	Relative Date	Volume Soil (L)	Volume Flot (ml)	Volume Botanical Remains (ml)	Sample Type
KBM17.B4.124.1028	Beaten earth surface	16th–17th c.	105.5	1000	8	Soil
KBM17.B4.124.1028	Beaten earth surface	16th–17th c.	0.25	N/A	N/A	Plaster
KBM17.B4.127.1037	Plaster surface	16th–17th c.	31	200	12	Soil
KBM17.B4.127.1037	Plaster surface	16th–17th c.	N/A	N/A	N/A	Plaster
KBM19.B4.127.1081	Plaster surface	16th–17th c.	0.25	N/A	N/A	Plaster
KBM19.B4.127.1082	Plaster surface	16th–17th c.	0.25	N/A	N/A	Plaster
KBM19.B4.127.1083	Plaster surface	16th–17th c.	0.50	N/A	N/A	Plaster
KBM17.B4.127a.1038	Plaster surface	16th–17th c.	87.5	250	11	Soil
KBM17.B4.127a.1038	Plaster surface	16th–17th c.	N/A	N/A	N/A	Plaster
KBM17.B4.127b.1039	Beaten earth surface	16th–17th c.	184	1300	20	Soil
KBM17.B4.127b.1039	Beaten earth surface	16th–17th c.	N/A	N/A	N/A	Plant-based fibers/ cloth fragment attached to coin object
KBM17.B4.127b.1039	Beaten earth surface	16th–17th c.	N/A	N/A	N/A	Plaster
KBM19.B4.128.1084	Final plaster floor	16th–17th c.	4	N/A	N/A	Plaster
KBM17.B4.129.1043	Burnt concentration in beaten earth surface L127b	16th–17th c.	6	30	4	Soil
KBM19.B4.129.1085	Soil below floor	16th–17th c.	4.5	30	10	Soil + Earthenware/cf. *Ṭābūn* fragment with straw/cereal chaff impressions
KBM17.B4.130.1044	Ash lens	16th–17th c.	4.5	40	2	Soil
KBM17.B4.130.1044	Ash lens	16th–17th c.	N/A	N/A	N/A	Earthenware/cf. *Ṭābūn* fragments with straw/ plant impressions
KBM17.B4.130.1048	Ash lens against W.122	16th–17th c.	2	25	3	Soil
KBM17.B4.155.1077	Fill in northwest sector of trench	ca. 16th c.	Dry sieved on site	No flot	1	Charcoal sample
KBM19.B4.167.1094	Ash concentration within plaster surface	15th–16th c.	8	10	7	Soil
KBM19.B4.176.2008	Hearth	15th–16th c.	9	30	12	Soil
KBM19.B4.183.2013	Plaster surface	15th–16th c.	3	15	2.5	Soil

Figure 2 Trench B4 archaeobotanical samples' context descriptions and sample metrics.

cloth. Because the disk did not feature any designs, etching, inscriptions, nor were any other features present, it was hypothesized to be a decorative object rather than a coin for exchange. The cloth, as it was made from seemingly plant-based fiber, may have originated from a piece of clothing (Figure 3). The thread that would have fastened the coin to the clothing was still intact and a few fibers were sub-sampled for further imaging and analysis. Fibers could be extracted while preserving the structure of the object, as only very few were needed for imaging. The fibers were imaged with a Scanning Electron Microscope (SEM) and analysed in the laboratory of Dr. Gregor Kirfel at the Institute for Cell Biology at the University of Bonn. The preliminary results are discussed below.

This surface (B4 L.127b B.1039) was located above a flagstone floor (L.128). Within this floor there were two noticeable breaks or shallow pits in which there was ashy material (L.130 B.1044 and L.130 B.1048). L.130 B.1044 was located within the center of the locus, while L.130 B.1048 was situated against W.122 in the northwest corner of the locus. The ashy material seems to be indicative of an ashy layer and therefore both areas, distinguished by different basket numbers, were sampled.

In the northwest extension of B4, the historical fill layers that were excavated likely relate to domestic use of the space in the different occupational phases of the site. No clear sealed contexts were identified in this section of the trench during the 2017 field season. This area was subsequently brought down to phase with the northeast extension of B4. A fill layer (L.126 B.1033) was removed above W.122 and sieved. This fill likely postdates the moment when the Early Ottoman structure went out of use and has therefore been relatively dated to the Middle Ottoman period (17th to 18th centuries). There was also a fill (L.134 B.1050) along W.122, stratigraphically underneath Locus 130 (the ashy lenses). This context has been relatively dated to between the late 16th and early 17th centuries, since from L.133 an early Ottoman pipe (*chibouk*) was recovered. A fill (L.135 B.1051) was removed in the western sector of B4 between W.118 and W.122. The western sector was located at a higher level than the eastern sector and consisted of much tumble (likely from W.122), suggesting a later use of the Early Ottoman domestic structure. The date

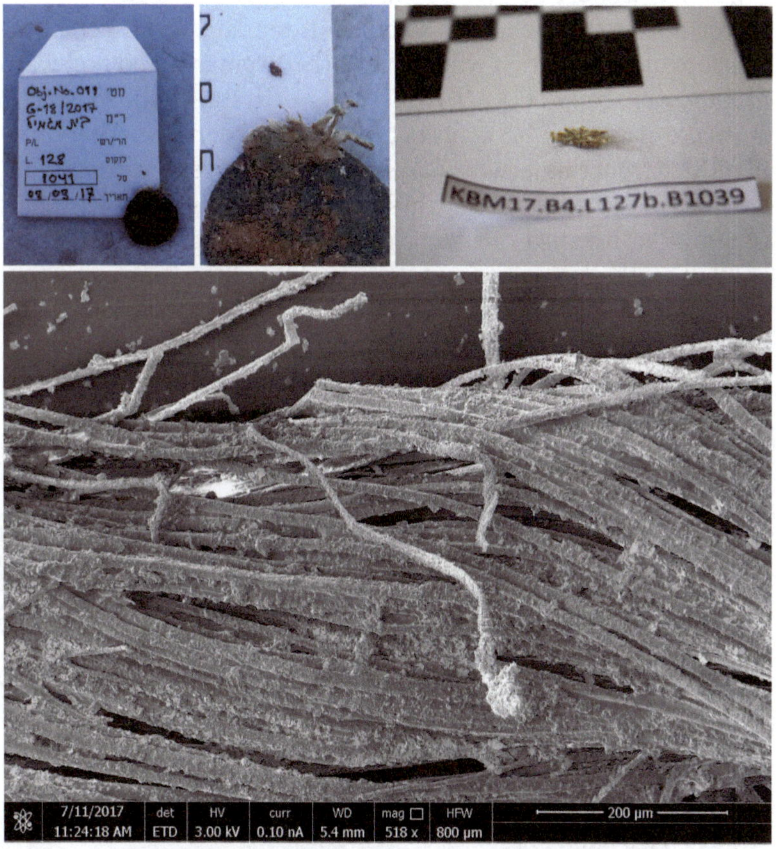

Figure 3 Top Left and Center: Disk with plant-based cloth photographed on site. Top Right: the object after cleaning (photo: A.M. Hansen 2017). Bottom: Scanning electron microscopy of plant-based fiber from object in Locus 127b, Basket 1039 (photo: G. Kirfel 2017).

suggested for this context is Middle Ottoman (17th to 18th centuries). A fill (L.141 B.1059) was removed from above W.122 and was also at a higher level than the eastern sector of B4. A Middle Ottoman date has also been suggested for this context. To further extend B4 north and connect it to B5, the topsoil of the balk was removed (L.143 B.1062). This material likely dates to the Mandate period of the 20th century. To better articulate W.122, the fill above it was removed (L.150 B.1069). This fill is associated with the later phases of use for the Early Ottoman domestic structure. Therefore, a date of late 16th to early 17th century is suggested.

A fill (L.155 B.1077) was excavated down to the lower flagstone floor (L.158) in the northwest sector of trench B4 and was sampled. It predates the cooking installation where the Early Ottoman *mangīrs* were found (L.127). A sherd of North Italian Ware of an early 16th century type has also been identified from this locus and therefore a tentative date of the 16th century is suggested for this locus (B. Dolinka *personal communication*). According to Walker, the loci until L.161, though difficult to date relatively, seem to represent a transitionary phase in the trench: this hypothesis is reinforced by the coarse wares that suggest a date between the 15th and 16th centuries, corresponding to the Late Mamluk period (B.J. Walker *personal communication*). Three contexts of interest were excavated below the flagstone floor (L.162), all of which were relatively dated to the 15th and 16th centuries (B.J. Walker *personal communication*). These consist of an ash concentration within a plaster surface (L.167 B.1094), a hearth (L.176 B.2008), and a plaster surface (L.183 B.2013).

Archaeobotanical results and observations

The summary of the identification and quantification of the archaeobotanical remains from contexts sampled in Trench B4 are presented in Figure 7. What is immediately clear from these results is that the surfaces, both beaten earth and plasters, are rich in archaeobotanical material indicative of both food and fuel remains. On the one hand, this shows a diverse range of cereals, pulses and fruits including free-threshing wheat, barley, lentil, pea, carob, white lupine, fig (*Ficus carica*, Figure 4), pomegranate, plum (*Prunus domestica*, Figure 4), bitter vetch, grape, olive and Christ's thorn, that were prepared and consumed within the space, likely over a span of multiple activities where the food waste was discarded or swept across the surface of the used space.

On the other hand, the presence of food waste and fuel waste within the matrices of the surfaces shows that these remains were likely used for another interesting economic activity: reuse as construction material, alongside cereal processing waste such as straw, in the formation of the surface itself. The omnipresence of diverse plant inclusions and impressions throughout the matrices of all the plaster surfaces are complementary to this hypothesis. The most common plant inclusions in this material are those of wood charcoal fragments, while the most common plant impressions are those of cereal straw (Figure 5).

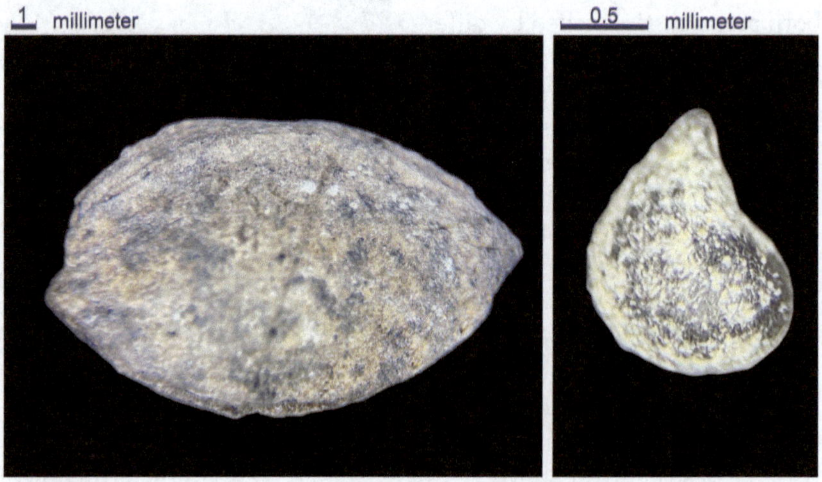

Figure 4 Left: *Prunus domestica* endocarp. Right: *Ficus carica* seed.

Figure 5 Top: Plaster surface Locus 127a, Basket 1038 with a free-threshing wheat seed, wood charcoal inclusions, and cereal straw impressions. Bottom: Plaster surfaces with inclusions of wood charcoal (left) and impressions of cereal straw (right).

In one plaster fragment, a unique, complete impression of a lentil with its hilum still visible was found in the internal matrix of L.127a B.1038 (Figure 6). Its position within the plaster matrix shows that it was a part of an intentionally added binding material when creating the floor plaster, and it is not a superficial plant remain that was deposited and secured after the plaster had been laid in place.

Within the same plaster floor, a well-preserved, charred free-threshing wheat seed was also present, alongside wood charcoal inclusions, and cereal straw impressions (Figure 5). These are telling examples of how plant remains, even after their primary or even secondary functions—for instance as a food and subsequently fuel source—have been carried out, can still not necessarily be considered a "final waste product." Moreover, they can also have a tertiary function.

Cereal processing waste, such as straw and chaff, or food and fuel waste may be recycled or "upcycled" to be used productively in other processes, in this case as temper. This is also attested in the case of the earthen ware ceramics or potential *ṭābūn*

Figure 6 Top Left and Right: modern lentils, showing the profile of the cotyledon and the lateral view of the embryo and hilum. Bottom: Impression of a lentil found within the plaster surface of L.127a, B.1038.

fragments uncovered in L.129 B.1085 and L.130 B.1044. The straw impressions are very clear in the *ṭābūn* rim fragment from L.130 B.1044 (Figure 8).

While historical sources do not describe such activities in detail, the early 20th century German ethnographer Gustaf Dalman in his seminal work *Arbeit und Sitte in Palästina* was very specific about how different fractions of cereal processing waste were collected, processed, and used as different products. Such uses included that of binders in clay for building *ṭābūn* ovens, and different types of dung cakes used for fuel resources, where they were a preferred material (Dalman 1935; Hansen *et al.* 2017). For plaster surfaces and floors, the principle is similar, but in these cases, as with mudbricks, less preferred sources of temper, such as

			Sample ID	KBM17. B4.124. 1028	KBM17. B4.127. 1037	KBM19. B4.127. 1081	KBM19. B4.127. 1082	KBM19. B4.127. 1083
Scientific Name	**Common Name**	**Plant Part**	**Preservation**					
cf. Agrimonia eupatoria	Common Agromony	Seed	Charred					
Ajuga cf. iva/ orientalis	Southern/ Eastern Bugle	Fruit (nutlet)	Charred					
Apiaceae (cf. Cuminum cyminum)	Apiaceae family (cf. Cumin)	Fruit	Charred					
Boraginaceae	Borage	Fruit	Mineralized					
Calendula sp.	Marigold	Seed	Charred					
Ceratonia siliqua	Carob tree	Seed	Charred		1			
Cerealia	Cereal Indeterminate	Grain kernal (quantifiable)	Charred	3				
Cerealia	Cereal Indeterminate	Grain kernal (fragments)	Charred	++				
Cerealia	Cereal Indeterminate	Culm node	Charred	4	1+			
Cerealia	Cereal Indeterminate	Culm base	Charred	1				
Fabaceae Indeterminate	Pulse Indeterminate	Seed	Charred	+	+			
Ficus carica	Fig	Seed	Charred					
Galium	Bedstraw*	Fruit	Charred	5				
Hulled Hordeum vulgare	Hulled Barley	Floret	Charred	1	3			
Hulled Hordeum vulgare	Hulled Barley	Floret (fragments)	Charred					
cf. Hulled Hordeum vulgare	Hulled Barley	Floret	Charred					
Hordeum vulgare	Barley	Rachis (node + internode)	Charred	1				
Hordeum vulgare	Barley	Rachis internode	Charred		1			
Lens culinaris	Lentil	Seed	Charred	10				
cf. Lens culinars	Lentil	Seed	Charred	1				
cf. Lupinus albus	White Lupine	Seed	Charred		1			

Figure 7 Trench B4 archaeobotanical assemblage, identifications, and quantifications.

KBM17. B4.127a. 1038	KBM17. B4.127b. 1039	KBM19. B4.128. 1084	KBM17. B4.129. 1043	KBM19. B4.129. 1085	KBM17. B4.130. 1044	KBM17. B4.130. 1048	KBM17. B4.155. 1077	KBM19. B4.167. 1094	KBM19. B4.176. 2008	KBM19. B4.183. 2013
1										
						1				
									1	
									1	
	1									
2										
6	4		3		1			3	2	
++	++			+		+		+		
27	3		1		1					
6						1				
+	9					2				
						1				
2										
	18		2	8	4	2		9	2	1
12								+	+	
					1					
1								1		
	1		1							
7	7		7			2				
									1	

(continued)

			Sample ID	KBM17. B4.124. 1028	KBM17. B4.127. 1037	KBM19. B4.127. 1081	KBM19. B4.127. 1082	KBM19. B4.127. 1083
Scientific Name	**Common Name**	**Plant Part**	**Preservation**					
Olea europaea	Olive	Endocarp (quantifiable)	Charred					
Olea europaea	Olive	Endocarp (fragments)	Charred	+				
Pisum sativum	Common Pea	Seed	Charred					
Poaceae	Grass family	Fruit	Charred	1				
Poaceae (cf. Bromus spp.)	Grass family cf. Brome genus	Fruit	Charred					
Prunus dulcis	Almond	Endocarp	Modern					
Prunus domestica	Plum	Endocarp	Charred					
cf. Punica granatum	Pomegranate	Seed	Charred		1			
T. a. aestivum	Bread Wheat	Rachis (node + internode)	Charred					
T. d. durum	Hard Wheat	Rachis (internode)	Charred					
T. d. durum	Hard Wheat	Rachis (node + internode)	Charred					
T. t. durum / T. a. aestivum	Hard / Bread Wheat	Grain kernal	Charred					
cf. T.t. durum / T. a. aestivum	Hard / Bread Wheat	Grain kernal	Charred					
Vicia ervilia	Bitter vetch	Seed	Charred					
cf. Vicia ervilia	Bitter vetch	Seed	Charred	1	1			
Vitis vinifera	Grape	Seed	Charred					
Wood Charcoal	N/A	Wood	Charred	3ml	15ml			
Ziziphus spina-christ	Christ's thorn	Endocarp	Charred					
Additional finds + Impressions								
Amorphous lump(s)	N/A	N/A	Charred		3			
Bead	N/A	N/A	N/A	1				
Cerealia	Cereal Indeterminate	Culm node	Modern					

Figure 7 Trench B4 archaeobotanical assemblage, identifications, and quantifications. (*continued*)

KBM17. B4.127a. 1038	KBM17. B4.127b. 1039	KBM19. B4.128. 1084	KBM17. B4.129. 1043	KBM19. B4.129. 1085	KBM17. B4.130. 1044	KBM17. B4.130. 1048	KBM17. B4.155. 1077	KBM19. B4.167. 1094	KBM19. B4.176. 2008	KBM19. B4.183. 2013
	1		1					1	2	
+	++					+			1.5ml	
					1					
2	1			8				7		
3						3				
			1							
1										
1										
1										
2	1									
14	21		6	4	2	2				1
					1	1		2		
9	6		8							
									1	
1										
4ml	8ml		0.75ml	× < 0.25ml		2ml	1/1ml	4.5ml	2.5ml	1ml
2										
						2		3	2	1
			1					1		
1										

(continued)

			Sample ID	KBM17. B4.124. 1028	KBM17. B4.127. 1037	KBM19. B4.127. 1081	KBM19. B4.127. 1082	KBM19. B4.127. 1083
Scientific Name	**Common Name**	**Plant Part**	**Preservation**					
Dung (sheep/goat)	N/A	N/A	Modern					
Earthenware fragment(s)	N/A	N/A	N/A					
Fruit endocarp	N/A	Endocarp	Charred	++	+			
Indeterminate	N/A	Seed/Fruit	Charred	15	7			
Indeterminate	N/A	Seed/Fruit	Mineralized		1			
Insect	N/A	N/A	Modern	+				
Sub-recent seeds/fruits	N/A	Seed/Fruit	Sub-recent		17			
Modern seeds/ fruits	N/A	Seed/fruit	Modern	+++	2			
Gossypium herbaceum/ arboreum	Levant Cotton/Tree Cotton	Fibers	Desiccated					
Plaster fragments w/chaff impressions	N/A	N/A	N/A			8	3	10
Plaster fragments w/hulled Hordeum vulgare charred seed	N/A	N/A	N/A				1	
Plaster fragment(s) w/wood charcoal	N/A	N/A	N/A					
Plaster fragment w/insect	N/A	N/A	N/A					
Slag	N/A	Slag	N/A	++	+			
Ṭābūn fragment(s)	N/A	N/A	N/A					

+ = present but not quantifiable 1–10 fragments

++ = 11–50

+++ = 51–100

KBM17. B4.127a. 1038	KBM17. B4.127b. 1039	KBM19. B4.128. 1084	KBM17. B4.129. 1043	KBM19. B4.129. 1085	KBM17. B4.130. 1044	KBM17. B4.130. 1048	KBM17. B4.155. 1077	KBM19. B4.167. 1094	KBM19. B4.176. 2008	KBM19. B4.183. 2013
1										
				++	350ml					
+	+							2		
14	8		6	28		5		30	10	
	1									
+	+									
									2	
	+									
	++	15								
++	++								4	
									1	

Figure 8 *Ṭābūn* rim fragment with cereal straw impressions from Trench B4, Locus 130, Basket 1044.

food and fuel waste, may often be added (Heinrich and Hansen 2020). The relative cost (i.e., the abundance and availability of high-quality temper in relation to needs for other, mutually exclusive, uses such as fuel or fodder) of the temper, the amounts required for the construction of an item/surface, and its importance/function are factors that codetermine decision making in temper selection (Heinrich and Hansen 2020). As the archaeological evidence at the site suggests that the many large surfaces present were regularly replastered (some surfaces show evidence of seven different sub-layers, see section Trench C6), the choice for waste/rest material as a low-quality alternative or additional temper seems likely.

An interesting find among the food and fuel waste remains within beaten earth surface L.127b B.1039 was a disk or coin-type object with fibers attached that were hypothesized to be from cloth, as described above. This object was likely discarded along with the other waste found within the context. The sub-sampled fibers were imaged using a Scanning Electron Microscope (SEM) in the laboratory of Dr. Gregor Kirfel at the University of Bonn Institute for Cell Biology. Based on the preliminary results, in comparison to reference materials, the fibers were determined to be plant-based and most closely resemble cotton (Figure 3; see Rast-Eicher 2016, 73–79). Further study with comparative material will help confirm this identification.

Some of the food remains found within Trench B4 are worth mentioning in more detail, particularly a fragmented white lupine seed (Figure 9). This represents only the second find of white lupine in the southern Levant. The first, which is a complete specimen, is from a 7th century context in Nessana, in the Negev Highlands; this taxon likely entered the region during the Roman to Early Byzantine period (Fuks *et al.* 2022). Pulses are not found frequently in the archaeobotanical record as they typically undergo various processing steps, most of which do not include exposure to fire—reducing the chances of accidental carbonization—before they are consumed. Boiling, roasting, grinding, if used in flour or in a porridge, are common processing steps for pulses. As lupine is also soaked for an extended period to remove certain antinutrients and is afterwards often pickled in brine, its chances for preservation are even lower, making this a rare and special find (Heinrich and Hansen 2019). For the Early Islamic period, the crop is attested in al-Muqaddasī's *Aḥsan al-Taqāsīm fī Ma'rifat al-Aqālīm* (10th century), who mentions that in the "province of Palestine" lupines are among the "thirty-six products not to be found together in any other land" (al-Muqaddasī, translated by Collins 2001,

151). Lupine can be very high in protein and is a versatile crop, as in addition to food it can also be used for green manuring. Their anti-nutritional effects can be easily counteracted through soaking and they can be pickled or roasted as a snack, be boiled and incorporated into dishes, or be dried and ground into flour to mix with cereal flours to make bread; lupines even have benign health effects and can be used as a medicinal plant and in traditional veterinary practices in ruminants (Fechner *et al.* 2011; Heinrich and Hansen 2019; Huyghe 1997; Ishaq *et al.* 2022; Islam and Ma 2016; Knecht *et al.* 2020; Landau *et al.* 2014). Studies on sustainable crops and agriculture moreover suggest that white lupine, beyond its nitrogen-fixing capabilities as a pulse, has the benefit of a unique cluster root structure only present in a small number of plant taxa that make them well adapted for nutrient acquisition even in nutrient-poor soils and they have even been shown to transfer these positive effects in cereal-pulse intercropping (Boudsocq *et al.* 2022; Shane and Lambers 2005). This trait may have attributed to the crop's success in the southern Levant and elsewhere in the past.

Other important pulse crops that were present in the assemblage were the common pea (*Pisum sativum*) and carob (*Ceratonia siliqua*) (Figure 9). Although the common pea is a founder crop, which came to the southern Levant during the Neolithic alongside many cereals and pulses, it can be a challenging taxon to identify after it has been carbonized. This specimen fractured along its long and narrow hilum (characteristic of the common pea) and cotyledon, emphasizing these features without breaking. Though it is a later find for the taxon, since it is rare to find in assemblages for the reasons described above, it is important to document. The carob was likely introduced to the region sometime during the Hellenistic to Roman period and it became an important food (Zohary *et al.* 2012). The fruit of the carob was primarily used or consumed, while the seed was often discarded. This is likely why we find the charred

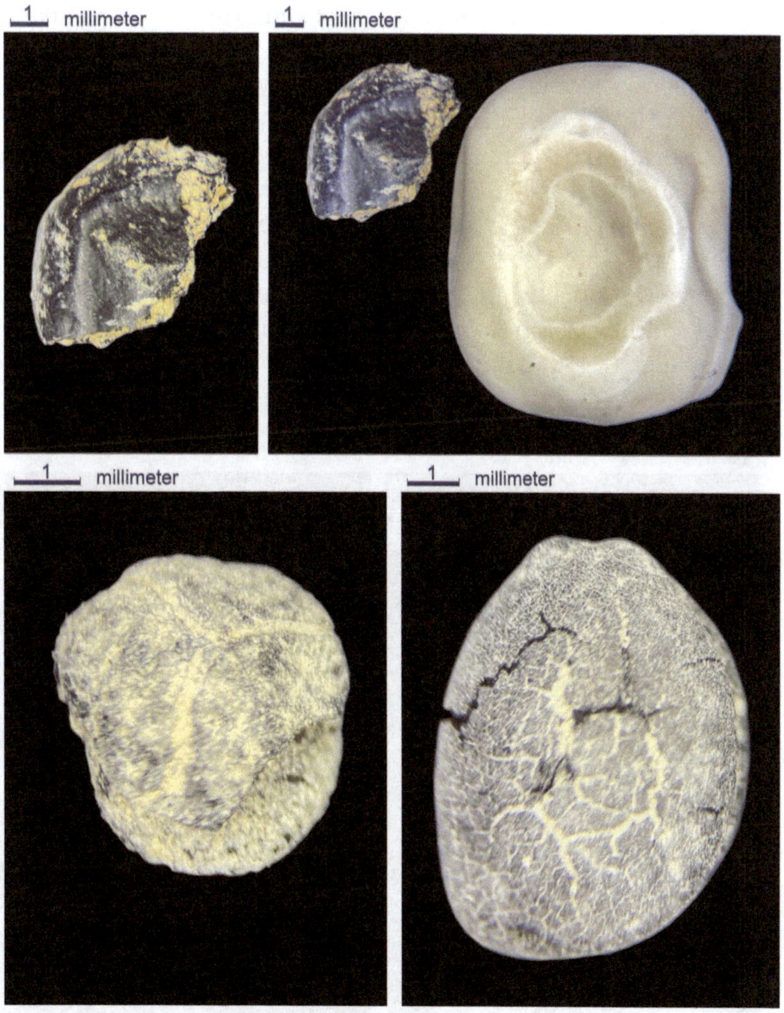

Figure 9 Top Left and Top Right: cf. *Lupinus albus*. Charred seed fragment (Top Left) and charred seed fragment next to modern *Lupinus albus* seed (Top Right). Bottom Left: *Pisum sativum* seed. Bottom Right: *Ceratonia siliqua* seed.

seeds within the cooking contexts. Carob is likewise mentioned in al-Muqaddasī's list of crops for the region in the 10th century (al-Muqaddasī, translated by Collins 2001, 152). This crop also has many medical applications and has been documented ethnographically in the southern Levant as used to treat gastrointestinal problems and diarrhoea, diabetes, as well as an ingredient in many veterinary treatments (Ali-Shtayeh *et al.* 2016; Dakia 2011; Jaradat *et al.* 2016; Landau *et al.* 2014). Carob seeds were found in cooking-related contexts in Trenches B4, B7, and C6, in both Mamluk and Ottoman period occupational layers.

Specimens of several taxa of common field weeds were found in the assemblage of B4, including marigold (*Calendula* sp., Figure 11), borage (*Boraginaceae*, Figure 10), bedstraw (*Galium* sp.), wild grasses (Poaceae), and brome grasses (*Bromus* sp.). Several of these taxa were also found in loci in Trench C6 (see below). A unique and interesting carbonized fruit of southern/eastern bugle was also found in an ashy lens (L.130) below a flagstone pavement (Figure 10). It was not possible to determine this specimen down to the species level; both *Ajuga iva* and *A. orientalis* are possibilities from among the local flora. Bugles are perennial plants and these particular taxa have been documented ethnographically in folk medicine in order to treat several ailments from allergies and metabolic issues to cardiovascular disorders and cancer (Bouyahya *et al.* 2020; Israili and Lyoussi 2009; Luan *et al.* 2019). This find is likely out of its original context, and it is not clear if it came from a cooking context.

Some taxa that were present in the assemblage may be indicative of the use of herbs or food additives or may represent medical pharmaceuticals. The best examples of these are common agrimony (*Agrimonia eupatoria*; B4 L.127a B.1038) and the spice cumin (*Cuminum cyminum*; B4 L.176 B.2008). Cumin was likely introduced during the Iron Age (Figure 11), though its seeds/fruits, likely due to food processing and the difficulty of their recovery, are rarely encountered in the archaeobotanical record (Frumin *et al.* 2015; Neef 1990).

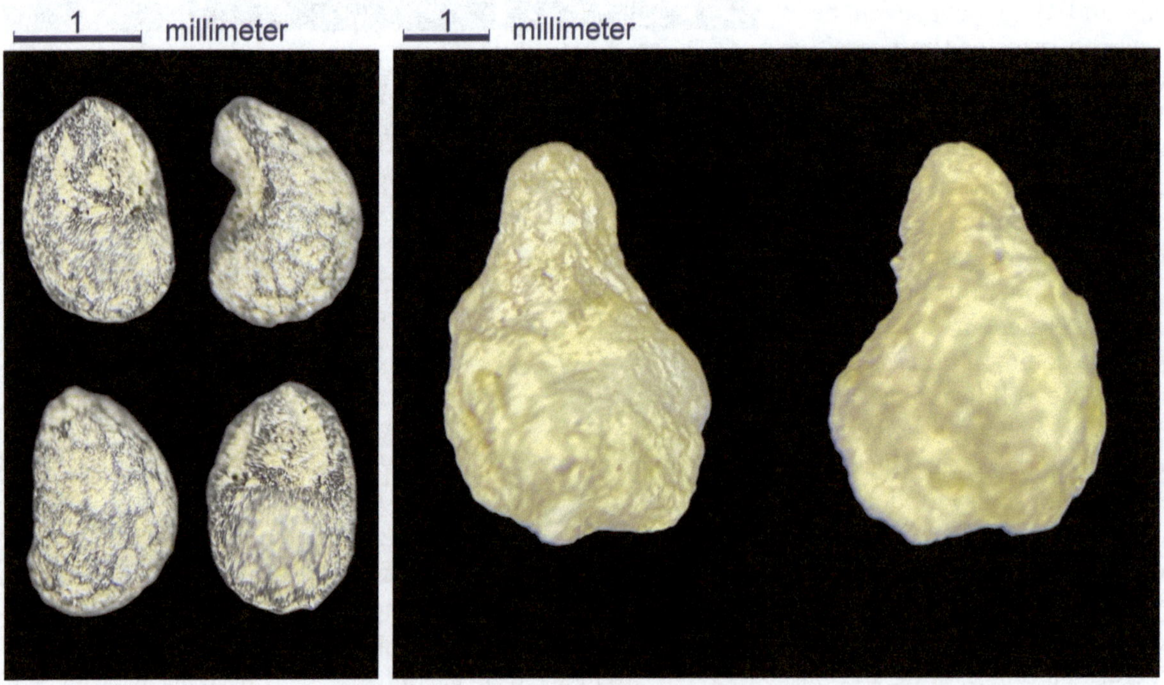

Figure 10 Left: *Ajuga* cf. *iva/orientalis* fruit. Center and Right: *Boraginaceae* fruit.

Figure 11 Top: *Calendula* sp. seed. Middle and Bottom: Carbonized Apiaceae cf. *Cuminum cyminum* fruit dorsal (middle) and ventral (bottom) sides flanked by modern fruits showing both dorsal (left) and ventral (right) sides.

Common agrimony (*Agrimonia eupatoria*, Figure 12) is a plant native to the region and comes from the Rosaceae family. It has a plump seed that is shaped like a semi-rectangular prism, the soft, rounded edges of which are lined with subtle ridges that begin at the basal end of the seed and draw towards the apical end of the seed into a slight, stubby point. The base of the seed is rounded. These remains were identified through comparison with the specimens depicted in Neef *et al.* (2012, 563) and Sabato and Peña-Chocarro (2021, 371). Danin and Zohary, in their respective *Flora Palaestina*, mention that it grows on damp sites. Zohary adds that in the past it was used as a "folk remedy," mainly as an astringent and vermifuge, among other medical applications (Danin 2004, 122; Zohary 1972, 13–14). It has also been documented in the sources of Kohen al-ʻAṭṭar (1940) and Ibn al-Bayṭar (ed. Bulaq 1874; ed. Ibn Murad 1989) from the 13th century that form part of the *materia medica* of medieval al-Shām, specifically to treat internal diseases, stomach and intestinal problems, animal bites, stings and poisons, and eye diseases (Lev 2002). It is also mentioned in the 10th century Cairo Genizah where it was used in all manners listed above, but also for liver ailments (Lev and Amar 2008). Modern ethnographic studies of traditional folk medicine also seem to confirm these applications. In traditional Arab medicine, it can be used to treat sore throat and diarrhea (Oumeish 1999). This is also documented in traditional applications of the plant in the Iberian Peninsula (Garcia-Oliveira *et al.* 2020), and a modern case study in Morocco that documents the use of common agrimony leaves to treat gastrointestinal problems (Teixidor-Toneu *et al.* 2016). Common agrimony also contains the chemical substance *agrimophol*, which can be used as an anthelmintic or antiparasitic medicine (Fabricant and Farnsworth 2001). It is additionally documented in folk medicine for its anti-inflammatory, analgesic, and hypotensive properties (Malheiros *et al.* 2022). From the evidence in the written and ethnographic sources, this seed is a rather curious find

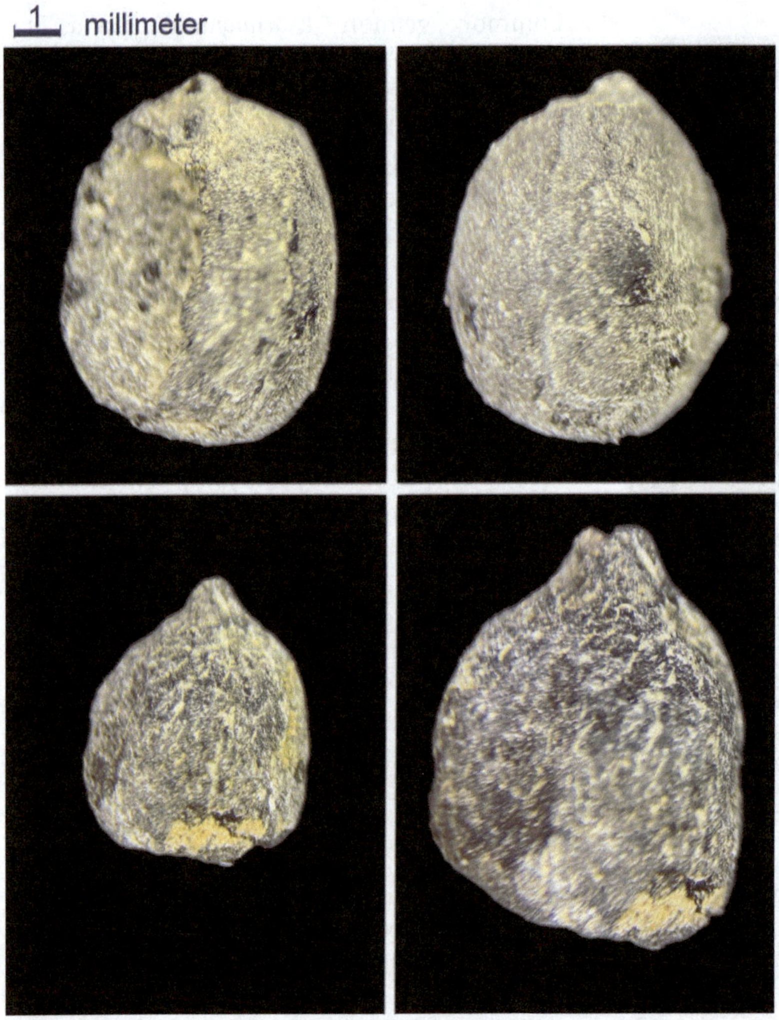

Figure 12 cf. *Agrimonia eupatoria* seeds.

within a domestic or cooking-related context; yet another seed of this taxon was found in a cooking installation in Trench C6 (L.226 B.2031). It seems likely that this plant was collected and its leaves, and potentially its fruit, were used in food preparation, as it is incorporated into an assemblage rich in cereals, pulses, and fruits. The seeds could have been discarded into the hearth or cooking installation as part of cooking/preparation waste. Placing this find into context with the other archaeobotanical finds is important as it shows that the distinction between medicinal plants and food plants, especially in the past, is far from straightforward. Often, plants with a medicinal function were also consumed as foodstuffs, herbs, or spices, while similarly medical preparations also included the addition of common foods. Although the taxon is native to the southern Levant, these are the first archaeobotanical remains documented for the region.

Between the occupational phases of Trench B4, from the Late Mamluk/Early Ottoman (Loci 155–183; 15th to 16th century) to Middle Ottoman periods (16th to 17th century), there are no significant differences in plant use and consumption within the space. This shows a degree of continuity in plant use throughout these periods at the site. Trench B4 gives a strong impression of local access to a diverse palette of plant foodstuffs, as well as insight into the flavourings that were incorporated into daily meals. The material from the trench furthermore illustrates the circular use of plant resources from farm to fork to fuel to building material. This is a trend that can be observed in some of the other trenches as well (for example, Section Trench C6).

Trench B5—transitional space between Trenches B4 and C7

Summary of the contexts

Bones were collected through dry sieving from transitional squares B5 and C6–7, where critical features of architecture were revealed: the topsoil in B5 that connected the square between B4 and C7 (L.138 B.1055) and a potential midden in the western sector of B5 (L.142 B.1061), situated to the west of W.126. The collection of bones from these squares was conducted to establish a more holistic view of wild and domesticated animal taxa in the later occupational phases of the site that were either exploited or simply present as a part of the site's natural fauna. For the complete discussion of the results of the zooarchaeological material and their implications for human-animal relationships and the local environment at Khirbet Beit Mazmīl, see Chapter 7.3 by Corbino. There were no sealed contexts within this trench, and therefore, no archaeobotanical samples were taken from this trench.

Trench B7—extramural structure

Summary of the contexts

Two soil samples were taken during the 2013 salvage excavation of the southern part of the site. This part of the site has now been built over by the modern community center. The samples originated from an extramural structure that has been tentatively identified as a stable due to the presence of metal objects associated with equines (B. Dolinka *personal communication*). The two soil samples came from the ashy fill layer immediately below a plaster floor (L.507). This locus was sampled in two areas labelled "E" for the eastern section (L.537 B.5240) and "W" for the western section (L.537 B.5236) (B. Dolinka *personal communication*). Architectural and ceramic evidence relatively dates this context to the Mamluk II period, or 15th century. In addition, a sample for radiocarbon dating was taken to provide an absolute date. The results of this analysis are discussed in detail by Walker and Pini in Chapter 4. The samples were subjected to floatation and the macrobotanical results from this context have shed light on fuel use in the southern part of the site. A brief description of the sampled contexts is provided in the following section. A summary data of the contexts sampled as well as a summary of the metrics of the processed and sieved samples from Trench B7 is presented in Figure 13.

Archaeobotanical results and observations

The data regarding the identification and quantification of the archaeobotanical remains from contexts sampled in Trench B7 are presented in Figure 13. Although the two samples taken were relatively small, they were rich in material and indicative of frequent food preparation, cooking activities, and fuel use, attesting to a wide range of taxa and plant parts. The crop remains include free-threshing wheat—both tetraploid (likely hard wheat) and hexaploid (likely bread wheat), with tetraploid being dominant—hulled barley, lentil (*Lens culinaris*, Figure 14), olive, pea, foxtail millet, bitter vetch (*Vicia ervilia*, Figure 14), common vetch (*Vicia sativa*, Figure 14), carob, and grape. While bitter vetch and common vetch are often considered low status foods or animal fodder, they were commonly consumed in the past. They are also drought-resistant crops and have a high protein content. Use as human food should therefore not be disregarded when encountered in cooking contexts in the archaeological record (Heinrich and Hansen 2019; Nguyen *et al.* 2020).

The presence of cereal straw (culm nodes and bases), barley, free-threshing tetraploid and hexaploid rachis (nodes and internodes, Figure 15), alongside several wild taxa including wild grasses, bedstraw, mallow, medick/alfalfa, and other small, wild legumes indicative of field weeds, strongly suggests the use of cereal crop processing waste as a fuel. Likewise, wood charcoal and a sheep/goat dung pellet were present within the context, indicating their use as fuel, likely during cooking activities within the space.

A high concentration of crushed olive endocarps (199 fragments making up a volume of 6 millimeters) was found in the sample from the western part of the ashy layer (L.537 [West] B.5236, Figure 15). This high concentration could indicate that olive pressings (also known as olive pomace or *jift*) were also being used as a fuel source (Neef 1990, fig. 5–6). Ethnographic examples of *jift* and sheep/goat dung used as fuel in *ṭābūn* ovens in Jordan were recorded in the village Suleikhat by Neef (1990). It appears that the practice of mixing multiple fuel sources, including sheep/goat dung and crop processing waste from cereal and olive,

		Sample ID	
		Context	
		Relative Date	
		Volume Soil (L)	
		Volume Flot (ml)	
Scientific Name	**Common Name**	**Plant Part**	**Preservation**
Ceratonia siliqua	Carob	Seed	Charred
Cerealia	Cereal Indeterminate	Grain kernal (quantifiable)	Charred
Cerealia	Cereal Indeterminate	Grain kernal (fragments)	Charred
Cerealia	Cereal Indeterminate	Culm node	Charred
Cerealia	Cereal Indeterminate	Culm base	Charred
Fabaceae indeterminate	Pulse Indeterminate	Seed	Charred
Galium	Bedstraw*	Fruit	Charred
Hulled Hordeum vulgare	Hulled Barley	Floret	Charred
Hordeum vulgare	Barley	Rachis (node + internode)	Charred
Lens culinaris	Lentil	Seed	Charred
Malva spp.	Mallow	Seed	Charred
Medicago spp.	Medick	Seed	Charred
Olea europaea	Olive	Endocarp (quantifiable)	Charred
Olea europaea	Olive	Endocarp (fragments)	Charred
Poaceae	Grass family	Fruit	Charred
Poaceae cf. Bromus spp.	Grass cf. Bromus genus	Fruit	Charred
Setaria italica	Foxtail millet	Fruit	Charred
Triticum aestivum ssp. aestivum	Bread wheat	Rachis (node + internode)	Charred
Triticum turgidum ssp. durum	Hard wheat	Rachis (node + internode)	Charred
T. t. durum / T. a. aestivum	Hard / Bread Wheat	Grain kernel	Charred
Vicia ervilia	Bitter vetch	Seed	Charred
Vicia sativa	Common vetch	Seed	Charred
Vitis vinifera	Grape	Seed	Charred
Wood Charcoal	N/A	Wood	Charred
Additional finds + Impressions			
Chrystalline lump	N/A	N/A	N/A
Dung (Sheep/Goat) w/culm, chaff inclusions	N/A	N/A	Charred
Fruit endocarp	N/A	Endocarp	Charred
Indeterminate	N/A	Seed/Fruit	Charred
Indeterminate	N/A	Seed/Fruit	Mineralized

+ = present but not quantifiable 1–10 fragments

++ = 11–50

+++ = 51–100

++++ = 101–150

+++++ = 151–200

Figure 13 Trench B7 archaeobotanical assemblage, context descriptions, identifications, and quantifications.

KBM13.B7.537 (West).5236	KBM13.B7.537 (East).5240
Ashy fill layer below plaster floor (Locus 507)	Ashy fill layer below plaster floor (Locus 507)
Mamluk II/15th c.	Mamluk II/15th c.
6.5	**5**
60	**30**
	3
4	1
+	+
2	18
1	6
1	1+
	3
2	1
2	2
	1
1	
	2
17	6
+++++/6ml	+
1	2+
3	12
1	
	6
	56
3	2
1	
1	
1	
2ml	3ml
1	
1	
	+
15	20
1	

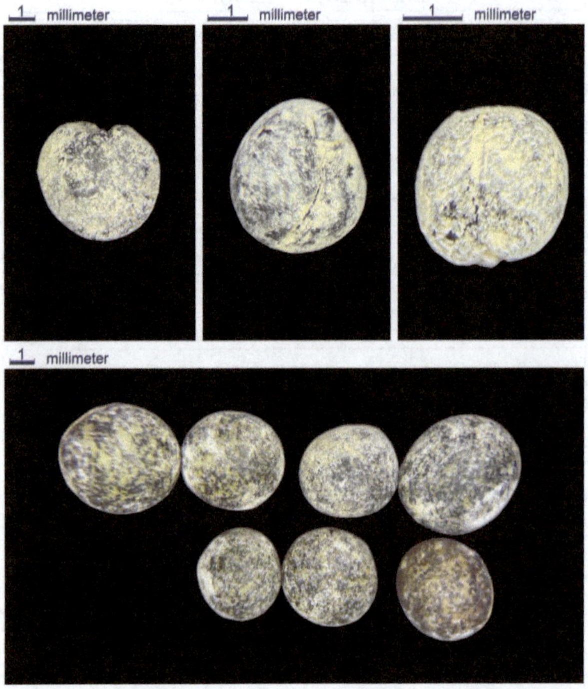

Figure 14 Top Left: *Lens culinaris* seed. Top Center: *Vicia ervilia* seed. Top Right: *Vicia sativa* seed. Bottom: Sub-recent/Modern *Lens culinaris* seeds from Trench E2, Locus 438, Basket 4058.

Figure 15 Top Left, Middle Left, and Middle Right: Free-threshing tetraploid wheat rachis. Top Right: Free-threshing hexaploid wheat rachis. Bottom Left, Middle, and Right: Fragmented olive endocarps originating from olive pomace or *jift* at Khirbet Beit Mazmīl.

was also practiced at Khirbet Beit Mazmīl. Crushed endocarps were found frequently in burning/fuel contexts throughout the site: they were present in 18 different loci in five of the seven trenches. The importance of *jift* as a fuel source at the site therefore cannot be overstated. As the cultivation and pressing of olives took place at or close by the olive groves, this confirms that olive cultivation for oil making must have been of importance to the economy of the village. As is apparent from ethnographic evidence (Dalman 1935, 201–255; Rowan 2019), olives are pressed in different phases, each with a subsequent increase in pressure to continue to press out the desired oily substance from the fruits. The first oil to be pressed is of the highest quality and much less pressure is applied than at the last stage of pressing. Therefore, when more pressure is applied, a lower quality of oil is extracted. The two final products from olive oil pressing are a liquid and a solid product. The liquid product is a black substance called *amurca*, which has various uses, especially as a pesticide for warding off insects on threshing floors, granaries, and other storage spaces for foodstuffs. The second product is a solid "cake" of compressed olive pomace, or *jift*. This material is composed of the fruit-flesh, (crushed) endocarps, and seeds remaining from this process. These cakes are used as fuel, and this is what the fragmented remains of olive in the archaeological context represent. Such cakes, or their remains, have been attested from at least the Bronze Age, and perhaps even the Chalcolithic, onwards at various sites across the Mediterranean and North Africa (Neef 1990). They are likewise attested in the southern Levant during the Bronze (Salavert 2008) and Iron Ages (Simchoni and Kislev 2006), and throughout the Roman (Kloner and Sagiv 2003), Late Antique (Fischer 1999), Early Islamic (Forste 2023), and Middle-Late Islamic periods (this chapter). As to the historical use of olive pomace, we know that in antiquity it was often used a fuel in (single-use) ceramic production at olive producing estates, in glass blowing, and it was also the preferred fuel source of bakers as it burns clean, leaves little ash, and does not have an unpleasant smell (Coubray *et al.* 2019; Rowan 2015; 2018; 2019). While the material has a high energetic value, it is not a good fire starter, which could explain its complementary use with cereal processing waste.

Today olive pomace is used as an industrial fuel and pressed into small pellets, as well as a biofuel (Ball 2022; Todorović 2021), while its potential role as a functional ingredient in bakery products such as bread, biscuits, and toast bread, as well as animal fodder is being explored (Di Nunzio *et al.* 2020; Difonzo *et al.* 2020; Sleim *et al.* 2020), although such uses would not apply to the past. The traditions as observed at Khirbet Beit Mazmīl are being reapplied, however, both due to the pressures of rising fuel costs and the movement towards sustainable energy use with lower carbon emissions, with olive pomace becoming visible again as an alternative fuel source to heat homes in Jordan, Syria, and Greece (Campbell and Horton 2021; al-Ketan 2012; Tawarah 2019; Vourdoubas 2016).

In the archaeobotanical literature, the smoothness or sharpness of the fractures of olive endocarps have been interpreted to correspond with either crushing before firing in the case of smoother edges, or following digenesis, excavation, and floatation in the case of sharper edges (Neef 1990; Rowan 2015). However, an experimental study of modern olive pomace and its broken endocarps and comparison with archaeological parallels found that these interpretations were far from clear cut. Fragments of endocarps that were charred after crushing could also have sharp edges, whereas the smoothing of the edges could also occur during post-depositional processes and post-excavation processing (Braadbaart *et al.* 2016). Although the authors conclude that consulting the form of the edges of charred archaeobotanical olive stones as a technique to recognize olive oil production is problematic and unreliable, the sheer volume of remains of olive stone fragments, especially from sealed contexts at archaeological sites, could still be considered as evidence for the activities of oil pressing and the crushing of the endocarps (Braadbaart *et al.* 2016; Coubray *et al.* 2019; Kaniewsky *et al.* 2012; Neef 1990). Similar to the archaeobotanical endocarp remains evaluated by Braadbaart (*et al.* 2016), the edges of the olive endocarps found within the ashy layer at Khirbet Beit Mazmīl were found to be both sharp and smooth and these features had no clear relationship to their historical post-depositional or post-excavation processing. Based on the sheer volume of remains within this context, however, there is indeed a high probability that these remains relate to olive oil processing at or near the site, and that these pressings were later used as a fuel in the excavated context.

It was also observed that the large concentration of olive pomace remains were on the west side of the trench, while the concentration of cereal rachis were on the east side of the trench. While the installations related to these ash concentrations were no longer intact, it is interesting to note these subtle differences between the two areas in the trench. Likewise, cereal culm (i.e., straw), was also found in noticeably higher concentrations in the east side of the trench than in the west side. These concentrations may be related to different cooking activities occurring within the trench, though without the installations or other related tools or ceramics present, it is not possible to explore this further.

Trench C6—domestic space

Summary of the contexts

Trench C6 revealed another domestic space with several plaster and beaten earth surfaces. A total of 12 samples were taken from this trench over the course of the 2017, 2018, and 2019 field seasons. The main objective in excavating this trench was to investigate potential cooking activities within the space. A brief description of the sampled contexts is provided in the following section. A summary of the contexts sampled and a summary of the metrics of the processed and sieved samples from Trench C6 is presented in Figure 16. Radiocarbon dates were provided by B.J. Walker (*personal communication*).

For a discussion on the chronology of the Mamluk phases of the site, please refer to Chapter 4.

L.220 B.2023 was an uneven, beaten earth surface where fragments of the flagstone were still visible. The dimensions of the locus were not clear, as it was partially merged with a plaster surface directly below it (L.221). Two areas of the locus were sampled: one area in the northeast corner of the locus, where there was a small concentration of charred material within the surface, and a small area in the southwest corner of the beaten earth surface, located above an ashy layer.

Located underneath the beaten earth surface was a relatively large plaster surface (C6 L.221 B.2024)

Sample ID	Context	Relative Date	Volume Soil (L)	Volume Flot (ml)	Volume Botanical Remains (ml)	Sample Type
KBM17.C6.220.2023	Beaten earth surface	ca. 16th c.	4	100	2.5	Soil
KBM17.C6.221.2024	Plaster surface	16th c.	187.5	1500	19	Soil
KBM17.C6.222.2025	Beaten earth surface with flagstone bits	ca. 16th c.	49	500	10	Soil
KBM17.C6.223.2026	Small charcoal concentration	ca. 16th c.	N/A	N/A	N/A	Charcoal sample
KBM17.C6.225.2030	In-situ jar and its soil fill	ca. 16th c.	9	5	1.5	Soil
KBM17.C6.226.2031	Cooking installation and its soil fill	16th c.	N/A	N/A	N/A	*Ṭābūn* fragments
KBM17.C6.226.2031	Cooking installation and its soil fill	16th c.	124.5	1400	49	Soil
KBM17.C6.226.2033	Cooking installation and its soil fill	16th c.	N/A	N/A	N/A	Charcoal sample
KBM18.C6.240.2050	Beaten earth surface	ca. 16th c.	46	100	10	Soil
KBM18.C6.240.2050	Beaten earth surface	ca. 16th c.	N/A	N/A	N/A	Plaster
KBM18.C6.241.2051	Plaster surface	15th c.	23	15	3	Soil
KBM18.C6.241.2051	Plaster surface	15th c.	N/A	N/A	N/A	Plaster
KBM18.C6.241.2051	Plaster surface	15th c.	N/A	N/A	N/A	Earthenware pottery
KBM18.C6.242.2052	Beaten earth surface	15th c.	32	10	3	Soil
KBM18.C6.242.2052	Beaten earth surface	15th c.	N/A	N/A	N/A	cf. *Ṭābūn* base fragment
KBM18.C6.242.2052	Beaten earth surface	15th c.	N/A	N/A	N/A	Plaster
KBM18.C6.243.2054	Plaster surface	15th c.	9	25	2	Soil
KBM18.C6.243.2054	Plaster surface	15th c.	N/A	N/A	N/A	Plaster
KBM18.C6.244.2055	Probe	14th/15th c.	18	40	10	Soil
KBM18.C6.244.2055	Probe	14th/15th c.	N/A	N/A	N/A	Plaster
KBM18.C6.244.2055	Probe	14th/15th c.	N/A	N/A	N/A	Earthenware pottery
KBM19.C6.259.2068	Fill of plastered Basin #2	13th/14th c. (1233–1391 CE)	10	55	5	Soil

Figure 16 Trench C6 archaeobotanical samples' context descriptions and sample metrics.

that was rich in archaeobotanical material. These plaster surfaces were smoothed against both stone walls to the north and south of the locus, demarcating the edges of this space. There were several plaster layers within this locus that had been laid on top of one another, totaling seven visible layers, though there could have been additional, more-subtle layers that were no longer distinguishable.

Underneath this multiple-plastered surface was a beaten earth surface that contained flagstone fragments, designated Locus 222 B.2025. A subsample, consisting of 25 percent of this surface, was taken, as it was a very large surface similar to the plaster surface above it and yielded many archaeobotanical remains.

While bringing the eastern section of C6 down to the level of L.222, several interesting installations were uncovered. An in situ jar filled with soil (likely a secondary fill) was discovered next to a potential cooking installation, where *ṭābūn* fragments were also present. Each of these sub-areas were designated separate locus numbers. The jar and the soil within and immediately outside of it, including ceramics and a cut glass fragment contained within the soil, were assigned locus number L.225 B.2030. The cooking installation and its earthenware fragments, potentially a *ṭābūn*, as well as the related fill—the latter being rich in archaeobotanical remains—were assigned locus number L.226 B.2031 (Figure 16). This fill represents the last use phase of the cooking installation (or potential *ṭābūn* oven). L.225 B.2031 was located directly above another flagstone floor (Locus 229), which could be traced along the full length of the trench and showed evidence of continuing underneath the southern balk. L.229 could also be traced all the way to the eastern edge of the trench, demarcated by Wall 125 and a thick plaster installation with several stone and ceramic inclusions that ran from Wall 125 to the vaulted arch/wall in the north part of the trench (Wall 104). This installation ran deeper than the flagstone floor. Within the eastern part of the trench there was an uneven plaster layer that was removed (Locus 233), lying directly above the flagstone floor (Locus 229) that had been removed to better articulate the floor, and continued underneath Wall 125. One sample was taken from the plaster, and it contained both metal slag and charred plant materials including one seed. There was a break in the flagstone floor directly in front of Wall 125. The flagstone floor (L.229) could also be traced all the way to the western edge of the trench following the wall (Wall 104) towards Square C5 and C4. It appears that the flagstone floor may have been a part of an earlier phase of domestic occupation that might have covered the entirety of the trench. Later, as indicated by demarcations in plaster surfaces above the flagstone floor that were smoothed flush against the wall, this large space appears to have been divided into smaller domestic spaces. The articulation of the demarcations within this space (C6) seem to indicate changes in its domestic use during the different occupational phases of the trench.

Another feature of interest, a circular installation (L.238) that abutted the north face of W.124, and its associated contexts were investigated to assess its potential function as a cooking installation. This feature appears to be contemporary with cooking installation L.225. Below L.238, there was a beaten earth surface (L.240 B.2050) that was situated above a plaster surface (L.241 B.2051), which was above another beaten earth surface (L.242 B.2052) that in turn was above a thin plaster surface at the bottom of the installation (L.243 B.2054). Each of these contexts was sampled for archaeobotanical analysis.

Objects dating to the 15th century, or Late Mamluk period, were recovered from both the plaster surface (L.241 B.2051) and beaten earth surface (L.242 B.2052). Stratigraphically the plaster surface (L.243 B.2054) should therefore also be dated to the 15th century. A soil sample was taken from a probe (L.244 B.2055) below the flagstone floor (L.229) where finds, including a Mamluk period lamp, glass fragments, and a horseshoe likely dating to the same period, were recovered (B.J. Walker *personal communication*). The last soil sample taken from Trench C6 was from the fill of installation L.259 B.2068, plastered basin #2, where samples for radiocarbon dating and OSL dating were also taken. The obtained radiocarbon date of 1233–1391 CE dates to the Mamluk period and therefore attests to an earlier occupation layer than the samples above it. Earthenware pottery fragments were found in L.241 and L.244. A potential *ṭābūn* oven base fragment, as indicated by its thick, rounded form, and flat base, was found in L.242. Each of these objects were assessed for plant impressions.

Archaeobotanical results and observations

The summary of the identification and quantification of the archaeobotanical remains from contexts sampled in Trench C6 are presented in Figure 20. The archaeobotanical assemblage from this trench is diverse, comparable to the assemblage of trench B4, and includes carob, a cucurbit seed, barley, grass pea/red pea (*Lathyrus sativus/cicera*, Figure 17), mallow, medick, olive, common pea, pomegranate (*Punica granatum*, Figure 17), foxtail millet, free-threshing wheat (both tetraploid and hexaploid), bitter vetch, faba bean (*Vicia faba*, Figure 17), and grape. *Lathyrus sativus* and *Lathyrus cicera* can be particularly challenging to distinguish from one another in a charred/carbonized state.

Both surfaces, such as L.221, and cooking installations, such as the potential *ṭābūn* oven area or cooking installation (L.226), were rich in archaeobotanical materials. In the former case, these are likely indicative of sweepings of previous cooking/food preparation activities and in the latter case of cooking activities themselves. The cucurbit seed was highly damaged and somewhat broken up during the carbonization process, and its seed coat was completely absent. However, the characteristic groove on the circumference of the seed remains visible despite extreme burning (Figure 17). Based on its dimensions, it seems to be closest to *Citrullus* sp., though because of the damage it is not possible to specify to the species level, either *lanatus* or *colocynthis*, because of the absence of other surviving morphological features. In terms of fuel sources, wood, cereal processing waste, and *jift* (as attested through the concentration of fragmented olive endocarps) were frequently found in Trench C6; for example, in L.244 B.2055 and L.226 B.2031.

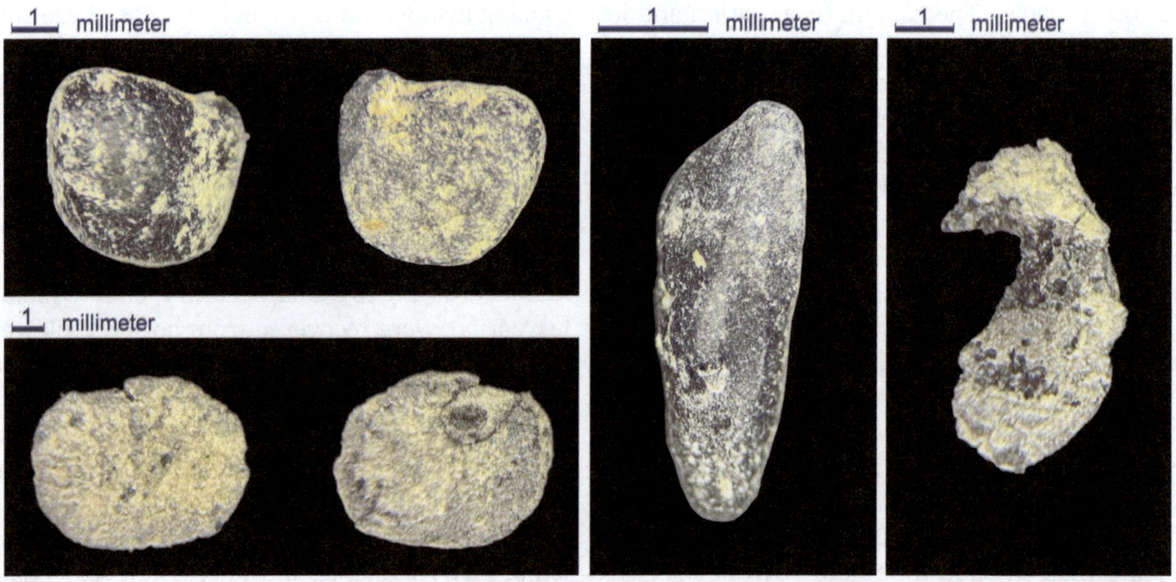

Figure 17 Top Left: *Lathyrus sativus/cicera* seed. Bottom Left: *Vicia faba* seed. Middle: *Punica granatum seed*. Right: Carbonized and damaged Cucurbitaceae seed from C6, Locus 226, Basket 2031.

A particularly important find from this trench is a fruit, also referred to in botanical literature as a caryopsis, of the summer crop foxtail millet (*Setaria italica*) in L.240 B.2050. A second specimen of this crop was found in Trench B7 (L.537 [West] B.5236). Though the fruit coats of both specimens were damaged by burning, its long and relatively broad scutellum, spanning over almost the full length of the fruit (as is characteristic for the taxon), is still visible. In one of the specimens the embryo had exploded out of the fruit, likely due to intense heat (Figure 18). While this may be a later find, these specimens are among the very few positive identifications of this taxon in the region. Caryopses identified to this taxon have been recorded from a context relatively dated to the mid-1st to 4th century—the Byzantine period—from al-Humayma (Ramsay 2013). Foxtail millet, along with the other "millets"—a commercial or practical term rather than taxonomic group of crops comprised of dozens of small-seeded grasses (Heinrich 2019; Heinrich and Hansen 2020)—is often challenging to identify in archaeological contexts. This is because of the manner in which it is processed, as its seed/fruit coat is particularly fragile and easily damaged during carbonization. These finds, however, point to a long and continued use of the foxtail millet in the southern Levant as an additional cereal crop. As elsewhere, it is underrepresented in both the archaeobotanical record and the written sources of the region. The extent to which this crop was used, and its economic importance, is therefore difficult to determine. As millets were often considered low status, emergency, or ascetic crops that made for "inferior foodstuffs," they were infrequently mentioned in more elite sources. This skewed view has also permeated the archaeological debate, often unjustly underestimating their importance to rural populations and within agricultural systems (Heinrich *et al.* 2021).

There were also some unique finds within the assemblage of the trench that have not yet been documented in the archaeobotanical record for the region, such as numerous specimens of carbonized flowers, all of the same type, which were found in L.240 B.2050 (four specimens), L.242 B.2052 (one specimen), and L.243 B.2054 (one specimen) (Figure 18). These appear to be the basal structures of the flowers, as the attachment points of the flowers are clearly visible. The features of the specimens

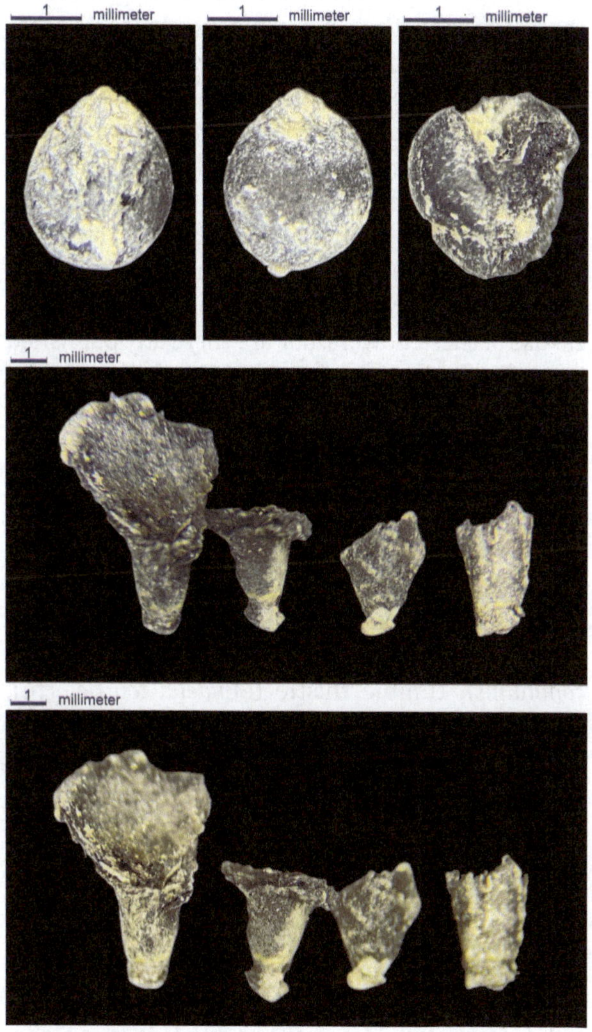

Figure 18 Top Left, Center, and Right *Setaria italica* caryopses. Top Left: Ventral side B7, Locus 537 (West), Basket 5236. Top Center: Dorsal side B7, Locus 537 (West), Basket 5236. Top Right: Ventral side C6, Locus 240, Basket 2050. Middle and Bottom: Carbonized flowers from C6, Locus 240, Basket 2050.

are very similar to the basal structure of a pomegranate flower, though it is difficult to identify more precisely in lieu of sufficient comparable reference material (Neef *et al.* 2012, 257). The presence of multiple carbonized flowers in more than one locus is significant, as such flowers are very unlikely to be preserved during burning, and it is therefore likely that they came into contexts regularly and/or in large numbers. Moreover, they must have come from a taxon that has a fairly robust flower, making pomegranate a feasible candidate. Their presence in three different loci may indicate a more intentional use of flowers in food preparation rather than a random and incidental signal of settlement noise. There is unfortunately no existing archaeobotanical literature on the archaeological use of these flowers, nor are there any published images of similar finds for the region with which to explore parallels to this material.

Consumption of flowers as a food, additive, or ingredient in medical preparations is documented ethnographically across the Mediterranean. The flowers of some taxa are frequently mentioned ethnographically and in the historical written sources for the southern Levant, including (but not limited to) taxa such as the giant fennel (*Ferula communis*), tumble thistle (*Gundelia tournefortii*), and pomegranate (*Punica granatum*) (Ali-Shtayeh *et al.* 2008; Hani *et al.* 2022; Lev and Amar 2008; Lev-Yadun and Abbo 1999; Motti *et al.* 2022). The tumble thistle (*Gundelia tournefortii*)—known as *a'kūb* (عكوب) in Arabic—is mentioned in the 10th century by al-Muqaddasī, though he is not explicit about the consumption of its flowers. Perhaps because of its conical flower, robust petals, and method of preparation, *a'kūb* has been sometimes mistranslated as artichoke (Lev-Yadun and Abbo 1999, 218, fig. 1; al-Muqaddasī, translated by Collins 2001, 152). The morphology of the flowers of giant fennel and tumble thistle, however, being from the Apiaceae and Asteraceae families respectively, are not consistent with the flowers found in the archaeobotanical assemblage at Khirbet Beit Mazmīl. The flowers of the pomegranate, alongside its seeds and peels, are mentioned in the 10th century Cairo Genizah as having practical medicinal uses, including creating a "medical soap," helping with "urinary complaints," and being useful when following a "medical diet." Pomegranate flowers additionally had several theoretical medicinal uses, including the treatment of "eye diseases, inflammatory conditions of the tongue and gums, hectic and septic fevers, cancer, erysipelas, soft and hard inflammatory swellings and elephantiasis" (Lev and Amar 2008, 29). While flowers are underrepresented in the archaeological record due to low visibility, the use of flowers as food and/or medicine in the past should not be underestimated.

Common field weeds are also present in Trench C6, many of which also appear in trenches B4 and B7. These include wild barley (*Hordeum spontaneum*, Figure 19), wild grasses (Poaceae), brome grasses (*Bromus* sp., Figure 19), including drooping brome (*Bromus* cf. *tectorum*, Figure 19), *Phalaris* sp. (Figure 19), bedstraw (*Galium* sp.), and mallow (*Malva* sp., Figure 19). Though some morphological features were damaged during carbonisation, the specimen of drooping brome (*Bromus* cf. *tectorum*, Figure 19) has a very distinct Y-shaped scutellum visible and its parallel ridges in the center and on each side of the caryopsis are well preserved too. The identification was made based on the parameters as specified by Nesbitt (2009, 81, fig. 78.32).

Trench C6–7—transitional area

Summary of the contexts

Square C6–7, similar to B5, is a trench connecting Squares C6 and 7 in order to further articulate Wall 104 in C7 and better assess its relationship with the domestic structures in C6. A significant amount of fill, both modern and historical, was removed from this area along with a modern almond tree. Situated underneath the tree was a semi-circular coping wall (W.125). Within this fill, an object first interpreted as a kohl kit and later confirmed as a *narghila* made of copper sheet (L.218 B.2020) was discovered. The *narghila* has been relatively dated to the 17th to 18th century. The soil from

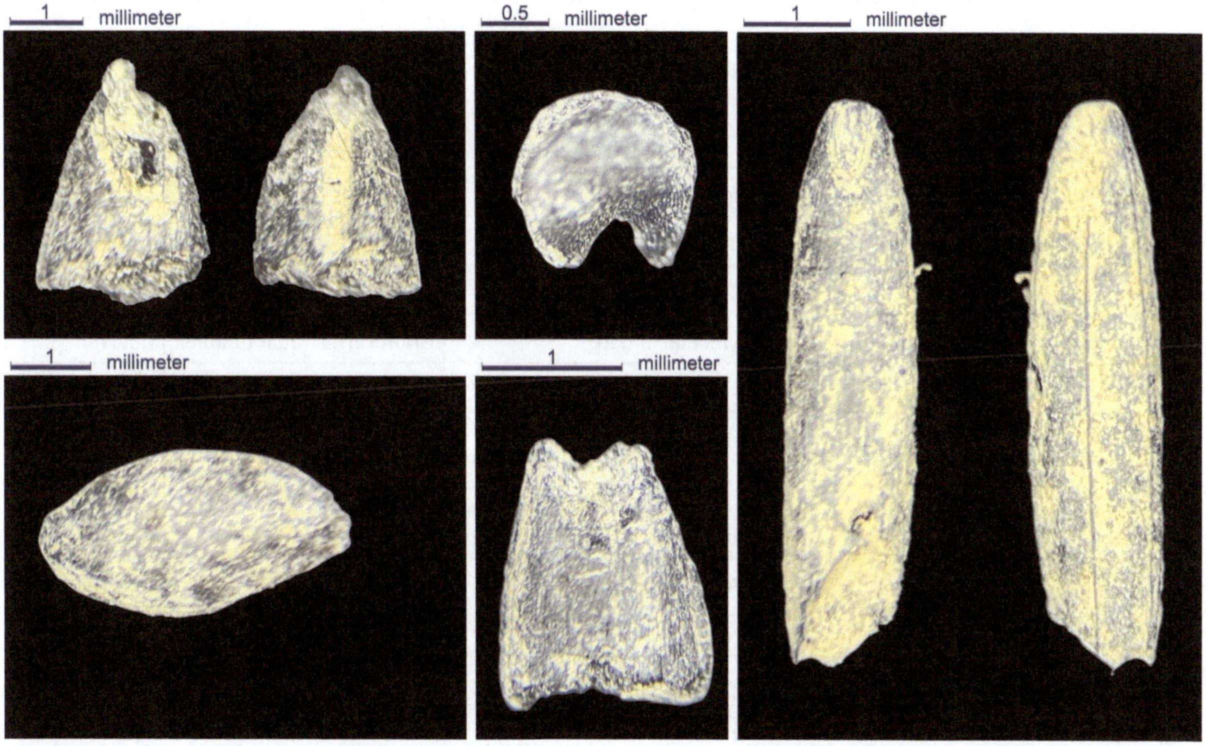

Figure 19 Top Left: Fragmented *Hordeum spontaneum* caryopsis dorsal (left) and ventral (right) sides. Top Middle: *Malva* sp. seed. Bottom Left: *Phalaris* sp. caryopsis. Bottom Middle: Fragmented *Bromus* sp. caryopsis dorsal side. Right: *Bromus* cf. *tectorum* caryopsis dorsal (left) and ventral (right) sides.

this context also contained an 18th century porcelain sherd (B.J. Walker *personal communication*). Significant amounts of charcoal, both compressed and fragmented, were found within the *narghila*. This material was sampled for later anthracological identification of the plant material and radiocarbon dating, as the *narghila* was sealed. The *narghila* was likely taken from its original, historical context and deliberately placed below the roots of the modern almond tree when it was (trans)planted to Khirbet Beit Mazmīl's orchard in the late 20th–early 21st century.

Archaeobotanical results and observations

The summary of the identification and quantification of the archaeobotanical remains from the contexts sampled in Trench C6–7 are presented in Figure 21. The material within the *narghila* appears to be very finely compressed ash that was mixed with a substance that, while resting inside the container, eventually hardened and crystallized over time into a large mass. The fragments are brittle and under 40× magnification the fine substance is a dark black-brown, glossy substance. The mixture was very firmly pressed into the container, since the larger fragments have impressions from the inside of the object (Figure 22). The original material that the ash was made from was already pulverized prior to being put into the container, presumably for use in smoking, as is attested by the absence of any recognizable morphological features of plant tissues, such as seeds/fruits, leaves, flowers, or wood charcoal. These pressings would then have carbonized during use.

		Sample ID	
Scientific Name	**Common Name**	**Plant Part**	**Preservation**
cf. Agrimonia eupatoria	Common Agromony	Seed	Charred
Bromus sp. cf. tectorum	Drooping Brome	Fruit	Charred
Ceratonia siliqua	Carob	Seed	Charred
Cerealia	Cereal Indeterminate	Grain kernal (quantifiable)	Charred
Cerealia	Cereal Indeterminate	Grain kernal (fragments)	Charred
Cerealia	Cereal Indeterminate	Culm node	Charred
Cerealia	Cereal Indeterminate	Culm base	Charred
Curcubitaceae cf. Citrullus sp.	Curcubit Family cf. Citrullus Genus	Seed	Charred
Fabaceae indeterminate	Pulse Indeterminate	Seed	Charred
Galium	Bedstraw*	Fruit	Charred
Hordeum spontaneum	Wild Barley	Fruit	Charred
Hulled Hordeum vulgare	Hulled Barley	Floret	Charred
Hulled Hordeum vulgare	Hulled Barley	Floret (fragments)	Charred
Hordeum vulgare	Barley	Rachis (node + internode)	Charred
Lathyrus sativus/cicera	Grass pea/Red pea	Seed	Charred
Lens culinaris	Lentil	Seed	Charred
Cf. Lens culinars	Lentil	Seed	Charred
Malva sp.	Mallow	Seed	Charred
Cf. Medicago sativa	Alfalfa	Seed	Charred
Olea europaea	Olive	Endocarp (quantifiable)	Charred
Olea europaea	Olive	Endocarp (fragments)	Charred
Phalaris sp.	Canary Grass	Fruit	Charred
Pisum sativum	Common Pea	Seed	Charred
cf. Pisum sativum	Common Pea	Seed	Charred
Poaceae	Grass family	Fruit	Charred
Punica granatum	Pomegranate	Seed	Charred
Setaria italica	Foxtail millet	Floret/Caryopsis	Charred

Figure 20 Trench C6 archaeobotanical assemblage, identifications, and quantifications.

KBM17. C6.220. 2023	KBM17. C6.221. 2024	KBM17. C6.222. 2025	KBM17. C6.223. 2026	KBM17. C6.225. 2030	KBM17. C6.226. 2031	KBM18. C6.240. 2050	KBM18. C6.241. 2051	KBM18. C6.242. 2052	KBM18. C6.243. 2054	KBM18. C6.244. 2055	KBM19. C6.259. 2068
					1						
					2						
										1	
2	20	4			8	1	2	1		5	
	+				++					+	
1					4					2	
	2	2			1			1			
					1						
					++	+					
					4		1				
					1						
	6	1			10	1	1	1		4	1
					1						
					1						
	1	1									
	4	7			22		1	1			1
	+	+									
					5						
							1				
	1	2			1			1		2	
	++	+			++		+	+	+	+++	+
					1						
					1						
					1						
	1	2			5	8	5	2			
						1					
						1					

(continued)

		Sample ID	
Scientific Name	**Common Name**	**Plant Part**	**Preservation**
T.a. aestivum	Bread Wheat	Rachis	Charred
T.t. durum	Hard Wheat	Rachis	Charred
T. t. durum / T. a. aestivum	Hard / Bread Wheat	Grain kernal	Charred
Vicia ervilia	Bitter vetch	Seed	Charred
Cf. Vicia ervilia	Bitter vetch	Seed	Charred
Vicia faba	Faba bean	Seed	Charred
Vitis vinifera	Grape	Seed	Charred
Wood Charcoal	N/A	Wood	Charred
Additional finds + Impressions			
Amorphous lump(s)	N/A	N/A	Charred
Dung	N/A	N/A	Charred
Earthenware fragment(s)	N/A	N/A	N/A
Flower base	Flower base	Flower	Charred
Fruit endocarp	N/A	Endocarp	Charred
Glass	N/A	N/A	N/A
Indeterminate	N/A	Seed/Fruit	Charred
Insect	N/A	N/A	Modern
Modern seeds/fruits	N/A	Seed/fruit	Modern
Obsidian fragment	N/A	N/A	N/A
Plaster fragments w/chaff impressions	N/A	N/A	N/A
Plaster fragments w/seed impression	N/A	N/A	N/A
Plaster fragments w/hulled Hordeum vulgare charred seed	N/A	N/A	N/A
Plaster fragment(s) w/wood charcoal	N/A	N/A	N/A
Plaster fragment w/insect	N/A	N/A	N/A
Slag	N/A	Slag	N/A
Ṭabun fragment(s)	N/A	N/A	N/A

+ = present but not quantifiable 1–10 fragments

++ = 11–50

Figure 20 Trench C6 archaeobotanical assemblage, identifications, and quantifications. (*continued*)

KBM17. C6.220. 2023	KBM17. C6.221. 2024	KBM17. C6.222. 2025	KBM17. C6.223. 2026	KBM17. C6.225. 2030	KBM17. C6.226. 2031	KBM18. C6.240. 2050	KBM18. C6.241. 2051	KBM18. C6.242. 2052	KBM18. C6.243. 2054	KBM18. C6.244. 2055	KBM19. C6.259. 2068
	3										
2					1	4	2				
	3	1		3	26	7+	1	4		1	1
	3	3		1	15	2	4	1	1		
	1										
					1						
	1				1						
0.25ml	10ml	7ml	1/0.50ml	0.50ml	45ml		1ml	1ml	1.5ml	6ml	1.5ml
					+	++	++	+	+	+	
						1					
						4		1	1		
						+					
						+					
4	7	1			21	16	1	1	3	3	12
		+			+			+		+	
	6				1	1	4		11	1	
											1
+						+			+		
1/+											
	+								+		1
											1
		+									

		Sample ID		KBM17.C6–7.L218.B2020
		Context		Charcoal from *narghila* Obj. No. 024
		Relative Date		17th–18th c.
Scientific Name	**Common Name**	**Plant Part**	**Preservation**	**Volume (ml)**
N/A	Compressed ash	Wood	Charred	25ml

Figure 21 Trench C6–7 archaeobotanical assemblage, context description, identifications, and quantifications.

Figure 22 Compressed ash from the inside of the *narghila*, showing impression from the vessel.

Trench C7

Summary of the contexts

Trench C7 was excavated to aid in the articulation of the architecture from the north tower, following W.123 and its springer W.126 towards trench B5 and B4 in the south and C6 in the east. A significant amount of fill and collapse were removed before historical discard and fill layers were reached (Locus 202–204). Later occupational phases (Locus 205) were relatively dated to the Early Mandate period, early 20th century, while earlier phases of the trench seem to go down to the Late Mamluk or Early Ottoman period, between the 15th and 16th centuries. No archaeobotanical samples were taken from this trench. There was one fragmented earthenware vessel collected from the final excavated surface (L.232) during the 2017 field season, however, where plant impressions and inclusions were visible.

Archaeobotanical results and observations

Although the earthenware vessel was rather brittle and did not have any discernible decoration or form to determine its function, it was rich in plant remains in terms of impressions and inclusions. These were mostly made up of microscopically visible plant remains from grasses (Poaceae) and macroscopic cereal by-products such as straw. Some of the inclusions are on the surface of the ceramic. These likely represent the use of cereal by-products during the formation of the ceramic that were then charred during firing. Cereal straw impressions were also present on the surface and within the fabric (Figure 23). This once again highlights the importance of cereal processing by-products/waste, such as straw, as a resource.

Figure 23 Plant impressions and inclusions from earthenware vessel. Top Left: Carbonized cereal straw inclusion. Top Right: Carbonized cereal straw inclusion (above) and cereal straw impression (below). Bottom Left: Impression of cereal straw. Bottom Right: Impression of cereal straw. Scale: 1 centimeter (photo: F. Weber and A. Legowski 2017).

Trench D1—water reservoir

Summary of the contexts

The investigation of Trench D1, a water reservoir, was intended to determine the approximate dating and period of use of this structure, and to assess if its use was contemporary with the use of other structures onsite. OSL dating of the structure and water system dated the last phase of use to circa 1330 CE, the Mamluk period, seemingly confirming the contemporaneity of use for the various structures on site. See Ze'evi-Berger and Ben Melech, Chapter 6, for a discussion on the use of OSL dating at the site and on the use of nearby contemporary agricultural terraces. One archaeobotanical probe sample was taken from the thin layer of soil under the flagstone floor, just above the bedrock, in the water reservoir (L.510 B.5015). Its context description and its sample metrics are summarized in Figure 24.

Archaeobotanical results and observations

The summary of the identification and quantification of the archaeobotanical remains from contexts sampled in Trench D1 are presented in Figure 24. As the remains were uncovered from underneath the floor, and do not seem to be related to a specific feature, it is hypothesized that this material is settlement noise that was blown into the context during the construction of the floor. There is much micro-charcoal present and some small seeds/fruits, likely representing fuel waste that was swept or blown away from its original context(s). One charred fruit from the Caryophyllaceae family was identified, which likely represents a wild native taxon that grew on or near the site. Its overall morphology was preserved as it had some of the features on the seed/fruit coat. Charring, however, did partially burn away most

		Sample ID		**KBM19.D1.L510.B5015**
		Context		**Soil under flagstone floor**
		Relative Date		**14th c.**
		Volume Soil (L)		**3.5**
		Volume Flot (ml)		**25**
		Volume Botanical Remains (ml)		**20**
Scientific Name	**Common Name**	**Plant Part**	**Preservation**	
Caryophyllaceae	Caryophyllaceae family	Fruit	Charred	1
Wood Charcoal	N/A	Wood	Charred	20ml
Additional finds + Impressions				
Indeterminate	N/A	Seed/Fruit	Charred	2

+ = present but not quantifiable 1–10 fragments

++ = 11–50

Figure 24 Trench D1 archaeobotanical assemblage, context description, identifications, and quantifications.

other features on the seed/fruit coat, making it difficult to identify to a lower taxonomic level (Figure 25).

Figure 25 Caryophyllaceae fruit. D1 Locus 510, Basket 5015.

Trench E1—potential storeroom with installations

Summary of the contexts

This space was excavated to assess a potential storeroom with some installations of interest. Based on radiocarbon dates from the lower layers, the earliest phase of the structure is dated to the Abbasid period. This is confirmed by the rich ceramics found, especially in Locus 433. As is clear from radiocarbon dating carried out on material from a later locus (Locus 428), the structure was reused during the Mamluk period, between the 14th and 15th centuries, with its use continuing into the 15th to 16th centuries, or Late Mamluk period. A total of five samples were taken from E1 (Figure 26). The first locus to be sampled from this trench was a hard-packed, orange-coloured, beaten earth surface (L.413 B.2022) directly below the fill layer (Locus 412) and above a flagstone floor (Locus 414). Based on the ceramic evidence, the loci date to the Late Mamluk period and imply a prolonged use of this space. The surface had plaster chunks throughout its matrix. In addition to the archaeobotanical remains, pottery, bone, and pieces of metal and slag were found, which were likely mixed into its matrix during construction.

Brick-like fragments (E1.W1 L.133 B.4007) were excavated from the E/W wall parallel to W.124 within the space and sampled to assess the potential presence of plant impressions, as these were first thought to be potential *ṭābūn* fragments built into the structure of the wall. While it is difficult to determine if these fragments were fired bricks or part of the thick base of a *ṭābūn* oven that was repurposed, it is clear from their position that these fragments were clearly incorporated into the wall's structure. The clay is also clearly fired. Small wood micro-charcoal remains, though only few, are present in the matrix of the fragments. Large plant remains and impressions are absent from these fragments.

Two installations were uncovered that date to an earlier usage phase of the space. The first was a stone-lined pit (L.140) located just outside of the storeroom in E1. Soil from within the stone-lined pit was sampled (L.428 B.4035). Within the room, a stone-lined installation (L.432) was uncovered, and a sample of the fill was taken (L.431 B.4037). A small sample was taken from the fill below the final flagstone floor (L.433 B.4038). This was located just above an ash and brick layer, which itself was not archaeobotanically sampled (L.434).

Archaeobotanical results and observations

The summary of the identification and quantification of the archaeobotanical remains from contexts sampled in Trench E1 is presented in Figure 26. The beaten earth surface (L.413 B.2022) contained some remains of food crops, such as free-threshing wheat (Figure 27) and lentil as well as olive endocarp and wood charcoal fragments. These are likely the sweepings or discarded materials of a former cooking area that were incorporated into the surface's matrix. The presence of broken olive endocarps alongside wood charcoal is consistent with Locus 428 and elsewhere on the site (Trench B7) and associated with fuel use. Upon closer inspection of the matrix of the brick-like fragments of W1 L.133 B.4007, the presence of wood micro-charcoal could be attested, but neither identifiable plant remains nor impressions were found. Within the fill of the stone-lined pit (L.428 B.4035), there were various plant remains, with some indicating either prepared food or food preparation, such as crushed grain kernels and barley floret/caryopsis specimens (Figure 27).

The remainder of the assemblage within trench E1 is representative of fuel use. There is a large amount of wood charcoal (23 ml), among which there are some larger fragments that could be identified anthracologically at a later stage, as well as cereal culm, crushed olive endocarps, and some wild flora including grass caryposes (Poaceae) and bedstraw (*Galium* sp.), which are common field weeds (Figure 28).

These field weeds likely came into the assemblage with cereal crop processing by-product, such as straw, which was collected for use as fuel. There was one fragment from a *ṭābūn* oven encountered, on which the structure of the base is clearly visible. The archaeobotanical material from this context is yet another example of crushed olive endocarps being used alongside other plant-based fuels such as wood and cereal processing by-products. These materials likely come from olive pressings that were used as fuel in a cooking context. Some modern material was blown into the sample during excavation and collection: these could be easily distinguished and separated from the archaeological material. A small sample was taken from a fill within another installation (L.431 B.4037), also capturing evidence of potential food processing and preparation. A few remains of cereals and fuel remains were present and indicate their use within the space. There was also a concentration of wood charcoal (6 ml) that further supports this interpretation. One seed closely resembling medick/alfalfa was present, though due to charring distortion this could not be

		Sample ID	
		Context	
		Relative Date	
		Volume Soil (L)	
		Volume Flot (ml)	
		Volume Botanical Remains (ml)	
Scientific Name	**Common Name**	**Plant Part**	**Preservation**
Cerealia	Cereal Indeterminate	Grain kernal (quantifiable)	Charred
Cerealia	Cereal Indeterminate	Grain kernal (fragments)	Charred
Cerealia	Cereal Indeterminate	Culm internode	Charred
Galium	Bedstraw*	Fruit	Charred
Hulled Hordeum vulgare	Hulled Barley	Floret	Charred
Lens culinaris	Lentil	Seed	Charred
Cf. Lens culinars	Lentil	Seed	Charred
Cf. Medicago sativa	Alfalfa	Seed	Charred
Olea europaea	Olive	Endocarp (quantifiable)	Charred
Olea europaea	Olive	Endocarp (fragments)	Charred
Poaceae	Grass family	Fruit	Charred
T. t. durum / T. a. aestivum	Hard / Bread Wheat	Grain kernal	Charred
cf. T.t. durum / T. a. aestivum	Hard / Bread Wheat	Grain kernal	Charred
Vitis vinifera	Grape	Seed	Charred
Wood Charcoal	N/A	Wood	Charred
Additional finds + Impressions			
Bead	N/A	N/A	N/A
Bricky material	N/A	N/A	N/A
Indeterminate	N/A	Seed/Fruit	Charred
Indeterminate	N/A	Other	Charred
Indeterminate	N/A	Other	Modern
Insect	N/A	N/A	Modern
Modern seeds/fruits	N/A	Seed/fruit	Modern
Ṭābūn fragment(s)	N/A	N/A	N/A

+ = present but not quantifiable 1–10 fragments

++ = 11–50

Figure 26 Trench E1 archaeobotanical assemblage, context descriptions, identifications, and quantifications.

KBM18. E1.413.2022	**KBM18. E1.W.133.4007**	**KBM19. E1.428.4035**	**KBM19. E1.431.4037**	**KBM19. E1.433.4038**
Beaten earth surface	***Ṭābūn* fragments included in wall construction**	**Soil inside stone-lined pit**	**Fill of installation**	**Fill below final flagstone floor in "storeroom" in E1**
15th–16th c.	**15th–16th c.**	**14th–15th c.**	**14th–15th c.**	**14th–15th c.**
44	**N/A**	**20**	**9**	**0.25**
8	**N/A**	**30**	**11**	**5**
5	**N/A**	**25**	**8**	**1**
1				
		+	+	
		1		
		1		
		1		1
1				2
				1
			1	
		1		
+		+		
		4	1	
			1	1
1				
				1
1ml		23ml	6ml	× < 0.1ml
1				
	+			
		33	2	3
		1		
		49		
+				
		2		
		+		

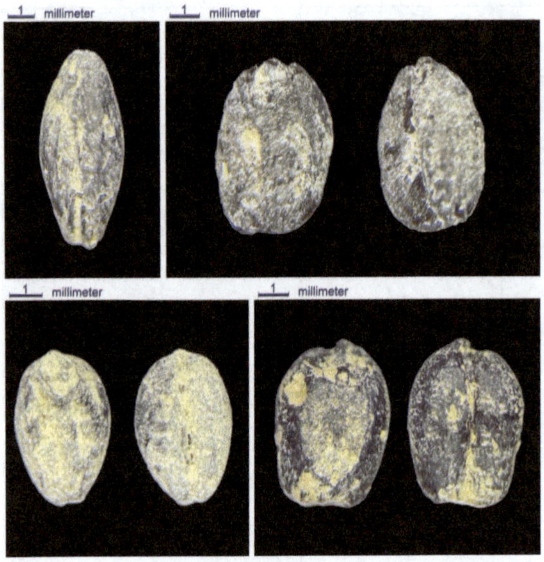

Figure 27 Top Left: Hulled *Hordeum vulgare* caryopsis ventral side. Top Center, Top Right, Bottom Left, Bottom Right: Free-threshing wheat seeds dorsal (left) and ventral (right) sides.

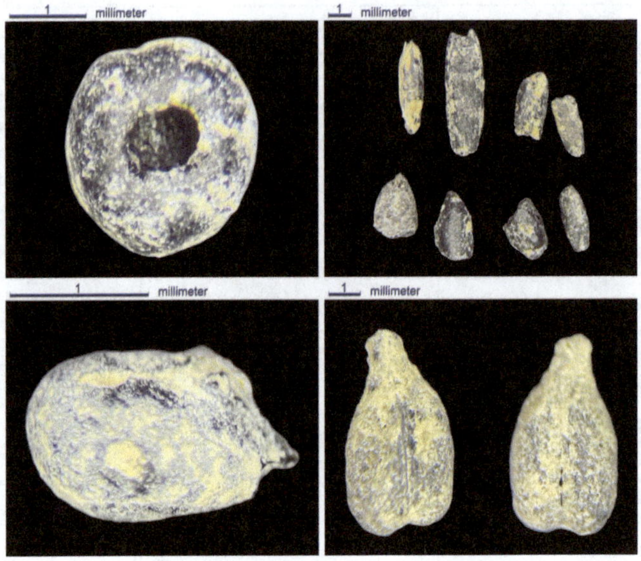

Figure 28 Top Left: *Galium* sp. fruit. Top Right: Poaceae caryopses. Bottom Left: *Medicago* sp. seed, Locus 431, Basket 4037. Bottom Right: *Vitis vinifera* seed ventral (left) and dorsal (right) sides, Locus 433, Basket 4038.

confidently confirmed (Figure 28). The sample from under the flagstone (L.433 B.4038) contained several crop species including hulled barley, free-threshing wheat, lentil, and grape (Figure 28), alongside wood charcoal, all of which are indicative of past food preparation and fuel use from one or more activities that were eventually included in the fill.

Trench E2—potential agricultural terrace

Summary of the contexts

Trench E2 was investigated to ascertain the presence of an agricultural terrace within the confines of the site. The northern half of the trench, located between Walls 138 and 135, was a long-term use midden situated in between the North and South Buildings that dated to the main Ottoman phase, during the 17th century. The southern part of Trench E2, south of Wall 135 and contemporary to the northern part of the trench, was an area that was hypothesized to be an agricultural terrace. This part of the trench is composed of soils brought from elsewhere, possibly off-site, and deposited. The pottery recovered from the soil suggests that the soil was a part of a midden or secondary midden (B.J. Walker *personal communication*). One archaeobotanical sample was taken from this potential agricultural terrace (L.438 B.4058), the context information and the sample metrics of which are summarized in Figure 29.

Archaeobotanical results and observations

The summary of the identification and quantification of the archaeobotanical remains from the contexts sampled in Trench E2 are presented in Figure 29. The overall archaeobotanical composition of the sample appears to support that the context is indeed a midden or originated from a midden. It contains almost exclusively food and fuel waste, alongside the pottery. The reuse of middens as additional material in the construction of an agricultural terrace is probable, although the archaeobotanical assemblage of this sample is likely random and representative of the archaeobotanical remains' secondary or even tertiary use. This would follow cooking/fuel use; deposition within the original midden; collection; and redeposition as a construction component within the potential terrace. It is important to note that the archaeobotanical assemblage represented within this sample is consequently a result of waste material from multiple activities, likely from different areas of the site, accumulated over a long period of time. The archaeobotanical assemblage from this trench can therefore neither attest to nor refute the hypothesis that the area had once been an agricultural terrace. During excavation, or in a sub-recent phase, there could have been a disturbance in the trench, as seven modern or sub-recent lentil seeds, a green variety with black spots, were recovered in this context (Figure 14).

The archaeological sample from the context contains seeds of free-threshing wheat, lentil, and bitter vetch, as well as crushed fragments of cereal grains and olive endocarps. As is noted elsewhere

		Sample ID		**KBM19.E2.L438.B4058**
		Context		**Fill of potential terrace**
		Relative Date		**17th c.**
		Volume Soil (L)		7
		Volume Flot (ml)		8
		Volume Botanical Remains (ml)		5
Scientific Name	**Common Name**	**Plant Part**	**Preservation**	
Cerealia	Cereal Indeterminate	Grain kernal (fragments)	Charred	++
Lens culinaris	Lentil	Seed	Charred	4
Lens culinaris	Lentil	Seed	Sub-recent	7
Olea europaea	Olive	Endocarp (quantifiable)	Charred	1
Olea europaea	Olive	Endocarp (fragments)	Charred	+
T. t. durum / T. a. aestivum	Hard / Bread Wheat	Grain kernal	Charred	1
cf. T.t. durum / T. a. aestivum	Hard / Bread Wheat	Grain kernal	Charred	1
Vicia ervilia	Bitter Vetch	Seed	Charred	2
Wood Charcoal	N/A	Wood	Charred	0.5ml
Ziziphus spp.	Ziziphus spp.	Endocarp	Mineralized	1
Additional finds + Impressions				
Amorphous lump(s)	N/A	N/A	Charred	2.5ml
Indeterminate	N/A	Seed/Fruit	Charred	24

+ = present but not quantifiable 1-10 fragments

++ = 11–50

Figure 29 Trench E2 archaeobotanical assemblage, context description, identifications, and quantifications.

on-site, the crushed olive endocarps likely represent waste material from the use of olive pressings as fuel. A number of small, charred seeds and plant remains were also uncovered, though their seed/fruit coats had sustained significant damage during the charring process, making them impossible to identify. Mineralized *Ziziphus* spp. fruit endocarps (Figure 30) were also encountered as well numerous carbonized amorphous lumps (Figure 30). While *Ziziphus lotus* is a native taxon of the area, *Ziziphus spina-christi* (Christ's thorn) cannot be ruled out based on the encountered specimen. The fruits of both taxa were often collected and eaten, and the fruits can also be used for medicinal purposes. Additionally, the endocarps of Christ's thorn tend to be rounded, similar to the remains encountered in the trench, whereas *Ziziphus lotus* endocarps tend to be more elongated and pointed towards the apical and basal ends. The internal structure that would hold the two seeds, however, did not survive during the mineralization process. Two carbonized specimens of *Ziziphus spina-christi* were also found in Trench B4 L.127a B.1038 (Figure 30).

The amorphous fragments encountered in the sample do not include any identifiable plant inclusions or tissues visible with a light microscope. Remains such as these are often associated with charred, processed food, or sometimes dung remains due to the presence of processed and morphologically altered plant remains (Figure 30). These remains are like those found in Trenches B4 and C6, and it is important that they are documented when they are found on an archaeological site. It is to be hoped that future studies will be able to identify these remains more precisely, at a higher magnification or using an SEM, as they have the potential to give more direct evidence of food and fuel processing (cf. Kabukcu *et al.* 2022). Microfragments of wood charcoal were also present in the sample, likely representing fuel waste.

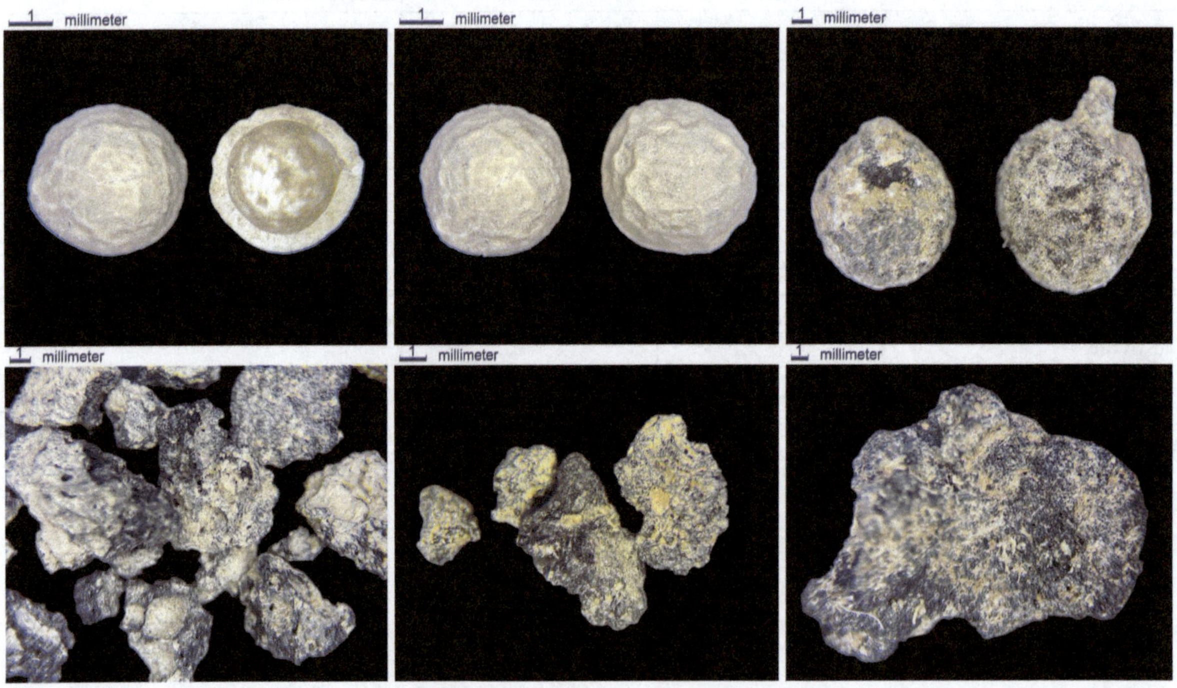

Figure 30 Top Left and Center: Mineralized *Ziziphus* spp. endocarp from E2, Locus 438, Basket 4058. Top Right: Carbonized *Ziziphus spina-christi* endocarps from B4, Locus 127a, Basket 1038. Bottom: Carbonized amorphous lumps.

Land use, diet, and food processing at Mamluk and Ottoman Khirbet Beit Mazmīl

This chapter has discussed the results of the archaeobotanical assessment of 56 samples from Khirbet Beit Mazmīl. What is clear from the results is that the inhabitants of the site had access to a wide and diverse range of food crops that were most likely cultivated at or near to the site, including both summer and winter cereals (foxtail millet, hulled barley, free-threshing wheats), pulses (lentil, broad bean, bitter vetch, common vetch, lupine, grass/red pea), fruit shrubs and trees (fig, plum, olive, Christ's thorn/local *Ziziphus* spp.), as well as locally available herbs and spices (cumin and common agrimony). In the case of finds of certain taxa, such as common agrimony, it is important to note that the distinction between food and medical or pharmaceutical plants was not as narrowly defined in the past as it is today. For those taxa non-food uses may also have been common. Fruit tree cultivation, especially that of olives, appears to have been especially important to both the food and fuel economy of the site, as processed olive pomace, or *jift*, was used as a major fuel source alongside wood charcoal and animal dung, including that from sheep and/or goats. This is supported by the historical evidence for the region during the Mamluk and Ottoman periods (Walker and Dolinka 2020).

In terms of plant-based food resources, as attested from the archaeobotanical evidence, the inhabitants of the village had access to a nutritionally complimentary and well-rounded selection of foodstuffs. The combination of cereals and pulses is particularly important nutritionally, as each category possesses some of the essential amino acids that the other lacks (Heinrich 2019; Heinrich and Hansen 2019). Some of the pulses encountered, such as bitter vetch and common vetch, were often considered fodder crops by both contemporary authors and historians, though their use as, albeit low status, human food is widely attested. This is also the case at Khirbet Beit Mazmīl, where they were found in explicit cooking contexts. Several other so-called low status crops, such as the other pulses and the cereals foxtail millet and barley were present in the village. While this is to be expected at a village site, it should certainly not be taken as sign of economic dearth or crisis, especially in the case of millets, as has become fashionable in some archaeological debates (Heinrich *et al.* 2021). Instead, especially for rural producer-consumers, it makes sense to cultivate these crops alongside each other in an agricultural regime. Due to their nitrogen fixing properties, pulses contribute to maintaining soil fertility while millets, being summer crops with a greater drought-resistance, allow for different cropping regimes, potentially allowing farmers to take advantage of seasonality and spread the demand for agricultural labour over a longer period and optimize a balanced regime with both subsistence and cash crop production. Moreover, cultivating a wide array of different cereals, pulses, and fruit trees suggests a high degree of risk-spreading strategies that would have imbued the community with resilience against adverse conditions.

Various amorphous lumps of carbonized organic remains were documented at the site. These are hypothesized to be either processed food remains or incompletely burnt remains of dung cakes. Future studies may be able to further their identification, using new methods at the cellular level to determine their composition (cf. Kabukcu *et al.* 2022). The presence of (potential) cotton cloth, attested by threads attached to a small, decorative disk encountered at the site, is also of note. It does not necessarily indicate that the cultivation and processing of cotton occurred at Khirbet Beit Mazmīl, although al-Muqaddasī specifically mentions cotton cultivation in the Jerusalem of the 10th century, confirming its cultivation in the city's hinterland (al-Muqaddasī, translated by Collins 2001, 151). If not locally produced, the inhabitants of Khirbet Beit Mazmīl could have obtained the material from elsewhere within the local network of nearby villages, or Jerusalem itself. If purchased in Jerusalem, the cloth could have of course come from farther afield.

Based on the archaeobotanical assemblages of the trenches, there appears to be a significant degree of spatial homogeneity and diachronic continuity in the crop regime of the site and the differences in the attestations of the various taxa do not appear to be significant. The range of wild flora present at the site clearly reflects arable weeds that both attest to local cultivation at or near the site, and the use of cereal crop processing waste as an important resource with multiple functions. The consistent presence in cooking contexts of some more unusual plant remains, such as the carbonized flowers and seeds of common agrimony, implies that these or other parts of the plant were consumed or otherwise used. The documentation and figures provided in this chapter will hopefully contribute to their future identification in other assemblages, facilitating a better understanding of their function at rural sites.

In terms of cereal cultivation, both hulled barley and free-threshing wheat were cultivated in similar proportions at or near Khirbet Beit Mazmīl. Cereal remains were present in all trenches, except for C6–7 and D1. In terms of the quantity of rachis, free-threshing wheat rachis (both tetraploid and hexaploid) were more dominant than barley rachis. Because of the bulkier structure of wheat rachis, however, this is to some degree to be expected in archaeological assemblages. Unfortunately, no barley rachis with preserved glumes were present in the archaeobotanical samples from Khirbet Beit Mazmīl. It is therefore not possible to determine the subspecies—either two-row or six-row barley—most commonly cultivated at or near the site. For the periods under study, however, it would not have been unlikely that hulled types of both subspecies of barley were cultivated in the region (Zohary *et al.* 2012). In terms of seed concentrations, there does not seem to be a significant difference between hulled barley and free-threshing wheat in the archaeobotanical assemblage at Khirbet Beit Mazmīl (Figure 31). There does appear to be a marked difference, however, in the ratio between free-threshing tetraploid wheat, the most likely candidate being *Triticum turgidum* ssp. *durum*, or hard wheat (which is sometimes also referred to as "pasta wheat" or "durum wheat"), and free-threshing hexaploid wheat, the most likely candidate being *Triticum turgidum* ssp. *aestivum*, or bread wheat (which is sometimes also referred to as "soft wheat"), in which the former seems to be the more dominant wheat sub-species cultivated at or near the site. This trend is consistent with what we see at sites from across the southern Levant and from the Roman to early Ottoman period.

The archaeobotanical assemblage at Khirbet Beit Mazmīl also provides useful information as to the site's fuel economy. In addition to cereal processing waste, dung—where cereal processing waste was present as a common inclusion, as it served both as a fodder and temper in dung cakes—and wood charcoal, *jift*, the processing waste of olive oil production, played an important role as a fuel source for food preparation and cooking at the

	Hulled Barley Florets	**Free-threshing wheat seeds**	**Barley Rachis**	**Tetraploid free-threshing wheat rachis**	**Hexaploid free-threshing wheat rachis**
B4	50	50	3	3	1
B7	3	5	4	56	6
C6	25	40	1	9	3
E1	2	2	—	—	—
E2	—	1	—	—	—

Figure 31 Quantification of hulled barley florets and rachis, free-threshing wheat seeds, tetraploid free-threshing wheat rachis, and hexaploid free-threshing wheat rachis.

site. The apparent abundance of this material also reflects the attested economic importance of fruit trees such as olive in the area.

The recycling and/or upcycling of by-products or rest products from crop production is not limited to *jift*. It appears to be a recurring theme in the agricultural economy of Khirbet Beit Mazmīl. Plant remains, especially cereal processing by-products or waste, such as straw and chaff, were crucial inputs into the processes of various productions, serving as construction materials, fodder, fuel, ceramic tempers, and as binders in plaster and dung cakes. This is illustrated by the ubiquity of cereal waste inclusions and impressions in objects, installations, and surfaces throughout the settlement, including plaster surfaces and earthenware ceramics, *ṭābūn* ovens, and bricks. While not as versatile as crop processing waste, food and fuel waste are also consistently encountered in archaeological contexts, acting as a type of binder in plaster surfaces and floors on-site and often representing a tertiary use of resources. This is illustrative of an efficient and circular use of resources.

Overall, the archaeobotanical assemblage at Khirbet Beit Mazmīl provides us with new data and insights into how village economies in the southern Levant functioned, and what factors shaped production and consumption in rural areas. The image that emerges is one of a versatile agricultural landscape managed by a resilient village community. These were a people that, on the one hand, applied a circular approach to agricultural rest products and employed risk reduction strategies, including the application of a diversified crop curriculum, to be self-sufficient in their basic needs of food, fuel, and construction materials. On the other hand, as both the apparent abundance of *jift* and the written sources indicate, the inhabitants of the village also produced cash crops—likely olives for olive oil, fruits, and surpluses of other crops, with which they entered into exchange with other communities, most likely Jerusalem itself. Finally, some of the plants uncovered at Khirbet Beit Mazmīl, such as lupine and foxtail millet, have only very rarely been documented in the southern Levant, while some other finds were the first of their taxon to be encountered archeologically in the region, including common agrimony, perhaps southern/eastern bugle, and flowers. Such finds, which are often lacking from urban contexts, alongside new insights into the complex relationships between rural and urban economies in the Late Medieval and Early Modern Islamic world, further illustrate the importance, value, and regionally still largely untapped potential of excavating rural settlements such as Khirbet Beit Mazmīl.

Acknowledgements

This chapter was completed with the help and feedback of an excellent team, for which I am extremely grateful. I would like to thank A.J. Bronkhorst (University of Groningen) for the excellent photo editing work on the archaeobotanical remains: his work did justice to the level of detail required of charred archaeobotanical plant remains. I would like to thank Aris Legowski and Dr. Felicitas Weber for taking photographs of some of the larger objects, ensuring I could capture the right features of the challenging *ṭābūn* and plaster fragments. I would like to thank Merit Hondelink (University of Groningen) for photographing the perfect example of a modern specimen of lentil for my composite figure. I would like to thank Dr. Daniel Fuks (University of Cambridge) for many enjoyable exchanges on crop introductions and diffusion into the southern Levant, within the context of our collaboration in the Crop History Consortium. I would especially like to thank the project directors, Professor Bethany Walker and Dr. Benjamin Dolinka, for involving me in the project and for the many valuable insights they shared concerning the ceramics, coins, and objects found on site, and Dr. Nicolò Pini for the valuable insights he provided about the site's building patterns. I would like to thank the German Israeli Foundation for Scientific Research and Development (GIF) for their support concerning transportation and accommodation, making the field work trips possible. I would like to thank Professor Oren Gutveld and his incredible team of students and excavators for our excellent collaboration on site and for lending me his team's floatation machine. I would like to thank the Israeli Antiquities Authority for their support and enthusiasm for our project. I would like to thank the Dutch Research Council (NWO) that supported a good part of my PhD work and provided funds for board and equipment during the field work campaigns. I would like to thank the AIAR staff, especially Nadia, Naual, Noor, Ahmad, Ashraf, and Hisham, as well as to the former director, Professor Matthew Adams, who always made me feel welcome and were very supportive of our work and let us use the garden for conducting floatation. I would also like to thank the AIAR chef, Hisham M'Farreh, for the many fruitful conversations concerning local agriculture, food, and food preparation, and reminiscing about local *ṭābūn* ovens in use. Hisham also kindly shared that following my floatation work, and the deposition of fertile sediments from the flotation tank, the fruits of the lemon and orange trees in the garden tasted better. Lastly, I would like to thank my family, especially my grandmother Grace, godmother Jill, and my husband and colleague Frits, who encouraged me to pursue my research.

Note

1. This chapter also forms a case study for the author's doctoral dissertation on trends in crop selection at various sites in wider Bilād al-Shām between the Roman and Ottoman periods. Additional figures from Khirbet Beit Mazmīl that could not be included within the confines of this chapter are included in the forthcoming dissertation.

References

Ali-Shtayeh, M.S., R.M. Jamous, J.H. Al-Shafie, W.A. Elgarabah, F.A. Kherfan, K.H. Qarariah, I.S. Khadair, I.M. Soos, A.A. Musleh, B.A. Isa, H.M. Herzallah, R.B. Khlaif, S.M. Aiash, G.M. Swaiti, M.A. Abuzahra, M.M. Haj-Ali, N.A. Saifi, H.K. Azem, and H.A. Nasrallah
2008. "Traditional knowledge of wild edible plants used in Palestine (Northern West Bank): A comparative study." *Journal of Ethnobiology and Ethnomedicine* 4(13): 1–13. Internet Edition: https://doi.org/10.1186/1746-4269-4-13.

Ali-Shtayeh, M.S., R.M. Jamous, and R.M. Jamous
2016. "Traditional Arabic Palestinian ethnoveterinary practices in animal health care: A field survey in the West Bank (Palestine)." *Journal of Ethnopharmacology* 182: 35–49. Internet Edition: http://dx.doi.org/10.1016/j.jep.2016.02.005.

Ball, M.
2022. "Movialsa, Spain." *EQTEC*, 07.29.2022.

Boudsocq, S., C. Cros, P. Hinsinger, and H. Lambers
2022. "Changes in belowground interactions between wheat and white lupin along nitrogen and phosphorus gradients." *Plant Soil* 476: 97–115. Internet Edition: https://doi.org/10.1007/s11104-022-05558-3.

Bouyahya, A., N. El Omari, N. Elmenyiy, F.-E. Guaouguaou, A. Balahbib, M. El-Shazly, and I. Chamkhi
2020. "Ethnomedicinal use, phytochemistry, pharmacology, and toxicology of *Ajuga iva* (L.,) schreb" *Journal of Ethnopharmacology* 258: 112875: 1–20. Internet Edition: https://doi.org/10.1016/j.jep.2020.112875.

Braadbaart, F., E. Marinova, and A. Sarpaki
2016. "Charred olive stones: experimental and archaeological evidence for recognizing olive processing residues used as fuel" *Vegetation History and Archaeobotany* 25: 415–430. Internet Edition: https://doi.org/10.1007/s00334-016-0562-2.

Brock, F., T. Higham, P. Ditchfield, and C.B. Ramsey
2010. "Current pretreatment methods for AMS radiocarbon dating at the Oxford Radiocarbon Accelerator Unit (ORAU)." *Radiocarbon* 52(1): 103–112. Internet Edition: https://doi.org/10.1017/S0033822200045069.

Campbell, M., and B.A. Horton
2021. "Olive waste is being turned into clean fuel to power homes in Syria." *Euronews*, 12.11.2021. Website: https://www.euronews.com/green/2021/11/12/olive-waste-is-being-turned-into-clean-fuel-to-power-homes-in-syria.

Cappers, R.T.J., R. Neef, and R.M. Bekker
2009. *Digital Atlas of Economic Plants*. Groningen: Barkhuis and University of Groningen Library.

Coubray, S., N. Monteix, and V. Zech-Matterne
2019. "Of olives and wood: baking bread in Pompeii." In *Fuel and Fire in the Ancient Roman World: Towards an Integrated Economic Understanding*, edited by R. Veal and V. Leitch, 121–133. Cambridge: McDonald Institute for Archaeological Research, University of Cambridge.

Dakia, P.A.
2011. "Chapter 35—Carob (Ceratonia siliqua L.) Seeds, Endosperm and Germ Composition, and Application to Health." In *Nuts and Seeds in Health and Disease Prevention*, edited by V.R. Preedy, R.R. Watson, V.B. Patel, 293–299. Cambridge: Academic Press. Internet Edition: https://doi.org/10.1016/B978-0-12-375688-6.10035-0.

Dalman, G.
1935. *Arbeit und Sitte in Palästina. Band IV: Brot, Öl und Wein*. Hildesheim: G. Olms Verlag.

Danin, A.
2004. *Distribution Atlas of Plants in the Flora Palaestina Area*. Jerusalem: Israel Academy of Sciences and Humanities.

Di Nunzio, M.G. Picone, F. Pasini, E. Chiarello, M.F. Caboni, F. Capozzi, A. Gianotti, and A. Bordoni
2020. "Olive oil by-product as functional ingredient in bakery products: Influence of processing and evaluation of biological effects." *Food Research International* 131:108940. Internet Edition: https://doi.org/10.1016/j.foodres.2019.108940.

Difonzo, G., M. Troilo, G. Squeo, A. Pasqualone, and F. Caponio
2020. "Functional compounds from olive pomace to obtain high-added value foods—a review." *Journal of the Science of Food and Agriculture* 101(1): 15–26. Internet Edition: https://doi.org/10.1002/jsfa.10478.

Fabricant, D.S., and Farnsworth, N.R.
2001. "The value of plants used in traditional medicine for drug discovery." *Environmental Health Perspectives* 109: Supplement 1: Reviews: 69–75. Internet Edition: https://www.jstor.org/stable/3434847.

Fechner, A., U. Schweiggert, K. Hasenkopf, and G. Jahreis
2011. "Lupine Kernel Fiber: Metabolic effects in human intervention studies and use as a supplement in wheat bread." In *Flour and Breads and their Fortification in Health and Disease Prevention*, edited by V.R. Preedy, R.R. Watson, and V.B. Patel, 463–473. Cambridge: Academic Press. Internet Edition: https://doi.org/10.1016/C2009-0-30556-5.

Feinbrun-Dothan, N.
1978. *Flora Palaestina. Part Three: Ericaceae to Compositae. 2 vols. Text and Plates*. Jerusalem: The Israel Academy of Sciences and Humanities.

Feinbrun-Dothan, N.
1986. *Flora Palaestina. Part Four: Alismataceae to Orchidaceae. 2 vols. Text and Plates*. Jerusalem: The Israel Academy of Sciences and Humanities.

Fischer, A.
1999. "Glass production activities as practised at Sepphoris, Israel (37 BC–AD 1516)." *Journal of Archaeological Science* 26: 893–905.

Frumin, S., A.M. Maeir, L.K. Horwitz, and E. Weiss
2015. "Studying ancient anthropogenic impacts on current floral biodiversity in the Southern Levant as reflected by the Philistine migration." *Scientific Reports* 5: 11308: 1–10. Internet Edition: https://doi.org/10.1038/srep13308.

Forste, K.M.
2023. "Arboriculture and viticulture as investment in the Early Islamic Levant: An archaeobotanical and historical investigation of the site of Ashkelon." *Journal of the Economic and Social History of the Orient* 66: 85–119.

Fuks, D., Y. Melamed, D. Langgut, T. Erickson-Gini, Y. Tepper, G. Bar-Oz, and E. Weiss
2022. "Unprecedented yet gradual nature of first millennium CE intercontinental crop plant dispersal revealed in ancient Negev desert refuse." *bioRxiv 2022.12.01.518650*: pre-print. Internet Edition: https://doi.org/10.1101/2022.12.01.518650.

Garcia-Oliveira, P., M. Fraga-Corral, A.G. Pereira, C. Lourenço-Lopes, C. Jimenez-Lopez, M.A. Prieto, and J. Simal-Gandara
2020. "Scientific basis for the industrialization of traditionally used plants of the Rosaceae family."

Food Chemistry 330: 127197: 1–16. Internet Edition: https://doi.org/10.1016/j.foodchem.2020.127197.

Hani, N., S. Baydoun, H. Nasser, T. Ulian, and N. Arnold-Apostolides

2022. "Ethnobotanical survey of medicinal wild plants in the Shouf Biosphere Reserve, Lebanon." *Journal of Ethnobiology and Ethnomedicine* 18: 73: 1–16. Internet Edition: https://doi.org/10.1186/s13002-022-00568-y.

Hansen, A.M., B.J. Walker, and F.B.J. Heinrich

2017. "'Impressions' of the Mamluk agricultural economy: Archaeobotanical evidence from clay ovens (*ṭābūn*) at Tall Ḥisbān (Jordan)." *Tijdschrift voor Mediterrane Archeologie* 56: 58–69.

Heinrich, F.B.J.

2019. "Cereals and bread." In *The Routledge Handbook of Diet and Nutrition in the Roman World*, edited by P. Erdkamp and C. Holleran, 101–115. London: Routledge.

Heinrich, F.B.J., A.M. Hansen, and P. Erdkamp

2021. "Roman isotopes and economic meaning: millets, manure, mobility, marine signals, and Malthus." *Archaeological and Anthropological Sciences* 13: Article number 44: 1–13. Internet Edition: https://doi.org/10.1007/s12520-021-01276-6.

Heinrich F.B.J., and A.M. Hansen

2019. "Pulses." In *The Routledge Handbook of Diet and Nutrition in the Roman World*, edited by P. Erdkamp and C. Holleran, 116–128. London: Routledge.

Heinrich, F.B.J., and A.M. Hansen

2020. "Mud bricks, cereals and the agricultural economy: Archaeobotanical investigations at the New Kingdom town." In *AcrossBorders 2 Living in New Kingdom Sai*, edited by J. Budka, 275–349; 480–488 (plates). Vienna: Austrian Academy of Sciences Press.

Huyghe, C.

1997. "White lupin (*Lupinus albus* L.)." *Field Crops Research* 53: 147–160. Internet Edition: https://doi.org/10.1016/S0378-4290(97)00028-2.

Ibn al-Baytar, ʿAbd Allāh ibn Aḥmad

1874. *Kitāb al-Jāmiʿ li-mufradāt al-adwiya wa-l-aghdhiya*, edited by Bulaq. Paris: [publisher not identified].

Ibn al-Baytar, ʿAbd Allāh ibn Aḥmad

1989. *Tafsīr Kitāb Diyāsqūrīdūs*, edited by I. ibn Murad. Beirut: Dār al-Gharb al-Islāmī.

Ishaq, A.R., H.A.S. El-Nashar, T. Younis, M.A. Mangat, M. Shahzadi, A.S. Ul Haq, and M. El-Shazly

2022. "Genus Lupinus (Fabaceae): a review of ethnobotanical, phytochemical and biological studies." *Journal of Pharmacy and Pharmacology* 74: 1700–1717. Internet Edition: https://doi.org/10.1093/jpp/rgac058.

Islam, S. and W. Ma

2016. "Lupine." In *Encyclopedia of Food and Health*, edited by B. Caballero, P.M. Finglas, and F. Toldrá, 579–585. Cambridge: Academic Press. Internet Edition: https://doi.org/10.1016/B978-0-12-384947-2.00432-3.

Israili, Z.H. and B. Lyoussi

2009. "Ethnopharmacology of the plants of genus Ajuga." *Pakistan Journal of Pharmaceutical Sciences* 22(4): 425–462.

Jaradat, N.A., O.I. Ayesh, and C. Anderson

2016. "Ethnopharmacological survey about medicinal plants utilized by herbalists and traditional practitioner healers for treatments of diarrhea in the West Bank/Palestine." *Journal of Ethnopharmacology* 182: 57–66. Internet Edition: http://dx.doi.org/10.1016/j.jep.2016.02.013.

Kabukcu, C., C. Hunt, E. Hill, E. Pomeroy, T. Reynolds, G. Barker, and E. Asouti

2022. "Cooking in caves: Palaeolithic carbonised plant food remains from Franchthi and Shanidar." *Antiquity* 97: 391: 12–28. Internet Edition: https://doi.org/10.15184/aqy.2022.143.

Kaniewsky, D., E. Van Campo, T. Boiy, J.-F. Terral, B. Khadari, and G. Besnard

2012. "Primary domestication and early uses of the emblematic olive tree: palaeobotanical, historical and molecular evidence from the Middle East." *Biological Reviews* 87: 885–899. Internet Edition: https://doi.org/10.1111/j.1469-185X.2012.00229.x.

al-Ketan, O.

2012. "Potential of using olive pomace as a source of renewable energy for electricity generation in the Kingdom of Jordan." *Journal of Renewable Sustainable Energy* 4: 063132: 1–8. Internet Edition: https://doi.org/10.1063/1.4769205.

Kloner, A., and N. Sagiv

2003. "Subterranean Complexes 44 and 45." In *Maresha Excavations Final Report. Vol. 1*, edited by A. Kloner,

D.T. Ariel, and I. Rashut ha-'atikot, 51–73. Jerusalem: Israel Antiquity Authority.

Knecht, K.T., P. Sanchez, and D.H. Kinder
2020. "Chapter 27—Lupine Seeds (Lupinus spp.): History of use, use as an antihyperglycemic medicinal, and use as a food plant." In *Nuts and Seeds in Health and Disease Prevention (Second Edition)*, edited by V.R. Preedy and R.R. Watson, 393–402. Cambridge: Academic Press. Internet Edition: https://doi.org/10.1016/B978-0-12-818553-7.00027-9.

Kohen al-'Attar, Abu-'l-Muna Dawud b. Abi Nasar
1940. "Minhādj al-dukkān wa-dustūr al-aʿyān fī tarkīb al-adwiya al-nāfiʿa li'l-abdān." Cairo: [publisher not identified].

Landau, S.Y., H. Muklada, A. Abu-Rabia, S. Kaadan, and H. Azaizeh
2014. "Traditional Arab ethno-veterinary practices in small ruminant breeding in Israel." *Small Ruminant Research* 119: 161–171. Internet Edition: http://dx.doi.org/10.1016/j.smallrumres.2014.01.004.

Lev, E.
2002. "Reconstructed *materia medica* of the medieval and Ottoman al-Sham." *Journal of Ethnopharmacology* 80:2–3: 167–179. Internet Edition: https://doi.org/10.1016/S0378-8741(02)00029-6.

Lev, E., and Z. Amar
2008. "'Fossils' of practical medical knowledge from medieval Cairo." *Journal of Ethnopharmacology* 119: 24–40. Internet Edition: https://doi.org/10.1016/j.jep.2008.05.042.

Lev-Yadun, S. and S. Abbo
1999. "Notes on Economic Plants: Traditional use of A'kub (*Gundelia tournefortii*, Asteraceae), in Israel and the Palestinian Authority area." *Economic Botany* 53: 217–219.

Luan, F., K. Han, M. Li, T. Zhang, D. Liu, L. Yu, and H. Lv
2019. "Ethnomedicinal uses, phytochemistry, pharmacology, and toxicology of species from the genus Ajuga L.: A systematic review." *The American Journal of Chinese Medicine* 47(05): 959–1003. Internet Edition: https://doi.org/10.1142/S0192415X19500502.

Malheiros, J., D.M. Simões, A. Figueirinha, M.D. Cotrim, and D.A. Fonseca
2022. "*Agrimonia eupatoria* L.: An integrative perspective on ethnomedicinal use, phenolic composition and pharmacological activity." *Journal of Ethnopharmacology* 296: 115498: 1–20. Internet Edition: https://doi.org/10.1016/j.jep.2022.115498.

Motti, R., B. Paura, A. Cozzolino, and B. de Falco
2022. "Edible flowers used in some countries of the Mediterranean Basin: An ethnobotanical overview." *Plants* 11: 3272: 1–22. Internet Edition: https://doi.org/10.3390/plants11233272.

al-Muqaddasī, Shams al-Dīn Abū ʿAbd Allāh Muḥammad ibn Aḥmad ibn Abī Bakr
2001. "The best divisions for knowledge of the regions" (*Aḥsan al-Taqāsīm fī Ma'rifat al-Aqālīm*), translated by B. Collins. Reading: Garnet Publishing.

Neef, R.
1990. "Introduction, development and environmental implications of olive culture: The evidence from Jordan." In *Man's Role in the Shaping of the Eastern Mediterranean Landscape*, edited by S. Bottema, G. Entjes-Nieborg and W. Van Zeist, 295–306. Rotterdam: A.A. Balkema.

Neef, R., R.T.J. Cappers, and R.M. Bekker
2012. *Digital Atlas of Economic Plants in Archaeology*. Groningen: Barkhuis and University of Groningen Library.

Nesbitt, M.
2009. *Identification Guide for Near Eastern Grass Seeds.* New York: Routledge. Internet Edition: https://doi.org/10.4324/9781315427096.

Nguyen, V., S. Riley, S. Nagel, I. Fisk, and I.R. Searle
2020. "Common vetch: A drought tolerant, high protein neglected leguminous crop with potential as a sustainable food source." *Frontiers in Plant Science* 11: Article 818: 1–7. Internet Edition: https://doi.org/10.3389/fpls.2020.00818.

Oumeish, O.Y.
1999. "Traditional Arabic medicine in dermatology." *Clinics in Dermatology* 17: 13–20. Internet Edition: https://doi.org/10.1016/s0738-081x(98)00068-6.

Pamuk, Ş.
2000. *A Monetary History of the Ottoman Empire*. Cambridge: Cambridge University Press.

Ramsay, J.
2013. "10.D Plant remains." In *Humayma Excavation Project, 2: Nabataean Campground and Necropolis Byzantine Churches, and Early Islamic Domestic Structures*, edited by J.P. Oleson and R. Schick, 351–358. Boston: American Schools of Oriental Research Archaeological Reports.

Rast-Eicher, A.
2016. *Fibres: Microscopy of Archaeological Textiles and Furs.* Budapest: Archaeolingua.

Rowan, E.
2015. "Olive oil pressing waste as a fuel source in antiquity." *American Journal of Archaeology* 119(4): 465–82. Internet Edition: https://doi.org/10.3764/aja.119.4.0465.

Rowan, E.
2018. "Sustainable fuel practices in Roman North Africa and the contemporary Mediterranean Basin." *Interdisciplinaria Archaeologica Natural Sciences in Archaeology (IANSA)* IX(2): 147–156. Internet Edition: http://dx.doi.org/10.24916/iansa.2018.2.2.

Rowan, E.
2019. "The utility of olive oil pressing waste as a fuel source in the Roman world." In *Fuel and Fire in the Ancient Roman World: Towards an Integrated Economic Understanding*, edited by R. Veal and V. Leitch, 109–120. Cambridge: McDonald Institute for Archaeological Research, University of Cambridge.

Sabato, D., and L. Peña-Chocarro
2021. *MARIS NOSTRI NOVUS ATLAS: Seeds and Fruits from the Mediterranean Basin.* Madrid: Ediciones Doce Calles.

Salavert, A.
2008. "Olive cultivation and oil production in Palestine during the Early Bronze Age (3500–2000 BC). The case of Tel Yarmouth, Israel." *Vegetation History and Archaeobotany* 17: 53–61. Internet Edition: https://doi.org/10.1007/s00334-008-0185-3.

Shane, M.W., and H. Lambers
2005. "Cluster roots: a curiosity in context." *Plant and Soil* 274: 101–125.

Simchoni O., and M.E. Kislev
2006. "Charred by-products of olive-oil production in the Iron Age." In *Excavations at Tel Beth-Shean 1989–1996. Beth-Shean Valley Archaeological Project Publication 1*, edited by A. Mazar and N. Amitai-Preiss, 679–685. Jerusalem: Institute of Archaeology, the Hebrew University of Jerusalem.

Sleim, K.A., W.Z. Badawy, and I. Smetanska
2020. "Utilization of olive pomace as a source of bioactive compounds in quality improving of toast bread." *Egyptian Journal of Food Science* 48(1): 27–40. Internet Edition: https://doi.org/10.21608/EJFS.2020.22871.1038.

Tawarah, K.M.
2019. "Specifications of exhausted olive pomace as an energy source: A statistical approach." *International Journal of Advanced Research in Chemical Science (IJARCS)* 6(1): 6–15. Internet Edition: http://dx.doi.org/10.20431/2349-0403.0601002.

Teixidor-Toneu, I., G.J. Martin, A. Ouhammou, R.K. Puri, and J.A. Hawkins
2016. "An ethnomedicinal survey of a Tashelhit-speaking community in the High Atlas, Morocco." *Journal of Ethnopharmacology* 188: 96–110. Internet Edition: http://dx.doi.org/10.1016/j.jep.2016.05.009.

Todorović, I.
2021. "Cooperative in Croatia supplies biomass power plant with olive tree waste." *Balkan Green Energy News*, 17.05.21. Internet Edition: https://balkangreenenergynews.com/cooperative-in-croatia-supplies-biomass-power-plant-with-olive-tree-waste/.

Vourdoubas, J.
2016. "Possibilities of creating zero CO^2 emissions olive pomace plants due to energy use in Crete, Greece." *Open Journal of Energy Efficiency* 5: 78–87. Internet Edition: http://dx.doi.org/10.4236/ojee.2016.53008.

Walker, B.J., and B. Dolinka
2020. "*Ḫirbet Bēt Mazmīl*. Investigations of Medieval Jerusalem's Hinterland: Interim report on the 2015 and 2017 Field Seasons" *Zeitschrift des Deutschen Palästina-Vereins* 136.2: 191–234.

Zohary, D.
1966. *Flora Palaestina. Part One: Equisetaceae to Moringaceae. 2 vols. Text and Plates.* Jerusalem: The Israel Academy of Sciences and Humanities.

Zohary, D.
1972. *Flora Palaestina. Part Two. Platanaceae to Umbelliferae. 2 vols. Text and Plates.* Jerusalem: The Israel Academy of Sciences and Humanities.

Zohary, D., M. Hopf, and E. Weiss
2012. *Domestication of Plants in the Old World: The origin and spread of domesticated plants in Southwest Asia, Europe, and the Mediterranean Basin (4th edition).* Oxford: Oxford University Press.

About the author

Annette M. Hansen, MSc (Oxon.) is an agricultural and food historian, and ethno-archaeobotanist specializing in the Islamic world. She is a senior archaeobotanist on different archaeological projects in Jordan, Israel, Egypt, and Sudan, and is also involved in historical and traditional farming experiments in the Netherlands and Belgium. She is the manager and co-founder of FOST-lab at the Vrije Universiteit Brussel, where she is also research associate at the Interdisciplinary Historical Food Studies (FOST) research group. She has published in peer-reviewed journals including the *Journal of Archaeological and Anthropological Sciences* and in monographs in Routledge, Austrian Academy of Sciences Press, and Palgrave Studies in Ancient Economies.

7.3: *The Zooarchaeological Remains*

Chiara A. Corbino

The study of animal bones from archaeological sites has become a well-developed research field. There remain, however, some chronological periods in specific geographic areas which need larger datasets to accurately reconstruct the interactions between humans and animals in the past.

The sample analyzed here represents a contribution to this issue and adds to the existing zooarchaeological studies concentrated on the Mamluk, Ottoman, and British Mandate periods in Israel and the surrounding countries. Khirbet Beit Mazmīl is a rural site on the outskirts of Jerusalem, which has provided clear evidence for the relationship between humans and animals. The faunal remains provide the kind of evidence about everyday life that is not usually described in the available written sources.

Materials and methods

The faunal remains analyzed here were collected over four excavation seasons that occurred in 2012, 2017, 2018, and 2019, and were recovered from squares B4–5–7, C6–7, D1, and E1–2 in both primary and secondary contexts. These included, for the Late Mamluk period, cooking areas and a hearth in B4 (L. 129, 176, and 183), the first use of E1 as a storeroom (L.428) in the Northern Sector, and a living surface (flagstone floor) in B7 of the Southern Sector (L.357). Early Ottoman contexts in the Northern Sector included the cooking installations of B4 (L.122–124, and 127) and C6 (L.220–224). Samples from the Late Ottoman and Mandate periods were collected from the midden fills (topsoil and subtopsoil) of B4 (L.126, 143, and 178) and B5 (L. 138 and 142) in the Northern Sector. In total, 462 bone fragments were collected.

The samples were dry and wet sieved to retrieve small bone elements and fragments and all the fragments recovered were cleaned and recorded. The identification of faunal remains relied on atlases (Barone 1976; Schmid 1972) and on comparisons with complete skeletons from the mammal collection of the Koenig Museum in Bonn (Germany) and the bone reference collection of the Department of Biology at the University of Florence (Italy).

Attempts were always made to identify specimens to species level, and, when this could not be done, to the lowest possible taxonomic level. Sheep and goat remains were divided using the published criteria (Boessneck *et al.* 1964) and when morphologic features were not preserved the elements were included in the sheep/goat group.

The number of identified specimens (NISP) and minimum number of individuals (MNI) were used to assess the relative abundances of different taxa. MNI counts were obtained by taking ontogenetic age and body size into account (Bökönyi 1969). The distribution of anatomical elements was based on the minimum number of skeletal elements (MNE) for each species considering body side (right or left), size variation, and age. Biometric measurements for both mammals and birds were taken according to von den Driesch (1976). The presence or absence of a spur on chicken tarsometatarsi was recorded to gain some insight into sex distribution. This was not considered a guaranteed sex indicator, however, as while they are usually present in males, in rare cases spurs can also occur in females with defective ovaries (Sadler 1991; West 1982).

Age estimation for mammals was based on long bone epiphyseal fusion (Reitz and Wing 1999) and tooth eruption and wear (Grant 1982; Greenfield and Arnod 2008; Hambleton 2001). Bird bones were divided into two broad age groups: "immature" for juvenile individuals and "mature" for adults, based on the observed stage of skeletal ossification.

Osteopathologies were recorded, providing the location and a general description of the alterations. Inferences about the type of disease indicated by the recorded evidence have been proposed according to data available in Bartosiewicz and Gál (2013). Different types of bone surface alterations have been tallied, such as weathering (Behrensmeyer 1978), root etching, gnawing by carnivores and rodents, and human-derived modifications. The latter includes butchery marks, burning, and worked bones.

Analysis and results

In total 152 faunal remains, including eggshell fragments, were identified. The complete list of the identified taxa includes *Bos taurus* (cattle), *Ovis/Capra* (sheep/goat), *Ovis aries* (sheep), *Capra hircus* (goat), *Canis familiaris* (dog), *Capra nubiana* (ibex), *Vulpes* sp. (fox), *Rattus* sp. (rat), small rodents other than rats, *Gallus gallus* (chicken), Passeriformes, sea mollusks (shells) and fish. This sample spans three chronological periods (Figure 1). The number of elements and the NISP both vary considerably across the periods. It was not possible to identify 310 small fragments of bone, 65 of which dated to the 13th–15th century, 230 to the 16th–18th century and only 15 from the 19th century contexts. Although the frequencies present in this sample are not statistically significant, they may still provide some suggestions about animal exploitation and human preferences at the site through time.

	13th–15th c.	16th–18th c.	19th–20th c.
	Mamluk	**Ottoman**	**Late Ottoman/British Mandate**
Cattle	1	1	11
Sheep/goat	2	13	12
Sheep			1
Goat		1	
Dog			1
Ibex			1
Fox		1	
Rat	5	9	
Rodents		6	
Chicken	1	4	1
Passeriformes		2	
Eggshell	6	57	3
Seashell	2	1	1
Fish	9		
TOTAL	26	95	31

Figure 1 Frequencies of taxa by period based on NISP.

Mamluk period (13th–15th century)

Only 26 animal remains were identified from this period. The Mamluk sample is dominated by fish, which are represented by small fragments of fine spines. These did not allow for species identification. Two very small fragments of seashells were also retrieved through flotation, but it was not possible to identify them. Sheep/goat (an upper molar and a scapula fragment; MNI = 1) and cattle (a first incisor; MNI = 1) are present along with a chicken (an adult phalange; MNI = 1) and a rat (five elements; MNI = 1). Eggshell fragments are likely of hen.

Ottoman period (16th–18th century)

A larger number of faunal remains were collected from the Ottoman contexts (NISP = 95) compared to the other periods. However, about 60 percent is composed of bird eggshells, likely belonging to hens. The domestic mammal sample is dominated by sheep/goat (13.7 percent of NISP; MNI = 2) followed by cattle (1.1 percent of NISP; MNI = 1). Chicken (MNI = 2) represents 4.2 percent of the NISP assemblage while Passeriformes just 2.1 percent. Rodent bones are very common in this period and are mostly composed of rats (9.5 percent). Fox (1.1 percent) and seashell (1.1 percent) are also included in this sample.

The frequencies of anatomical elements (Figure 2) show the occurrence of cranial bones of sheep/goat, among which one horncore was attributed to goat, and of meat bearing portions (humerus, radius, and tibia). The lower limb portion is attested by the presence of metatarsus and first phalanges. Cattle remains show only a pelvis fragment of an adult, while other anatomical parts are completely absent. The chicken is represented by a coracoid, a scapula, one femur, and a tarsometatarsus.

The age at death for the three sheep/goat individuals was estimated at 1–2 years, 3–4 years, and more than 42 months.

Passeriformes are represented by the diaphysis of an ulna and a fragment of sternum, and fox by a juvenile maxilla. The chicken remains belong to

	Cattle		Sheep/goat	
	16th–18th c.	19th–20th c.	16th–18th c.	19th–20th c.
skull			2	
mandible		1		2
scapula		1		
humerus		3	1	
radius/ulna		1	1	2
metacarpus		1		
pelvis	1			
femur				
tibia		1	1	1
calcaneus		1		
metatarsus		1		2
I phalange			3	1
III phalange		1		

Figure 2 Anatomical frequencies of cattle and sheep/goat by period.

one mature and one immature individual. A tarsometatarsus shows the spur scar and, although not completely reliable, this may indicate the presence of a male.

Late Ottoman/British Mandate period (19th–20th century)

This period has only 31 identified faunal remains (Figure 1). Twelve of sheep/goat (38.7 percent of NISP; MNI = 3), of which it was possible to assign one to sheep, and 11 of cattle (35.5 percent of NISP; MNI = 2). The dog is represented by a humerus of an adult while the chicken only by a femur skeletally mature. Three eggshells, likely of chicken, and one fragment of seashell are also present. One element which is particularly worthy of mention is the third phalange of an ibex.

The anatomical elements recovered for sheep/goat are related to areas of the body which provide both a medium quantity of meat (radius and tibia), and a low quantity in the case of mandible, metatarsus and a first phalange. The anatomical frequencies of cattle also show, besides meat-bearing portions, elements related to lower quantities of meat such as mandible and tibia as well as the third phalange, which is usually discarded during the initial butchery process. The available ontogenetic evidence for sheep/goat indicates the presence of one individual younger than two years and two of 3–4 years, while cattle is represented by one juvenile and one adult. Most of the cattle remains from this period belong to one very young individual collected from a single context (Square B4, Locus 178, Basket 2011) (Figure 3, left). The unfused epiphyses of the identified anatomical elements (distal scapula, distal humerus, proximal radius, and calcaneus) and the degree of ossification of the cattle bones collected from this context indicate an individual of less than seven months. Biometric and morphological comparisons with cattle fetal bones (Prummel 1988) suggest that this individual probably died in the first month of life. Cut and/or cooking marks are both absent, suggesting that this animal may have died from natural causes rather than as a result of human will. However, the presence of only select, non-articulated elements of the skeleton may suggest the slaughtering of this individual, which would have had tender and valuable meat due to its age.

The only pathological alteration found was noted on a right chicken femur: the bone callus formation on the middle of the shaft likely corresponds to a healed fracture that did not recover properly (Figure 3, right). Only the distal portion of this bone was recovered, suggesting that the proximal part may have been discarded elsewhere.

Taphonomic evidence

A relatively wide range of modifications are present on the bones from this site (Figure 4). Alterations caused by root-etching are most common in

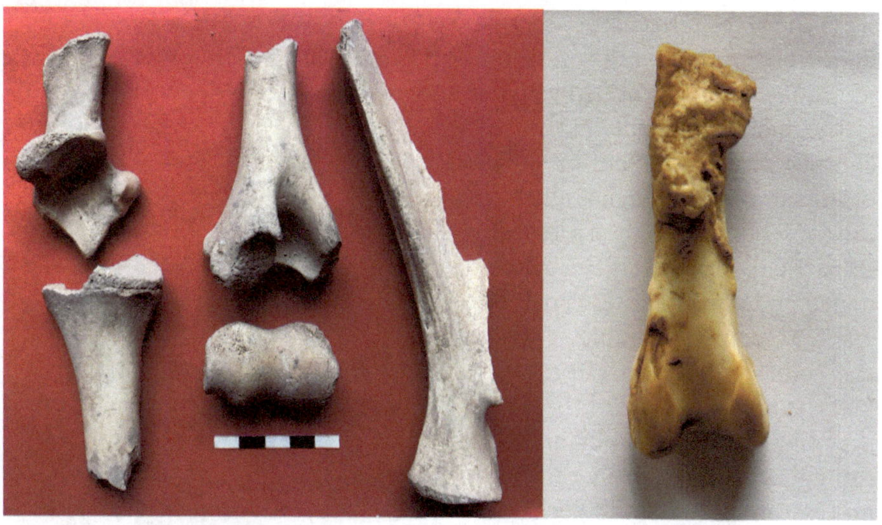

Figure 3 Ottoman period. Young calf remains from Square B4, Locus 178, Basket 2011 (left). Pathological right femur of chicken (Square B4, Locus 114, Basket 1039) (right).

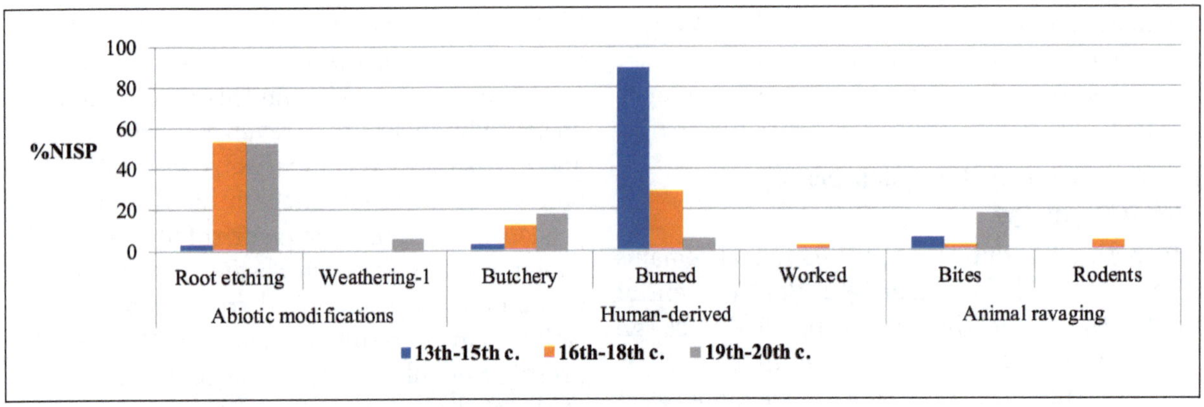

Figure 4 Frequencies of the taphonomic evidence by period.

the 16th–18th and 19th–20th century samples (53 percent of each sample respectively displaying these modifications). The extent of butchery evidence increases through time. Of particular note is a cattle tibia diaphysis which shows two chop marks possibly associated with the ossobuco meat-cut (Figure 5). This sample was recovered from balk topsoil (B4.143.1062) dated to the Mandate era. Another interesting example of human modification on bone from this site is the recovery of a pointed bone object, possibly a needle or hairpin, from the 16th–18th century sample. Evidence of burning was present on both identified and unidentified remains and is the most abundant in the 13th–15th century sample (present on 89 percent of the sample), although it also occurs in the subsequent periods to a lesser extent. Of particular interest is the burning evidence on the dog humerus dated to the 19th–20th century sample. Considering the colors of the bones (brown/black) and the limited extent of the damage, it may have been caused by an open fire. Animal bite marks, likely related to dog and cat, are present in all periods but are marginal compared to other alterations. Evidence of rodent gnawing was only recorded in the 16th to 18th century.

Discussion

Sheep/goat are present in all periods. This taxon is the most abundant domestic animal in all contexts. It was impossible to assess whether there was a preference for one of the two species (sheep or goat) comprising this group. It appears, however, that the goat remains were collected from the 16th–18th century samples, while the sheep came from the 19th–20th century.

Caprines provide not only meat but also milk and wool for textile production. They are quite

Figure 5 Cattle tibia fragment with chop marks (Square B4, Locus 143, Basket 1062).

adaptive and perfectly in line with the religious prohibitions and climatic conditions of that geographic area. The ontogenetic data for this taxon indicates a preference for young adult and adult individuals in the 16th–18th century and for adults in the 19th–20th century. These animals provided good quality meat but may also have been exploited for their fleeces.

Cattle are present in all the analyzed contexts as well. Its frequency seems to increase in the later periods but that could be due to the small sample size. Of interest is the identification of some elements of a very young calf from the 19th–20th century, which may indicate the consumption of high-quality meat. Regardless of whether the cause of death was natural or due to human selection, the presence of this animal in the sample indicates the local breeding of this taxon at the farmstead or in the nearby areas.

Anatomical frequencies of sheep/goat and cattle suggest a preference for meat-rich elements in the Ottoman and Mandate periods, although less palatable elements—such as phalanges—have also been recorded. The complete absence of second and third phalanges in the Ottoman contexts suggests that butchery practices occurred elsewhere, since these elements are usually discarded at the beginning of the butchery process. The presence of cranial and mandible fragments may indicate the consumption of brain and tongue.

Chicken remains, bones and eggshells, were also collected from all periods. The considerable number of eggshells at the site suggests that hens were likely kept alive for egg production. Eggshells are a particularly important component of the Ottoman contexts, indicating a high consumption of eggs and a well-planned management strategy to obtain this product. Indeed, in medieval Cairo cuisine, bird eggs were very popular and among them chicken eggs were the most consumed (Lewicka 2011, 207).

The anatomical elements of chickens from the Ottoman period include a tarsometatarsus with spur scar that may be associated with the meat consumption of young adult roosters. The pathology displayed by the chicken femur (Figure 3, right), dated to the 19th–20th century, could indicate poor health and living conditions for the chickens consumed at the site which were likely to have been kept in bad conditions or intensively bred to maximize meat and egg production.

A taste for bird meat in Ottoman-era Beit Mazmil is also suggested by a few remains of Passeriformes which occur only at that time. In medieval Cairo all social classes consumed bird meat (Lewicka 2011, 198). Fried chickens, sand grouses, small pigeons, and sparrows were all sold in the streets of Cairo (Lewicka 2011, 199). Among them, fried sparrows seem to have been particularly popular and sometimes they were cooked together with chicken meat or eggs (Lewicka 2011, 206–207).

Hunting activities are attested at the site by the recovery of fox and ibex remains. The latter was likely included in the diet while the former, usually not particularly appreciated for its meat, may have been used for its fur. It was also possibly consumed, as its young age may have made it more palatable.

The frequencies of the main domestic animals from the site—sheep/goat, cattle, and chicken—were compared to those of other settlements located in Israel and Jordan dated to the same periods. The complete list of the selected sites with chronological data and references is presented in Figure 6. Figure 7 shows the general abundance of sheep/goats ranging from about 60 percent to more than 90 percent at the Jordanian sites, and from 50 percent to more than 70 percent in Israel. The high frequencies of this taxon are due to religious rules as well as to environmental conditions at the sites that favored caprine husbandry. Moreover, they provide not only meat but also wool, milk, and leather. In the Israeli sites, besides sheep/goats, a considerable number of cattle remains have been noted, particularly from the 16th to 20th century. The site analyzed in this paper shows similarly high frequencies of cattle based on NISP and a considerable increase in the 19th–20th century context of Khirbet Beit Mazmīl.

Site	Country	Period	Reference
Aqaba castle	Jordan	Mamluk	De Cupere et al. 2017
Karak castle	Jordan	Mamluk, Ottoman	Brown, Rielly 2016
Shobak castle	Jordan	Mamluk, Ottoman	Brown, Rielly 2010
Ḥisban (citadel and villlage)	Jordan	Mamluk	Corbino 2020
Dhiban	Jordan	Mamluk	Fatkin et al. 2011
Al-Watta	Israel	Mamluk	Bar-Oz, Raban-Gerstel 2019
Jerusalem (Jewish Quarter)	Israel	Mamluk, Ottoman	Bouchnick 2018
Majdal Yābā	Israel	Late Ottoman/Mandatory	Taxel, Glick, Pines 2017

Figure 6 List of faunal assemblages included in Figure 7.

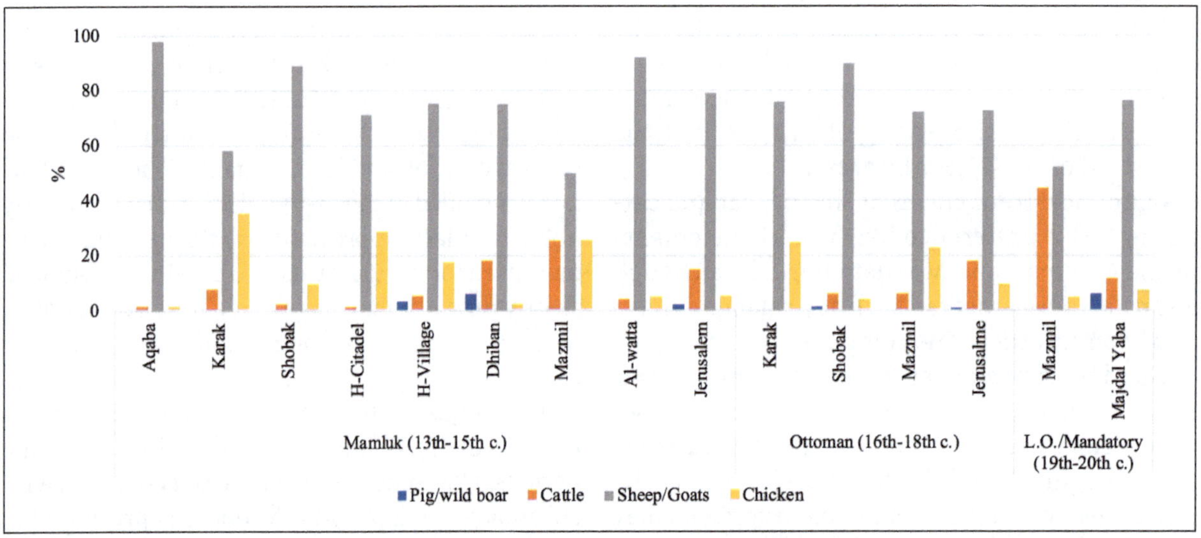

Figure 7 Frequencies of the main domestic animals at Israeli and Jordanian sites dated from the Mamluk period to the 20th century.

North African sites, dated from the 10th to 15th centuries, show the increase of cattle remains in western Maghreb Islamic sites when compared to those in eastern North Africa. This trend was linked to the higher demand of urban sites for cattle meat as further indicated by mortality profiles (MacKinnon 2017). Beef meat was permitted by the Muslim religion and is mentioned seven times in the Qur'an (MacKinnon 2017).

Chicken is attested to in all samples. It was an important source of meat, eggs, and feathers, which may have replaced the pig as the household source of meat (Redding 2015). Although absent at Khirbet Beit Mazmīl—and in most of the other sites considered—pig/wild boar remains frequencies are also included in Figure 7 to evaluate the contribution of this species. In general, it is a rare occurrence that does not seem to be linked to high status contexts. Its presence, when found, could be related to the presence of small Christian communities (pig) or to hunting activities (wild boar), though the attribution to earlier phases of occupation at the sites cannot be completely ruled out in some cases.

Rat remains occur frequently in the Mamluk and Ottoman periods, while they are completely absent from the 19th–20th century contexts. Rats are commensal animals that will have lived at the site, feeding on human food waste. They are an intrusive presence that are, however, linked to human behavior. A campaign to sanitize the cities took place not only in the Ottoman Empire but also in other parts of the world in the 19th century. The apparent disappearance of this species in this period at Khirbet Beit Mazmīl can be linked to such changes in human behavior (Mikhail 2014, 65–99).

Contact with the coast and the exploitation of marine resources are indicated by the occurrence of some fragments of seashells and fish; in particular in the Mamluk period sample. These could belong to the Mediterranean coast, just 60 kilometers from the site, or even from the Red Sea. Trade routes linking the Red Sea to northern territories are attested in Jordan by the recovery of parrotfish remains from sites quite distant from Aqaba, such as in the Petra Valley, Karak, Dhibān, and Ḥisban. From the 13th to 16th centuries, however, the consumption of this taxon decreased considerably. It was not an elite food and generally not particularly appreciated in the Mamluk period (Brown and Reilly 2016).

Human modifications—that is, burning and butchery—are present in all periods as is abiotic modification, in particular root-etching. Animal ravaging, mostly due to dogs and cats, indicates the presence of some commensal animals at the site in all periods.

Dogs may have been involved in hunting activities. They were important companions in hunting expeditions and, together with birds of prey, were considered among the most important animals for hunting and consequently often feature in Mamluk hunting treatises and iconographic sources (Shehada 2013, 44–45). It is worth noting, however, that wild animals are a marginal occurrence at the site. The evidence of burning on the dog bone is quite puzzling. The dog remains were collected from subtopsoil (B4.126.1033), which was part of refuse-disposal in the Mandate era. Human-dog relationships in the Islamic word have been largely misunderstood due to conventional thinking that this animal is ritually impure. Dogs were considered an important "productive" species in earlier Islamic communities, both rural and urban. They were involved in a number of activities, including hunting, herding, companionship, protection, and even as consumers of human waste. Ottoman officials actively encouraged the increase and maintenance of dog populations, providing food and water to these animals and punishing any who committed violence against them (Mikhail 2014, 67–69). Dogs become a problem in the Ottoman Empire only from the early 19th century, as they did throughout the world in this period. Dog killing campaigns were undertaken by the central powers in Cairo and Istanbul to eradicate this animal from their crowded cities at that time (Mikhail 2014, 88–95).

Conclusions

In the analyzed sample, sheep/goats were primarily exploited for meat production. However, the prevalent consumption of adult individuals indicates that they would have been exploited for secondary products as well. In general, the presence of adult sheep/goats also suggests that some individuals were exploited for breeding purposes and wool production. Cattle meat made a significant contribution to the diet of the site in all periods. Chicken was largely consumed in all periods, likely because its meat was enjoyable, reasonably priced, and its cooking required less time than red meat (Lewicka 2011, 203). This taxon was an important component, not only in the Khirbet Beit Mazmīl contexts, but also at the other sites discussed above (Figure 7).

The meat-diet of the inhabitants of this site was mostly based on mammals and chicken. Meat-eating was even encouraged by Islam, which considered meatless food not "real food" (Lewicka 2011). In the Islamic medieval cuisine of Cairo, meat was prepared in several ways: fried, roasted, baked, or cooked in the oven. However, the most popular cooking techniques were stewing and frying (Lewicka 2011, 189, 196). Roasted and grilled meat may have played a considerable role in the diet at Khirbet Beit Mazmīl as well, as suggested by the evidence of burning.

The sample presented in this contribution, though numerically limited, provided interesting hints to the nature of the human-animal relationship in the Jerusalem region from the Mamluk period to the 20th century, which only the analysis of further samples with those chronologies will clarify.

Acknowledgments

The access to the skeletons of the mammal reference collection of the Koenig Museum in Bonn (Germany) was kindly granted by curator Jan Decher and Eva Bärmann. Many thanks to Professor Jacopo Moggi Cecchi and Luca Bachechi for granting access to the bone reference collection of the Department of Biology at the University of Florence (Italy). Funding for the analysis in Bonn was provided by the DFG-funded Annemarie Schimmel Kolleg of the University of Bonn when the author was a Research Fellow there in February-March 2016 and January-July 2017. The author also gratefully acknowledges the logistical support of the university's Islamic Archaeology Research Unit during her fellowship in Bonn.

References

Barone, R.

1976. *Anatomia comparata dei mammiferi domestici.* Osteologia. Edagricole: Bologna.

Bartosiewicz, L., and E. Gál

2013. *Shuffling Nags, Lame Ducks: The Archaeology of Animal Disease.* Oxford: Oxbow Books.

Behrensmeyer, A.K.

1978. "Tophonomic and ecologic information from bone weathering." *Paleobiology* 4: 150–162.

Boessneck, J., H.H. Muller, and M. Teichert

1964. "Osteologische Unterscheidungsmerkmale zwischen Schaf (*Ovis aries* Linne) und Ziege (*Capra hircus* Linne)." *Kühn-Archiv*, 78: 1–29.

Bökönyi, S.

1969. "Archaeological problems and methods of recognizing animal domestication." In *The Domestication and Exploitation of Plants and Animals*, edited by P. Ucko and G. Dembley, 219–229. London: Duckworth.

Brown, R., and K. Reilly

2016. "Faunal remains from Mamluk and Ottoman occupations in the Middle Islamic period palace at Karak Castle (Qal'at al-Karak)." In *The Archaeology of Agro-pastoralist Economies in Jordan*, edited by Kevin M. McGeough, 91–134. Boston: The American Society of Overseas Research.

von den Driesch, A.

1976. *A Guide to the Measurement of the Animal Bones From Archaeological Sites.* Cambridge MA: Peabody Museum Bulletins 1.

Grant A.

1982. "The use of tooth wear as a guide to the age of domestic ungulates." In *Ageing and Sexing Animal Bones From Archaeological Sites*, edited by R. Wilson, C. Grigson, and S. Payne, 91–108. Oxford: BAR British Series 109.

Greenfield, H.J., and E.R. Arnold

2008. "Absolute age and tooth eruption and wear sequences in sheep and goat: determining age-at-death in zooarchaeology using a modern control sample." *Journal of Archaeological Science* 35: 836–849.

Hambleton, E.

2001. "A method for converting Grant Mandible Wear Stage to Payne style Wear Stages in sheep, cow and pig." In *Archaeological Sciences '97. Proceedings of the Conference Held at the University of Durham 2nd–4th September 1997*, edited by A. Millard, 103–108. Oxford: BAR International Series, vol. 939.

Lewicka, P.B.

2011. *Food and Foodways of Medieval Cairenes: Aspects of Life in an Islamic Metropolis of the Eastern Mediterranean.* Leiden: Brill.

MacKinnon, M.

2017. "Animals, acculturation and colonization in ancient and Islamic North Africa." In *The Oxford Handbook of Zooarchaeology*, edited by U. Albarella, 466–478. Oxford: Oxford University.

Mikhail, A.
2014. *The Animal in Ottoman Egypt*. New York: Oxford Academic.

Prummel, W.
1988. "Atlas for Identification of foetal skeletal elements of Cattle, Horse, Sheep and Pig / Part 3." *Archaeozoologia*, Vol. II/1.2: 13–26.

Redding, R.W.
2015. "The pig and the chicken in the Middle East: Modeling human subsistence behavior in the archaeological record using historical and animal husbandry data." *Journal of Archaeological Research* 23: 325–368. Internet Edition: https://doi.org/10.1007/s10814-015-9083-2.

Reitz, E., and E. Wing
1999. *Zooarchaeology*. Cambridge: Cambridge University Press.

Sadler, P.
1991. "The use of tarsometatarsi in sexing and ageing domestic fowl (*Gallus gallus* L.) and recognising five-toed breeds in archaeological material." *Circaea* 8:1: 41–48.

Schmid, E.
1972. *Atlas of Animal Bones*. Amsterdam, London, New York: Elsevier Publishing Company.

Shehada, H.
2013. *Mamluks and Animals: Veterinary Medicine in Medieval Islam*. Leiden: Brill.

West, B.
1982. "Spur development: recognizing caponized fowl in archaeological material." In *Ageing and Sexing Animal Bones From Archaeological Sites*, edited by R. Wilson, C. Grigson, and S. Payne, 255–262. Oxford: British Archaeological Reports, British Series 109.

About the author

Chiara A. Corbino is a postdoctoral fellow at the Institute for Heritage Science (ISPC-CNR) based in Naples and senior zooarchaeologist of the Pompeii Archaeological Park. Her research is focused on human-animal interactions through time. She investigates animal management and exploitation, ritual activities, biometry, taphonomy and environmental changes. Particular attention is dedicated to the study of human-animal interactions from the Classical period to the Middle Ages in Europe and, since 2005, from the Ayyubid-Crusader to the Ottoman period in the Levant. She cooperates with national and international archaeological teams as senior zooarchaeologist in Italy, England, Greece, Jordan, and Israel.

— 8 —

Market Gardening in Medieval Jerusalem

Bethany J. Walker

It is rare to find a rural site of the pre-modern period as well documented as Beit Mazmil. Although textually invisible until the 14th century, it leaves behind a rich historical record thereafter.[1] In fact, one can trace Beit Mazmil's history as an agriculturally productive village for 600 years: from the 14th through 19th centuries. These documents, moreover, tell us what crops were grown, how much the land was worth, and even the names of people who lived there through the centuries. It is beyond the scope of this work, which is primarily an archaeological report, to survey these texts in detail; this will be done in other publications.[2] What this chapter offers, however, is an overview of the sources that describe the village, its residents, and its land, in a way that ties together the fieldwork at the site and in the terraces and helps to reconstruct possible factors behind peasant decision-making and practices that do not leave a material trace. In the process, we come to understand more the agrarian world in which the people of Beit Mamzil lived, and what their relationship was with the imperial states of the day.

Critiquing the sources

The historian of the Mamluk and Ottoman sultanates has at her disposal a wealth of textual sources, and of varied form. The historiography of both periods, which together spans the 13th through early 20th centuries, is well developed and today engages with spatial and material culture studies. The nature of the written sources available, in part, makes this possible. The texts most-relevant to a study of rural society and its farmland, however, remain in manuscript form. The texts are challenging to locate and access, and even more difficult to read. It is perhaps for this reason that archaeological studies of the later Islamic periods have not engaged the written record as they should have. For the study of Mamluk Beit Mazmil, we rely here on endowment documents (*waqfiyyāt*) and farming manuals (*filāḥah* treatises), each of which present their own methodological and interpretive challenges.

Waqfiyyāt are the documents registered in court to donate income-producing property that one owns. This is usually done, ostensibly, for charitable purposes, though the management of the endowment generally stays in the hands of the original proprietor and his/her family. The documents tend to be quite detailed in their description of properties, and when this regards farmland, the texts mention the boundaries of the land, its relationship to the road system and other settlements, its infrastructure and resources, and occasionally what kind of land it was (how it was used). Many Mamluk endowment records have been reproduced in Ottoman tax and court files, as proof of land status and to support claims. The originals, if they have survived, are usually preserved in scroll form. Locating a particular village or field is like "looking for a needle in a haystack." One has to read a lot of scrolls before one finds the property of interest (if it exists in written form). The names of rural places are often misspelled, or the place-name has changed over time.

As a textual genre, *filāḥah* treatises are a cross between encyclopedias and farmers' manuals. They grew out of the translation movement of the Abbasid period, when Byzantine farmer's manuals were translated into Arabic. Manuals in Arabic first appeared in the 10th century in Iraq and really blossomed in al-Andalus. Over time they inspired the development of a tradition in Ottoman Turkish and Persian, and continued to be produced through at least the 18th century.[3] The pre-Mamluk treatises are encyclopedic, transmitting bits and pieces of written knowledge about crops, soils, water, and climate that were copied from earlier manuscripts, from different times and different places. Generally, one should not read them as descriptions of contemporary practice to explain the archaeological record. The genre, however, developed over time. Syrian manuals of the late Mamluk and early Ottoman periods (15th and 16th centuries) began to incorporate quite detailed descriptions of what appear to be agrarian practice of the day. As the compilers/authors of these texts at times acknowledged in their writing, they often watched or interviewed the peasants themselves to better understand farming practice—asking them what they were doing and why. In interpreting ancient landscapes, these are the most useful texts, if, and only if, we can read them as historians and tease out what was contemporary from what was curated knowledge. These late medieval manuals were copied in larger quantities and special-ordered for sultanic libraries in Cairo and Istanbul, at a time of growing interest in agriculture. Cuttings and seedlings of fruit trees were exchanged along with the books (Shopov 2020). As to who read such manuals, it was probably not peasants, but a new class of landed elites, who administered endowed agricultural land or were taking up market farming for the first time. This included the clerics (*ulamā'*), as well as sultans.

For Ottoman Beit Mazmil we have a century-long series of tax registers (the *tapu defterleri*, or "TD") in Ottoman Turkish,[4] and a 300-year series of detailed court files (*sijillāt*) of the Jerusalem Sharia court, in Arabic. Read together, and studied as a series, we learn how administrative practice (documented in the *defters*) intersected with the rhythms of peasant daily life (as recorded in the *sijills*).[5]

The *tapu tahrir defterleri* were the Ottoman's survey registers of population and estimates of expected annual revenue. Population is presented as numbers of households (and single men living alone), and the distribution of revenues by administrative category. They give us invaluable information about taxable assets, demographic trends, and cultivation, on a general level. As these surveys were only done "as needed" (and by sultanic order), and the recording did not keep pace with the collection of data, the registers could represent data that was a year or two (or more) out-of-date (Singer 1994, 17). Surveys of Jerusalem were conducted periodically throughout the 16th century, and we have six registers that cover the years 932–934 H/1525–1526 CE to 1005 H/1596–1597 CE. This series of registers have been published in annotated volumes (al-Bakhīt and al-Sawariyyah 2005–2012). By the beginning of the 17th century these periodic land surveys had come to an end (Ze'evi 1996, 7). As a result, we are generally limited to the 16th century when studying the Ottoman's taxation of Jerusalem District.

The study of these registers is further challenged by two factors. The first is that, as is common with hand-written records, the spelling of some village names is irregular. Beit Mazmil, for example, has been read by the editor as Farīl, Farmīl, Farnal, Fāsīn, Farfīl, Fasīl, and Farīk in the Arabic transliterations. The م and ف, as well as ر and ز, in placenames can look the same, particularly in the small hand of the *siyāḳat* script used by Ottoman scribes for such record-keeping. Scribal error may account for this: not understanding what people were saying when they were asked for information, for example, or a misreading of the original

Mamluk documents on which they were based (Kumakura 2015). The second is related to the Ottoman's fiscal categories, which diversify over the course of the century. Olives and olive oil are taxed as separate commodities. All fruit-bearing trees, except for olive trees, are combined into a single taxable category. The tax officials did not distinguish between different kinds of wheat and different kinds of barley (although there is plenty of archaeological evidence that peasants did) (Walker *et al.* 2017). The marriage and market taxes are combined into one-line entries. There are many questions we cannot answer, then, on the basis of these registers when read on their own.

By far the richest textual corpus of the period are Jerusalem's Sharia court registers (*sijillāt*), which provide detailed information on anything brought to court for filing.[6] These include a wide range of document types: sales and leases, debt claims and foreclosures, receipts, *timar* agreements, complaints against neighbors, and so forth. The Jerusalem series spans essentially the entirety of Ottoman rule; that is, 388 years from 1529 to 1917 CE. There are 82 *sijill*s alone from Jerusalem in the 16th century, covering the years 1538 to 1600. They provide "the most complete historical archive record for any city in Bilād al-Shām during this century" (al-Sabagh 2014, 40).

The privatization of land under the Mamluks

Iqṭāʿ*s and* waqf

Like many, if not most, rural sites across Bilād al-Shām, Beit Mazmil does not appear in textual sources until the 14th century, with the rapid expansion of the endowment of agricultural lands under the Mamluks. For centuries, most of Jerusalem's farmland, as throughout the sultan's domains, was state land, the tax revenues of which were distributed mainly to military officers as temporary grants for their military service (*iqṭāʿ*s). Technically, they could not be bequeathed, as they were not private property, and could not be sold or leased. In the late 14th century, the cash-strapped Mamluk state began to sell off this land, which was originally purchased by its own military and civilian officers, clerics, and merchants, and converted to religious endowments (in part as a form of tax shelter). A century later, rural elites began to do the same. Medieval Muslim endowments—in which the legal tenure over income-producing property is donated for a charitable purpose in perpetuity—left a paper trail. The donation had to be registered in court, signed by witnesses, and that property and the management of the institution it financially supported described in detail. With the privatization of former state lands from the 14th century, which included the liquidization of *iqṭāʿ* lands, much agricultural land was converted into *awqāf*, largely managed by the religious elite, and generating for them vast incomes.

A landscape of hilltop quṣūr

The relationship between land tenure, settlement typologies, the spatial organization of field systems, and landscape transformation has been acknowledged by archaeologists working in different parts of the globe and specializing in different periods of human history (Palmer 1999; Rippon *et al.* 2006; Wilkinson 2010). The Mamluk *iqṭāʿ* and endowment systems created in southern Bilād al-Shām, and perhaps in other regions of the sultans' realms, a landscape of castle-like estates, walled gardens, and terraced fields that would leave a permanent imprint in the highland regions until modern times. What began as *amiral* agricultural estates, based on *iqṭāʿāt*, developed over time into endowed orchards and gardens producing for the market. These could be readily recognized by a particular typology of architecture and field systems defined by agricultural terraces and plot walls.

The legal and financial status of the village of Beit Mazmil, and its farmland, stood at the border between *iqṭāʿ* and *waqf*, serving the interests of the political and religious elites alike. The village was surrounded by endowments in this period, and it was eventually absorbed by one in Hebron. Additionally, while there are no textual references confirming its status at any point in the Mamluk period as an *iqṭāʿ*, there is some archaeological evidence that it may have been so.

The Jerusalem-based historian Mujīr al-Dīn al-ʿUlaymī describes the rural outskirts and suburbs of his hometown in the 15th century in the following manner: "As for the orchards outside the city: the city is surrounded by them in every direction, and in them are all kinds of fruit, but mostly grapes . . . And in most (of the orchards) one can find '*quṣūr*,' built on the order of the state" (al-ʿUlaymī 1999: 81–83). The same term was used to describe suburban residences with walled orchards that were a familiar sight outside Damascus in the same period. "Qaṣr Abi Shāmah" and "Qaṣr Abi Riyās," located south of the walled city, and the vast zone of orchards located to the north and called "Qaṣr al-Labbād," were all part of the endowment for the Umayyad Mosque (Eychenne *et al.* 2018). The suburban landscape of Jerusalem in the Mamluk period must have been punctuated by such hilltop residences, perched atop terraced fields, and interspersed with a series of watchtowers of various functions.[7]

There is an assumption on the part of archaeologists that these *quṣūr* comprised the built environment of amiral estates (*iqṭaʿāt*), though they have had, until now, little evidence to support it. The officers to whom were assigned the tax revenues of these estates visited them on occasion (such as during harvests) and may have resided in them for short periods in the summer, but most did not live there year-round; military officers were otherwise resident in garrisons or urban houses or travelling on military campaigns. On the other hand, the historical sources lead us to believe that their estate managers (*wakīls*), many recruited from the local population, may have lived there, and had the most direct contact with local peasants and village sheikhs (Walker *et al.* 2017).

As noted in Chapter 5, hilltop residences called *quṣūr* became one of the most familiar components of the highland landscapes of Ottoman Palestine and Transjordan. Some of these structures may have developed out of the Mamluk *quṣūr*, as their standing remains sit on top of, and in many places reuse, components of earlier structures. The term generally became synonymous with a particular kind of landscape and terraced agriculture. From an archaeological perspective, the *quṣūr* are a key to understanding the way rural life and land-use developed in Jerusalem's hinterland in the Mamluk and Ottoman periods.

Khirbet Beit Mazmīl in the Mamluk period was more than a farmstead. The site appears to have been an amiral estate, a *qaṣr*. We have no parallels for such estates in this period, and outside of al-Ulaymī's account, no historical descriptions of what they may have looked like or where they were located. Nonetheless, the material culture of the site—equestrian accoutrements, weaponry (daggers, stone ballistae), and ceramic imports (some from Egypt)—and its location (a walled hilltop complex amidst orchards) makes it a likely option. The recovery of many sherds, and one complete bowl, of a special kind of glazed relief ware is key to understanding how this site may have functioned in this period (Figure 1). The molded decoration of the bowls included registers of Arabic inscriptions and, in some cases, heraldic designs.

Figure 1 Glazed Relief Ware bowl from the Southern Sector. The inscription is a variant of the formula "*al-ʿizz wa iqbāl*" (courtesy of Benjamin Dolinka and the IAA).

The inscriptions on this kind of ware are formulaic and replicate phrases used in the "registration of office" (*tawqīʿ*), when a Mamluk amir was promoted.[8] According to these certificates, in return for their *iqṭāʿ*s and their military ranks, the amirs were to provide "law and order" in the village, often intervening in communal disputes and responding to petitions against state officials. The state also expected them to guarantee the productivity of the income-generating properties entrusted to them, by renewing once-neglected lands and investing in critical infrastructures. The certificates also list the titles and salutations that were to be used for provincial and district governors in official correspondence (and possibly also in public ceremonies). They include such phrases as "may his name and rank be displayed throughout the province: 'everlasting glory and prosperity'" (*al-marqūm bi-ismihi wa-rasmihi ʿalaa arjāʾ al-wilāyāti: ʿizzu yadūmu wa iqbālu'*), and "and praise for his effort will bring 'achievement of hopes and attainment of dignity'" (*bulūghu-l-āmāli wa-idrāku-l-muḥābi*). They express more than well wishes; the use of the inscription may have been a privilege of office. Glazed relief wares bearing these kinds of inscriptions and blazons had a very limited distribution: southern Syria, from the Mediterranean coast to central Jordan, according to published excavation reports. Most of the archaeological sites are known from chronicles and *waqf* documents as military garrisons, administrative centers, and the private land or endowments of amirs—all places where officers spent some time or had financial interests, such as agricultural estates (the "country" residences of the Mamluk elite), replete with baths and gardens, as well as sugar mills. Amirs and governors regularly visited the mills during the harvest and often stayed there for several months. To my knowledge, glazed relief ware has not been found anywhere in Egypt (Walker 2023b).

We know little about the daily activities of the people who lived on these estates or how they were managed. Mamluk historians have long assumed that the officers who were granted the *iqṭāʿ*s took little active interest in them and delegated their day-to-day running to estate managers. However, recent studies of administrative practice in Ayyubid and Mamluk Egypt have provided evidence of a more active involvement of *iqṭāʿ*-holders in their estates, suggesting that they also spent long periods of time in residence there. The offices (*diwāns*) of the amirs collecting revenues from these estates were likely physically located on those estates, generating their own paperwork and keeping something akin to archives (Hirschler 2016). Documents of the period related to the management of *iqṭāʿ*s by lower-ranking amirs gives us a glimpse of the role these officers played in settling local disputes and administering their assets (Livingston 2020). These documents, kept in the Austrian Library of Vienna and still covered in dirt, may have originated in archaeological sites in the Fayyum, perhaps from an estate (Walker 2023a). Developing the estates was in the financial interest of the amirs, and their relationship to them may be closer than once imagined. For example, a trusted officer of the Sultan al-Nāṣir Muḥammad had a Sufi college (a *zawiyyah*) built in the 1330s on his estate, which he called "his village" (Behrens-Abouseif 1985, 62, 342, notes 18, 19; Little 1979, 398). He ordered a special minaret for the building, and this required the work of a foreign architect.

Al-ʿUlaymī's account suggests that the hilltop *quṣūr* of the Jerusalem area were residential, at least for part of the year. If they were, indeed, the rural residences of lower-ranking amirs (and their staff), then these complexes would have served a host of other functions, as described in the investiture documents. Their *diwāns*, which we suspect operated in the *quṣūr* described by al-ʿUlaymī, may have been the instruments for managing agricultural estates. In this way, these hilltop buildings would have been the locus of nascent agro-businesses (Walker 2023a).

A landscape of endowments and terraces

Khirbet Beit Mazmīl in the Mamluk period was likely one of these hilltop *quṣūr*, and it was surrounded by endowed land. It appears for the first time in written sources in a *waqfiyyāt* dated to 720 H/1320 CE. The manuscript is a later copy in Istanbul of a 13th century original (Massignon 1953, 92, note 1).[9] According to the text, Jerusalem's farmland had suffered from many years of neglect. The endowment of the lands of larger Ein Karem by the grandson of Shuʿayb Abū Madyān al-Maghribī—the well-known mystic who died in Tlemcen (Algeria) in 594 H/1197 CE—was meant to promote revival of these lands to create revenues for a Maghribi *zāwiyyah* in Jerusalem. The land included unworked soil and uncultivated lands; neglected fields; buildings (some of which were vacant); and gardens, fruit orchards, and olive trees irrigated with local spring water (Weigert 1990). A place called "Beit Mazmil" (though not actually called a "village" in the document) marked the easternmost border of this endowed area (Massignon 1953, 41), but was apparently not part of the endowment. The other borders were Qalunyā (to the north), al-Malḥa (to the south), and ʿEin Shaqīq (to the west).

Although the farmland earmarked in this document technically belonged to the Maghribi institution, it was leased and released many times, and sold off through extrajudicial means in the centuries following, so that by the 17th century increasingly more land was alienated from the original endowment (Massignon 1951). As with many other endowments, its legal status was disputed over the course of Ottoman rule and during the period of the British Mandate, as documented clearly by a series of decrees and legal complaints (Massignon 1953). The lands of Beit Mazmil also fell under the control of a religious endowment, in this case for the financial support of al-Ḥaram al-Ibrāhīmī Mosque in Hebron (al-Khatīb 2013, 119), at a time when Mamluk sultans and officials were actively financing religious institutions there (Frenkel 2014, 20).[10] The Ottoman state recognized the claims of the Hebron endowment, and taxed it; it appears as a line entry in the tax records of the 16th century. The Ottomans also maintained the Maghribi endowment, assigning the revenues of the village of Ein Karem, and neighboring lands, to it (Singer 1994, 71–74).[11] Legal challenges to the rights to endowment revenues aside, the Ein Karem region appears to have been prospering and agriculturally productive. By the 17th century, this region of Jerusalem's western hinterland was filled with countless vineyards; olive groves; and orchards of figs, apples, quince, lemons, carobs, and pomegranates. The Maghribi endowment clearly had achieved its purpose: the terraced lands had been revived, and Jerusalem's countryside was flourishing.

While the Mamluk regime sought to control revenues, neither it nor *waqf* managers interfered in cultivation. The decision on what to plant, how to enrich soils and share water, and the protocol for rotating fields among different families in lands traditionally held communally (*mushāʿ*) were generally left to the peasants themselves. The state relied on their know-how to best manage natural resources and maintain the productivity of agricultural land (Rapoport 2018). Why peasants chose to plant the way they did was a balance of need and greed, always fluctuating between subsistence and market agriculture. The documentary sources do not allow us to distinguish among the wide range of crops cultivated at the time: different kinds of wheats and barleys (sometimes planting them in the same fields), a diverse range of fruits and nuts, and specialization in vegetable production, all supplemented with animal husbandry. Botanical analysis from domestic contexts has opened new windows on peasant decision-making in this regard (Walker *et al.* 2017). The building of terraces, as well, appears to have been a choice made by the peasants themselves. The backbreaking work in the construction and maintenance had long-term goals in mind (Walker *et al.* 2020).

Though they literally covered the highlands of central Palestine—they were, in fact, the most distinguishing characteristic of the local landscape—agricultural terraces are virtually invisible in the textual record. Perhaps because they were so familiar and commonplace, the geographers of the time felt no need to describe them, and neither did travelers to the region.[12] As they were not a taxed commodity, they were not included in tax registers. Nor did they appear to attract the attention of the authors of the *filāḥah* manuals. The manuals describe soil constructions and the placing of stones in pits when planting individual trees (al-Nābulūsī 1882, 38),[13] but instructions on how to build or repair a terrace wall are not to be found, even though this should have been a critical skill for the gardeners that read such texts. The terms *ḥālūl*, *ḥabūl* (pl. *ḥabā'il*), *al-mudarrijah*, *maṣṭabah* (pl. *maṣāṭib*) and *judrān*, however, do occasionally appear in *waqfiyyāt* and *fatwa* manuals from Mamluk and Ottoman Syria.[14] Like cisterns, which were built by local initiative and locally owned, terraces are part of the human development of the landscape that raised the value of the land, though they do not always belong to the endowment.[15] In the *waqfiyyāt* of the Umayyad Mosque in Damascus—a Mamluk-era endowment text preserved in an Ottoman-era document—terraces (*ḥālūl*) merely form the boundary of the endowed property.[16] As the legal status of terraces, and the responsibility for their maintenance, was never clear, they became a flashpoint of conflict in rural communities. According to the 14th century scholar al-Subkī, agricultural terraces on *waqf* land presented a particular legal challenge. Terraced fields, in a legal sense, were comprised of three different kinds of property: the soil in them, the plants cultivated on them, and the terrace walls themselves. It seems from the sequence of question-and-answer in al-Subkī's manual that, in popular understanding, only the original fruit trees, soils, and wall belonged to the endowment, and that during seasonal maintenance any new soils brought in, and new trees planted, did not (Walker 2022, 348–349). For al-Subkī, however, the endowment had full legal claim to the entire terraced system (soils-plants-terrace wall), from its original form through all its seasonal repairs (al-Subkī, *Fatāwā* 1: 487).[17]

Perhaps the greatest impact the Mamluks had on rural Syria, then, was the *waqfization* of the countryside. It drove important investments in land development and gave an impetus for the exchange of agrarian knowledge. Plot walls, for example, physically marked out the borders of endowed lands. Terracing, on the other hand, brought more land in hilly zones under cultivation. Terracing could have, moreover, spread with grafting, as both supported the cultivation of fruit trees. Knowledge about grafting and buying cuttings and seedlings went hand-in-hand, as documented in the *filāḥah* manuals. They were practices adopted by a new group of landowners that emerged with the privatization and endowment of farmland from the later decades of Mamluk rule. The leasing of endowed land, which began in late Mamluk times, and the subdivision of lands within the family under Islamic law may have necessitated the further building of field walls.[18] Under the Mamluks, the Jerusalem highlands were beginning to literally form a landscape of endowments. We should understand the development of terracing as the physical expression of changes in land tenure, which are particular to the economic systems of the day.

Market gardening under the Ottomans

The market gardening that was supported by privatization of land, endowments, and terrace-building during the Mamluk era grew exponentially under the Ottomans. Cuttings of fruit trees were in demand, and *filāḥah* manuals were being copied at an increased rate and translated into other languages. The status of endowed farmland fell under scrutiny, as it was among the most profitable. Because so much land in Syria and Egypt was tied up in endowments, which were tax-free under the Mamluks, the Ottomans began, soon after the conquest of this region, to tax endowments at the rate of the tithe (10 percent). A rural real estate market developed with the leasing of endowed lands in the 17th century and the eventual replacement of the *timar* system (the Ottoman version of the Mamluk *iqṭā*ʿ system) with *iltizām* (tax-farming) by the 18th century.[19] Much of the land in the immediate vicinity of Jerusalem was cultivated as orchards, olive groves, and vineyards. Although some of these went up for sale, it was their produce, as well as the harvest from cereal fields, that were the object of active buying-and-selling in shares from the 17th century, a form of market exchange made possible by the emergence of new sources of credit from the first century of Ottoman rule. In the 16th century, credit became available even to peasants through a new financial practice called the "cash *waqf*," which was an innovative use of the endowment system to loan money (al-Sabagh 2014). Tribal sheikhs eventually came to lend money to peasants, as well.

A tax perspective (Figure 2)

Our main textual sources that allow us to trace these developments are tax and court registers. The Beit Mazmil entries in the extant registers appear in Figure 2; we will summarize the most important details here. Demographically, the entirely Muslim village grew significantly over the course of the 16th century. With only three households, Beit Mazmil was a small village community during the early years of Ottoman control. It had grown tenfold by the end of the century, and its lands appear to have been devoted to grain cultivation. It is important to note, however, that the village was never "empty," not even in the decades immediately following the Ottoman conquest.[20] The archaeological site of Khirbet Beit Mazmīl, on the other hand, has a clear abandonment phase in the 16th century. The site, then, was only a part of the larger village, and some other portion of it must have continued to be occupied.

In the earliest register, revenues are set aside for one policeman and a bondsman, who is appointed to guarantee the work of the policeman. This was a time of some insecurity, when the Ottoman state had to rely on local forces to keep law and order, until its own security apparatus could be put in place. The listing of individual taxes becomes more detailed over the course of century in the *defters*, and we see the people of Beit Mazmil paying taxes on a comfortably diverse range of crops and animals: wheat and barley, orchards (seeming to increase in taxable importance over time), and goats. As for the allocation of the revenues, those of Beit Mazmil were divided between a *timariot* and the Hebron *waqf*.

In comparison to other villages in the District of Jerusalem, Beit Mazmil was small at the beginning of the 16th century and had reached a middling size at the century's end. Its taxable revenues were similar, existing in the mid-range of what other villages in the district produced. The size of its fields (10 *feddān*) could be described in the same manner. A few entries make the village stand out from many of the others. The first is the market tax (*bac-ı bazar*) that was levied on the village, reaching 100 *akçe*, a not-negligible amount. While in one entry it is included with the marriage tax, in one (TD 516) it is listed on its own. In other words, Beit Mazmil hosted a market. Another detail that would seem to raise the status of the village is that it had an imam, which means there would have been a congregational mosque in the village.

Furthermore, its relationship to other income-producing lands suggests the village had a more complex occupational history and that it could have occupied a higher status in a ranking of village importance. Two isolated fields (*mazrāʿahs*) are listed as "belonging" to Beit Mazmil. The first is Wādī al-Badawīyyah, and the registers mention that it was located near Beit Mazmil. The revenues of 1,400 *akçe* were assigned to it, as part of the sultanic *has* (*khāṣṣ* in Arabic—the largest grant). As the name suggests, Ṭarish Laymūn was probably a lemon orchard. It is frequently listed after Wādī al-Badawīyya, and "belonged to" Beit Mazmil. For the entirety of the 16th century, this orchard was taxed at the rate of *kharāj*, the rate assigned to fruit trees and a tax paid in cash.[21] The revenues of 700–800 *akçe* were assigned as *has shāhī* (a sultanic estate).

ʿAyn Jāwud was a *mazraʿah* located close to Beit Mazmil, and apparently cultivated by peasants there, but did not belong administratively to the village. The nomenclature used for ʿAyn Jāwud is inconsistent, the different terms seemingly

T.D.#	DATE	POPULATION	WHEAT tax	BARLEY tax	TREES tax	SH/G tax	MARKET tax	*TIMAR/ ZIʿĀMAT/ KHĀṢṢ/*	*WAQF*	Notes
427	932–4/1525–6–1525/6	3 households 3 bachelors (1 guard, 1 bondsman, incl.)						(*timar*) unnamed—8 shares = 833	Hebron—16 shares = 1677	total tax (in kind) 2500
131	932–8/ 1525–1531–2							(*timar*) Murād (935/1528) = 2533		
1015 Farīl	945/1538–9	7 households 1 imām	2.5 sacks = 1200	2 sacks = 520	550			(*khāṣṣ* Amīr al-Liwāʿ) 8 shares of goats = 932	Hebron—12 shares = 1666	
289	961/1553–4									Beit Mazmīl left out of the register
516	970/1562	34 households	11 sacks = 5280	5 sacks = 1300	195	255	85	(*timar*) Tawkhatu Bey-8 shares = 2370	Hebron—16 shares = 4750	10 *feddān* taxed land at 7155 total; Wadi al-Badawiyya is its *mazrāʿah* (1400), a sultanic *khāṣṣ*
515	1005/1596–7	31 households	5 sacks = 2500	10 sacks = 2800	1800	300	100	(*ziʿāmat*) Ibrahim Mīrālay of Jerusalem; (*timar*) ʿĀli and Qayṭās Amīr/Mīr al-ʿIlm = 8 shares = 2500		10 *feddān* taxed land at 7500 total; Ṭārish Laymūn belongs to it (800) as Sultan's *khāṣṣ*

Figure 2 Chart of *tapu defter* entries mentioning Beit Mazmīl.

reflecting the uncertainty of the official doing the survey and not a change in the tax status of the place over time. In the 1520s (TD 427) it was listed as an abandoned village (al-Bakhīt and al-Sawarīyya 2005–2012, 24); in other registers it appears as a village (for example TD 131, entry #746—al-Bakhīt and al-Sawarīyya 2007, 338).[22] By the 1530s it is consistently referred to as a *mazraʿah*. In 1553–1554 (TD 289, entry #349) and again in 1596–1597 (TD 515, entry #309) it is registered as a *mazraʿah* of Beit Safāfa (al-Bakhīt and al-Sawarīyya 2010, 96; al-Bakhīt and al-Sawarīyya 2005–2012, 104). TD 515 (1596–1597), on the other hand, specifically mentions its location as a *mazraʿah* near Beit Mazmil, without saying it belonged to that village (al-Bakhīt and al-Sawarīyya 2005–2012, 308). Earlier, in the 15th century, part of its revenues were set aside as *waqf* for a Quran reader by a high-ranking Mamluk officer by the name of Ibn Muzhar.[23] The Ottomans maintained this endowment (as 16 shares worth 1400 *akçe*), but assigned the remainder (8 shares worth 700 *akçe*) as a *timar*. As for what this land produced, the registers mention only dry measure and the tithe (10 percent tax usually paid on grains). Moreover, by the end of the century, it was given the right to pasture animals and paid a tax on sheep and goats. According to the tax registers, this was not an orchard, but land devoted to cereal production and pasturage. On the other hand, the court registers document olive cultivation there in the same period, as we will see below.

Unfortunately, the size of these plots of land is not provided in these registers, and we cannot estimate the scale of production on the basis of the data provided in them. What was important was the estimated revenues from the annual fruit yield. There seems to be a correlation between orchards and *mazraʿahs*. The fiscal status of these lands alludes to demographic change here and in Beit Mazmil over time. While there is some scholarly debate about the meaning and significance of the term, a *mazrāʿah* is generally understood as land that continued to be cultivated after the people who used to live there left for some other place. Beit Mazmil may have absorbed some of the peasants who originally lived in Wādī al-Badawīyyah and Ṭārish Laymūn and maintained their old orchards there. However, in reading the *defters*, one could argue that some were simply orchards fiscally separated from the supporting villages but administered as part of them. They suggest a specialization of cultivation by village. The *defters* list hundreds of *mazraʿahs*, which suggests these were a fiscal category. In the 1530s (TD 1015) Jerusalem District had 263 of them, and Hebron District 245, which is twice the number of registered villages.[24] The village of Beit Mazmil was taxed primarily in cereals, with additional revenues from the raising of sheep and goat and a market tax; its *mazraʿahs*, however, were orchards. While this phenomenon needs to be systematically studied across the registers, comparing villages to one-another, a preliminary reading suggests some level of specialization in cultivation by village. For Beit Mazmil, barley and fruits were the highest-revenue local products. The botanical record from the site documents the same pattern of cultivation (see Chapters 7.2 and 7.2). On the basis of the excavations, the botanical record, and the tax and court registers, we cannot say what accounts for this apparent pattern of specialization across the region. One needs to study the registers as a whole and compare villages. Equally important is further fieldwork, to determine if topography, soil quality, and water resources determined a spatial patterning of cultivation.[25] If they do not, then the decision to crop in a particular manner would have been market-driven and could have been coordinated in a collective manner.

A legal perspective (Figure 3)

In the 400 Jerusalem *sijillāt* there are some 1500 entries alone that make reference to Beit Mazmil, with information related to the village, the land attached to it, or the people living there.[26] Covering a documentary series so vast must be tabled to its own monograph. Summarized in Appendix 7 are representative entries contemporary to the Early Ottoman period of occupation at the site of Khirbet Beit Mazmīl. As with the *defters*, it appears that the village name was spelled different ways. For the purposes of this study, we use only examples where the placename can be read with some confidence as "Beit Mazmil." Where summaries are provided, as in most of the IRCICA publications, we have checked them against the digitized manuscript. There are some misreadings in the summaries: what is provided in Appendix 7 are the corrected versions.

Important trends related to land tenure, cultivation, and social networks emerge from these entries, which largely relate to leasing contracts, debts, and purchases of harvest. The *sijills*, by and large, confirm the picture of specialization in cultivation by village that is suggested by the *defters*. They do not, however, give us information on the scale of production. Cereals remain the foundation of the cultivation strategy in Beit Mazmil, but

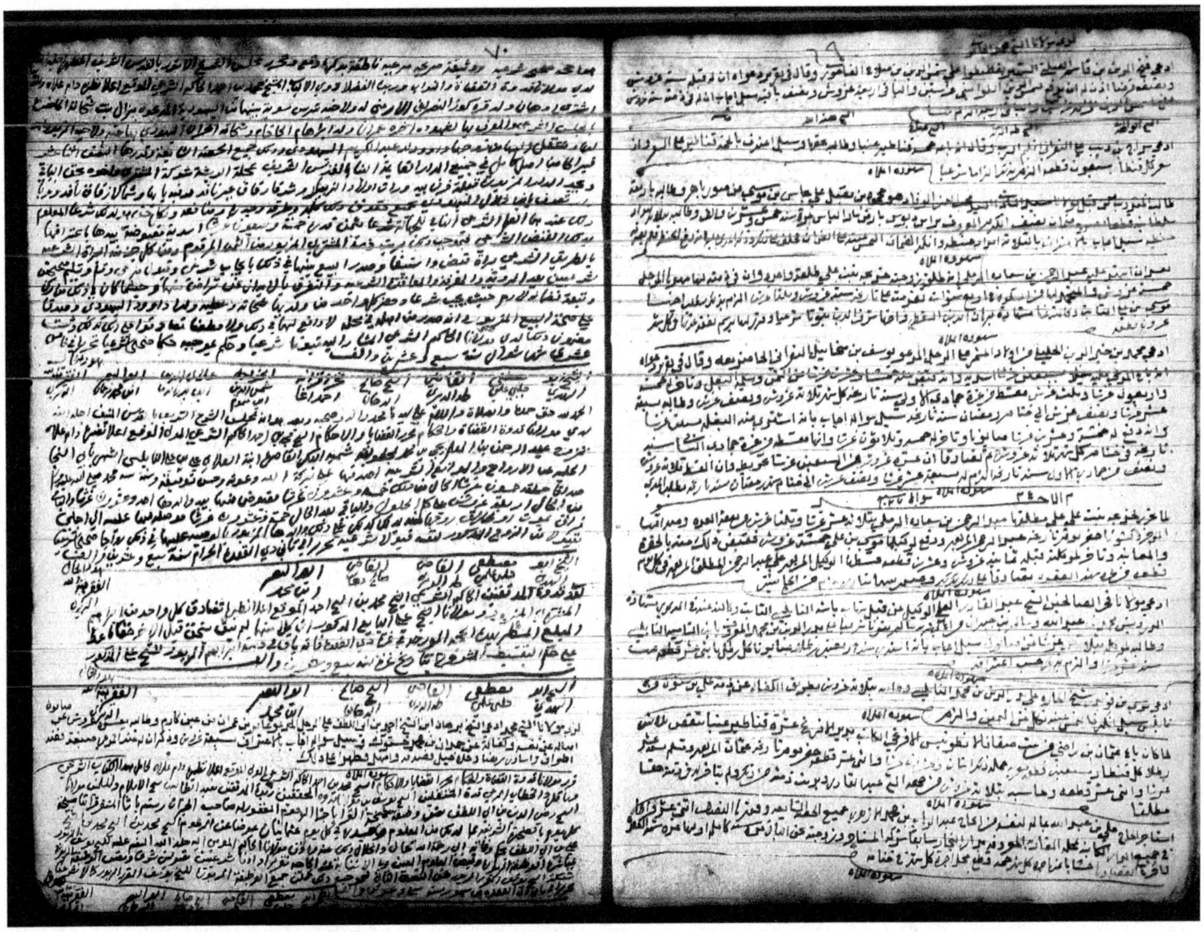

Figure 3 Page of Jerusalem *sijill* 101, 1/122, p. 70 (1027 H/1618 CE): a demand for payment of fruit from Wādī al-Badawiyya (courtesy of the IRCACA).

many of its residents maintain orchards in nearby *mazraʿahs*, or what we might call "satellite farms," and had the right to sell their harvests. In the court registers these are ʿEin Jāwat/Jawūd,[27] for grapes, figs, olives, and flowers, and Wādī al-Badawiyya, for apples, figs, and grapes. Al-Malḥa, the closest village to the south of Beit Mazmil, had vineyards. The kinds of crops grown at each village remain consistent over time. People from Beit Mazmil were actively involved in cultivation and buying and selling crops in these places. Overall, one gets the impression that local peasants had good knowledge about the agriculture in their district and knew what was going up for sale and when.[28]

Crops were not the only commodities bought and sold. We read of building plaster and donkeys, camels, and cattle, all of which have left a trace in the archaeological record. The Early Ottoman buildings at Khirbet Beit Mazmīl had heavily plastered floors and walls, which represents a local building style. Cattle appear in the faunal records in a late phase of occupation at the site (see Chapter 7.3), when its marrow may have been consumed.

The registers do not only provide information on agriculture. They are rich in details about local society and social networks. Several men from Beit Mazmil are named as village representatives in court and at the threshing field, where the harvest division of cereals with the *timariot* took place.[29] Most of the court cases were related to disputes between the Hebron *waqf* manager and the *timariot*, both of whom had claims to the revenues from Beit Mazmil. *Timar* contracts and *muqata'ah* (tax farming) agreements with the endowment were often renegotiated, or confirmed, as a result. Occasionally a group of men from the village took collective action against *timariots* and endowment managers.[30] Some villagers were simply frequent visitors to the Jerusalem court, either because of debt, or social conflict, or to register a sale or purchase. One of the most visible in the register for this period was one Su'ayfan b. Yusūf, who, at different times: filed a complaint against a man from al-Malḥa, who threw a rock at him; stood as a loan guarantor for a man from Deir Yāsīn, who was buying a cow; and was sued for not having paid for fruit he bought in Wādī al-Badawiyya. As for the harvest division, villagers were required to be at the threshing floor to vouch for the amount of the harvest, and make sure they delivered the *timariot's* share. Among these regular attendees was one Aḥmad b. Ḥasan, referred to as "the eminent farmer of Beit Mazmil" (Rec. 36 in Appendix 7).

The business of buying and selling crops, as well as leasing land, brought the people of Beit Mamzil into regular contact with absentee landowners, businessmen (and women), and judges in Jerusalem. Their business dealings also took them further afield. In one entry, dating to 972 H/1564 CE, al-Ḥājj b. Khalīl al-Salṭī demanded money back from a man in Beit Mazmil ('Abd al-Nabī b. Aḥmad) for having paid a security for him on a loan (Rec. 46 in Appendix 7). If we are to understand the *niṣba* "al-Salṭī" as referring to a place of origin, the plaintiff either lived in the town of Salṭ, in Transjordan, or his family hailed from there.

Beit Mazmil also had ties to the Egyptian community in Jerusalem. A Coptic woman by the name of Ḥana bint Mikhael al-Naṣrānī al-Qibṭī al-Najjār, a resident of Jerusalem, was particularly active in buying local produce, and on a large scale. She bought all the crops of 'Ein Jāwut in 1006 H/1597 CE from one Jābir b. Khalīfeh from Beit Mamzil for four Egyptian *sulṭānīs*, a gold coin (Rec. 78 in Appendix 7). A few months earlier, in 1005 H/1597 CE, Sa'īd b. Ramadhān, also from Beit Mazmil, sold her grapes, figs and olives from the same *mazra'ah* for 12 Egyptian *sulṭānīs* (Line 42 in Appendix 7). Not included in the Appendix are other references to what are clearly her business interests in the village and its associated farms. In 1005 H/ 1597 CE she also purchased from Sa'īd b. Ramadhān (as above) 12 *qirāṭ* of grapes, figs, and olives from the same *mazra'ah*, Ein Jāwut. The price paid was two Egyptian *sulṭānīs* (*sijill* 78, doc. 3, in Rabāyi'ah vol. 7, 158). The following year she bought the entire harvest of olives from Beit Mazmil from Jābir b. Khalīfa (as above) for four Egyptian *sulṭānīs* (*sijill* 78, doc. 3, in Rabāyi'ah vol. 7, 336).

If we explore registers dated to the later 17th and 18th centuries, we can identify efforts to further develop the region's agricultural land and signs of growth. In the 1660s some of Beit Mamzil's land was uncultivated/in ruins (*khirbah*). The manager (*mutawwalī*) of the Hebron endowment leased to the *sipāhī* of Jerusalem those lands in 1077 H/1667 CE, for a period of 90 years at the bargain annual rent of four *qursh* (copper coins), arguably to put the land "under the plow" and make it profitable in the future (*sijill* 167, doc. 1—Rabāyi'ah vol. 8, 166). For the 18th century, I cite one entry from 1114 H/1702 CE, one year before the Naqīb al-Ashrāf revolt. In that year, a man from the village of Beit Mazmil (Shāhīn Salāḥ al-Dīn al-Quṣāṣ) sold the olive harvest from 'Ein Jawūt to a man from Ein Karem (Sālim b. Suleymān) for 100 silver coins (*sijill* 201, doc. 2—in Rabāyi'ah vol. 11, p. 263). The value of the crops from this *mazra'ah* continued to rise, and demand stayed high.

Prosperity in times of abandonment?

The *sijills* continue to record economic activity at Beit Mamzil, and legal action involving people from there or living there, throughout its series. In other words, Beit Mazmil was an economically and socially viable place long after the archaeologically documented abandonment of the site. In this sense, there is a disconnect between the textual record and the archaeological one. Two factors could account for this. The first is that the archaeological site was only one part of the village. Any other buildings and housing complexes that once stood on this hill and its slopes have long since disappeared, their remains buried under modern concrete. Therefore, we can only reconstruct part of what must have been a rich occupational history from the archaeological ruins that were excavated between 2011 and 2019. Alternatively, the residents of the Beit Mazmil did, indeed, leave the village for a time, relocating to another village nearby, but continued to cultivate their old lands. Peasant flight pushed Ottoman officials to desperate measures to keep peasants in their villages and on their land, and forcefully return those that left (Singer 1996).

On a social level, the movement of peasants from one village to another, and often back again, was commonplace. It has been argued that peasant migration did not have a demonstrable impact on the overall regional economy, as lands continued to be cultivated, and the contraction of one village simply meant the growth of a nearby one (Marom 2021; Marom 2022; Marom and Taxel 2023; Marom *et al.* 2023). These kinds of "fluid landscapes" have also been identified in contemporary Greece, where peasant migration created a complex diversity of settlement forms, fields without villages were not exceptional, and tightly knit rural social networks lent cohesiveness to what appears, from our perspective, to be a patchwork landscape of settled and abandoned villages and fragmented fields (Sutton 2000).

The archaeological history of Beit Mazmil, then, is best interrogated from this larger landscape perspective; the fragments of the settlement studied in conjunction with the cultivated plots we can associate with the place textually and know were contemporary with the site through OSL dating of the agricultural terrace walls. It should also be examined as part of a network of social ties and economic exchange that bound one village to another in Jerusalem District.

Some patterns of cultivation practices and landscape development, reflecting the choices peasants living in Beit Mazmil made about their land, can be partially reconstructed, and on a general level, through the combination of archaeological, textual, and botanical approaches—with tight chronological control provided by OSL dating. The transformation of Jerusalem's farmland over the course of Mamluk and Ottoman rule—from primarily cereal to garden cultivation, from state to private tenure, and from subsistence farming to market cultivation—can literally be traced "on the ground" in the rapid expansion of terracing. The construction of plot walls appears to have been tied directly to the *waqf* system, delineating the lands of one endowment from another. Beit Mazmil was bordered on the west by the lands of the Maghribi endowment, for example, and that of Sheikh Jaljīl, and between one third and one half of its own lands were controlled by the Hebron *waqf*. Such endowments came to control a large part of the farmland outside Jerusalem, and by the 17th century increasingly more of it was leased. These changes in land tenure facilitated the shift to market gardening and may explain the "green belts" of villages that encircled Jerusalem from the Mamluk period, and which have been documented archaeologically (Yeger 2017).[31] The village of Beit Mazmil was a central participant in the long-term "greening" of Jerusalem's immediate hinterland.

Landscape archaeology, however, is not best-positioned to give us details on rural networks: its participants, the nature and intensity of exchange, and the nature of the social ties between villages. The 300 years' series of court registers can do this, however, but such a study is well beyond the scope of the current monograph. A preliminary survey of the registers for the 16th and 17th centuries, however, does suggest patterns that cannot be reconstructed through archaeological methods alone. Beit Mazmil, unsurprisingly, had close ties (and conflicts) with its immediate neighbors: namely, the villages of al-Malḥa and Ein Karem. It also maintained cultivation of and paid taxes on orchards in the vicinity, which were technically not village land, such as Wādī al-Badawīyya, ʿAyn Jāwud, and Ṭarish Laymūn. According to the tax and court registers, some of these gardens were once villages in their own right; their inhabitants likely relocated to Beit Mazmil for reasons that cannot be reconstructed from the sources. The village regularly sent representatives to the court in Jerusalem to represent their interests, and to buy and sell produce, shares of harvests, and leases of land. It had an imam and a marketplace. Beit Mazmil appears to have been a local center for social networking and business. This movement of people and goods continued after the site we excavated was abandoned, for a time, in the early part of the 18th century.

Notes

1. Though the site was clearly settled in the Abbasid period, it does not appear in the most important geographic account of the time for this region (al-Muqaddasī 2002).
2. This is the goal of the project "TERRSOC: 'Reading' Ancient Landscapes: Peasant Decision-Making and Terraced Agriculture in Central Palestine over *la Longue Durée*," funded by the German Research Foundation (project number 493699283), and will be covered in the author's forthcoming monograph on peasant society in Mamluk Syria.
3. An excellent resource on medieval Arabic *filāḥa* manuals is the Filāḥa Texts Project, found online at: www.filaha.org.
4. The Ottoman Turkish manuscripts have been published in print form and supplemented with Arabic commentaries by the Jordanian historian Muhammad al-Bakhīt (al-Bakhīt and al-Sawariyyah 2005–2012). We use these volumes for the current study.
5. We build here on the groundbreaking work of A. Singer (1994), who was the first to combine Ottoman tax and court registers to write a history of peasant-state relations in 16th-century Jerusalem.
6. These documents have never been published in edited form or translated. They are found in dispersed collections. The most important is that of the Endowments Ministry (*Wizārat al-Awqāf*) in Abū Dīs, Jerusalem. A full series is also found in the archives of Istanbul: 346 extant Jerusalem *sijill*s covering the years 936 H/1529 CE–1280 H/1863 CE can be studied at ISAM (the Turkish Religious Foundation's Islamic Research Center) (al-Sabagh 2014, 39). The IRCICA (Istanbul Research Center for Islamic Culture and Arts) has been publishing annotated catalogues of these, along with CD-ROMs of the digitized documents (Rabāyiʿah 2010–2018). Microfilms of part of the manuscript series can also be found at the Givat Ram branch of the National Library of Israel. Microfilms of the series can also be found in the Centre for Archives and Manuscripts at the University of Jordan; partial indexes and summaries have been published (al-Bakhīt *et al.* 1991). This chapter relies mostly on the digitized registers of the IRCICA.
7. On the function and layout of Jerusalem's Mamluk-era watchtowers (*mezadem*), see Yeger 2017.
8. These are collected in administrative manuals of the period. For an in-depth study of them and a full bibliography of the Arabic sources, see Walker 2023b.
9. To my knowledge, the original Arabic document has never been published. The studies of the text, which have been reproduced by Wiegert and Massignon, were published in Hebrew and French, respectively. According to Massignon, a copy of the document was sent to Jerusalem. The original must have been kept in Tlemcen.
10. This endowment should not be confused with that of Tamīm al-Dārī in Hebron, the validity of which was actively challenged in court in the 15th century, but upheld by the Ottomans (Frenkel 2014, 31).
11. According to the tax registers, the tax (tithe of 10 percent) on that endowment was assigned to a *timar*-holder.
12. European travelers to the region in the late 19th and early 20th century, though, did describe such terracing and took an interest in what role they played in local societies (Dalman 2013, vol. I/1, 5, 216–217).

13. This practice has been identified in our terrace excavations and has been explained as a technique to facilitate drainage (see Chapter 6).
14. The terms *ḥabalah* and *jisr* are used in Palestinian Arabic dialect today. The terms may also refer to plot walls, and this is not always clear from their use in the texts.
15. On the cisterns in Mamluk-era *waqfiyyāt*, see Walker (2017).
16. أسهم من أربعة وعشرين سهما 65 ومن ذلك جميع الحّصة الشائعة ومبلغها سّ وهي الربع شائعا ذلك من ستة
جميع أرض الجنينة لغير وشربها من الماء من نهر يزيد
ّبا الخراجية التي من جملة أراضي النيرب وتعرف بجنينة
العميان خلا غراسها فإنه مختص من ماصية البريكة ح
ّق معلوم وهو في كل النهار إلى الظهر حّدها من القبلة نهر
ثورة ومن الشرق ّل يومي خميس وجمعة من أوّ الباب
وتمامه جنينة ابن أبي العيش يفصل بينهما حائط ومن الشام
بستان البريكة ومن الغرب ملك لأربابه يفصل بينهما الحالول
(original Arabic text courtesy of M. Eychenne).
17. Similar debates existed in Jewish law (in the Mishneh) regarding ownership of terraces (s. *madrigah*, in Hebrew) and responsibility for repairing them. One concerns the collapse of a terrace: when one collapses, who owns the materials of the old terrace that falls downslope—the owners of the upper terrace or the lower one? I am grateful to R. Havilio for this observation.
18. Generally, however, this inheritance practice is expressed in the tax and court registers as shares of and revenues, rather than physical fragmentation of the land.
19. The Ottomans divided most of the land of Palestine into *timar*s (small estates), *zi'amet*s (a large *timar*), and *khāṣṣ* (the imperial share and the biggest estate). These were grants to military officers and high-level administrators of the taxable revenues, and not land assignments. The average *timar* consisted of one or several villages, was worth up to 20,000 *akçe*, and was generally assigned to a *sipāhī* (cavalryman). The *zi'amet* was worth 20,000–100,000 *akçe* and was assigned to higher ranking officers or officials. The largest "estate" was the *khāṣṣ*, valued at over 1,000,000 *akçe*, and distributed among district and provincial governors or senior officials of the provincial administration (Zeʿevi 1996, 117).
20. There is a tax category for an abandoned settlement: *khālī*.
21. Taxes on cereals were paid in kind and converted into cash in the brokers markets of Jerusalem; taxes on trees were paid in cash; peasants had to sell the produce to obtain the revenues to pay their taxes on fruit.
22. Its mapping coordinates are 35°05 N-S and 31°32 E-W (al-Bakhīt and al-Sawarīyya 2005–2012, 275, note 245).
23. The editor of the register, al-Bakhīt, provides biographical information on the donor, citing al-Ulaymī's chronicle on Jerusalem (*al-Uns al-Jalīl*). Zayn al-Dīn Muhammad b. Ahmad b, Muzhar al-Anṣārī al-Dimashqī was from Egypt and served at different times under the Mamluk sultan Qaytbay as *wālī kātib al-sirr* (head scribe), *nāzir al-jaysh* (Inspector of the Army), and head of the *Diwān al-Inshā* (Chancery) in Egypt (al-Bakhīt and al-Sawarīyya 2005–2012, 278, note 426). He also endowed a university (*madrasah*) in Damasus in 885 H/1480–1481 CE.
24. On the basis of numbers provided in the same register (TD 1015), the District of Jerusalem had 133 villages.
25. This work is being done as part of the ongoing project TERRSOC, which continues the work begun with the study of Beit Mazmil's terraced lands.
26. This is an estimate based on published and manuscript registers (R. Marom, *personal communication*).
27. There was no reference to fruit trees at ʿEin Jawūd in the *defters*. This may be due to a change in cultivation over time, or a confusion on the part of the surveyor about placename.
28. On peasant market knowledge in Palestine and Transjordan in the Mamluk and Ottoman periods, see Walker 2007 and Walker 2021.
29. Among them, in the 16th century, were Khalaf b. ʿAlī, Muḥammad b. Iyād, and ʿAlī b. Maṭār (Rec. 35 in Appendix 7).
30. Names mentioned in the registers in this regard in the 16th century include Radwān b. Saʿādeh, Ḥamdan b. Mūsā, Khatāb b. ʿUmayd, and Mirʿib b. Sāliḥ (Rec. 47 in Appendix 7).
31. At the time of writing this monograph, Yeger's MA thesis has produced the only comprehensive

catalogue of Mamluk sites around Jerusalem. It is based on the internal and published reports of the IAA and includes mostly salvage excavations.

References

Primary sources

al-Bakhīt, M.ʿA., N.R. Hmoud, and A.ʿA. Khuraysāt, eds.
1991. *Al-Quds al-Sharīf: Wathāʾiquhā wa-Sijillātuhā al-Muṣawwarah fī Markaz al-Wathāʾiq wa-l-Makhṭūṭāt fī Jāmiʿah al-Urduniyyah (936H/1529M 1404 H/1984 M).* Amman: Centre for Archives and Manuscripts, University of Jordan.

al-Bakhīt, M.ʿA. and N.R. al-Sawarīyyah, eds.
2005–2012. *The Detailed Defter of the Liwa' of Noble Jerusalem.* Vols. 1–6: Tapu Defterleri No. 427 (932–934 A.H./1525–26–1527–28), 131 (932–938/1525–1531–2), 1015 (945/1538–9), 289 (961/1553–4), 516 (976/1562), and 515 (1005/1596–7). Amman/London: al-Furqān.

Dalman, G.
2013. *Work and Customs in Palestine.* Volume I/1 and 2, translated by N. Abdulhadi-Sukhtian. Ramallah: Dar Al Nasher.

al-Khatīb, M.
2013. *al-Awqāf al-Islamīyah fī Filasṭīn fī-l-Aṣar al-Mamlūkī.* Irbid: Yarmouk University.

al-Muqaddasī, Shams al-Dīn Abū ʿAbdallah Muḥammad ibn Aḥmad ibn Abi Bakr
2002. *Aḥsan Taqāsīm fī Maʿrifat al-Aqālīm*, edited by Muḥammad Amīn al-Dunāwī. Beirut: Dār al-Kutub al-ʿIlmiyyah.

al-Nābulūsī, ʿAbd al-Ghanī ibn Ismāʿīl
1882. *Kitāb ʿalam al-milāḥah fī ʿilm al-filāḥah.* Damascus: Nahaj al-Suwāb.

al-Subkī, Taqī al-Dīn
n.d. *Fatāwā.* Beirut: Dār al-Maʿrifah.

Sijillāt Maḥkamat al-Quds al-Sharʿīyah
a) National Library of Israel, Givat Ram branch (microfilms).
b) Rabāyiʿah, I., ed. 12 vols. Istanbul: IRCICA, 2010–2018.

al-ʿUlaymī, Mujīr al-Dīn
1999. *al-Uns al-Jalīl bi-Tārīkh al-Quds wa-Khalīl*, edited by M.A. al-Kaʿābinat. Amman: Maktabat Dandīs.

Secondary sources

Behrens-Abouseif, D.
1985. *The Minarets of Cairo: Islamic Architecture from the Arab Conquest to the End of Ottoman Empire.* London/NY: I.B. Tauris.

Eychenne, M., A, Meier, and E. Vigouroux
2018. *Le waqf de la mosquée des Omeyyades de Damas: Le manuscrit ottoman d'un inventaire mamelouk établi en 816/1413.* Damascus: Institut Français du Proche-Orient.

Frenkel, Y.
2014. *Ḍaw' al-sārī li-maʿrifat Ḫabar Tamīm al-Dārī: On Tamīm al-Dārī and his Waqf in Hebron.* Leiden: Brill.

Hirschler, K.
2016. "From archive to archival practices: Rethinking the preservation of Mamluk administrative documents." *Journal of the American Oriental Society* 138.1: 21–26.

Kumakura, W.
2015. "Who handed over Mamluk land registers to the Ottomans? A study on the administrators of land records in the Late Mamluk period." *Mamluk Studies Review* 18: 279–298.

Little, D.P.
1979. "Notes on Aitamiš, a Mongol Mamlūk." In *Die islamische Welt zwischen Mittelalter und Neuzeit: Festschrift für Hans Robert Roemer zum 65. Geburtstag*, edited by U. Haarmann and P. Bachmann, 387–401. Wiesbaden: Franz Steiner.

Livingston, D.
2020. "The Paperwork of a Mamluk *Muqṭaʿ*: Documentary life cycles: Archival spaces, and the importance of documents lying around." *Al-ʿUṣūr al-Wusṭā* 28: 346–375.

Marom, R.
2021. "The Abu Hameds of Mulabbis: An oral history of a Palestinian village depopulated in the Late Ottoman period." *Journal of Middle Eastern Studies* 50.1: 87–106.

Marom, R.
2022. "Jindās: A history of Lydda's rural hinterland in the 15th to the 20th centuries CE." *Lod, Lydda, Diospolis: Journal of the History, Archaeology and Heritage of Lod* 1: 2–31.

Marom, R., and I. Taxel
2023. "Ḥamāma: The historical geography of settlement continuity and change in Majdal ʿAsqalan's hinterland, 1270–1750 CE." *Journal of Historical Geography* 82 (October): 1–16.

Marom, R., Y. Tepper, and M.J. Adams
2023. "Lajjun: forgotten provincial capital in Ottoman Palestine." *Levant* 55.2: 218–241.

Massignon, L.
1951. "Documents sur certains waqfs des lieux saints de l'Islam: principalement sur le waqf Tamimi à Hébron et sur le waqf tlemcénien Abû Madyan à Jérusalem." *Revue Études Islamiques* (1951): 73–120.

Massignon, L.
1953. "Documents sur certains waqfs des lieux saints de l'Islam: principalement sur le waqf Tamimi à Hébron et sur le waqf tlemcénien Abû Madyan à Jérusalem." (Additions et corrections à Rev. Et. Isl., 1951, pp. 73–120), *Revue Études Islamiques* (1953): 39–42.

Palmer, C.
1999. "'Whose land is it anyway?' An historical examination of land tenure and agriculture in northern Jordan." In *The Prehistory of Food: Appetites for Change*, edited by C. Gosden and J. Hather, 300–302. London: Routledge.

Rapoport, Y.
2018. *Rural Economy and Tribal Society in Islamic Egypt: A Study of al-Nābulusī's "Villages of the Fayyum."* Turnhout: Brepols.

Rippon, S.J., R.M. Fyfe, and A.G. Brown
2006. "Beyond villages and open field: The origins and development of a historic landscape characterized by dispersed settlement in southeast England." *Medieval Archaeology* 50: 31–70.

al-Sabagh, M.H.
2014. "Before banks: Credit, society, and law in sixteenth-century Palestine and Syria." Santa Barbara: PhD dissertation, Department of History, University of California–Santa Barbara.

Shopov, A.
2020. "The vernacularization of sixteenth-century Ottoman agricultural science in its economic context." In *Living with Nature and Things: Contributions to a New Social History of the Middle Islamic Periods*, edited by B.J. Walker and A. Al Ghouz, 639–681. Göttingen: V&R Unipress.

Singer, A.
1994. *Palestinian Peasants and Ottoman Officials: Rural administration around sixteenth-century Jerusalem.* Cambridge: Cambridge University Press.

Sutton, S.
2000. *Contingent Countryside: Settlement, Economy, and Land Use in the Southern Argolid Since 1700*, edited by S. Sutton. Stanford: Stanford University Press.

Walker, B.J.
2007. "Regional markets and their impact on agriculture in Mamluk and Ottoman Transjordan." In *On the Fringe of Society: Archaeological and Ethnoarchaeological Perspectives on Pastoral and Agricultural Societies*, edited by B. Saidel and E. van der Steen, 117–125. Oxford: BAR International Series 1657.

Walker, B.J.
2017. "The struggle over water: Evaluating the 'water culture' of Syrian peasants under Mamluk Rule." In *Developing Perspectives in Mamluk History*, edited by Y. Ben-Bassat, 287–310. Leiden: Brill.

Walker, B.J.
2021. "Peasants, rural economy and cash crops in medieval Islam." In *Markets and Exchanges in Pre-Modern and Traditional Societies*, edited by J.C.M. García, 91–113. Oxford: Oxbow Books.

Walker, B.J.
2022. "The changing face of agricultural 'estates' in 15th and 16th-Century Palestine: The commercialization of Khirbet Beit Mazmīl." In *The Mamluk-Ottoman Transition: Continuity and Change in Egypt and Bilad al-Sham in the Sixteenth Century*, Vol. II, edited by S. Conermann and G. Şen, 325–357. Bonn: V&R Press.

Walker, B.J.
2023a. "Did the *qaṣr* stimulate market agriculture?; the case of Khirbet Beit Mazmīl and Mamluk Jerusalem." In *Festschrift for Hillel Geva on the Occasion of his 75th Birthday*, edited by O. Gutfeld, A. Mazar, O. Peleg-Barkat, and S. Weksler-Bdolah, in print. Eretz Israel series, vol. 35.

Walker, B.J.
2023b. "Made on order for the Amir: What was the function of Mamluk "Barracks Wares?" In *Material Entanglements in the Islamic World: New Approaches to Islamic Archaeology & Ceramics*, edited by J. Vroom, in print. Medieval and Post-Medieval Mediterranean Archaeology Series. Tournhout: Brepols.

Walker B.J., S. Laparidou, A. Hansen, and C. Corbino
2017. "Did the Mamluks have an environmental sense?: Natural resource management in Syrian villages." *Mamluk Studies Review* 20 (2017): 167–245.

Walker, B.J., Y. Gadot, Y. Elgart-Sharon, and O. Ze'evi
2020. "Agricultural terracing and rural revival in Late Medieval Palestine." In *Proceedings of the 11th International Conference on the Archaeology of the Ancient Near East*, vol. II, Islamic Archaeology section, edited by L. Korn and A. Heidenreich, 651–661. Wiesbaden: Harrossowitz.

Weigert, G.
1990. "A Maghribi religious endowment in fourteenth-century Jerusalem: On Abū Madyan al-Maghribī's endowment in Jerusalem." *Cathedra* 58: 25–34.

Wilkinson, T.
2010. "The tell: Social archaeology and territorial space." In *Development of Pre-State Communities in the Ancient Near East*, edited by D. Bolger and L.C. Maguire, 55–62. Oxford: Oxbow.

Yeger, D.
2017. "The Mamluks' Watchtowers around the city of Jerusalem: A reevaluation." Jerusalem: Unpublished MA thesis, Institute of Archaeology, The Hebrew University of Jerusalem.

Zeʿevi, D.
1996. *An Ottoman Century: The District of Jerusalem in the 1600s*. Albany: State University of New York Press.

About the author

Prof. Dr. Bethany J. Walker (PhD 1998, University of Toronto) is Professor of Islamic Archaeology at the University of Bonn, the Director of the Research Unit of Islamic Archaeology at the same institution, and former Co-Speaker of the Annemarie Schimmel Kolleg of Mamluk Studies in Bonn. A historically trained archaeologist, Walker directs field projects in Israel and Jordan. Among her monographs and edited and co-edited works are *Jordan in the Late Middle Ages: Transformation of the Mamluk Frontier* (Chicago 2011), *Reflections of Empire: Archaeological and Ethnographic Studies on the Pottery of the Ottoman Levant* (Boston 2009), *Living with Nature and Things: Contributions to a New Social History of the Middle Islamic Periods* (Bonn 2020), and the *Oxford Handbook of Islamic Archaeology* (Oxford 2020). She is the senior and founding editor of the Journal of Islamic Archaeology, and her research focuses on peasant society, land use and tenure, and landscape transformation.

— 9 —

What Makes a Village?

Social Networks and Resilience

Bethany J. Walker

What kind of place was Beit Mazmil?

There is much we will never know about Beit Mazmil, either archaeologically or historically, and many questions will remain unanswered. We cannot quantify production and consumption, as has been done for Roman or Byzantine estates. The kind of Annalistic research that brought the small community of medieval Montaillou to life for its readers (Le Roy Ladurie 1978)—through the village's "mentalities" and individual stories—is not possible for a late/post-medieval village like Beit Mazmil. We are missing the kinds of written sources that allow the historians of medieval Europe to quantify, map, and explore the larger questions of microhistory. Our textual record, as rich as it is, is spotty and specialized, and does not include European-style censuses, cartography, locally written narrative sources (letters), or ego-documents (diaries). The archaeological record from a site located in the middle of an urban building zone, where the largest part of the original settlement was never studied and is buried under concrete, is naturally partial and fragmented. Much of the fieldwork at the site took the form of salvage projects, leaving us with non-contiguous areas of excavations, and different methods of excavation and recording. What we have is a kaleidoscope of images, giving color to a place, but offering limited clarity.

We cannot say how large the village was (spatially), where its boundaries were, or how many buildings constituted the built environment. Neither the archaeological record nor the textual sources give us this information. We have material traces of and textual references to a market in the village in the early Ottoman period, but we do not know how large it was, how large a region it served, what goods it specialized in, and how often it was held (weekends, during pilgrimage, and so on). We cannot estimate how much land was devoted to the cultivation of particular crops, as information on field sizes appears only fleetingly in the registers; likewise, while our study of terraces can be used to determine field size, one cannot determine cropping practices from the fields themselves, or connect specific fields to those documented in the registers. The issue of crop specialization was brought up in the previous chapter: to determine what degree soil, hydrology, and topography impacted cropping practices requires a different kind of fieldwork than was done for this project.

From a historical perspective, perhaps the biggest lacuna is our inability to evaluate the economic importance of this village to the town of Jerusalem. There is no clear distinction between subsistence-based and market-oriented cultivation practices. Archaeological studies of subsistence agriculture tend to focus on risk-management strategies, such as diversification (Marston 2011; Marston 2015), and those of market agriculture on specialization, cash crops, and imperial interventions (Walker *et al.* 2017; García 2021). Beit Mazmil appears to have cropped for both objectives.

To assess market capabilities, one needs to know exactly how much of all locally consumable commodities were produced, after the *timariots* and *waqf* managers took their shares. One needs population figures, in detail, for all periods. Animal products are underrepresented in both the archaeological and textual sources for this village, yet chicken, chicken eggs, and sheep and goat were important food items sent to towns. Syrian chronicles of the Mamluk and early Ottoman periods attest to the regular transport of such items from village to town (see Ibn Qāḍī Shuhbah, *Tārīkh*, 3:164), and contemporary recipe books demonstrate the degree to which urban cuisine was dependent on a reliable supply of fresh meat, fruits, and vegetables from the countryside (Lewicka 2011). However, even in the imperial capitals of Cairo and Istanbul chickens were raised and urban gardens kept in the courtyards, and possibly roofs, of crowded apartment complexes (Quickel 2015; Shopov 2022). The question remains: to what degree did a village like Beit Mazmil actually contribute to "feeding" Jerusalem? How important was it to the economy of the town?

These outstanding questions aside, we learn four important lessons from this study. The first concerns our expectations about what a village should be and how it should look. The hilltop villages of Ottoman Palestine took the form of *quṣūr*: walled complexes of beehive-shaped towers (*manāṭīr*). Since they appear in woodblock drawings of the 17th century, we know that their history is much longer than the 19th and 20th century structures still standing in the central highlands and documented by RIWAQ. Whether these structures continue a type known from Mamluk times has yet to be confirmed. A single such complex may have been all that architecturally defined the village. The village, however, was much more than its built environment, and included fields, gardens, and pastureland, all tied together by field walls and paths. The settlement was part of an agricultural and pastoral landscape. We should not be surprised by the small scale of the built environment, as that was not the only space of socialization and work.

The second lesson deals with Jerusalem as a social actor and a consumer, and how important the village of Beit Mazmil and the Holy City were to one another; this is discussed below. The third is relational: what do we learn about the Mamluk and Ottoman states from their relationships with this village? Because of its longevity, its rich textual record, and the preservation of the field systems (cultivated by the people of Beit Mazmil and their neighbors), Beit Mazmil provides us with a unique perspective on empire-building in the later Islamic centuries. Finally, we learn lessons about peasant resilience in times of political uncertainty and economic change from Beit Mazmil's long history.

Relations with Jerusalem

The court registers, and to a lesser degree the ceramic record, document the close relationship the village had with Jerusalem in the Ottoman period. Villagers were regular visitors to the Sharia court there, had business partnerships with Jerusalemites, and likely bought and sold in the town markets, as well as that in the village. Further social ties, such as through marriage, have yet to be confirmed through the registers, and there is no proof—textual or archaeological—that Jerusalemites spent any time in Beit Mazmil or that their familiarity with locals extended beyond loaning money, collecting revenues from *waqf* and *timar* land, and buying and selling crops. We have no archaeological evidence, thus far, of village products—outside of crops and animals—making their way to the markets in town. However, if Beit Mazmil was like other villages of the day, regular contact with individual merchants and consumers took place outside the formal marketplaces. Olives, for example, were pressed in the villages, the oil frequently transported by the villagers themselves to the buyers' homes (Cohen 1989, 70). Once receiving notice about the ripening and harvest of cereals, town-based grain brokers would visit the village to inspect the crops and make purchases.[1]

These contacts can be described and categorized, but not quantified. The question remains how important a village like Beit Mazmil was to the economy of Jerusalem. To what degree did this village help to feed the town? To determine how much of Beit Mazmil's production was market-oriented, one needs to know its local needs and scale of production: how much was surplus after paying taxes? To estimate the contributions of the village to Jerusalem's annual food consumption, we need to know the town's population, and how much was supplied from other sources. One cannot get this kind of fine-tuned information about a single village from the textual sources, and the archaeological ones do not allow us to quantify in this manner. Nonetheless, studies of the court registers for reconstructing some aspects of Jerusalem's economy in the 16th and 17th centuries give us a broad picture of the economic relationships of its villages as a group with the local markets. Population estimates for Jerusalem at the beginning of Ottoman rule vary from 5,000–6,000 (Zeʿevi 1996, 3) to 9,000 (Cohen 1989, 24). The population of the District of Jerusalem as a whole, was about 200,000, the majority living in the villages (Zeʿevi 1996, 3). Estimated population numbers peaked by around 1560 at 16,000 individuals (Cohen 1989, 56), and they fell from that point on. In the 17th century, there were no more than 10,000 people living in the town (Zeʿevi 1996, 23). Bread was the main food staple, but as the urban population grew, there was increasing demand for meat. Nearby villages could not satisfy Jerusalem's demand for meat (Cohen 1989, 44).[2] As a result, sheep (the preferred source of protein) were transported "on the hoof" from villages all over Palestine, and from as far away as northeast Syria and Anatolia. Cohen has estimated that the town required at least 24,000 sheep a year to meet mutton demand in the 16th century (Cohen 1989, 56). On the other hand, the town was surrounded by a "green belt" of villages with orchards, vineyards, and gardens, which appear to have met Jerusalem's requirements for fruit and vegetables. The active business of buying and selling produce from these fields documented in the *sijill*s suggests it was a dependable source of extra income for those with the legal rights of cultivation. There appears to have generally been a surplus of cereals, so much so that villagers took what was not needed at home (and after paying taxes) to the open marketplaces in town. There was usually a surplus of olive oil, as well, which villagers could sell at profit at public sales (Cohen 1989, 94). Beit Mazmil, which is remembered today as having some of the most splendid gardens of the region, likely played an important role in supplying Jerusalem with fruit and fruit by-products.

Relations with the Mamluks and Ottomans

Arguably the most important reason why the archaeological study of this site is valuable is for the information it provides about imperial-village relations. Beit Mazmil appears to have held a special status in the administration of rural Jerusalem. While not the largest or wealthiest village, its proximity to Jerusalem, hilltop vantage point, and location along a well-travelled Christian pilgrimage route connecting Ein Karem and Bethlehem with the Holy City gave it a locational advantage.

There is sufficient archaeological evidence to identify the Mamluk phase of occupation at Khirbet Beit Mazmīl as an amiral estate. While there is no textual evidence that the village of Beit Mazmil was an amiral *iqṭāʿ*—the Mamluk-era *iqṭāʿ* lists having been lost, arguably during the Ottoman conquests—it is highly likely that it was. We know little about how these estates were run and what relationship *muqṭāʿ*s had with these properties. The archaeological site appears to have been the estate house, the only one identified as such to date. The village associated with it has been lost with modern urban development. What we learn from this site is just how embedded Mamluk officials were, in fact, in the local fabric of rural society. Excavations of small "castle towns," such as Ḥisban in central Transjordan, have identified ways in which the modest Mamluk garrisons became part of local society, in terms of water and food distribution, shared participation in the local marketplace, and so on (Walker 2011). What is most remarkable about the site of Khirbet Beit Mazmīl is its industrial facilities—the press, kilns, evidence of glass and metalworking—and a ceramic assemblage made up of both imports from across the Mediterranean (obtained from the Jerusalem markets) and local products. The recovery of what have been identified at other sites as rattles (used either to pacify infants or as a magical device thought to ward off the "evil eye")—produced in a clay tentatively identified as a Nile silt—gives a more intimate view of this place as a residence of an official from Egypt. The site is our first window on what such an estate could have looked like, and how it may have functioned. The written record cannot do as much.

As for its relationship with the Ottoman state, Beit Mazmil, as we know from the registers, had its own imam, so arguably its own mosque and its own marketplace. We do not know the nature of this local market—though it was clearly monetarized—but it does suggest that the village was a central place for local exchange. The villagers, moreover, were in and out of Jerusalem on a regular basis, knew how to use the courts, and were intimately familiar with the markets and buying and selling. They were among the first in the area to begin smoking tobacco. They were active participants in the globalizing markets opened with the Ottoman conquest. Beit Mazmil was incorporated from the start into the imperial Ottoman system.

A lesson on peasant resilience

On a final note, this study of a single village and its people paints a picture, however blurry the details, of a resilient peasant community. Like many villages in pre-modern Palestine, the place name of Beit Mazmil has remained the same for hundreds of years; it is a constant presence in the *sijills*; and the site of Khirbet Beit Mazmīl has left a visible trace in the archaeological record, with a few periods of interruption, from at least the Early Islamic period. Was Beit Mazmil typical, in this regard, or were there factors that set this place apart? Given its longevity, was it uniquely resilient?

Resilience, from a systems perspective, refers to changes made within a system (or in this case a social group or community) that allow it to survive as a system under periods of extreme stress. Resilience is a survival strategy: it keeps the group together by adapting to outside pressures (and opportunities). The survival of the community is a measure of the success of these adaptations. A community can retain its integrity as a social unit by migrating to a new place as a group and leaving the homeland behind. This is a kind of communal resilience. The place is sacrificed for the survival of the community. There are limits, however, to the degree to which any social unit (or individual) can adapt, accommodate, or sacrifice before it ceases to exist as a group or person. That is a "system threshold." When these thresholds are exceeded, the system can no longer survive in that form, and either dissolves (dies) or changes form (Redman 2005). For a social group or community, this could take several different forms, including full-scale abandonment of villages/sites and migration, site dispersal, "bedouinization," and the disappearance of that community as a recognizable group. Partial, and temporary, abandonment of Khirbet Beit Mazmīl in the early 18th century did not lead to the dissolution of the community or abandonment of its farmland. The court records continue to document the economic and legal activities of the people of Beit Mazmil through the 18th and 19th centuries. The archaeological site was reoccupied multiple times and not abandoned for good until 1948.

Market accommodation is another form of communal resilience. The traditional form of cultivation on the village land was cereal production; in the 17th century this changed, as Beit Mazmil became part of Jerusalem's "green belt." One *sijill* entry of 1065 H/1655 CE is of special interest.[3] It describes the purchase of a large share of the grape, fig, olive, apple, and quince harvest for that year from the garden of al-Dawāyih in the village of Beit Mazmil by one Ḥajj Khayr al-Dīn b. Muḥammad al-Shaybeh. In this entry the location of the garden is actually described: it was surrounded by vineyards on three sides. By the mid-18th century, this area appears to have shifted to garden agriculture, vineyards occupy much of the space in the immediate vicinity of the village, and the village land now includes orchards of diverse cultivation, the crops of which are owned in full by local villagers. This trend would develop with increasing force through the 18th and 19th centuries. Altogether, this exemplifies social resilience: the ability to adapt to changing markets, to survive as a community after political turmoil, and to continue to cultivate the land, regardless of where one lives. It reflects knowledge of the local and urban markets, effective social networks, and a dramatic shift in cultivation that is suggested by the expansion of terracing in the region and documented by the botanical record at the site.

Was the village of Beit Mazmil special, or does it only provide us with a case study of medieval Jerusalem's countryside? In terms of its architecture and archaeological assemblages, Beit Mazmil is a rather typical village of Ottoman Palestine. However, the exceptional preservation of a portion of the site, its botanical record, and its agricultural fields, combined with a centuries-long series of documents related to the village, its fields, and its people allow us to follow the long life of this small peasant community that is not possible in the same way anywhere else. It is not only longevity that makes this village stand out from others of the time, however, but its special relationship with the imperial states of the day. We learn as much about Mamluk and Ottoman statehood—how these sultanates penetrated rural society and functioned on the local level—as we do about daily life on the farm, all from this archaeological encounter with the people of Beit Mazmil.

Notes

1. "Future deals," where the produce was purchased even before the crop ripened, was also done with olives (Cohen 1989, 75).
2. Sheep, goats, camels, water buffalo, and cattle were slaughtered for consumption, in that order of preference.
3. Sijill no. 155, p. 36, Sharia court al-Quds (Jerusalem), photocopy kept at the National Library of Israel. I am grateful to Yehoshua Frenkel for directing me to this resource.

References

Cohen, A.
1989. *Economic Life in Ottoman Jerusalem*. Cambridge: Cambridge University Press.

García, J.C.M.
2021. *Markets and Exchanges in Pre-Modern and Traditional Societies*, edited by J.C.M. García. Oxford: Oxbow Books.

Ibn Qāḍī Shuhbah, Taqī al-Dīn Abū Bakr b. Aḥmad.
1977–1997. *Tārīkh Ibn Qāḍī Shuhbah*, edited by ʿA. Darwīsh. Damascus: L'Institut Français du Proche Orient.

Le Roy Ladurie, E.
1978. *Montaillou: The Promised Land of Error*. New York: G. Braziller.

Lewicka, P.B.
2011. *Food and Foodways of Medieval Cairenes: Aspects of Life in an Islamic Metropolis of the Eastern Mediterranean*. Leiden: Brill.

Marston, J.M.
2011. "Archaeological markers of agricultural risk management." *Journal of Anthropological Archaeology* 30.2: 190–205.

Marston, J.M.
2015. "Modeling resilience and sustainability in ancient agricultural systems." *Journal of Ethnobiology* 35.3: 585–605.

Quickel, A.
2015. "Farm to fork: Cairo's food supply and distribution during the Mamluk sultanate (1250–1517)." Cairo: MA thesis, Arab and Islamic Civilizations Department, American University of Cairo.

Redman, C.L.
2005. "Resilience theory in archaeology." *American Anthropologist* 107.1: 70–77.

Shopov, A.
2022. "When Istanbul was a city of Bostāns." In *A Companion to Early Modern Istanbul*, edited by S. Hamadeh and Ç. Kafescioğlu, 279–307. Leiden: Brill.

Walker, B.J.
2011. *Jordan in the Late Middle Ages: Transformation of the Mamluk Frontier. Chicago Studies on the Middle East 8.* Chicago: Middle East Documentation Center, University of Chicago.

Walker, B.J., S. Laparidou, A. Hansen, and C. Corbino.
2017. "Did the Mamluks have an environmental sense?: Natural resource management in Syrian villages." *Mamluk Studies Review* 20: 167–245.

Ze'evi, D.
1996. *An Ottoman Century: The District of Jerusalem in the 1600s.* Albany: State University of New York Press.

About the author

Prof. Dr. Bethany J. Walker (PhD 1998, University of Toronto) is Professor of Islamic Archaeology at the University of Bonn, the Director of the Research Unit of Islamic Archaeology at the same institution, and former Co-Speaker of the Annemarie Schimmel Kolleg of Mamluk Studies in Bonn. A historically trained archaeologist, Walker directs field projects in Israel and Jordan. Among her monographs and edited and co-edited works are *Jordan in the Late Middle Ages: Transformation of the Mamluk Frontier* (Chicago 2011), *Reflections of Empire: Archaeological and Ethnographic Studies on the Pottery of the Ottoman Levant* (Boston 2009), *Living with Nature and Things: Contributions to a New Social History of the Middle Islamic Periods* (Bonn 2020), and the *Oxford Handbook of Islamic Archaeology* (Oxford 2020). She is the senior and founding editor of the Journal of Islamic Archaeology, and her research focuses on peasant society, land use and tenure, and landscape transformation.

Appendices

App. 1: Full site plan

Khirbet beit Mazmil
Full plan of the archaeological site

Elaboration by Nicolò Pini

App. 2: Plan of North Building of the Northern Sector

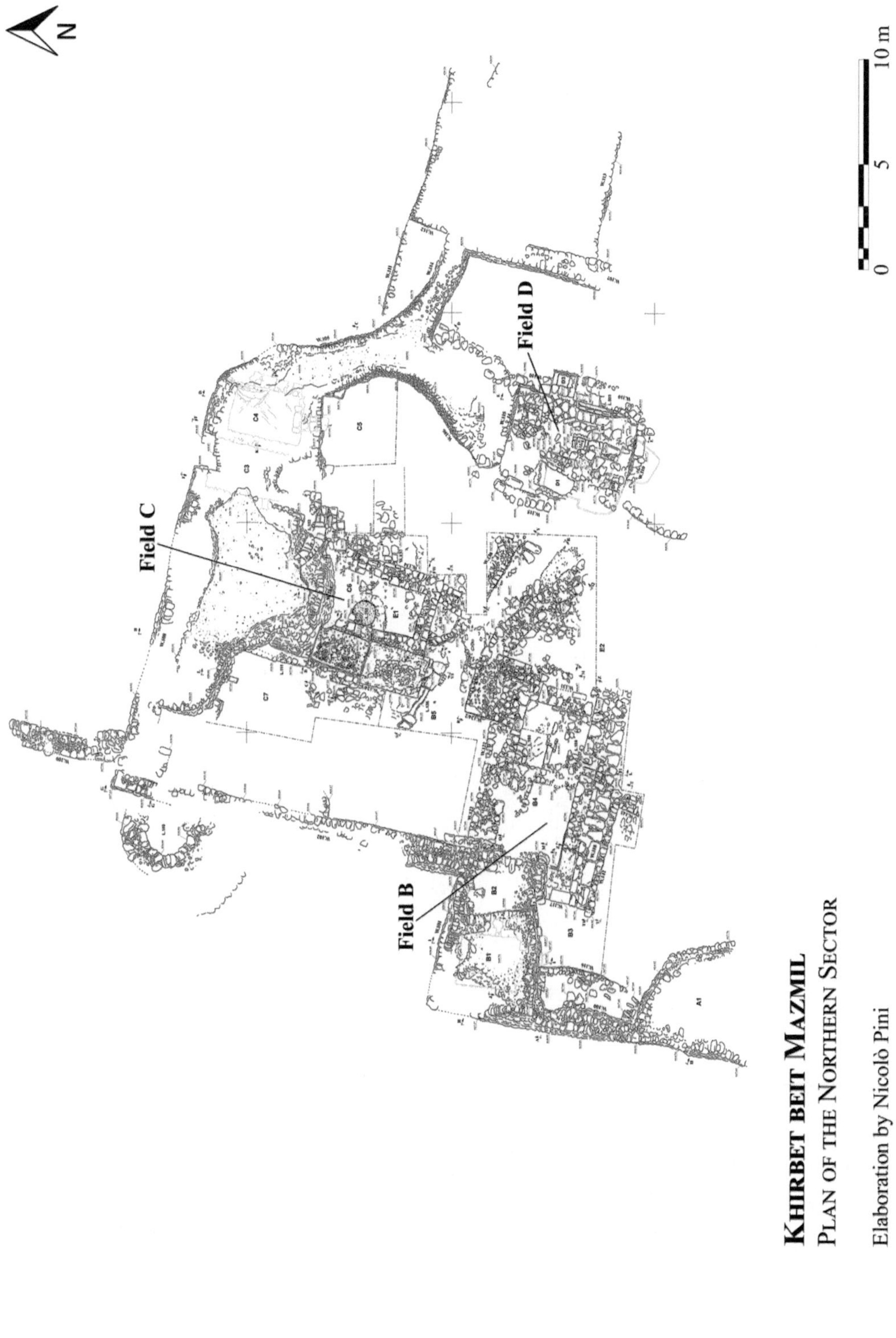

App. 3: Plan of Strata VI–IX (pre-Mamluk)

Appendix 3
Khirbet beit Mazmil
Plan of Strata VI–IX - Pre-Mamluk

Elaboration by Nicolò Pini

App. 4: Plan of Strata III–IV (Mamluk)

APPENDIX 4
KHIRBET BEIT MAZMIL
PLAN OF STRATA III–IV - MAMLUK

Elaboration by Nicolò Pini

App. 5: Plan of Stratum II (Early Ottoman)

Appendix 5
Khirbet beit Mazmil
Plan of Stratum II - Early Ottoman

Elaboration by Nicolò Pini

App. 6: Plan of Stratum I (Late Ottoman / Mandate)

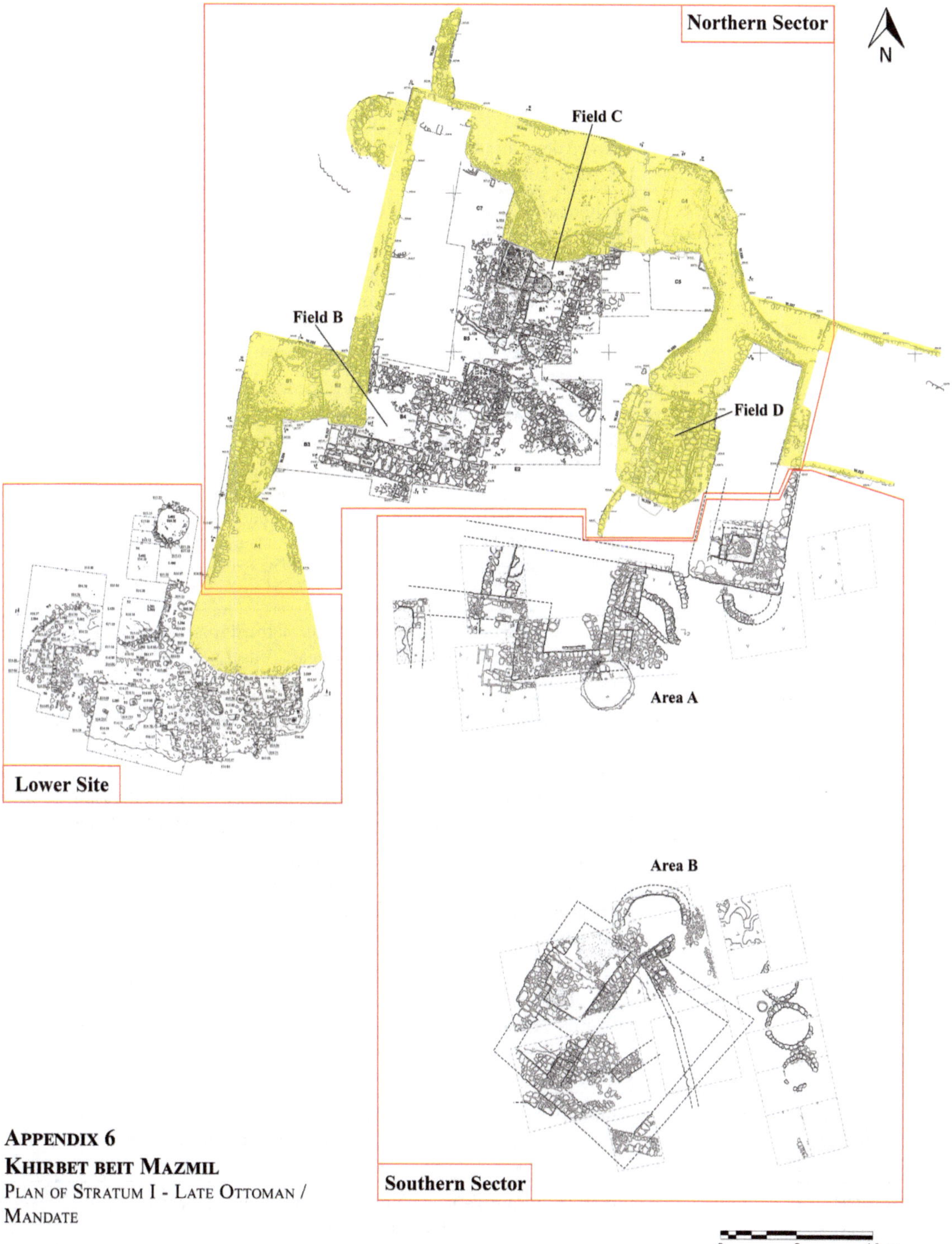

Appendix 6
Khirbet beit Mazmil
Plan of Stratum I - Late Ottoman / Mandate

Elaboration by Nicolò Pini

App. 7: Summaries of representative entries on Beit Mazmīl in the Jerusalem *sijills*

Rec.	Date	Subject	Case summary
1	28.11.936 H / 24.07.1530 CE	Land lease	The lease of all the land bearing fig trees, vines, and apple trees along with the house that is on the property. The term of the lease is 30 years. The court record describes the land and mentions that, to the north of it, al-Qaṣab bridge is located, which is held (used?/leased?) by Beit Fāsīn villagers.
16	15.07.951 H / 03.10.1544 CE	Debt claim	Yahūd bn Mousa al-Ṣayrafiy (the money changer) demanded a debt payment from Saʿed bn Khaṭāb from Biet Mazmīl.
16	12.02.952 H / 04.05.1545 CE	Foreclosure	The plaintiff, al-Sheikh Shihāb al-Dīn Ahmad bn Muhmmad al-Qudsiy (the Jerusalemite), in his capacity as the legal guardian of the orphans of Khalīl bn Zayneh, claimed that Mousa bn ʿbed al-Hadiy from al-Malḥah village had illegally seized the whole of the crops at Beit Mazmil including the half that belongs the orphans.
19	06.06.954 H / 24.07.1547 CE	Compliant against Beit Mazmīl village	The *sipāhī* Ahmad Jalbiy bn Ghanim filed a complaint against the villagers of Beit Mazmīl. He argued that the villagers intentionally neglect cultivating the Miri lands every year. As representative of Beit Mazmil villagers, Ibrahīm bn Zallum, Ahmad bn Zeyadeh, Saʿādeh bn Khalaf, ʿUthman bn Khalaf, Mousa bn Ghānim, Deyab bn Ghānim, and Ramadhan bn Saʿed were present before the court and they argued that the work in the Miri lands was suspended for agricultural reasons.
28	21.08.961 H / 22.07.1554 CE	Payment claim	Mousa bn Abed al-Karīm bn Shamāḥah claimed that Yousef bn Khalifeh from Beit Mazmīl owed him 10 *muds* of wheat. Yousef paid the full sum and was issued a deed of release.
28	24.09.961 H / 23.08.1554 CE	*Timar* agreement	The Jerusalemite Sipahi, Ali Bey bn Ramdhan the Soubashi, has drawn an agreement with Saʿed bn Khatāb, Deyāb bn Ghānim, Ḥussein al-Khūliy, and Ḥasan bn Ramdhan—all from Beit Mamzīl—stating that the latter party has a share in the crops of wheat and barley that they will be cultivating for Ali Bey on his *timar* farm. In return, they are obliged to pay Ali Bey 33 *Dinār Sulṭaniy.*
28	23.10.961 H / 21.09.1554 CE	Receipt	Alaʾ al-Dīn Ali bn Sinān, one of the eminent *sipāhīs* of Jerusalem, has received from al-Sheikh Zain al-Dīn Abed al-ʿAẓīm bn Shihāb al-Dīn Ahmad al-Maʿariy, one of Jerusalem's eminent scholars, in his capacity as the attorney of Deyāb bn Ghānim, Saʿādeh bn Saʿed, and Ramadhan bn Saʿed—all from Beit Mazmīl—the sum of [. . .]. The entry states that a third of Mazmīl village is Alaʾ al-Dīn Ali Timār.
28	22.11.961 H / 19.10.1554 CE	Receipt	The *sipāhī* Alaʾ al-Dīn Ali bn Sinān received from Saʿādeh bn Saʿed, Deyāb bn Ghānim, and Ramadhan bn Saʿed—all form Beit Mazmīl—16 golden Sulṭaniys and 26 Silver *Suleimanīs* for the *timar* of Beit Mazmil village. A third of village of Beit Mazmīl is Alaʾ al-Dīn Ali's *timar.*
36	04.07.965 H / 22.04.1558 CE	Guarantee	Ibrahīm bn Ali bn Argoun demanded from Ibrahīm bn Ḥajīj the fulfilment of his responsibilities as the guarantor of Ḥasan bn al-Khūliy from Beit Mazmīl.
36	21.08.965 H / 08.06.1558 CE	Guarantee	Ibrāhīm bn Zain al-Dīn Zallum form Beit Mazmīl pledged to be the guarantor of Khalīfeh bn Khalaf, who is also from Beit Mazmīl. Latter owes al-Ḥāj Ahmad al-Malāḥ and his children—Shiḥādeh, Khatāb, and Sit-al-Banāt—six golden *Sulṭaniy.*

Individuals	Articles involved	Locales	Case no./P.	Sum. Page no.
[1] Shamis al-Dīn bn al-Shamis Majīd Abu al-Ḥasan. [2] Hishihab-al-Dīn Ahamd bn al-Ḥusamī, aka "Ibn al-Qaṣīr" [2] Hishihab-al-Dīn Ahamd.	[1] Fruit-bearing land. [2] Bridge.	[1] Wadi al-Badawayeh [2] al-Qaṣab bridge. [3] Beit Fāsīn	X	99
[1] Sa'ed bn Khaṭāb [2] Yahūd bn Mousa al-Ṣayrafiy	Debt	Biet Mazmīl	4/43	23
[1] al-Sheikh Shihāb al-Dīn Ahmad bn Muhmmad al-Qudsī [2]the orphans of Khalīl bn Zayneh [3] Mousa bn 'bed al-Hadī from al-Malḥah village	Crops (not specified)	[1] Biet Mazmīl [2] al-Malḥah village	1/463	224
[1] Sipahi Ahmad Jalbiy [2] Ibrahīm bn Zallum [2] Ahmad bn Zeyadeh [4] Sa'ādeh bn Khalaf [5] 'Uthman bn Khalaf [6] Mousa bn Ghānim [7] Deyab bn Ghānim [8] Ramadhan bn Sa'ed	Miri lands of Beit Mazmīl village.	Beit Mazmīl	5/207	87
[1] Abed al-Karīm bn Shamāḥah [2] Yousef bn Khalifeh	10 *muds* of wheat	Beit Mazmīl	4/425	108
[1] Sipahi Ali Bey bn Ramdhan [2] Sa'ed bn Khaṭāb [3] Deyāb bn Ghānim [4] Ḥussein bn al-Khūliy [5] Ḥasan bn Ramdhan	*Timar* Farm	Beit Mazmīl	3/177	139
[1] Sipahis Ala' al-Dīn Ali bn Sinān [2] al-Sheikh Zain al-Dīn Abed al-'Aẓīm al-Ma'ariy [3] Deyāb bn Ghānim [5] Sa'ādeh bn Sa'ed [6] Ramadhan bn Sa'ed	*Timar*	Beit Mazmīl	3/11	165
[1] Sipahis Ala' al-Dīn Ali bn Sinān [2] Deyāb bn Ghānim [3] Sa'ādeh bn Sa'ed [5] Ramadhan bn Sa'ed	*Timar*	Beit Mazmīl	2/68	186
[1] Ibrahīm bn Ali bn Argoun [2] Ibrahīm bn Ḥajīj [3] Ḥasan bn al-Khūliy	Debt	Beit Mazmīl	5/89	47
[1] Ibrāhīm bn Zain al-Dīn Zallūm [2] Khalīfeh bn Khalaf [3] al-Ḥāj Ahmad al-Malāḥ	Debt	Beit Mazmīl	3/181	96

(continued)

App. 7: Summaries of representative entries on Beit Mazmīl in the Jerusalem *sijills* (*continued*)

Rec.	Date	Subject	Case summary
36	11.10.965 H / 27.07.1558 CE	Crops transaction	Jalāl al-Dīn bn Makiy claimed that Ibrahīm bn Mar'iy sold his father, Yousf, a quarter of the crops of grapes and [. . .], which are in Beit Mazmīl. Jalāl al-Dīn transferred the value of the crops to al-KhawajahTaqī al-Dīn al-'Eynabusiy in payment for the rice he bought from him.
36	12/10.965 H / 28.07.1558 CE	Conciliation agreement	In the presence of 'Abeddullah al-Qāboujiy, the representative of the prince of Jerusalem Qīṭās Bey, the Soubashi Murād bn the *sipāhī* 'Abeddullah vouched for Younis bn Khalīfeh from Beit Mazmīl as part of the conciliation agreement drawn between him and his companions.
36	24.12.965 H / 07.10.1558 CE	*Timar* division	*Soubashi sipāhī* Ala' al-Dīn Ali bn Sinān attended; along Ramadhan bn Sa'ed, and Deyāb bn Ghānim, Ahmad bn Hasan—the eminent farmers of Beit Mazmīl; the division of his *timar* harvest. The harvest was divided into 5 *muds* of wheat, 3 *muds* mixed, and 11 *muds* of barley.
46	15.04.972 H / 21.11.1564 CE	Purchase	'Iz al-Dīn bn 'bed al-Qudūs demanded from 'Ali bn Maṭar from Beit Mazmīl the price of 32 *qinṭār* of plaster/stucco.
46	22.04.972 H / 27.11.1564 CE	Arrest	'Uthmān bn Khalaf from Beit Mazmīl was arrested for defaulting on his debt.
46	22.04.972 H / 27.11.1565 CE	Arrest	'Abed al-Nabiy bn Ahmad from Beit Mazmīl was arrested for defaulting on the payment of 13 *qirāṭ* of plaster/stucco he purchased from Zain al-Dīn bn 'Abed, the Jerusalemite merchant.
46	06.05.972 H / 10.11.1564 CE	Payment demand	al-Khawaja Shamis al-Dīn bn Jalal al-Dīn claimed that Ramadhan bn Sa'ed Beit from Mazmīl owes him 525 *Ḥalabeyeh* and crops (not specified). 250 *Ḥalabeyeh* are for half of the sheep herd and rest for rice and olive oil.
46	17.05.972 H / 20.12.1564 CE	Payment demand	Ḥusein bn al-Khuliy Ali from Beit Mazmīl owes Jerjis bn Yuḥanā 2 *Sulṭaniy Suleimāniy Jadīd* and 2 oil jars.
46	22.05.972 H / 25.12.1564 CE	Payment demand	al-Ḥāj 'Ali bn Khalīl al-Ṣaltiy claimed that 'Abed al-Nabiy bn Ahmad from Beit Mazmīl owed him 107 *Sulṭaniy Suleimāniy Jadīd* by way of sponsorship.
46	28.05.972 H / 31.12.1565 CE	Arrest	By the order of Mahmūd Jalbiy, the clerk of al-Ḥarameyn Waqf, Ramadhan bn Sa'ed from Beit Mazmīl was arrested for owing Waqf al-Khalīl 100 *muds* of barley.
46	02.06.972 H / 05.01.1565 CE	Payment demand	Younis bn Khalifah from Beit Mazmīl sold Khaṭāb bn Ghānim from Ein Kārem a camel and demand full payment from Khaṭāb.
46	02.06.972 H / 05.01.1566 CE	Rent payment	al-Mu'alim Ḥusein bn 'Ali bn Nimer owned half a camel while the other half was owned by Younis bn Khalifah from Beit Mazmīl. Younis rented the camel out and Ḥusein demand the payment of his share in the rent revenue.
46	17.07.972 H / 18.08.1565 CE	Purchase of crops	al-Mu'alim Ḥusein bn 'Ali bn Nimer bought from Khalīfah bn Khalaf from Beit Mazmīl 6 *qirāṭ* of the crops of grapes, figs, and pears that are from the land of Ein Jāwat.
46	20.08.972 H / 01.04.1565 CE	Debt claim	Deib bn Ḥankash al-Mardawiy has registered with the court the debt Mahmoud bn Ahmad from Beit Mazmīl owes him. The sum of the deby is *Suleimāniy* silver.
54	01.03.979 H / 23.08.1571 CE	Payment demand	Muhmmad bn 'Aby Dāghir demand the payment of 5 *Sulṭaniy Suleimāniy* for the price of a donkey he sold to 'Abed al-Nabiy bn Ahmad from Beit Mazmīl.

Individuals	Articles involved	Locales	Case no./P.	Sum. Page no.
[1] Jalāl al-Dīn bn Makiy [2] Ibrahīm bn Marʿiy [3] al-KhawajahTaqī al-Dīn al-ʿEynabusiy	Crops	Beit Mazmīl	2/256	139
[1] ʿAbeddullah al-Qāboujiy [2] Soubashi Murād bn the Sipahi ʿAbeddullah [3] Younis bn Khalīfeh	Conciliation agreement	Beit Mazmīl	3/258	140
[1] Sipahis Alaʾ al-Dīn Ali bn Sinān [2] Ramadhan bn Saʿed [3] Deyāb bn Ghānim [4] Ahmad bn Hasan	*Timar*	Beit Mazmīl	6/348	194
[1] ʿIz al-Dīn bn ʿbed al-Qudūs [2] ʿAli bn Maṭar	Plaster/stucco	Beit Mazmīl	5/53	45
ʿUthmān bn Khalaf	Debt	Beit Mazmīl	4/61	49
[1] ʿAbed al-Nabiy bn Ahmad [2] Zain al-Dīn bn ʿAbed	Plaster/stucco	Beit Mazmīl	7/61	50
[1] al-Khawaja Shamis al-Dīn bn Jalal al-Dīn [2] Ramadhan bn Saʿed	[1] Sheep [2] Rice [3] Olive oil	Beit Mazmīl	6/79	60
[1] Ḥusein bn al-Khuliy [2] Jerjis bn Yuḥanā	[1] *Sulṭaniy Suleimāniy Jadīd* [2] Oil	Beit Mazmīl	5/100	79
[1] al-Ḥāj ʿAli bn Khalīl al-Ṣaltī [2] Abed al-Nabiy bn Ahmad	107 *Sulṭaniy Suleimāniy Jadīd*	[1] al-Ṣalṭ [2] Beit Mazmīl	5/107	84
[1] Mahmūd Jalbiy [2] Ramadhan bn Saʿed	100 *muds* of barley	[1] Beit Mazmīl [2] Waqf al-Khalīl	2/120	93
[1] Younis bn Khalifah [2] Khaṭāb bn Ghānim	Camel	[1] Ein Kārem [2] Beit Mazmīl	1/126	97
[1] al-Muʿalim Ḥusein bn ʿAli bn Nimer [2] Younis bn Khalifah	Rented camel	Beit Mazmīl	2/127	98
[1] al-Muʿalim Ḥusein bn ʿAli bn Nimer [2] Khalīfah bn Khalaf	Crops of grapes, figs, and pears	[1] Beit Mazmīl [2] Ein Jāwat	1/160	119
[1] Deib bn Ḥankash al-Mardawiy [2] Mahmoud bn Ahmad	75 *Suleimāniy* silver	Beit Mazmīl	7/216	156
[1] Muhmmad bn ʾAby Daghir [2] ʿAbed al-Nabiy bn Ahmad	5 *Sulṭaniy Suleimāniy* for the price of a donkey	Beit Mazmīl	3/311	144

(continued)

App. 7: Summaries of representative entries on Beit Mazmīl in the Jerusalem *sijills* (*continued*)

Rec.	Date	Subject	Case summary
54	26.04.979 H / 17.09.1571 CE	Marriage rights breach	ʿŪbeyd bn Saʿed from Beit Mazmīl filed a complaint against his sister-in-law, Shaqra bnt Yousef from al-Waljeh village, for taking his wife out of his house without his permission.
54	03.06.979 H / 23.10.1571 CE	Dowry claim	Muhmmad bn Khatāb bn Ibrahīm from Beit Mazmīl demanded from Ṣālih Bn Sweyd al-Kurdiy, the attorney of Mohmmad's daughter, Saʿādeh, the payment of her dowry.
54	06.06.979 H / 24.10.1571 CE	Payment demand	Mousa bn Dāghir al-Saʿediy demand the price of grapes from Shiḥādeh bn Rafiʿ from Beit Mazmīl.
54	09.06.979 H / 28.10.1571 CE	Granting power of attorney	Fāṭimah bnt Marwan from Deir Yasīn village granted her husband, Deyāb bn Ghanim from Beit Mazmīl, power of attorney to represent her in a land dispute against Muhmmad al-Kābuliy, ʿAlwan bn al-Ḥāj ʿAli al-Hindi, and Ḥasanāt bn Thabit—all from al-Malḥah village.
54	09.06.979 H / 28.10.1572 CE	Land dispute hearing	The land dispute alluded to above was resolved. The second party, represented by al-Kābuliy admitted to wrongdoing and paid 100 *Sulṭaniy Suleimāniy* as compensation.
54	15.08.979 H / 02.01.1572 CE	Inheritance	The court has determined that Faṭimah Khatūn bnt Ali Jalbiy has inherited from her husband, al-Sheikh Sharaf al-Dīn Mousa al-Dairiy al-Ḥanafiy, the Imam of the Dome of the Rock, among other things, a plot of land in Beit Mazmīl.
54	28.10.979 H / 14.03.1572 CE	Debt payment	Taj al-Dīn Khalifah bn Khadhir, the guardian of Ṣafeyeh bnt Musṭafā al-Qarmaniy, claimed that ʿAbed al-Nabiy bn Ahmad from Beit Mazmīl owed her father 8 *Sulṭaniy Suleimāniy* for the price of half a donkey. Nabiy bn Ahmad clarified that he bought half the donkey for 7 not 8 and that he paid 6 already to Ṣafeyeh's father. He was made to pay the rest, which is 1 *Sulṭaniy*.
67	08.06.995 H / 18.05.1587 CE	Fixed sum agreement (مقاطعة)	al-Sheikh ʿAli bn Rajab Jalbiy, the guardian of al-Khalīl Waqf, has drawn an agreement with Khalaf bn ʿAli, Muhmmad bn Zeyād, and ʿAli bn Maṭār from the village of Beit Mazmīl. Two thirds of Beit Mazmīl belong to al-Khalīl Waqf. The Waqf will receive 140 *muds* of barley while the other party has the right to benefit from the village's winter and summer crops.
67	09.01.996 H / 10.12.1587 CE	Compliant	ʿAli bn Maṭar, Ṭahir bn al-Dawayeh, Shahīn bn al-Qaṣās, and Shiḥādeh bn Zallum—all from the Beit Mazmīl—filed a complaint against Suʿeyfān bn al-ʿAtāl from al-Jūrah village. Suʿeyfān bought a camel from the governor (حاكم السياسة) on behalf of his village, Beit Mazmīl, and al-Malḥah. The price of the camel was 30 *Sulṭaniy*. Each village should pay the governor a third. However, the camel was not delivered to any of the villages. Suʿeyfān said the camel had died.
67	10.01.996 H / 08.02.1588 CE	Purchase	al-Sheikh Shīhāb al-Dīn Ahmad bn Sheikh al-Ḥaram Taqī al-Dīn bought from Shāhīn bn Khalīl al-Qaṣās from Beit Mazmīl 12 *qirāṭ* of grape and fig crops, which originate from Beit Jāwat land/farm.
67	04.05.996 H / 13.06.1588 CE	Purchase	al-Sheikh Shīhāb al-Dīn Ahmad bn Sheikh al-Ḥaram Taqī al-Dīn bought from ʿAli bn Khalaf from Beit Mazmīl 12 *qirāṭ* of olive, grape, fig, other crops, which originate from Beit Jāwat land/farm. al-Sheikh Shīhāb paid 8 *Sulṭaniy Shāmiy*.

Individuals	Articles involved	Locales	Case no./P.	Sum. Page no.
[1] ʿŪbeyd bn Saʿed [2] Shaqra bnt Yousef		[1] Beit Mazmīl [2] al-Waljeh	4/334	155
[1] Muhmmad bn Khatāb bn Ibrahīm [2] Sālih Bn Sweyd al-Kurdiy [3] Saʿādeh bnt Muhmmad	Dowry	Beit Mazmīl	8/377	176
[1] Mousa bn Dāghir al-Saʿediy [2] Shiḥādeh bn Rāfiʿ	Price of grapes	Beit Mazmīl	5/379	177
[1] Fāṭimah bnt Marwan [2] Deyāb bn Ghanim [3] Muhmmad al-Kābuliy [4] ʿAlwan bn al-Ḥāj ʿAli al-Hindi [5] Ḥasanāt bn Thabit	A plot of land on a mountain facing Beit Arzā	[1] Deir Yasīn [2] Beit Mazmīl [3] Malḥah village [4] Beit Arzā	1/383	179
[1] Fāṭimah bnt Marwan [2] Deyāb bn Ghanim [3] Muhmmad al-Kābuliy [4] ʿAlwan bn al-Ḥāj ʿAli al-Hindi [5] Ḥasanāt bn Thabit	A plot of land on a mountain facing Beit Arzā	[1] Deir Yasīn [2] Beit Mazmīl [3] Malḥah village [4] Beit Arzā	3/383	179
[1] Faṭimah Khatūn bnt Ali Jalbiy [2] al-Sheikh Sharaf al-Dīn Mousa al-Dairiy al-ḥanafiy	A plot of land in Beit Mazmīl	Beit Mazmīl	1/341–342	207
[1] Taj al-Dīn Khalifah bn Khadhir [2] Ṣafeyeh bnt Mustafā al-Qarmaniy [3] ʿAbed al-Nabiy bn Ahmad	Half a donkey	Beit Mazmīl	1/529	244
[1] al-Sheikh ʿAli bn Rajab Jalbiy [2] Khalaf bn ʿAli [3] Muhmmad bn Zeyād [4] ʿAli bn Maṭār	140 *muds* of barley	[1] Beit Mazmīl [2] Waqf al-Khalīl	3/44	31
[1] ʿAli bn Maṭar [2] Ṭahir bn al-Dawayeh [3] Shahīn bn al-Qaṣās [4] Shiḥādeh bn Zallum [5]Suʿeyfān bn al-ʿAtāl	Camel	[1] al-Jūrah [2] Beit Mazmīl [3] al-Malḥah	2/60	39
[1] al-Sheikh Shīhāb al-Dīn Ahmad [2] Shāhīn bn Khalīl al-Qaṣās	Grape and fig crops	[1] Beit Mazmīl [2] Ein Jāwat	2/107	68
[1] al-Sheikh Shīhāb al-Dīn Ahmad [2] ʿAli bn Khalaf	12 *qirāṭ* of olive, grape and fig crops	[1] Beit Mazmīl [2] Ein Jāwat	2/247	149

(continued)

App. 7: Summaries of representative entries on Beit Mazmīl in the Jerusalem *sijills* (*continued*)

Rec.	Date	Subject	Case summary
67	22.08.996 H / 18.07.1588 CE	Transaction	al-Sheikh Ahmad bn Mansour, the Sheikh of al-Maghāribah Waqf, sold to Muhmmad bn Younis from Beit Mazmīl a quarter of a vineyard's produce that is a *Waqf* for the benefit of al-Maghāribah. The vineyard is located in Kabssā's plot of land.
67	13.09.996 H / 07.08.1588 CE	Guarantee	Mahmoud bn Ḥasan from Beit Mazmīl guaranteed to pay what Khaṭāb bn Zallum Muhān' obligation to al-Khalīl Waqf in case of his defaulting.
78	06.05.1005 H / 26.12.1596 CE	Payment demand	Ahmad bn Wali-al-Dīn al-Dairiy, the attorney of Muhmmad al-Dairiy, has demanded from Muhmmad bn Zeydeh from Beit Mazmīl the price of an amount of olive fruit that originates from Wadi al-Badawayeh. The amount is worth 40 *raṭil* of good olive oil. Muhmmad bn Zeydeh claimed that the amount he bought was only worth 25 *raṭil* and that he already paid 19 *Sulṭaniy* for it.
78	23.09.1005 H / 10.05.1597 CE	Purchase	Ḥanā bn Mikha'īl al-Naṣraniy al-Qibṭiy al-Najār bought 12 *qirāṭ* of grape, fig, and olive crops that originate from Ein Jāwat farm near Jerusalem from Sa'ed bn Ramadhān from Beit Mazmīl. The buyer paid 12 Egyptian *Sulṭaniy*.
78	05.10.1005 H / 21.051597 CE	*Waqf* agreement	Ali Jawīsh, the deputy of the Khalīl Waqf custodian, Hasan Bey, has reached an agreement with Radhwān bn Sa'ādeh and Khalaf bn Ali from Beit Mazmīl on 2 thirds of the barley harvest from Beit Mazmīl. The agreement stipulates that Radhwān bn Sa'ādeh and Khalaf bn Ali should pay the fixed sum (مقاطعة) of 160 *muds* of barley to al-Khalīl Waqf.
78	18.11.1005 H / 03.07.1597 CE	Price of a donkey	Ahmad bn 'Abed al-Razāq claimed that Ḥasan bn Ali—both from Beit Mazmīl—owed him the price of half a donkey.
78	23.12.1005 H / 07.08.1597 CE	Deferred sale	Moḥsin bn Ḥasan from Beit Mazmīl has registered in the court 'Ali bn Mirān's commitment to sell him 4 *muds* of wheat from his coming harvest.
78	03.03.1006 H / 15.10.1597 CE	Capital	'Abed al-Moḥsin bn Ḥasan from Beit Mazmīl owes al-Ḥāj Mohmmad bn Qazān al-Ramliy, who lives in Jerusalem, 4 *Sulṭaniy* and a half and 23 *raṭil* of oil, both are the capital of a commercial venture.
78	01.04.1006 H / 11.11.1597 CE	Purchase	Ḥanā bn Mikha'īl al-Naṣraniy al-Qibṭiy al-Najār bought all the harvest of the crops that originate from Ein Jāwat farm near Jerusalem from Jabir bn Khalifeh from Beit Mazmīl. The buyer paid 4 Egyptian *Sulṭaniy*.
78	04.04.1006 H / 14.11.1598 CE	Complaint against the *timar* collectors	Radhwān bn Sa'ādeh, Ḥamdān bn Mousa, Khatāb bn 'Ubeyd, and Mir'ib bn Saliḥ—all from Beit Mazmīl—filed a complaint against the *timar* collectors of the village, the *sipāhīs*: Ibrahīm bn Jalbiy, Qīṭās Jalbiy, and 'Ali bn Murād. They said that two thirds of the village are part of al-Khalīl Waqf while one third goes to the *sipāhīs*. They said that they owed 160 *muds* of barley to the *Waqf* and the *sipāhīs* wanted half of them. The court ruled against the *sipāhīs*.
78	01.06.1006 H / 25.09.1597 CE	Purchase	al-Sheikh Muhmmad Shamis al-Dīn bn Ahmad Shihāb al-Dīn al-Sororiy bought from Ṣaleh and Shahīn, the sons of Khalīl al-Qaṣās, from Beit Mazmīl 24 qirāṭ of grapes, figs, apples, olive fruits, and everything else that originate from Khirbet Ein Jāwat. al-Sheikh Muhmmad paid 19 Egyptian *Sulṭaniy*.
78	05.03.1006 H / 16.10.1597 CE	Loan	'Abed al-Moḥsin bn Ḥasan from Beit Mazmīl testified that he borrowed 4 *Sulṭaniy* and a half form Mohmmad bn Qazān al-Ramliy.

Individuals	Articles involved	Locales	Case no./P.	Sum. Page no.
[1] al-Sheikh Ahmad bn Mansour [2] Muhmmad bn Younis	Vineyard's produce	[1] al-Maghāribah [2] Beit Mazmīl [3] Kabssā	6/331	204
[1] Mahmoud bn Ḥasan [2] Khaṭāb bn Zallum Muhān	An obligation to al-Khalīl Waqf	[1] Beit Mazmīl [2] al-Khalīl Waqf	5/374	231
[1] Ahmad bn Wali-al-Dīn al-Dairī [2] Muhmmad bn Zeydeh	The price of olive fruit	[1] Wadi al-Badawayeh [2] Beit Mazmīl	2/14	13
[1] Ḥanā bn Mikha'īl al-Naṣraniy al-Qibṭiy al-Najār [2] Sa'ed bn Ramadhān	Grape, fig, and olive crops	[1] Beit Mazmīl [2] Ein Jāwat [3] Egypt	2/158	99
[1] Ali Jawīsh [2] Radhwān bn Sa'ādeh [3] Khalaf bn Ali	Fixed sum (مقاطعة) of 160 *muds* of barley	[1] Beit Mazmīl [2] al-Khalīl Waqf	6/186	112
[1] Ahmad bn 'Abed al-Razāq [2]Ḥasan bn 'Ali	Half a donkey	Beit Mazmīl	4/235	142
[1] Moḥsin bn ḥasan [2]'Ali bn Mirān	4 *muds* of wheat	Beit Mazmīl	2/274–275	165
[1] Abed al-Moḥsin bn Ḥasan [2] al-Ḥāj Mohmmad bn Qazān al-Ramliy	[1] 4 *Sulṭaniy* and a half [2] 23 *raṭil* of oil	Beit Mazmīl	1/310	182
[1] Ḥanā bn Mikha'īl al-Naṣraniy al-Qibṭiy al-Najār [2] Jabir bn Khalifeh	Crops (not specified)	[1] Beit Mazmīl [2] Ein Jāwat [3] Egypt	3/336	194
[1] Radhwān bn Sa'ādeh [2] Ḥamdān bn Mousa [3] Khatāb bn 'Ubeyd [4] Mir'ib bn Saliḥ [5]Ibrahīm bn Jalbiy [6]Qīṭās Jalbiy [7] 'Ali bn Murād	160 *muds* of barley	[1] Beit Mazmīl [2] al-Khalīl Waqf	2/338	195
[1] al-Sheikh Muhmmad Shamis al-Dīn bn Ahmad Shihāb al-Dīn al-Sororiy [2] Saliḥ bn Khalīl al-Qaṣās [3] Shahīn bn Khalīl al-Qaṣās	Crops of grape, fig, apple, olive fruit, and other	[1] Beit Mazmīl [2] Ein Jāwat [3] Egypt (?)	8/419	231
[1] 'Abed al-Moḥsin bn Ḥasan [2] al-Ḥāj Mohmmad bn Qazān al-Ramliy	4 *Sulṭaniy* and a half	Beit Mazmīl	2/423	244

(continued)

App. 7: Summaries of representative entries on Beit Mazmīl in the Jerusalem *sijills* (*continued*)

Rec.	Date	Subject	Case summary
88	20.12.1016 H / 06.04.1608 CE	Assault	Suʿayfan bn Yousef from Beit Mazmīl village accused Muhmmad bn Muhmmad from al-Malḥah village with assaulting him with a rock.
88	24.06.1017 H / 05.09.1608 CE	Foreclosure	al-Sheikh al-Sayed Maḥmoud al-Wafa'iy claimed that Salim bn Suleiman from al-Malḥah village has been illegally using a land in Beit Mazmīl for four years and demanded from him a compensation from the harvest. Salim admitted to two years of use only and agreed to pay a *mud* and half of barley for each year. Then, Maḥmoud claimed that Salim has been illegally using from the vineyard that is a *Waqf* for the orphans of al-Sayed Ahmad. Salim admitted to two years of use and claimed that he has paid a compensation save for 50 pieces. Then, Salim was accused of illegally benefiting from the cucumber harvest. Salim admitted to that and paid 5 *Gurush*.
96	18.01.1024 H / 17.02.1615 CE	Debt acknowledgement	Khaṭāb bn ʿUbeyed from Beit Mazmīl is owes the orphans of al-Sheikh Ahmad bn al-Sheikh ʾAkram the sum of 12 *Gurush* for the crops of grapes, figs, and apples, which exist in Wadi al-Badawayeh.
96	08.04.1024 H / 07.05.1615 CE	Affidavit	ʿAmir bn Ẓahir from Beit Mazmīl has given an affidavit confirming that he has no rights in the crops of figs and flowers of the land located in Ein Jāwat. It has been confirmed by the court that the crops are the property of the deceased Muhmmad bn Ahmad bn Qāzān al-Ramliy, whose heirs are his wife, Faṭimah bnt ʿAli, and his sister, Assen.
96	06.04.1024 H / 05.05.1615 CE	Payment demand	ʿEmad bn al-Sheikh Ismāʿīl al-Maʿariy demanded the rest of the payment of a cow's price from the guarantor of Nazāl bn Deib from Deir Yasīn village, Suʿeyfān bn Yousef from Beit Mazmīl.
101	16.10.1027 H / 06.10.1618 CE	Payment demand	al-Sheikh Shamis al-Dīn al-Sororiy claimed that Deyāb bn Mohmad bn Deyāb from Beit Mazmīl is indebted to him for the crops of the farm of Ein Jāwat.
101	26.10.1027 H / 22.10.1618 CE	Debt payment	In his capacity as the guardian of al-Sheikh ʿAbed al-Wāḥid's orphan daughters, al-Sheikh Muhmmad al-Sororiy demanded from Ibrahīm bn Muhmmad bn Zeyadeh from Beit Mazmīl that he fulfil his duties as the guarantor of ʿAbed al-Moḥsin bn Qureish from the aforementioned village, and pay the 14 and half *Gurush* he owes.
101	27.10.1027 H / 17.10.1618 CE	Trees recount	al-Sheikh ʿAbed al-Qadir al-Wafʾiy demanded a recount of the vine trees in the Ḥusaineyeh farm in al-Malḥah village from Ibrahīm bn Muhmmad from Beit Mazmīl.
101	29.11.1027 H / 17.11.1618 CE	Payment demand	al-Sheikh Ahmad bn Taqiy al-Dīn al-Ghanimiy demanded from Suʿeyfān bn Yousef from Beit Mazmīl the price of the apple, fig, and grape crops, which originate from Wadi al-Badawayeh farm.
101	01.02. 1028 H / 18.01.1619 CE	Acknowledgement of debt	Muhmmad bn ʿUbeyed from Beit Mazmīl owes al-Sheikh ʿIneyet al-Ghaziy 5 Sulṭaniy for the *Waqf* of al-Sheikh Nur-Allah bn Jamāʿah.

Individuals	Articles involved	Locales	Case no./P.	Sum. Page no.
[1] Suʿayfan bn Yousef [2] Muhmmad bn Muhmmad		[1] Beit Mazmīl [2] al-Malḥah	8/15	16
[1] l-Sheikh al-Sayed Maḥmoud al-Wafaʾiy [2] Salim bn Suleiman	[1] A plot of land in Beit Mazmīl [2] Waqf vineyard [3]cucumber harvest	[1] Beit Mazmīl [2] al-Malḥah	6/231	126
[1] Khaṭāb bn ʿUbeyed [2] al-Sheikh Ahmad bn al-Sheikh ʾAkram	The price of sold crops	[1] Wadi al-Badawayeh [2] Beit Mazmīl	12/161	100
[1] ʿAmir bn Ẓahir [2] Muhmmad bn Ahmad bn Qāzān al-Ramliy [3] Faṭimah bnt ʿAli [4] Assen bnt Ahmad	[1] Fig's crop [2] Flower's (ورد) crop	[1] Beit Mazmīl [2] Ein Jāwat	2/224	145
[1] ʿEmad bn al-Sheikh Ismāʿīl al-Maʿariy [2] Nazāl bn Deib [3] Suʿeyfān bn Yousef	Cow's price	[1] Beit Mazmīl [2] Deir Yasīn	7/233	151
[1] al-Sheikh Shamis al-Dīn al-Sororiy [2] Deyāb bn Mohmad bn Deyāb	Crops (not specified)	[1] Beit Mazmīl [2] Ein Jāwat	2/54	36
[1] al-Sheikh Muhmmad al-Sororiy [2] al-Sheikh ʿAbed al-Wāḥid's orphan daughters [3] Ibrahīm bn Muhmmad bn Zeyadeh [4] ʿAbed al-Muḥsin bn Qureish	Debt	Beit Mazmīl	4/79	50
[1] al-Sheikh ʿAbed al-Qadir al-Wafʾiy [2] Ibrahīm bn Muhmmad	Vine trees	[1] Beit Mazmīl [2] al-Malḥah village [3] al-Ḥusseineyeh farm	1/81	51
[1] al-Sheikh Ahmad bn Taqiy al-Dīn al-Ghanimiy [2] Suʿeyfān bn Yousef	Crops of apple, fig, and grape	[1] Beit Mazmīl [2] Wadi al-Badawayeh	1/122	70
[1] al-Sheikh ʿIneyet al-Ghaziy [2] Muhmmad bn ʿUbeyed	*Waqf* of al-Sheikh Nur-Allah bn Jamāʿah	Beit Mazmīl	5/261	132

App. 8: Chart of radiocarbon and OSL dates

COL#	Sample ID	Sample type	Location	Core	Depth [m]	Origin	Sample weight [mg]	Total Organic carbon [%] if available	Age estimate (calBP)
leave empty	C6.L239 #1	Other	Khirbet Beit Mazmīl, Jerusalem, Israel		<1m	Terrestrial			300–600
	C6.L239 #2	Other	Khirbet Beit Mazmīl, Jerusalem, Israel		<1m	Terrestrial			300–600
	L411.B4014	Plant/Wood	Khirbet Beit Mazmīl, Jerusalem, Israel		<1m	Terrestrial			300–600
	B4.L127a Plaster	Other	Khirbet Beit Mazmīl, Jerusalem, Israel		<1m	Terrestrial			300–600
	B Reserv hydrolic plaster		Khirbet Beit Mazmīl, Jerusalem, Israel		<1m	Terrestrial			300–600
	B 16thCt house	Other	Khirbet Beit Mazmīl, Jerusalem, Israel		<1m	Terrestrial			300–600
	B4.L111 mortar	Other	Khirbet Beit Mazmīl, Jerusalem, Israel		<1m	Terrestrial			300–600
	C6 Cistern	Other	Khirbet Beit Mazmīl, Jerusalem, Israel		<1m	Terrestrial			300–600
	Wall123i	Other	Khirbet Beit Mazmīl, Jerusalem, Israel		<1m	Terrestrial			300–600
	Wall123o	Other	Khirbet Beit Mazmīl, Jerusalem, Israel		<1m	Terrestrial			300–600
	Wall124i	Other	Khirbet Beit Mazmīl, Jerusalem, Israel		<1m	Terrestrial			300–600
	Wall124o	Other	Khirbet Beit Mazmīl, Jerusalem, Israel		<1m	Terrestrial			300–600
	Wall128i	Other	Khirbet Beit Mazmīl, Jerusalem, Israel		<1m	Terrestrial			300–600
	Wall128o	Other	Khirbet Beit Mazmīl, Jerusalem, Israel		<1m	Terrestrial			300–600
	B7.L357 East	Plant/Wood	Khirbet Beit Mazmīl, Jerusalem, Israel		<1m	Terrestrial			300–600
	B7.L357 West	Plant/Wood	Khirbet Beit Mazmīl, Jerusalem, Israel		<1m	Terrestrial			300–600
	C6–7.L218.B2020	Charcoal	Khirbet Beit Mazmīl, Jerusalem, Israel		<1m	Terrestrial			300–600

Sample pretreatment*	AMS-Lab-ID	$F^{14}C$	+/-	Age (yr BP)	+/-	C (µg)	Pretreatment	Sample mat.	Remarks
flotation									plaster
flotation									plaster
flotation									burnt olive pits
									plaster with botanics
									plaster
									plaster (floor+wall)
									mortar (from courses above vaults)
									Cistern Interior plaster (behind *ṭābūn*)
									wall plaster—innermost
									wall plaster—outermost
									wall plaster—innermost
									wall plaster—outermost
									wall plaster—innermost
									wall plaster—outermost
									Olea europea (endocarp)
									Olea europea (endocarp)
									charcoal from kohl kit

(*continued*)

App. 8: Chart of radiocarbon and OSL dates (*continued*)

COL#	Sample ID	Sample type	Location	Core	Depth [m]	Origin	Sample weight [mg]	Total Organic carbon [%] if available	Age estimate (calBP)
	C6.L222.B2025	Plant/Wood	Khirbet Beit Mazmīl, Jerusalem, Israel		<1m	Terrestrial			300–600
	C6.L226.B2031	Plant/Wood	Khirbet Beit Mazmīl, Jerusalem, Israel		<1m	Terrestrial			300–600
	B4.L127b.B1039	Plant/Wood	Khirbet Beit Mazmīl, Jerusalem, Israel		<1m	Terrestrial			300–600
	C6.L221.B2024	Plant/Wood	Khirbet Beit Mazmīl, Jerusalem, Israel		<1m	Terrestrial			300–600
	B4.L129.B1043	Plant/Wood	Khirbet Beit Mazmīl, Jerusalem, Israel		<1m	Terrestrial			300–600
	B4.L127a (.1038?)	Plant/Wood	Khirbet Beit Mazmīl, Jerusalem, Israel		<1m	Terrestrial			300–600
	B4.L130.B1048	Plant/Wood	Khirbet Beit Mazmīl, Jerusalem, Israel		<1m	Terrestrial			300–600
	E1.433.4038B	sed/soil	Khirbet Beit Mazmīl, Jerusalem, Israel		?	Terrestrial		-5.9	669–972
	E1.433.4038A	sed/soil	Khirbet Beit Mazmīl, Jerusalem, Israel		?	Terrestrial		-10.1	893–1155
	C6.251.2061	sed/soil	Khirbet Beit Mazmīl, Jerusalem, Israel		?	Terrestrial		-17.4	891–1030
	B4.183.2013	plant/wood	Khirbet Beit Mazmīl, Jerusalem, Israel		?	Terrestrial		-27.1	–
	E1.448.4065	sed/soil	Khirbet Beit Mazmīl, Jerusalem, Israel		?	Terrestrial		-23.7	1040–1220
	C6.259.2068	sed/soil	Khirbet Beit Mazmīl, Jerusalem, Israel		?	Terrestrial		-24.5	1233–1391
	C6.252.2062	plant/wood	Khirbet Beit Mazmīl, Jerusalem, Israel		?	Terrestrial		-14.9	-
	E1.428.4035	charcoal	Khirbet Beit Mazmīl, Jerusalem, Israel		?	Terrestrial		-24.0	1312–1442
	D1.510.5015	charcoal	Khirbet Beit Mazmīl, Jerusalem, Israel		?	Terrestrial		-20.7	1529–heute

Sample pretreatment*	AMS-Lab-ID	$F^{14}C$	+/-	Age (yr BP)	+/-	C (µg)	Pretreatment	Sample mat.	Remarks
									Olea europea (endocarp)
									Hordeum vulgare (hulled)
									Hordeum vulgare (hulled)
									Hordeum vulgare (hulled)
									Hordeum vulgare (hulled)
									T.t. durum/ T.a. aestium
									T.t. durum/ T.a. aestium
	COL6634.1.1	0.859	0.006	1217	57	663		AAA	Probe besteht aus Putz
	COL6635.2.1	0.879	0.005	1032	45	744		AAA	
	COL6636.2.1	0.876	0.004	1067	38	994		AAA	Probe besteht aus Putz.
	COL6638.1.1	1.052	0.005	> modern	–	1000		AAA	
	COL6639.2.1	0.894	0.004	899	37	994		AAA	Probe besteht aus Putz.
	COL6640.2.1	0.916	0.004	706	37	994		AAA	Probe besteht aus Putz. Ziegelfragment entfernt.
	COL6641.1.1	1.057	0.005	> modern	–	1000		AAA	Wurmsekret + Schimmel—> Ultraschall (Hex:DCM:MeOH) + Usltraschall NaClO2 Lsg
	COL6643.1.1	0.935	0.005	542	39	988		AAA	eingereicht als sed/soil
	COL6644.1.1	0.974	0.005	215	37	1000		AAA	

(continued)

App. 8: Chart of radiocarbon and OSL dates (*continued*)

COL#	Sample ID	Sample type	Location	Core	Depth [m]	Origin	Sample weight [mg]	Total Organic carbon [%] if available	Age estimate (calBP)
	E2.438.4065	charcoal	Khirbet Beit Mazmīl, Jerusalem, Israel		?	Terrestrial		-18.0	1022–1201
	B4.184.2014	charcoal	Khirbet Beit Mazmīl, Jerusalem, Israel		?	Terrestrial		-23.8	650–819

$F^{14}C$: see Reimer et al., 2004, Radiocarbon

$\delta^{13}C$: measured with AMS, not comparable to stable isotope ratio-MS measurement due to different isotopic fractionation

cal AD: 95.4% probability

Pretreatment: calculated with OxCal v. 4.4.1 (Bronk Ramsey 2020), Calibration curve: IntCal13 (Reimer et al. 2020)

AAA—Acid-Alkali-Acid extraction; A—Acid extraction

full preparation method see Rethemeyer et al. 2019: https://doi.org/10.1017/RDC.2019.16

Sample ID	Sample type	Location	Probe	Depth [m]	Origin
BMZ 1	OSL	Nahal HaYovel, Jerusalem, Israel	A1	0.25	Terrace Wall A/F201
BMZ 2	OSL	Nahal HaYovel, Jerusalem, Israel	A2	0.70	Stone Pile A/F203
BMZ 8	OSL	Nahal Ein Karim, Jerusalem, Israel	C1	1.10	Terrace C/F202
BMZ 7	OSL	Nahal Ein Karim, Jerusalem, Israel	C2	0.30	Plot Fence C/F201
BMZ 6	OSL	Nahal Ein Karim, Jerusalem, Israel	C2	0.60	Terrace C/F203
BMZ 13	OSL	Nahal Ein Karim, Jerusalem, Israel	C4	0.10	Terrace C/F203
BMZ 14	OSL	Nahal Ein Karim, Jerusalem, Israel	C4	0.50	Terrace C/F203 and Plot Fence C/F201
BMZ 12	OSL	Nahal Ein Karim, Jerusalem, Israel	C4	0.70	Terrace C/F203
BMZ 15	OSL	Nahal Ein Karim, Jerusalem, Israel	C5	0.85	Terrace C/F211
BMZ 124	OSL	Nahal Ein Karim, Jerusalem, Israel	C7	0.30	Plot Fence C/F213
BMZ 152	OSL	Nahal Ein Karim, Jerusalem, Israel	C8	0.60	Terrace C/F218
BMZ 145	OSL	Nahal Ein Karim, Jerusalem, Israel	C9	0.15	Terrace C/F219
BMZ 138	OSL	Nahal Ein Karim, Jerusalem, Israel	C9	0.85	Terrace C/F219
BMZ 135	OSL	Nahal Ein Karim, Jerusalem, Israel	C9	1.15	Terrace C/F219
BMZ 133	OSL	Nahal Ein Karim, Jerusalem, Israel	C10	0.60	Terrace C/F220 and Plot Fence C/F215
BMZ 116	OSL	Nahal Ein Karim, Jerusalem, Israel	C13	0.20	Terrace C/F224
BMZ 111	OSL	Nahal Ein Karim, Jerusalem, Israel	C13	0.70	Terrace C/F224
BMZ 106	OSL	Nahal Ein Karim, Jerusalem, Israel	C15	0.05	*manṭarah* C/F206
BMZ 108	OSL	Nahal Ein Karim, Jerusalem, Israel	C16	0.20	Stone-Steps C/F225
BMZ 107	OSL	Nahal Ein Karim, Jerusalem, Israel	C16	0.35	Terrace C/F226

Sample pretreatment*	AMS-Lab-ID	$F^{14}C$	+/-	Age (yr BP)	+/-	C (µg)	Pretreatment	Sample mat.	Remarks
	COL6645.1.1	0.888	0.004	951	38	994		AAA	
	COL6646.1.1	0.850	0.004	1307	39	1000		AAA	

Age (years before 2022)	Age range (CE)	Remarks
1920 ± 140	-40–240	
2330 ± 140	-450–-70	
520 ± 30	1470–1530	
670 ± 60	1290–1410	
850 ± 50	1120–1220	OSL age matches two coins dating to the early 13th century
280 ± 30	1710–1770	
600 ± 40	1380–1460	
1100 ± 80	840–1000	
1360 ± 80	580–740	
670 ± 110	1240–1460	
830 ± 60	1130–1250	
670 ± 50	1300–1400	
180 ± 10	1830–1850	taken next to a Monoglazed Mamluk vessel
1830 ± 90	100–280	
490 ± 40	1490–1570	
380 ± 30	1610–1670	
1690 ± 70	260–400	
340 ± 40	1640–1720	
800 ± 70	1150–1290	
1420 ± 110	490–710	

App. 9: List of critical loci

Chapter 3

Early Roman (Strata IX)

Northern Sector: D1 water system

Miqveh (no locus number)

Southern Sector: throughout

Stone quarries (in bedrock) (no locus number)

Late Roman (Strata VIII)

Northern Sector: D1 water system

Cistern (no locus number)

Late Byzantine (Stratum VII)

Southern Sector: A1

W111 + 143—wide walls forming exposed corner of this structure

L.134—beaten earth floor with flat-lying Byzantine sherds

L.135—soil under floor L.134

Lower Site: S1

Mosaic floor (no locus number)

Early Islamic (Abbasid) (Stratum VI)

Northern Sector: E1

L.433—modified bedrock

Northern Sector: C4

L.307—modified bedrock (= L.433)

L.309—fill of robber's trench along east wall of room

Northern Sector: B5

L.163—modified bedrock (= L.433)

Southern Sector: A11

Cisterns I and II—modified bedrock (no locus number assigned, similar to C4.307)

Crusader/Ayyubid (Stratum V)

Northern Sector: C4 vaulted room

W120—wall stub visible on outside of east wall of C4

Northern Sector: E1 furnace room

L. 434—ashy soil sealed by pebble floor of Pit 140

L.447—fill sealed by flagstone floor L.445

L. 448—soil under first flagstone floor, in doorway W147 (dated by radiocarbon to 1040–1220 CE)

Northern Sector: B4 domestic space

L.178—soil against exterior (south) face of W118, on bedrock

Southern Sector: Area A

W143—wall running north-south in southwest corner of A2

W405—wall running east-west across north end of A2–4–6

A2: L.132—fill on bedrock

A6: L.128, 136, 138—part of plaster floor 114, on bedrock

B8: L.543 + 544—floors on bedrock

"Pre-Mamluk" (sometime before Stratum IV)

Northern Sector: Wall numbers

W101 + 102—walls that delineate site on northwest side (heavily rebuilt in Early and Late Ottoman periods)

W107—wall that delineates site on east side (northernmost part of wall is likely Mamluk)

W120—pre-Mamluk wall in C5, under the 19th century W105

W123—pillar of the cloister vault (C6/E1)

W124/1—"pillar" in Area C6

W124/2—"pillar" in Area B5

Chapter 4

Mamluk (Strata IV–III)

Northern Sector: B4 Domestic Space

W118—southern wall in B4
L.161—soil layer associated with the stairs on the final flagstone floor
L.162—stairs on the final flagstone floor
L.165—soil layer under W127
L.180—semi-circular hearth in the northwest sector
L.186—water channel (not fully excavated)

Northern Sector: C6 Press Room

L.229—flagstone floor beneath the Early Ottoman cooking installation (L.225)
L.233—fill on the L.229 flagstone floor
L.238—oval cooking installation
L.244—probe below the L.229 flagstone floor
L.245—rectangular plaster-lined basin in the press room
L.246—small semi-circular channel cut into the plaster on the E floor of L.245
L.247—stairs leading into the E1 storeroom
L.256—corridor leading to the south that connected the press room with the E1 storeroom
L.258—semi-circular plaster-lined settling basin to the east of L.245 connected by the L.24 channel
L.259—contents of plaster-lined settling basin (L.25); radiocarbon sample taken (dated to 1233–1391 CE)
L.263—doorway leading to (unexcavated) room to the east of L.258

Northern Sector: E1 Storeroom

L.419—first flagstone floor in E1
L.420—plaster floor sealed under flagstone floor L.419
L.423—soil layer below L.420 plaster floor
L.428—contents of W140; radiocarbonsample taken (dated to 1312–1443 CE)

Northern Sector: C6 Press Room / E1 Storeroom (wall numbers)

W121—springer of the barrel vault abutting against the cloister vault
W126—cloister vault
W128—continuation of W136
W136—wall abutting against W124/1 on its eastern side
W140—circular stone-lined pit in the northeast corner

Northern Sector: D1 water system

L.510—soil on bedrock and sealed by flagstone floor L.504 (dated by OSL to c. 1329 CE)

Southern Sector: Area A

L.104—soil layer on plaster floor in A8
L.106—soil layer below the plaster floor in A8
L.114—soil layer on the plaster floor in the northwest sector of Square A6
L.117—soil layer on plaster floor in the southwest sector of Square A6
L.121—soil layer on plastered floor in Square A4
L.126—soil layer below plaster floor L.116 in A6
L.133—Mamluk-era robber's trench cut into the Byzantine plaster floor in Square A4

Southern Sector: Area B

L.507—fill above plaster floor in B7
L.512—fill above Circular Installation 1 in Squares B1/2
L.515—fill above Circular Installation 2 in Squares B2/3
L.528—fill under flagstone floor and above the bedrock in B7
L.530–532–541—flagstone floor which covered most of the building in B8
L.534—*ṭābūn* contents
L.536—fill under flagstone floor
L.537—fill below plaster floor in B7
L.539—circular Installation 3 contents in Squares B2/3

Chapter 5

Early Ottoman (Stratum II)

Northern Sector: B4 domestic space

L.118, 112–124, 127—cooking installation

Northern Sector: C6 "kitchen"

L.220–225, 228—cooking installations

Northern Sector: E2 "midden"

W135—terrace wall forming south end of courtyard midden

L.439—courtyard midden fill (= E1.439)

Late Ottoman/Mandate (Stratum I)

Northern Sector: Wall numbers

W105 + 106 + 108—drystone walls built around the *'ezbeh* and on top of earlier walls

W109 + 110 + 111 + 113 + 114—drystone terrace on upper level of walls around the *'ezbeh*

App. 10: Sketchfab—access to the three-dimensional models and instruction for use

The three-dimensional documentation of the site of Khirbet Beit Mazmīl is available on a dedicated page on Sketchfab, an online three-dimensional modelling platform website that allows the user to publish, share, and access three-dimensional contents. The platform is compatible with all forms of support—smartphone, tablets, and computers—but it is strongly recommended to use a computer for the sake of visibility and ease of use. It is also strongly recommended to have a stable and strong internet connection to avoid dramatic quality downgrades of the contents. All contents are free to access and to share but are not downloadable.

The following guideline is tailored for PC and laptop access:

1. To access the dedicated page with the three-dimensional contents, connect to the following web address: https://sketchfab.com/URBHUB/collections
2. Select the collection "Khirbet Beit Mazmīl" to open the list of all available three-dimensional models.
3. Click on the three-dimensional content to open it. The interface that opens is the same for all contents of the site, and strongly resembles those of other online platforms for videos: in the top left is the three-dimensional content's window; directly below is the content's title, description, and the comment section; on the right is a list of further suggested contents.
4. The basic navigation tools for the three-dimensional content are:
   a. Orbit around: Left click + drag, or "One finger drag" (touch)
   b. Zoom: Double click on model, or scroll anywhere, or "Pinch" (touch)
   c. Pan: Right click + drag, or a "Two fingers drag" (touch)
5. To activate "Fullscreen" view: F
6. To switch between "Standard quality" and "High definition," hover on the bottom right corner of the three-dimensional content's window and click on the "Gear" symbol of the menu, then select the desired option.
7. For further instructions and a complete controls list, hover on the bottom right corner of the three-dimensional content's window and click on the "Question mark" symbol of the menu.
8. Numbered buttons on the three-dimensional content allow one to open information tabs (with detailed descriptions and detail photos). To open them, click on the desired button and a pop-up window will automatically open.
9. It is possible to scroll between the different annotations automatically by hovering at the bottom centre of the three-dimensional contents window and clicking on the two arrows beside the name of the currently opened information tab.
10. It is possible to browse directly between the different three-dimensional contents by using the left and right arrows on the margins of the windows or on the keyboard.

Index

About the Cover Design

When the archaeologists responsible for this book first saw the site of Khirbet Beit Mazmīl, they were confronted by a picturesque scene of ruination: the site was lost beneath an increasingly wild orchard of almond trees and grapevines. This pastoral vision of the site has left a lasting impression upon all involved in its excavation over the subsequent years. Nikolaus Hochstein Cox has tried to honor this first impression of Khirbet Beit Mazmīl in his cover design. Inspired by Ottoman floral art, he has created a fictional plant with the flowers of the almond tree (*Prunus amygdalus*) and the leaves of the grapevine (*Vitis*) that grows across the book: representative of the site and the generations of people enmeshed within its history that called it home.